Engineering Solutions for CO_2 Conversion

Engineering Solutions for CO_2 Conversion

Edited by

Tomas R. Reina
José A. Odriozola
Harvey Arellano-Garcia

WILEY-VCH

Editors

Dr. Tomas R. Reina
University of Surrey
Department of Chemical & Process Engineering
388 Stag Hill
GU2 7XH Guildford, Surrey
United Kingdom

Prof. José A. Odriozola
Universidad of Sevilla
Inorganic Chemistry Department
4 San Fernando Street
41004 Sevilla
Spain

Prof. Harvey Arellano-Garcia
University of Surrey
Department of Chemical & Process Engineering
388 Stag Hill
GU2 7XH Guildford, Surrey
United Kingdom

Library of Congress Card No.:
applied for

British Library Cataloguing-in-Publication Data
A catalogue record for this book is available from the British Library.

Bibliographic information published by the Deutsche Nationalbibliothek
The Deutsche Nationalbibliothek lists this publication in the Deutsche Nationalbibliografie; detailed bibliographic data are available on the Internet at <http://dnb.d-nb.de>.

Print ISBN: 978-3-527-34639-4
ePDF ISBN: 978-3-527-34650-9
ePub ISBN: 978-3-527-34651-6
oBook ISBN: 978-3-527-34652-3

Typesetting SPi Global, Chennai, India

Printed on acid-free paper

10 9 8 7 6 5 4 3 2 1

Contents

1

CO_2 Capture – A Brief Review of Technologies and Its Integration

Mónica García[1], Theo Chronopoulos[2], and Rubén M. Montañés[3]

[1] *International Energy Agency- Greenhouse Gas R&D Programme (IEAGHG), Pure Offices, Hatherley Lane, Cheltenham GL51 6SH, United Kingdom*
[2] *128/15 Hoxton Street, N1 6SH, London, United Kingdom*
[3] *Energy Technology, Chalmers University of Technology, Department of Space, Earth and Environment, Hörsalsvägen 7E, Gothenburg SE-412 96, Sweden*

1.1 Introduction: The Role of Carbon Capture

The Intergovernmental Panel for Climate Change (IPCC) recently released the special report on 1.5C [1] and pointed out the need to implement all available tools to cut down CO_2 emissions. Energy efficiency, fuel switching, renewables, and carbon capture represent the largest impact on CO_2 emission reduction in power and industrial sectors. Carbon capture represents a contribution of 23% in the "Beyond 2 degrees scenario" (B2DS) modeled by the International Energy Agency (IEA)[1] and has other interesting characteristics that increase its value beyond its cost: (i) easiness to retrofit current power plants or industrial facilities,[2] (ii) simplicity to integrate that in the electricity grid and offer an interesting tool to cover the intermittency of renewables, (iii) ideal to cut down industrial process emissions that otherwise cannot suffer deep reductions, and (iv) current carbon budgets rely on negative emissions to compensate the use of fossil fuels [1]. Carbon capture combined with bioenergy (BECCS) can provide negative emissions at large scale in an immediate future.

CO_2 capture (also called CO_2 sequestration or carbon capture) involves a group of technologies aiming to separate CO_2 from other compounds released during the production of energy or industrial products, obtaining a CO_2-rich gas that can be stored or used for the obtention of valuable products. The main classification of CO_2 capture technologies relies on where in the process the CO_2 separation occurs. For the power sector, it can be divided into pre-, oxy-, and post-combustion. For the industrial sector, the classification is similar, although their integration would be different. In addition, other new arrangements are emerging.

1 https://www.iea.org/etp/explore/ (visited in January 2019).
2 Under specific arrangements.

Engineering Solutions for CO_2 Conversion, First Edition.
Edited by Tomas R. Reina, José A. Odriozola, and Harvey Arellano-Garcia.

1.2 CO_2 Capture Technologies

1.2.1 Status of CO_2 Capture Deployment

GCCSI reported in 2018 23 large-scale CCS facilities in operation or under construction globally, summing up 37 $MtCO_2$ per year. This wide range of facilities shows the versatility of CO_2 capture processes[3].

In the power sector, the United States is leading the implementation deployment, although Europe has the highest CO_2 capture capacity. The Boundary Dam project (Canada) and Petra Nova (USA) are pioneers in reaching commercial scale. Moreover, based on the successful results of the Boundary Dam project, a CO_2 capture facility has been planned for the Shand power facility (Canada), incorporating not only learnings from the Boundary Dam but also enhanced thermal integration and tailored design. The results show a significant cost reduction [2]. Also in Canada, the Quest project completes the list of Canadian CCS projects in operation [3] and The National Energy Laboratory (NET) power project recently appeared in the United States as a potential significant reduction on CO_2 capture costs [4].

In the industrial sector, cement, steel, refining, chemicals, heavy oil, hydrogen, waste-to-energy, fertilizers, and natural gas have been identified by the Carbon Sequestration Leadership Forum (CSLF; https://www.cslforum.org) as the main intensive emitter industries. As it is highlighted, the Norcem Brevik plant [5, 6], LEILAC [7] (cement production), and Al Redayah (steel production) are on the way to start running carbon capture systems in industrial facilities at pilot and large scales.

1.2.2 Pre-combustion

Pre-combustion systems can be applied to natural gas combined cycles (NGCC) or integrated gasification combined cycle (IGCC) (Figure 1.1), where a syngas, comprising mainly CO and H_2, feeds a gas turbine (GT) combined cycle system to produce electricity. The potential advantages are higher conversion efficiencies of coal to electricity and cheaper removal of pollutants [8]. The syngas, based on the water shift reaction, can be converted into CO_2 and H_2O. This mixture is typically separated with physical solvents (as described in Section 1.2.4), membranes, or sorbents. However, hybrid technologies can also be used. Depending on the technology, further post-treatment would be needed to avoid degradation and loss of efficiency.

The main theoretical advantage of pre-combustion is the production of hydrogen, which will add value to the business model, and a lower energy penalty compared to using the traditional chemical absorption within a post-combustion configuration. However, large projects demonstrated that this difference is only 1–2%, as reported by National Energy Technology Laboratory (NETL) [9].

The most notable pre-combustion project was the Kemper County IGCC plant in the United States, which stopped its operation in 2017.This demonstration facility

3 (The Global Status of CCS, GCCSI 2018 https://indd.adobe.com/view/2dab1be7-edd0-447d-b020-06242ea2cf3b).

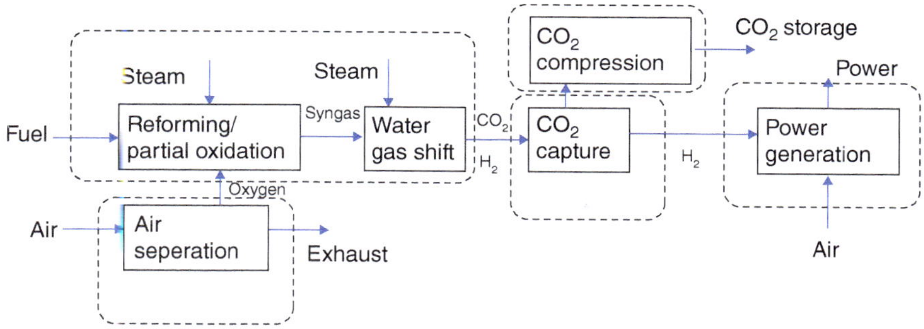

Figure 1.1 Diagram of pre-combustion capture for power generation in IGCC. Source: Adapted from Jansen et al. [72].

would place this arrangement at high TRL, while other testing campaigns would reach up to a TRL of 6.

1.2.3 Oxyfuel

In the oxyfuel process, the air is split into nitrogen and oxygen, generally using an air separation unit (ASU), for the combustion of fuel with nearly pure oxygen. The consequence is a higher flame temperature and a highly concentrated CO_2 stream (60–75%, wet and might contain impurities and incondensable components) that can be further purified to meet the final use specifications. The CO_2-rich gas is typically recirculated to manage the unstable flame and its high temperature. Nowadays, the progress on oxyfuel combustion is focused on the reduction of air separation costs and the enhancement of process configuration to reduce capture costs. Further information can be found, for example, in Ref. [10]. Based on the current progress, the most advanced arrangements can be assessed as TRL 7.

An advanced oxyfuel process, called the Allam cycle (Figure 1.2), is being tested at large scale as part of the NET Power project in the United States [4]. This involves oxyfuel combustion and a high-pressure supercritical CO_2 working fluid in a highly recuperated Brayton cycle, aiming to reduce CO_2 capture costs and prove stable operation. Based on that, there is a potential to progress to a TRL of 7 once the facility is fully operational.

1.2.4 Post-combustion

Post-combustion refers to the group of technologies able to separate CO_2 from the flue gas emitted during the fuel combustion and/or other reactions in the industrial sector. This indicates that those systems are mainly installed as additional equipment downstream in new plants or during the retrofitting of the existing facilities. The latter represents the main advantage of post-combustion technologies compared to pre- or oxy-combustion, as a fundamental redesign or complex integration with the existing facilities would be minimal.

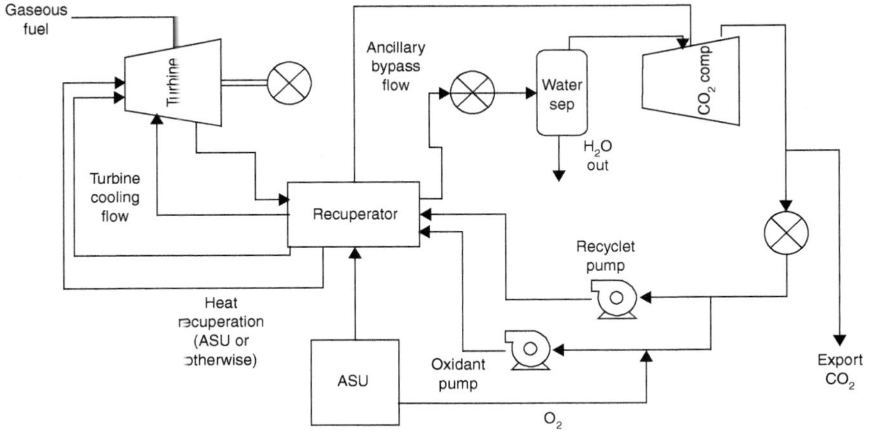

Figure 1.2 Process schematic of a simplified commercial scale natural gas Allam cycle. Source: Adapted from Allam et al. [4].

1.2.4.1 Adsorption

Adsorption refers to the uptake of CO_2 molecules onto the surface of another material. Based on the nature of interactions, adsorption can be classified into two types: (i) physical adsorption and (ii) chemical adsorption. In physical adsorption, the molecules are physisorbed because of physical forces (dipole–dipole, electrostatic, apolar, hydrophobic associations, or van der Waals) and the bond energy is 8–41 kcal mol^{-1}, while in chemical adsorption, the molecules are chemisorbed (chemical bond; covalent, ionic, or metallic) and the bond energy is about 60–418 kcal mol^{-1} [11].

A theoretical advantage of adsorption against other processes is that the regeneration energy should be lower compared to absorption because the heat capacity of a solid sorbent is lower than that of aqueous solvents. However, other parameters, such as working capacity and heat of adsorption, should also be considered [12]. The higher the heat of adsorption, the stronger the interaction between the CO_2 molecules and adsorbent-active sites and thus the higher the energy demand for the regeneration. The potential disadvantages for adsorbents include particle attrition, handling of large volumes of sorbents, and thermal management of large-scale adsorber vessels.

Solid sorbents can be classified according to the temperature range where adsorption is performed. Low-temperature solid adsorbents (<200 °C) include carbon-based, zeolite-based, metal–organic framework (MOFs)-based, several alkali metal carbonate-based, and amine-based solid adsorbents. Intermediate-temperature (200–400 °C) solid adsorbents include hydrotalcite-like compounds or anionic clays, while high-temperature (>400 °C) sorbents refer to calcium-based adsorbents and several alkali ceramic-based adsorbents.

Usually, adsorption takes place in packed or fluidized beds, as can be seen in Figure 1.3. For the packed bed case, the adsorbent is loaded into a column, the flue gas flows through the void spaces between the adsorbent particles, and CO_2

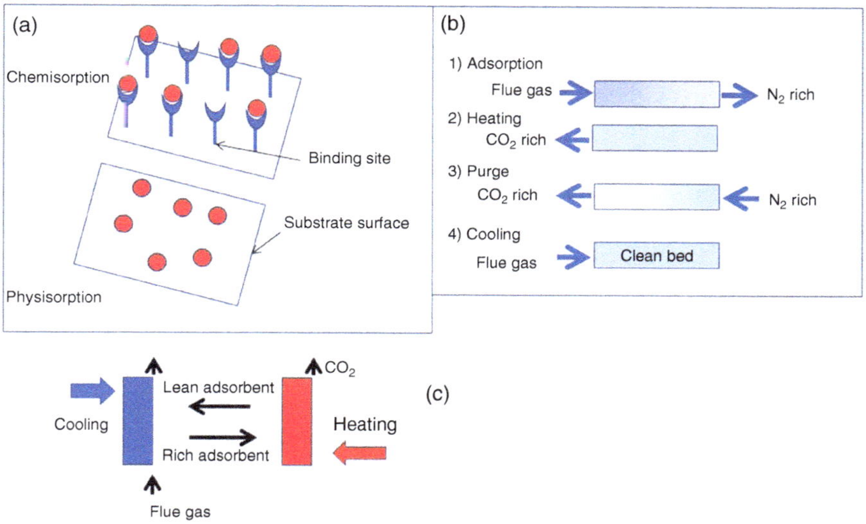

Figure 1.3 The adsorption process: (a) difference of physisorption and chemisorption, (b) a packed bed configuration, and (c) a fluidized bed configuration. Source: Adapted from Global CCS Institute (https://www.globalccsinstitute.com/archive/hub/publications/29721/co2-capture-technologies-pcc.pdf).

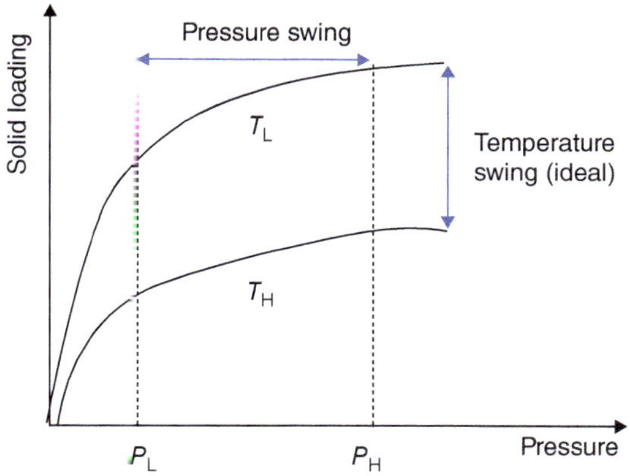

Figure 1.4 Comparison of TSA and PSA for the regeneration of solid adsorbents. H = high; L = low. Source: Adapted from Rackley [73].

gets adsorbed onto the surface of the particles. In fluidized beds, the flue gas flows upward through a column above the minimum fluidization velocity and the adsorbent particles are as such suspended in the gas flow. Regardless of the process configuration, the adsorbent selectively adsorbs CO_2 from the flue gas and is subsequently regenerated to complete the cyclic adsorption process.

Cyclic adsorption processes alternate between the adsorption and desorption modes of operation. Based on the intensive variable that is cycled, the adsorption processes are broadly classified as pressure swing adsorption (PSA) or temperature swing adsorption (TSA), as can be seen in Figure 1.4. If the cycle switches between adsorption at atmospheric pressure and desorption under vacuum, then it is called vacuum swing adsorption (VSA). Pressure vacuum swing adsorption (PVSA) cycles have an adsorption step above atmospheric pressures and desorption under vacuum [13].

In a packed bed configuration, regeneration is accomplished by heating the CO_2-loaded adsorbent to liberate CO_2. During this time, the flue gas is diverted to a second packed bed, which continues to adsorb CO_2 from the gas. By alternating the flue gas between two packed beds that alternatively undergo absorption and regeneration in a cycle, CO_2 can be continually removed from the flue gas. In a fluidized bed, the sorbent is circulated between an absorber vessel where it contacts the flue gas and a regenerator vessel where it is heated to liberate gaseous CO_2.

Usually, the PSA process is preferred to other cyclic operations when the process is carried out at elevated pressures. Otherwise, when the concentration of the adsorbate is low (0–15 vol%), or when the process is at low pressure, other options such as TSA may need to be considered. For a low-concentration adsorbate, the PSA technology may result in a much longer desorption step, whereas for low-pressure processes, the installation should also include additional vacuum pumps and compressors, both resulting in a more complicated process, increased capital cost, and reduced efficiency [8]. A potential option that could overcome these issues is vacuum pressure swing adsorption (VPSA).

TSA can work both for low and elevated pressures; however, it is usually preferred when the adsorption step is carried out at a low temperature. Consequently, the main advantage of TSA over PSA is its ability to separate efficiently strong-bonded adsorbates onto adsorbents, as for the case of chemisorption. However, a major drawback of TSA is the high energy intensity of the desorption process compared to PSA. Other alternatives to TSA include microwave swing adsorption (MSA) [14] and electric swing adsorption (ESA) [15] that could offer potential energy savings and faster heating rates; however, these technologies are still at low technology readiness level (TRL).

Generally, TSA is usually preferred for post-combustion CO_2 capture at low temperature and atmospheric pressure, while PSA usually is the right choice for pre-combustion CO_2 capture at elevated temperatures, as in the case for an IGCC plant configuration. As a post-combustion arrangement, PSA and TSA are assessed as TRL 6.

Adsorption equilibria, kinetics, and regeneration ability are key factors to evaluate the performance of an adsorbent. Fast adsorption/desorption kinetics, influenced by functional groups present, as well as the pore size and distribution in the support, are essential for an efficient CO_2 adsorption process and control of the cycle time and the required amount of adsorbent. Other selection criteria include high CO_2 selectivity, mechanical strength after multi-cycling, chemical stability/tolerance to impurities, high availability, and, lastly, low costs.

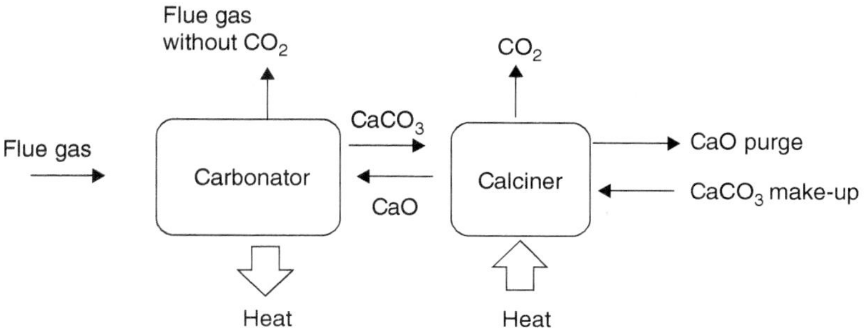

Figure 1.5 Calcium looping system as post-combustion configuration. Source: Adapted from Abanades [16].

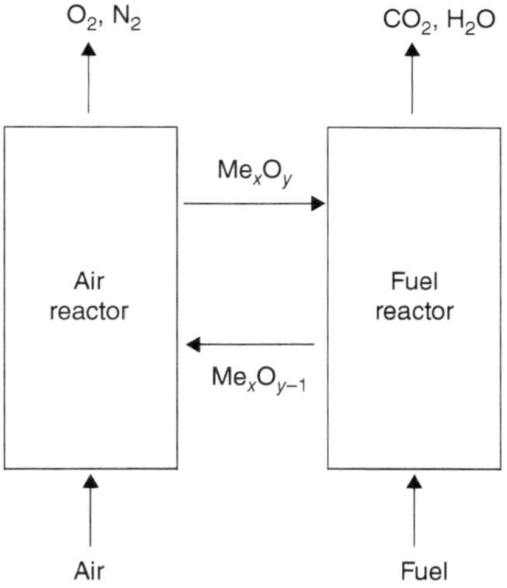

Figure 1.6 Chemical looping combustion. Me_xO_y/Me_xO_{y-1} denotes the recirculation oxygen carrier material. Source: Adapted from Abanades et al. [17]. © Elsevier.

1.2.4.2 High-Temperature Solids Looping Technologies

The most common types of high-temperature solids looping technologies are calcium and chemical looping combustion. Calcium looping uses CaO as a sorbent, which produces $CaCO_3$ at approximately 650 °C (Figure 1.5). Chemical looping is a two-step conversion process where the fuel reacts with almost pure O_2 as in the oxyfuel process, while a metal oxide acts as an oxygen carrier and reacts with the fuel, obtaining CO_2 and water (Figure 1.6). In both cases, the metal oxide or CaO is regenerated.

Note that calcium looping can be considered as post-combustion or pre-combustion, while chemical looping can be considered as oxy-combustion or pre-combustion depending on the configuration [16].

Because of the high operation temperature, the advantage of this process is the potential recovery of energy for steam production, which can be used for additional power production and reduce the efficiency penalty in the power plant.

Calcium looping has shown a significant evolution over the past 15 years from lab scale to pilot testing, reaching a TRL of 6. The main research focus to cut down the costs over the next years is on the sorbent, reactors (configurations and interconnections), and process designs [17]. If used in the industrial sector, calcium looping can be beneficially integrated in the cement production facility because of the use of solids from the capture system in the production. In this regard, the CLEANKER project aims to scale up a calcium looping process in a cement production environment, which will increase the TRL of this technology up to 7[4].

Chemical looping has reached a TRL of 6 as oxyfuel arrangement while a TRL of 3 as pre-combustion system. The main research areas on chemical looping are focused on the reactor design, oxygen carrier development, and prototype testing. Moreover, more than a thousand materials have been tested at the laboratory scale. At a larger scale (0.3–1 MW), the accumulated operational experience is more than 7000 hours [17]. A detailed review of the main process routes under development within the chemical looping systems is included in Ref. [17].

1.2.4.3 Membranes

Membranes are porous structures able to separate different gases at different rates because of their different permeation [8]. These can be used not only in post- and pre-combustion processes but also in oxyfuel for oxygen separation. In post-combustion, the main interest in these systems is their low energy requirements compared to the traditional chemical absorption process.

The energy needs are reduced to those from the compressor and vacuum pump. Moreover, membrane systems are easy to start and operate, have no emissions associated, and are modular, offering installation advantages [8]. However, the separation mechanism of membranes is based on the difference of CO_2 partial pressure. In post-combustion, because of the relative low CO_2 concentration in the flue gas to

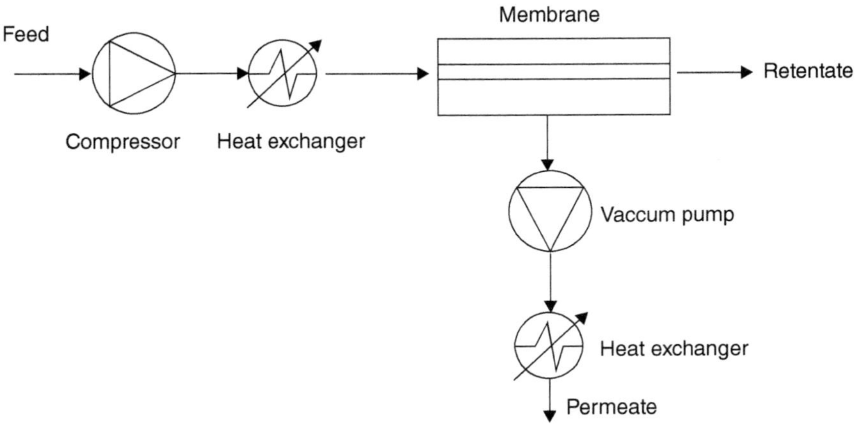

Figure 1.7 Scheme of a single-stage membrane system. Source: Adapted from Mores et al. [18].

4 (http://www.cleanker.eu).

Table 1.1 Advantages of each type of membrane [21].

Type of membrane	Advantages
Ceramic	Good selectivity–permeability
	Easier to manufacture larger areas
Polymeric	Good thermal stability and mechanical strength
Hybrids	Aiming to show the advantages of both ceramic and polymeric membranes

Source: Adapted from Wang et al. [21].

be treated (approximately 4–12% for power plants), this driving force would not be enough to achieve high CO_2 capture ratios through simple configurations. However, membranes could offer advantages for partial capture arrangements and generally more complex arrangements are used to reach a full capture rate (90%). In pre-combustion, because of the higher partial pressure of CO_2 in the gas to be treated, membranes can be more effective. In any case, the gas containing CO_2 must be cooled down to meet the temperature limitations of the membrane [18] and that could be a drawback (Figure 1.7).

There are two main characteristics to define a membrane material for CO_2 capture: permeability, which will impact on the CO_2 separation ratio and selectivity, which will define the CO_2 concentration in the output gas. From a techno-economic perspective, the optimum values for selectivity and permeability would be a function of the gas to be treated, as studied in Ref. [19]. The ratio of the permeability to the thickness of the membrane will be of high importance as that will characterize the permeance (commonly measured as gas permeation units [GPU]). To maximize the permeance without impacting the mechanical stability, the membranes are typically a dense layer supported by a porous layer [20].

The membrane materials can be divided into ceramic, polymeric, and hybrid (Table 1.1). Moreover, the design of the membrane-based system will be a key factor on the separation process. Firstly, the membrane module will be the key factor. The main modules for polymeric membranes are described as a spiral wound, a hollow fiber, and an envelope [21].

The majority of the membranes used currently for post-combustion are based on polymeric materials [20], and a large list of polymers have been studied in the literature, including polyimides, polysulfones, and polyethylene oxide. The most advanced processes have reached currently a TRL of 6. Because of the modularity membranes offer, although sometimes predicted, it is not clear if there will be a fast development toward higher TRLs [21].

1.2.4.4 Chemical Absorption

The basic configuration of chemical absorption (Figure 1.8) includes the reaction of a liquid solvent with CO_2 in a column called absorber at a relatively low temperature, 40–60 °C, and its desorption in another column called desorber or stripper, generally at a high temperature, 100–140 °C. It must be noted that process modifications and solvent enhancements might modify those process conditions.

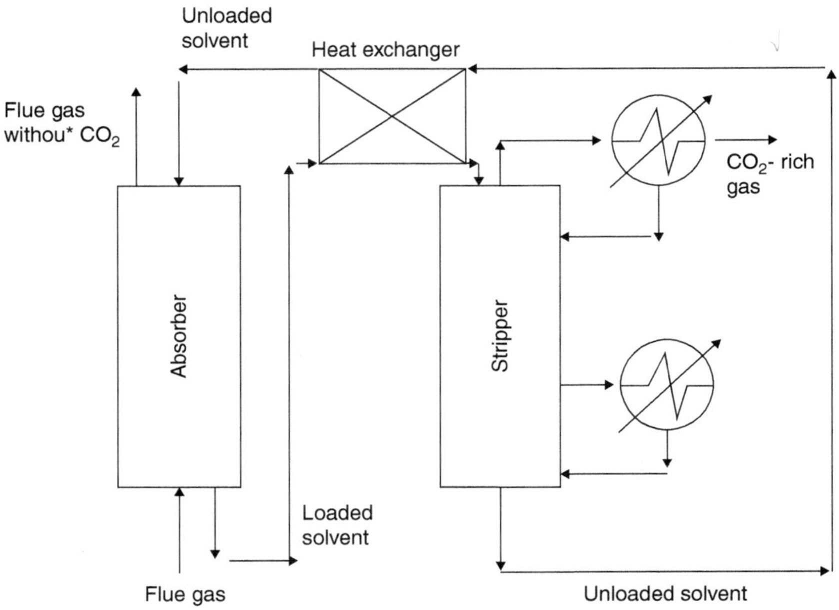

Figure 1.8 General chemical absorption configuration

The absorption of CO_2 into liquid solvents takes place by three phenomena: chemical reaction, physical absorption, and diffusivity. Depending on the compound and the conditions, one phenomenon will be predominant over the others.

Chemical solvents are more attractive candidates for typical post-combustion processes, with relatively low partial pressures of CO_2 (10–15% in coal power plants and 4–8% for gas-fired power plants). Chemical absorption follows a standard configuration such as in Figure 1.8. However, new configurations have appeared to enhance the process, increase the efficiency, and/or decrease the capture costs.

Chemical absorption with amines is by far the most advanced carbon capture process and the only one that reached a TRL of 9 [2]. The most tested solvent is aqueous monoethanolamine (MEA) solution, although it does not represent any more the benchmark solution as consolidated alternatives show enhanced properties. Two large-scale facilities have used enhanced systems, the Boundary Dam Capture plant [2] and Petra Nova. One of the main pathways to get more efficient chemical absorption processes and cut down costs is the development of new solvents. However, many solvents are emerging and only few have been tested at large scale, maintaining the TRL of other new systems still low. A review of commercial solutions and relevant projects can be found, for example, in Ref. [22]. The main criteria for the selection of a solvent are included in Table 1.2.

Primary amines are of high interest because of their fast reaction with CO_2. However, the main drawback is their high energy consumption for the solution regeneration. Several alternatives are emerging to decrease such penalty, the most common one being the use of tertiary amines. However, the CO_2 absorption in tertiary amines

Table 1.2 Desired solvent properties and its impact on the absorption process [75].

Solvent property	Impact on the absorption process
High capacity and low heat of absorption	It is linked to the energy requirements per ton of CO_2, but the absorption capacity is connected to heat (thermodynamics) and independent variation is limited
High mass transfer and chemical kinetics	It reduces equipment size or the capacity by operating near the equilibrium limit
Low viscosity	It reduces the pumping costs and potentially increases the mass transfer and the heat transfer rate
Low degradation tendency	It reduces the solvent make-up and the regenerator can operate at higher pressure/temperature, increasing the thermal efficiency
Low toxicity/environmentally friendly	It becomes more important if toxic by-products are released during volatility losses
Cost and availability	It will impact on reaching commercial scale
Low fouling tendency	It will impact on the operation

Source: Adapted from Mathias et al. [75].

is much slower. Consequently, other alternatives are emerging, such as the use of blends combining primary and tertiary amines (commonly called "promoted tertiary amines" or "activated tertiary amines"). Numerous alternatives have emerged during the past years; perhaps it is difficult to establish the best alternative.

A potential substitute of traditional solvents is the use of compounds that, at unloaded or loaded conditions, separate into two phases, called biphasic solvents. There are two types of biphasic solvents, namely, liquid–liquid or solid–liquid, depending on the phases in solution. The main advantage is that only one phase needs to be regenerated, and consequently, the stripper size is reduced, and the energy consumption is potentially lower. Consequently, numerous biphasic solvents have been studied in the literature (e.g. in Ref. [23]).

Another strategy is to add enzymes, such as carbonic anhydrase (CA) [24]. CA increases the kinetic constant of the absorption of CO_2 in aqueous amine and dilute carbonate solutions by catalyzing the CO_2 hydration. The impact will depend on the compounds in solution, as the regeneration of the enzyme regeneration rate will vary. The challenges enzymes offer are their pH and thermal stability, lifetime, and sensitivity to pollutants such as SO_x and NO_x.

At lower development stage, solvents can be encapsulated in thin polymer shells and be considered as a bed of capsules containing the solvent. Capsules must be permeable enough to allow carbon dioxide to get in contact with the solvent but strong enough to resist the high regeneration temperatures during a number of cycles [25]. The benefit of this configuration is to increase the surface area of the solvent in contact with the flue gas and avoid issues related to viscosity and precipitation.

Recently, ionic liquids (ILs) are of great interest. These are composed of ions and are at liquid state below 100 °C. If the melting point is below the room temperature, these are referred as room temperature ionic liquids (RTILs). These solvents are recognized by their low vapor pressure, high thermal and chemical stability, nonflammability, and high viscosity. These properties open new possibilities for the solvent regeneration at different pressures and temperatures, which can be optimized accordingly. Some ILs show a high absorption capacity, although the viscosity could be decremental for the mass transfer.

Physical solvents are characterized for the high physical solubility of CO_2 in these and are especially interesting for flue gas with high CO_2 content [26]. There are commercial processes based on this principle, such as Rectisol®, Selexol®, Purisol®, Morphysorb®, and Fluor®, particularly effective at high concentrations of acid gas, high pressure, and low temperature [27] and are characterized by their low vapor pressure, low toxicity, and low corrosion [15].

An emerging pathway is the use of hybrid solvents, solutions containing amine/s and organic compound/s with or without the presence of water, the former called as water-lean solvents. The goal is to maintain an enhanced physical absorption by substituting partial/totally the water content and maintaining a considerable chemical reaction by keeping the amine in the solution. It is known that at low concentration of the amine(s), the physical solubility plays an important role and the diffusivity can also become an important factor in viscous solutions. The enhanced solubility of CO_2 in organic solvents, compared to water, has been widely studied in the past [28–31], and this presents advantages in its application in chemical absorption. During the desorption, the main energy penalty is due to the water evaporation. Decreasing the water content will decrease this energy penalty. Partial and total substitution of water by organic solvents has been considered as an alternative to decrease the steam consumption in the desorber. However, as studied in Ref. [32], the absorption kinetics would just be favorable, compared with aqueous amine solutions, at certain conditions of pressure and temperature in the absorber. The total substitution of water in water-lean solvents will limit the reactions that take place in solution: hydrolysis will not occur and the carbamate and bicarbamate ions will be nonexistent [33]. However, the net benefit in the energy consumption when using water-lean solvents is not yet clear, as discussed in Ref. [34].

1.2.4.4.1 Advances on Process Configurations

As mentioned previously, chemical absorption is the most advanced technology, reaching commercial status (TRL 9). However, there are still barriers that slow down its application in industrial and power sectors. Cost is one of the challenges to overcome and energy consumption has a strong contribution. The development of new solvents and improvements on the process flow sheet and/or its integration in the industrial or power facility could reduce this energy consumption.

The common process modifications can be divided as in Ref. [35]: (i) absorption enhancement, (ii) heat integration, and (iii) heat pumps. Perhaps these can also be classified by their purpose, as in Ref. [36]: (i) increase of rich solvent loading, (ii) reduction of the specific reboiler duty, or (iii) combination of both. The enhancement on the absorption and desorption processes and its impact on costs will depend on other factors such as the solvent and the facility. The modifications on the stripper to reduce energy consumption are being considered for the next generation of post-combustion processes' configurations with advanced solvents (e.g. as in Ref. [37]).

1.2.5 Others CO_2 Capture/Separation Technologies

Other CO_2 capture/separation technologies such as electrochemical, cryogenic separation, liquefaction, microbial/microalgae, or direct air separation are described in the literature.

Hybrid technologies have been studied in the past years, aiming to achieve higher capture rates and/or sum up the advantages of each CO_2 capture technology. The hybrid processes can be classified into absorption-based, adsorption-based, membrane-based, and cryogen-based hybrid processes. The integration of membranes into the absorption process (such as in the membrane contractor arrangement), catalysis process, and cryogenic process has progressed over the past years. However, the majority of the results are based on simulations or small-scale testing campaigns, and the real value of using two technologies is not clear [38].

Within the range of emerging technologies, electrochemical separation has had a fast development over the past years and, potentially, will continue in this pathway. The following Section 1.2.5.1 will be focused on fuel cells because of the growing expectation on this electrochemical separation technology for its integration in power plants.

1.2.5.1 Fuel Cells

Fuel cells convert chemical energy of a gaseous fuel directly into electricity and heat. The fuel is oxidized electrochemically, which leads to lower exergy losses compared to direct combustion. In general, fuel cells are classified by the electrolyte material and their operating temperature (Figure 1.9). Low-temperature fuel cells (100–250 °C) include alkaline fuel cells (AFCs), phosphoric acid fuel cells (PAFCs), and proton exchange membrane fuel cells (PEMFCs), while high-temperature fuel cells (600–900 °C) refer to Molten carbonate fuel cells (MCFCs) and solid oxide fuel cells (SOFCs). Because of the high temperature at which MCFCs and SOFCs operate, natural gas reformation and the subsequent shift reaction can be performed in the fuel cell itself. MCFCs and SOFCs are most appropriate for stationary power production at scales ranging from a few hundred kilowatts up to a few megawatts because of their high electrical efficiencies and the ability for cogeneration of electricity and heat [39] Moreover, SOFCs and MCFCs are more fuel flexible and are not poisoned by carbon monoxide and carbon dioxide.

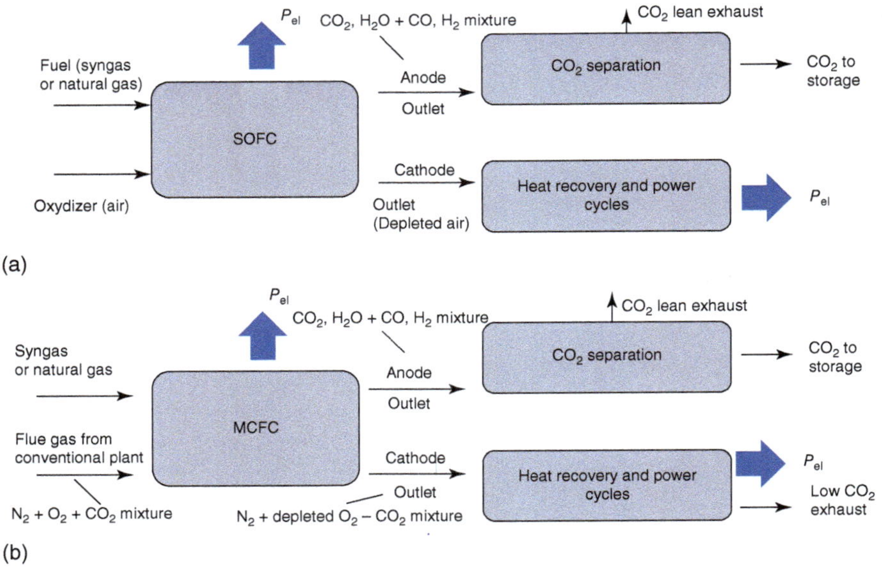

Figure 1.9 Two main options for CO_2 capture using fuel cells. (a) The FC oxidizes a fuel taking oxygen from air and later separating CO_2 from the anode effluent. (b) The MCFC concentrates the CO_2 in flue gas from a conventional power plant from the cathode inlet to the anode outlet, while also oxidizing a portion of additional fuel. Source: Adapted from [11].

When MCFCs/SOFCs are fueled with natural gas or syngas, CO_2 capture can be implemented at different points, for example, after the fuel cell ("post-anode capture"). Alternatively, H_2 can be produced by reforming/partial oxidation of natural gas or coal gasification upstream the fuel cell and CO_2 can be removed after syngas is shifted by means of physical solvents, membranes, or adsorbents – "pre-anode CO_2 capture," similar to pre-combustion.

Fuel cells generally operate with an approach that is similar to the "oxyfuel" concept, oxidizing fuel with oxygen extracted from air while generating power and releasing concentrated effluents at the anode outlet (Figure 1.9). This kind of power cycles generally require an integration with custom-tailored gas turbine cycles, often operating at unconventional turbine inlet temperatures and pressure ratios, either using natural gas as a fuel or coal through integrated gasification fuel cell (IGFC) concepts. Because most fuel is oxidized in the fuel cell to allow a high CO_2 capture efficiency, the fuel cell (FC) generates the majority of the cycle power output. The alternative option offered by MCFCs is shown at the bottom of Figure 1.9, where the fuel cell can operate "draining" CO_2 from the cathode inlet stream, receiving the flue gases of a conventional power plant. In this configuration, the fuel cell operates with a post-combustion approach, although also oxidizing a minor portion of additional fuel with the same "oxyfuel" features discussed above.

The parameters affecting the selection of operating conditions of the SOFC/MCFC are stack size, heat transfer rate, voltage output and cell life, load requirement, and cost. The main operating conditions are pressure, fuel utilization factor at the anode and O_2/CO_2 utilization factor (for SOFC and MCFC cases, respectively) at the cathode, voltage, current density, and temperature. The optimization of the process configuration in conjunction with optimal operating parameters is critical to minimize stack degradation, which directly impacts the performance and life of the FC.

Currently, the main challenges for stationary fuel cells are cost and cell durability. For the IGFC system, the gas cleaning process adds another energy barrier to its power generation.

1.2.5.1.1 Solid Oxide Fuel Cells (SOFCs)

Adams et al. [40] divided SOFC systems for CO_2 capture into first- and second-generation systems as a function of the operating pressure of the SOFC. Low-pressure, first-generation SOFC systems are the most promising option for SOFC commercialization at large scale (100 MW or greater) in the short term. Several process configurations and design options are possible (Figure 1.10), although those generally follow the same pattern and offer some flexibility to select the optimum combination of variables such as gas clean-up/reforming, water gas shift (WGS), CO_2 capture technology, and heat recovery.

Second-generation SOFC systems are high-pressure SOFCs with separate streams for the anode and cathode exhausts. This arrangement promotes the use of an SOFC system that captures and compresses CO_2 at significantly reduced costs and minimum complexity via "pre-anode" and/or "post-anode" capture.

In the pre-anode CO_2 capture process, syngas is generated at high pressure through high pressure coal gasification or by reforming the natural gas available from a natural gas pipeline at high pressure. Similar to the above cases, the syngas can be optionally shifted using the WGS reaction, creating a stream of steam, H_2, and CO_2 Up to about 90% of the CO_2 can then be recovered from the syngas (or shifted syngas) using absorption or adsorption technologies.

The post-anode CO_2 capture has been extensively studied in SOFC IGCC and natural gas cycles. A simple IGFC system is similar to an IGCC system, but the gas turbine (GT) power island is replaced by a FC island. Some system configurations still have a gas or steam turbine to utilize the extra heat. "Post-anode" CO_2 capture can be applied via CO_2 separation from H_2O via H_2O condensation (or via cooling, knockout, and additional drying) and can effectively result in a 100% CO_2 removal. A separation system that uses condensation followed by a cascade of flash drums can be used to produce CO_2 at high enough purity for pipeline transport at the SOFC anode exhaust pressure.

1.2.5.1.2 Molten Carbonate Fuel Cells (MCFCs)

The MCFC can be used to separate CO_2 thanks to the functional reactions that occur inside the cell. By sending flue gas from a power plant to the cathode, the CO_2 from the flue gas is selectively separated and concentrated at the anode, in a mixture of water and small amounts of unreacted hydrogen and methane. The "cleaner flue gas" is delivered to the atmosphere with up to 70% less CO_2 content, which is transferred to the MCFC anode exhaust stream where it can be separated much more

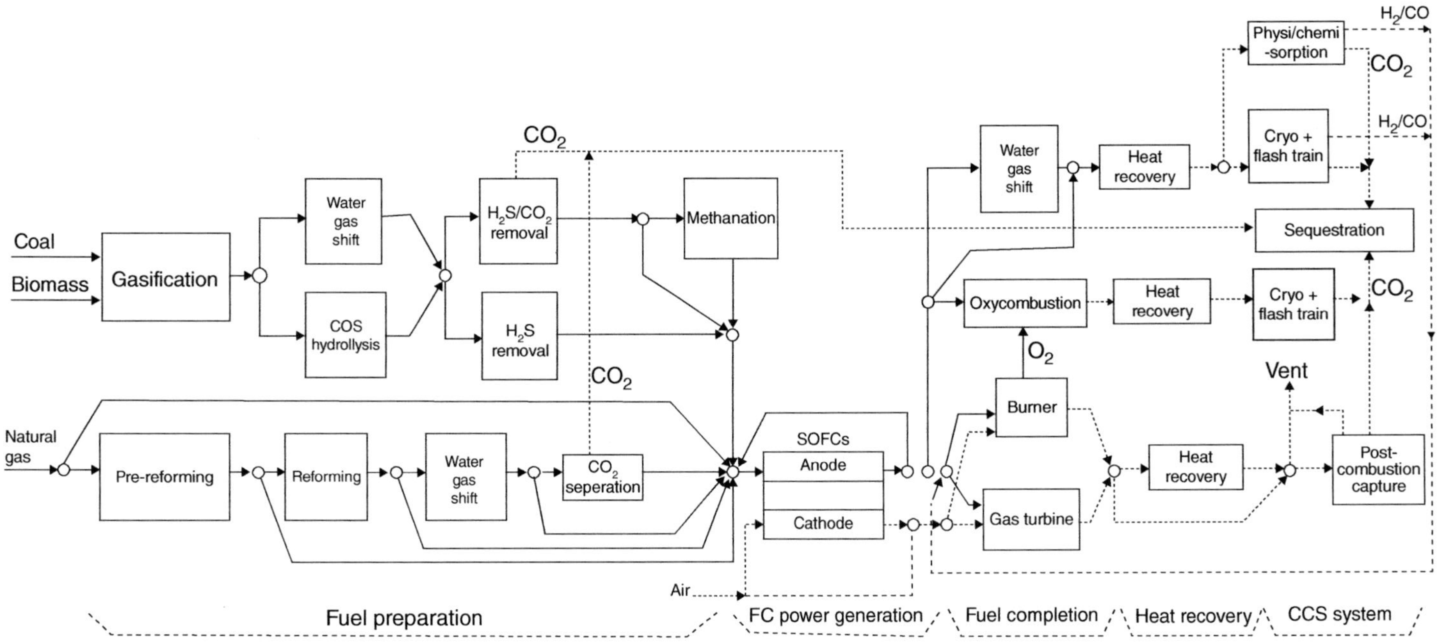

Figure 1.10 Superstructure of SOFC – CO_2 capture process configurations. Source: Adams et al. [40].

effectively, resulting in a high-purity CO_2 flow. The main advantage in this process is that extra power is generated because the MCFC will be fueled and operated normally to carry out the separation, and it increases the overall efficiency of the power plant and compactness of the post-combustion unit, while reduces the energy penalty. The modularity feature of MCFC systems allows to tailor the installation to the capture needs or gradually increases the size of the capture unit.

One example of an MCFC and CO_2 capture system was developed by Fuel Cell Energy (FCE), namely, the Combined Electric Power and Carbon-dioxide Separation (CEPACS). In the process of capturing >90% CO_2. In this configuration, the system can generate up to 351 MWe additional power (net AC), after compensating for the auxiliary power requirements of CO_2 capture and compression[5].

1.3 Integration of Post-combustion CO_2 Capture in the Power Plant and Electricity Grid

A key aspect of thermal power plants is their carbon intensity (CO_2 emitted per unit of energy generated, generally expressed as kg CO_2/MWh). Nowadays, the global average is around 500 kgCO_2/MWh, which must be reduced to 100 kgCO_2/MWh by the late 2030s to be consistent with a 2 °C climate pathway [36]. Even if combined cycle thermal power plants can be considered as low carbon alternatives in some scenarios, in the mid-to-long term, it might be required to further decarbonize the existing units by retrofitting them with CCS or by building novel designs with low CO_2 emissions. As demonstrated at commercial scale, post-combustion CO_2 capture can significantly reduce the carbon intensity of thermal power plants [2]. Table 1.3 compares the carbon intensity of thermal power plants with and without CCS.

1.3.1 Integration of the Capture Unit in the Thermal Power Plant

In principle, the key integration aspects of the power plant and the capture unit are the flue gas, emitted by the power plant and sent to the capture unit, and the energy requirements of the chemical absorption/desorption process, provided by the power plant to the capture unit (Figure 1.11). Figure 1.11 shows a simplified schematic of a power plant integrated with a post-combustion CO_2 capture system. The main energy and mass integration flows are described. Fuel and air are used in the combustion process, providing heat to produce steam in the power cycle. The flue gas from the combustion is sent to the CO_2 capture unit and leaves it lean in CO_2. A CO_2 rich stream is produced in the CO_2 capture plant and sent to conditioning, transport, and storage. Heat in the form of steam is provided from the power plant and is returned back as water condensate. Electricity from the power plant is utilized to run the auxiliary systems of the capture unit, including the flue gas fan, cooling, and solvent circulation pumps. Higher levels of process integration between the power plant and the capture unit can be considered, as explained in [41].

5 (https://www.netl.doe.gov/project-information?p=FE0026580).

Table 1.3 Low heat value (LHV) efficiency, carbon intensity, and energy penalty in coal- and gas-based thermal power plants with CCS [43, 54, 76].

	Carbon intensity (kg CO_2/MWh)	LHV efficiency (%)	Efficiency reduction (%)
Pulverized coal subcritical	700–1000	30–47	—
Combined cycle	350–450	56–62	—
Pulverized coal with CCS at 90% capture rate	130	25–42	5–7
Combined cycle with CCS at 90% capture rate	40–50	50–54	6–8

Source: Adapted from Adams and Mac Dowell [43], Gonzalez-Salazar et al. [54], Kvamsdal et al. [76].

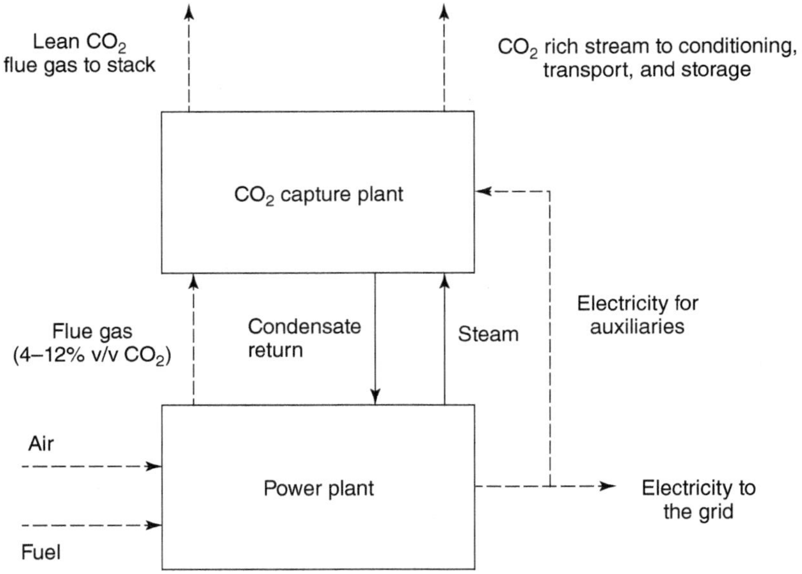

Figure 1.11 Schematic integration of a power plant with a post-combustion CO_2 capture system.

The energy required to run the chemical absorption–desorption process in the capture unit process is mainly due to the (i) mechanical work to drive the flue gas fan to compensate the pressure drop induced by direct contact cooler, absorber column, and water wash sections and ducting; (ii) mechanical work to drive the pumps for cooling water and solvent circulation pumps; (iii) steam for solvent reclaiming because of its degradation in order to keep the solvent fresh and contaminant free; and (iv) steam to feed the reboiler duty: regenerate the solvent, generate stripping vapors, heat up the solvent to saturation conditions, and evaporate the water.

The flue gas is sent to the capture unit and additional pressure drop is imposed, which is a function of the thermal power plant unit, its equipment for emissions control, and the boiler type. In gas plants, the heat recovery steam generator in the exhaust gas generally imposes the additional pressure drop downstream the gas turbine (in the order of 20–40 mbar [8]), and in boilers (coal/oil/gas fired), a fan is commonly used to keep it under slightly sub-ambient pressure. The main flue gas line equipment inducing pressure drop would be as follows [42]:

- The particle removal system (electrostatic precipitator [ESP]).
- The flue gas desulfurization (FGD) unit (if existing).
- The NO_x scrubber (if required) 2–7 mbar [42].
- The bypass stack and damper if installed to bypass the capture unit.
- Flue gas recirculation ducting and/or bypass (if utilized).
- Direct contact cooler, absorber column, and water washes 6–80 mbar [43].
- Absorber duct and stack.

In general, a fan will be required to overcome those pressure drops, whose size will depend on the volumetric flow and pressure drop. In combined cycle thermal power plants, most of the pressure drop might be overcome by raising the back pressure of the gas turbine. However, in boilers, the pressure drop is generally overcome by one or more fans [43].

The extraction of steam from the power plant steam turbine could cover the requirements in the capture system. This strategy will reduce the power output at a lower degree than the amount of heat extracted because the exergy content of the steam is just a fraction of the heat [41]. When the extracted steam is superheated (typically at higher steam pressure), it is normally cooled down by high-pressure water injection. The heat content of the steam can be fully utilized or part of it is returned via the condensate recirculation to the power plant. The steam extraction design depends on the specific configuration and power plant unit, level of process integration, and steam requirement in the reboiler (also function of the solvent characteristics), which has been extensively discussed in the literature [44–46].

In order to compensate for the efficiency reduction in power generation introduced by the CO_2 capture and conditioning processes, several studies have been conducted to increase the efficiency of the integrated process. Increasing process integration within the capture unit itself might lead to reduced specific reboiler duty, for example, by using lean vapor recompression, absorber intercooling, or solvent split flow to stripper [41, 47]. In addition, studies have shown the potential to reduce the specific reboiler duty by using a technique called exhaust gas recirculation (EGR), which consists of recirculating part of the CO_2-rich stream to increase the partial pressure of the exhaust [48] using supplementary firing to increase the partial pressure of CO_2 [49] and/or even integrating part of the reboiler duty in the power plant [50]. These options can lead to lower capital and operational costs at the expense of higher integration between the power plant and the capture unit under operation.

1.3.2 Flexible Operation of Thermal Power Plants in Future Energy Systems

Thermal power plant operation is highly coupled to the operation of power systems and power markets. Flexible operation of thermal power plants has become an important issue in the past decades because the increased integration of renewables CCS must operate accordingly (e.g. [51]). In decarbonized power systems, thermal power plants must be operated in cycling mode in order to cope with variability in demand and generation [52, 53], following the main patterns:

- *Efficiency at part load*: In power systems with high penetration of renewables, it is expected that thermal power plants will be operated during a significant number of hours at part load [52, 54, 55]. However, at part load, the efficiency of thermal power generation is generally reduced and specific emissions at minimum compliant load increase. Thermal power plant developers are striving to reduce minimum compliant load level (to minimize economic losses at times when marginal costs of operation are higher than electricity prices) and increase part load efficiency. Design and operation should take into consideration the part load performance of thermal power plants with CCS. An important aspect is to keep minimum specific reboiler duty and an economically suitable capture rate in the capture unit over the whole load range [43, 44, 56, 57].
- *More frequent changes in load*: Faster ramping can be valuable for thermal power plants in order to be more competitive in day-ahead power markets and balancing markets [54, 55] and the different time scales required for ramping the power plant load and the capture plant will be the key. Generally, thermal power plant load change is characterized by stabilization times in the order of 5–10 minutes, while the capture unit can take up to several hours [54] to stabilize under load changes because of the inertia of the chemical process [58–60]. Efforts are being made to develop operational and control strategies to improve the stabilization time and reduce the specific reboiler duty under transient conditions [57, 61–63].
- *More frequent start-up and shutdown events*: The start-up and shutdown increase CO_2 emissions during start-up and fuel utilization without any significant power output from the power plant. Efforts are being made in order to reduce the start-up time to provide power on demand and/or reduce emissions during start-up [64]. Because the start-up of amine-based post combustion CO_2 capture is time and energy intensive, minimizing the start-up time and emissions during the start-up sequence might be relevant.

Several operational strategies are proposed to operate thermal power plants with CCS in flexible operation mode, being the main purpose to change the power output of the power plant by changing the operational conditions of the integrated process. The main goal from the power operator perspective is to maximize the profits, while from the power system operator perspective, the power plant would be providing variation management to the power system to accommodate the variability of renewables. In addition, flexible operation of post-combustion capture might be required when integrated within industrial processes because of the inherent variability of the industrial process operations, such as in primary steelmaking [65]. The main strategies proposed for power plants with a carbon capture system can be summarized as follows:

- Allowing the thermal power plant to follow load changes. The capture unit follows the power plant load change [58, 59].
- Varying the CO_2 capture rate, depending on CO_2 costs and electricity prices [51]. In such case, the solvent regeneration is variable, using the large amount of loading capacity and large inventories of solvent as CO_2 storage [66]. At times with high electricity prices, the steam is used for power production, while the regeneration takes place at low electricity prices.
- Turning on-and-off the capture unit or flue gas bypass. The flue gases sent to the capture unit are bypassed to the stack of the power plant so that partial or no CO_2 is being captured. Part of the flue gas is vented to the atmosphere. This allows part of the steam used for solvent regeneration to be used for power production in the steam turbine. This option might be viable in scenarios in which CO_2 emission costs or prices are low.
- Providing solvent storage to decouple plant operation from the capture unit. The capture rate is kept constant and the solvent is stored in tanks. The regeneration energy is shifted to times when electricity prices are low. Solvent storage can incur in significant capital expenditure required for solvent storage, which could be favorable in scenarios with high CO_2 emission costs.

1.4 CO_2 Capture in the Industrial Sector

The industrial sector was responsible for almost 25% of the CO_2 emissions in 2014. CO_2 is emitted on the fuel combustion, intrinsic reactions and indirectly on the use of electricity. IEA predicted a required reduction on the CO_2 emissions of 3–6 Gt/yr to achieve the 2 degrees scenario (2DS) or B2DS. Although other measures such as increasing energy efficiency, developing new production process, using renewable energy or fuel switching, will reduce CO_2 emissions, still there is a significant amount of CO_2 from the process that can be only reduced through CO_2 capture [20]. To achieve the B2DS, the contribution of CCS is estimated as 23%.

All the available CO_2 capture technologies can be potentially installed in industrial facilities. However, while certain industries would have similar or even more favorable characteristics for the implementation of carbon capture utilisation and storage (CCUS) compared to power plants, the design of CO_2 capture systems must be tailored for each facility. The heat and energy integration will be site specific and, together with the composition and CO_2 emission stacks, will impact on the optimum capture rate and the CO_2 avoidance cost.

An exhaustive description of the integration of certain CO_2 capture technologies in the cement sector can be found, for example, in Refs. [67, 68]. A large scale chemical absorption system will be installed in the Norcem Brevik facility, after other technologies (solid sorbents and membranes) were tested at smaller scale [6]. Oxy-fuel has been included in the Front End Engineering Design (FEED) studies within the European climate research alliance (ECRA) project and the LEILAC project will test the Calix technology (direct separation) [7]. Other technologies, such as chilled ammonia, membrane-based capture combined with liquefaction, and calcium looping were studied, for example, in the CEMCAP project at modeling scale [69]. Moreover, partial capture configurations for several industries are being studied by the CO2STCAP project [70] and the CLEANKER project will scale up the calcium looping up to a TRL of 7[6].

The peculiarity of the steelmaking sector is the heterogeneity of production processes that will be more or less dependent on the electricity grid. At large scale, the most significant project is the Al Reyadah in Abu Dhabi, where CO_2 is captured in the steam methane reforming (SMR) for H_2 production to be used in a direct reduction iron (DRI) process. A recent cost review identified promising CO_2 capture solutions for this sector, perhaps at lower TRL and potentially with less accurate cost figures [71]. Other projects are advancing on CO_2 capture technologies applied to the steelmaking sector. For example, the C4U project will test high-temperature solid sorbents, aiming to reach a TRL of 7 once the demonstration facility is fully operational. Additionally, the STEPWISE project will advance on the testing of the sorption- enhanced water gas shift technology, reaching a TRL of 7 once it operates successfully, while the 3D project will test an advanced solvent in a steel mill[7].

Other sectors such as refining, hydrogen, natural gas, heavy oil, fertilizer productions, and waste-to-energy are important and are being considered for further study, for example, by the CSLF.

1.5 Conclusions

In this chapter, the main CO_2 capture systems applied to the industrial and power sectors have been described, covering a wide range of TRLs. Chemical absorption as post-combustion arrangement was further discussed, including advanced process configurations and its integration in the power plant and electricity grid.

Based on the information from the literature, Figure 1.12 aims to provide an overview of the current TRLs of the different CO_2 capture technologies applied to the power and industrial sectors. Note that differences on the TRL definitions from different sources can impact on the TRL assessment. Additionally, several systems can vary and it would be reflected in their TRL. For example, chemical absorption systems have reached their maximum TRL when commercial solvents are used. However, emerging solvents might be at a much lower TRL. Similar limitations of those estimations can be seen, for example, in the use of different absorbents, different types of membranes or using novel O_2 separation process for oxyfuel. Moreover, in the case of the industrial sector, the TRL is also dependent on the industry. For example, while a system has been tested within a cement production facility, it might not have been used in the iron and steel production environment.

6 (www.cleanker.eu).
7 https://3d-ccus.com/

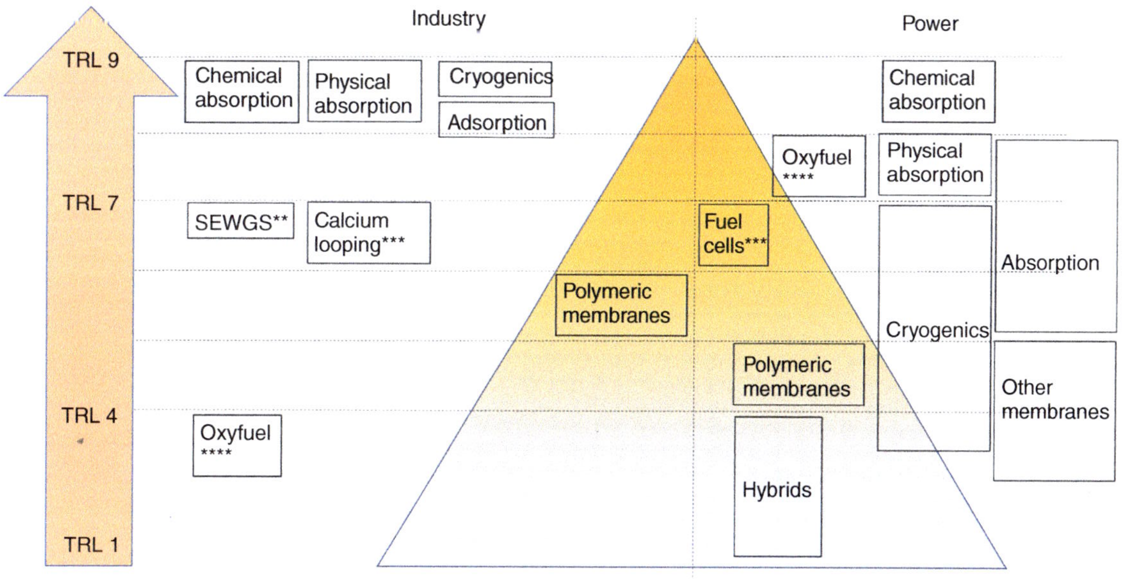

Figure 1.12 Review of current TRL of different CO_2 capture technologies. *The prediction of the TRL of fuel cells is based on the project implemented in Alabama by ExxonMobil and Fuel Cell Energy partnership using MCFC. **SEWGS = Sorption-enhanced water gas shift. This prediction is based on the expected outcome of the STEPWISE project. ***The prediction of the calcium looping technology on the industrial sector is based on the expected outcome of the CLEANKER project. ****Oxyfuel is considered here as the combustion with almost pure oxygen. Other configurations of, for example, chemical looping, cryogenics, membranes (oxygen separation), among others, can be considered as part of the oxy-combustion technologies.

In addition, in some industries, there could be a wide range of production processes, which impact on the CO_2 emitted and composition of the flue gas, and will be considered when assessing the TRL at the relevant environment.

References

1. Masson-Delmotte, V., Zhai, P., Pörtner, H.-O. et al. (eds.) (2018). IPCC, 2018: summary for policymakers. In: *Global Warming of 1.5°C. An IPCC Special Report on the Impacts of Global Warming of 1.5°C Above Pre-industrial Levels and Related Global Greenhouse Gas Emission Pathways, in the Context of Strengthening the Global Response to the Threat of Climate Change*. Geneva, Switzerland, 32 pp.: World Meteorological Organization.
2. Giannaris, S., Jacobs, B., Srisang, W. et al. (2019). Heat integration analysis and optimization for a post combustion CO_2 capture retrofit study of SaskPower's Shand Power Station. *Int. J. Greenhouse Gas Control* 84: 62–71.
3. Rock, L., McNaughton, C., Black, A. et al. (2017). Assessment of CO_2 levels prior to injection across the quest sequestration lease area. *Energy Procedia* 114: 2836–2846.
4. Allam, R., Martin, S., Forrest, B. et al. (2017). Demonstration of the Allam cycle: an update on the development status of a high efficiency supercritical carbon dioxide power process employing full carbon capture. *Energy Procedia* 114: 5948–5966.
5. Knudsen, J.N., Bade, O.M., Askestad, I. et al. (2014). Pilot plant demonstration of CO_2 capture from cement plant with advanced amine technology. *Energy Procedia* 63: 6464–6475.
6. Bjerge, L.-M. and Brevik, P. (2014). CO_2 capture in the cement industry, Norcem CO_2 Capture Project (Norway). *Energy Procedia* 63 (1876): 6455–6463.
7. Hills, T.P., Sceats, M., Rennie, D., and Fennell, P. (2017). LEILAC: low cost CO_2 capture for the cement and lime industries. *Energy Procedia* 114: 6166–6170.
8. Herzog, H.J. (2018). *Carbon Capture*. The MIT Press Essential Knowledge Series.
9. Fout, T., Zoelle, A., Keairns, D. et al.. Cost and Performance Baseline for Fossil Energy Plants Volume 1b: Bituminous Coal (IGCC) to Electricity. NETL-PUB-22638. https://netl.doe.gov/projects/files/CostAndPerformanceBaselineForFossilEnergyPlantsVol1BitumCoalAndNGtoElectBBRRev4-1_092419.pdf.
10. Wu, S., Kukoski, A., Jin, P. et al. *Development of Oxyfuel Combustion Technology for Existing Power Plants*, 1–6. Hitachi Power Systems America, Ltd www.hitachipowersystems.us.
11. Younas, M., Sohail, M., Kong, L.L. et al. (2016). Feasibility of CO_2 adsorption by solid adsorbents: a review on low-temperature systems. *Int. J. Environ. Sci. Technol.* 13 (7): 1839–1860.
12. Samanta, A., Zhao, A., Shimizu, G.K.H. et al. (2012). Post-combustion CO_2 capture using solid sorbents: a review. *Ind. Eng. Chem. Res.* 51 (4): 1438–1463.

13 Cherbański, R. and Molga, E. (2009). Intensification of desorption processes by use of microwaves: an overview of possible applications and industrial perspectives. *Chem. Eng. Process. Process Intensif.* 48 (1): 48–58.

14 Chronopoulos, T., Fernandez-Diez, Y., Maroto-Valer, M.M. et al. (2014). CO_2 desorption via microwave heating for post-combustion carbon capture. *Microporous Mesoporous Mater.* 197: 288–290.

15 Yu, C.H., Huang, C.H., and Tan, C.S. (2012). A review of CO_2 capture by absorption and adsorption. *Aerosol Air Qual. Res.* 12 (5): 745–769.

16 Abanades, J.C. (2013). Chapter 21-Calcium looping for CO_2 capture in combustion systems. In: *Fluidized Bed Technologies for Near-Zero Emission Combustion and Gasification*, 931–970. Woodhead Publishing Series in Energy.

17 Abanades, J.C., Arias, B., Lyngfelt, A. et al. (2015). Emerging CO_2 capture systems. *Int. J. Greenhouse Gas Control* 40: 126–166.

18 Mores, P.L., Arias, A.M., Scenna, N.J. et al. (2019). Cost-based comparison of multi-stage membrane configurations for carbon capture from flue gas of power plants. *Int. J. Greenhouse Gas Control* 86: 177–190.

19 Roussanaly, S. and Anantharaman, R. (2017). Cost-optimal CO_2 capture ratio for membrane-based capture from different CO_2 sources. *Chem. Eng. J.* 327: 618–628.

20 Lockwood, T. (2016). Next-generation carbon capture technologies for coal. *IEA Clean Coal Centre, CCC/265, London, United Kingdom.*

21 Wang, Y., Zhao, L., Otto, A. et al. (2017). A review of post-combustion CO_2 capture technologies from coal-fired power plants. *Energy Procedia* 114: 650–665.

22 Abu-Zahra, M.R.M., Sodiq, A., and Feron, P.H.M. (2016). *Commercial Liquid Absorbent-Based PCC Processes.* Elsevier Ltd.

23 Xu, Z., Wang, S., Zhao, B., and Chen, C. (2013). Study on potential biphasic solvents: absorption capacity, CO_2 loading and reaction rate. *Energy Procedia* 37: 494–498.

24 Gladis, A., Gundersen, M.T., Fosbøl, P.L. et al. (2017). Influence of temperature and solvent concentration on the kinetics of the enzyme carbonic anhydrase in carbon capture technology. *Chem. Eng. J.* 309: 772–786.

25 Aines, R.D., Spaddaccini, C.M., Duoss, E.B. et al. (2013). Encapsulated solvents for carbon dioxide capture. *Energy Procedia* 37: 219–224.

26 Zheng, D.-Q., Ma, W.-D., Wei, R., and Guo, T.-M. (1999). Solubility study of methane, carbon dioxide and nitrogen in ethylene glycol at elevated temperatures and pressures. *Fluid Phase Equilib.* 155 (2): 277–286.

27 The DOW Chemical Company (2015). The DOW Chemical Company Product Safety Assessment SELEXOL TM Solvents. pp. 1–5.

28 Takahashi, S., Song, K.Y., and Kobayashi, R. (1984). Experimental vapor-liquid equilibriums in the carbon dioxide-diethylene glycol-water and carbon dioxide-triethylene glycol-water systems at feasible absorption temperatures and pressures. *J. Chem. Eng. Data* 29 (1): 23–28.

29 Jou, F.-Y., Otto, F.D., and Mather, A.E. (2000). Solubility of H_2S and CO_2 in diethylene glycol at elevated pressures. *Fluid Phase Equilib.* 175: 53–61.

30 Lin, H., Wu, T., and Lee, M. (2003). Isothermal vapor – liquid equilibria for binary mixtures of carbon dioxide with diethylene glycol (diethyl, butyl, hexyl, or dibutyl) ether at elevated pressures. *Fluid Phase Equilib.* 209: 131–145.
31 Henni, A., Tontiwachwuthikul, P., and Chakma, A. (2008). Solubilities of carbon dioxide in polyethylene glycol ethers. *Can. J. Chem. Eng.* 83 (2): 358–361.
32 Garcia, M., Knuutila, H.K., Aronu, U.E., and Gu, S. (2018). Influence of substitution of water by organic solvents in amine solutions on absorption of CO_2. *Int. J. Greenhouse Gas Control* 78: 286–305.
33 Guo, C., Chen, S., Zhang, Y., and Wang, G. (2014). Solubility of CO_2 in non-aqueous absorption system of 2-(2-aminoethylamine)ethanol + benzyl alcohol. *J. Chem. Eng. Data* 59: 1796–1801.
34 Yuan, Y. and Rochelle, G.T. (2019). Lost work: a comparison of water-lean solvent to a second generation aqueous amine process for CO_2 capture. *Int. J. Greenhouse Gas Control* 84: 82–90.
35 Le Moullec, Y., Neveux, T., Al Azki, A. et al. (2014). Process modifications for solvent-based post-combustion CO_2 capture. *Int. J. Greenhouse Gas Control* 31: 96–112.
36 Bhattacharyya, D. and Miller, D.C. (2017). Post-combustion CO_2 capture technologies – a review of processes for solvent-based and sorbent-based CO_2 capture. *Curr. Opin. Chem. Eng.* 17: 78–92.
37 Rochelle, G.T., Wu, Y., Chen, E. et al. (2019). Pilot plant demonstration of piperazine with the advanced flash stripper. *Int. J. Greenhouse Gas Control* 84: 72–81.
38 Song, C., Liu, Q., Ji, N. et al. (2018). Alternative pathways for efficient CO_2 capture by hybrid processes – a review. *Renewable Sustainable Energy Rev.* 82: 215–231.
39 EG&G Technical Services, Inc (2004). *Fuel Cell Handbook*, 7e). US-DOE NETL Under Contract No. DE-AM26-99FT40575.
40 Adams, T.A., Nease, J., Tucker, D., and Barton, P.I. (2013). Energy conversion with solid oxide fuel cell systems: a review of concepts and outlooks for the short- and long-term. *Ind. Eng. Chem. Res.* 52 (9): 3089–3111.
41 Amrollahi, Z., Ystad, P.A.M., Ertesvåg, I.S., and Bolland, O. (2012). Optimized process configurations of post-combustion CO_2 capture for natural-gas-fired power plant – power plant efficiency analysis. *Int. J. Greenhouse Gas Control* 8: 1–11.
42 Bolland, O. (2010). *Thermal Power Generation*, 32. NTNU.
43 Adams, T. and Mac Dowell, N. (2016). Off-design point modelling of a 420 MW CCGT power plant integrated with an amine-based post-combustion CO_2 capture and compression process. *Appl. Energy* 178: 681–702.
44 Jordal, K., Ystad, P.A.M., Anantharaman, R. et al. (2012). Design-point and part-load considerations for natural gas combined cycle plants with post combustion capture. *Int. J. Greenhouse Gas Control* 11: 271–282.
45 Lucquiaud, M., Chalmers, H., and Gibbins, J. (2009). Capture-ready supercritical coal-fired power plants and flexible post-combustion CO_2 capture. *Energy Procedia* 1 (1): 1411–1418.

46 Pan, M., Aziz, F., Li, B. et al. (2016). Application of optimal design methodologies in retrofitting natural gas combined cycle power plants with CO_2 capture. *Appl. Energy* 161: 695–706.

47 Karimi, M., Hillestad, M., and Svendsen, H.F. (2011). Capital costs and energy considerations of different alternative stripper configurations for post combustion CO_2 capture. *Chem. Eng. Res. Des.* 89 (8): 1229–1236.

48 Diego, M.E., Bellas, J.-M., and Pourkashanian, M. (2017). Process analysis of selective exhaust gas recirculation for CO_2 capture in natural gas combined cycle power plants using amines. *J. Eng. Gas Turbines Power* 139 (12): 121701.

49 Gonzalez Díaz, A., Sanchez, E., Gonzalez Santalób, J.M. et al. (2014). On the integration of sequential supplementary firing in natural gas combined cycle for CO_2-enhanced oil recovery: a technoeconomic analysis for Mexico. *Energy Procedia* 63: 7558–7567.

50 Jonshagen, K., Sipöcz, N., and Genrup, M. (2010). A novel approach of retrofitting a combined cycle with post combustion CO_2 capture. *J. Eng. Gas Turbines Power* 133 (1): 011703.

51 IEAGHG (2012). Operating Flexibility of Power Plants with CCS.

52 Brouwer, A.S., van den Broek, M., Seebregts, A., and Faaij, A. (2015). Operational flexibility and economics of power plants in future low-carbon power systems. *Appl. Energy* 156: 107–128.

53 Johnsson, F., Odenberger, M., and Göransson, L. (2014). Challenges to integrate CCS into low carbon electricity markets. *Energy Procedia* 63: 7485–7493.

54 Gonzalez-Salazar, M.A., Kirsten, T., and Prchlik, L. (2018). Review of the operational flexibility and emissions of gas- and coal-fired power plants in a future with growing renewables. *Renewable Sustainable Energy Rev.* 82: 1497–1513.

55 Montañés, R.M., Korpås, M., Nord, L.O., and Jaehnert, S. (2016). Identifying operational requirements for flexible CCS power plant in future energy systems. *Energy Procedia* 86 (1876): 22–31.

56 Rezazadeh, F., Gale, W.F., Akram, M. et al. (2016). Performance evaluation and optimisation of post combustion CO_2 capture processes for natural gas applications at pilot scale via a verified rate-based model. *Int. J. Greenhouse Gas Control* 53: 243–253.

57 Hauger, S.O., Flø, N.E., Kvamsdal, H. et al. (2019). Demonstration of non-linear model predictive control of post-combustion CO_2 capture processes. *Comput. Chem. Eng.* 123: 184–195.

58 Lawal, A., Wang, M., Stephenson, P., and Obi, O. (2012). Demonstrating full-scale post-combustion CO_2 capture for coal-fired power plants through dynamic modelling and simulation. *Fuel* 101: 115–128.

59 Montañés, R.M., GarÐarsdóttir, S., Normann, F. et al. (2017). Demonstrating load-change transient performance of a commercial-scale natural gas combined cycle power plant with post-combustion CO_2 capture. *Int. J. Greenhouse Gas Control* 63: 158–174.

60 Gardarsdóttir, S., Montañés, R.M., Normann, F. et al. (2017). Effects of CO_2-absorption control strategies on the dynamic performance of a supercritical pulverized-coal-fired power plant. *Ind. Eng. Chem. Res.* 56 (15): 4415–4430.

61 Bui, M., Gunawan, I., Verheyen, V. et al. (2016). Flexible operation of CSIRO's post-combustion CO_2 capture pilot plant at the AGL Loy Yang power station. *Int. J. Greenhouse Gas Control* 48: 188–203.

62 Montañés, R.M., Flø, N.E., Dutta, R. et al. (2017). Dynamic process model development and validation with transient plant data collected from an MEA test campaign at the CO_2 Technology Center Mongstad. *Energy Procedia* 114 (1876): 1538–1550.

63 Panahi, M. and Skogestad, S. (2012). Economically efficient operation of CO_2 capturing process. Part II. Design of control layer. *Chem. Eng. Process. Process Intensif.* 52: 112–124.

64 Marx-Schubach, T. and Schmitz, G. (2018). Dynamic simulation and investigation of the startup process of a postcombustion-capture plant. *Ind. Eng. Chem. Res.* 57 (49): 16751–16762.

65 Martinez Castilla, G., Biermann, M., Montañés, R.M. et al. (2019). Integrating carbon capture into an industrial combined-heat-and-power plant: performance with hourly and seasonal load changes. *Int. J. Greenhouse Gas Control* 82: 192–203.

66 Mechleri, E., Fennell, P.S., and Mac Dowell, N. (2017). Optimisation and evaluation of flexible operation strategies for coal- and gas-CCS power stations with a multi-period design approach. *Int. J. Greenhouse Gas Control* 59: 24–39.

67 IEAGHG (2013). Iron and Steel CCS Study (techn-economics integrated steel mill). Report 2013/04.

68 IEAGHG (2013). Deployment of CCS in the cement industry, 2013/19.

69 CEMCAP (2018). D 4.6 – Comparative techno-economic analysis of CO_2 capture in cement plants, Deliverable 4.6, 2018.

70 Skagestad, R., Normann, F., Garðarsdóttir, S.Ó. et al. (2017). CO_2 stCap – cutting cost of CO_2 capture in process industry. *Energy Procedia* 114 (1876): 6303–6315.

71 IEAGHG (2018). Cost of CO_2 capture in the industrial sector: cement and iron and steel industries, 2018/03.

72 Jansen, D., Gazzani, M., Manzolini, G. et al. (2015). Pre-combustion CO_2 capture. *Int. J. Greenhouse Gas Control* 40: 167–187.

73 Rackley, S.A. (2009). *Carbon Capture and Storage*. Butterworth-Heinemann.

74 IEA. *CO_2 Capture and Storage: A Key Carbon Abatement Option*. Paris: OECD Publishing.

75 Mathias, P.M., Reddy, S., Smith, A., and Afshar, K. (2013). A guide to evaluate solvents and processes for post-combustion CO_2 capture. *Energy Procedia* 37: 1863–1870.

76 Kvamsdal, H.M., Ehlers, S., Kather, A. et al. (2016). Optimizing integrated reference cases in the OCTAVIUS project. *Int. J. Greenhouse Gas Control* 50: 23–36.

2

Advancing CCSU Technologies with Computational Fluid Dynamics (CFD): A Look at the Future by Linking CFD and Process Simulations

Daniel Sebastia-Saez[1], *Evgenia Mechleri*[1], *and Harvey Arellano-García*[1,2]

[1] *University of Surrey, Department of Chemical and Process Engineering, Guildford GU2 7XH, United Kingdom*
[2] *Brandenburgische Technische Universität Cottbus-Senftenberg, LS Prozess- und Anlagentechnik, D-03046, Cottbus, Germany*

2.1 Sweep Across the General Simulation Techniques Available

The application of simulation techniques to the study of carbon capture, storage and utilization (CCSU) underpins significant advantages such as gaining detailed insight into the underlying phenomena and reducing the experimental load and its associated risks. Simulation methods of interest for CCSU applications include those dealing with quantum mechanics, molecular dynamics, continuum mechanics (computational fluid dynamics, also known as CFD), and process engineering simulations, each with different length and time scales as illustrated in Figure 2.1.

At the smallest time and length scales, quantum mechanics focuses on solving the equations of quantum theory, whereas molecular dynamics is based on the application of Newton's laws of motion to a limited set of molecules. Quantum mechanics and molecular dynamics thus provide information on the phenomena taking place at the subatomic and molecular level, respectively.

The application of quantum mechanics and molecular dynamics to the field of CCSU provides valuable data on fractional free volume (space between particles), diffusion coefficients, and a better understanding of the role of hydrogen bonds on the absorption of CO_2 into amino acid ionic liquids (AAILs) [1]; the mechanism for carbon dioxide sequestration within porous rocks [2]; the study of the phase change mechanism of biphasic solvents for CO_2 capture [3]; and the adsorption mechanism including selectivity of different gas molecules in porous matrices [4]. In conclusion, quantum mechanics and molecular dynamics provide a visualization of the diffusion and chemical bonding at the subatomic and molecular level and are of interest for the development and study of new solvents and solid sorbents and the evaluation of sequestering media both in terms of diffusion time and depth.

In some other instances, however, researchers might need to obtain a broader vision of the phenomena taking place at larger time and length scales. Having

Engineering Solutions for CO_2 Conversion, First Edition.
Edited by Tomas R. Reina, José A. Odriozola, and Harvey Arellano-Garcia.

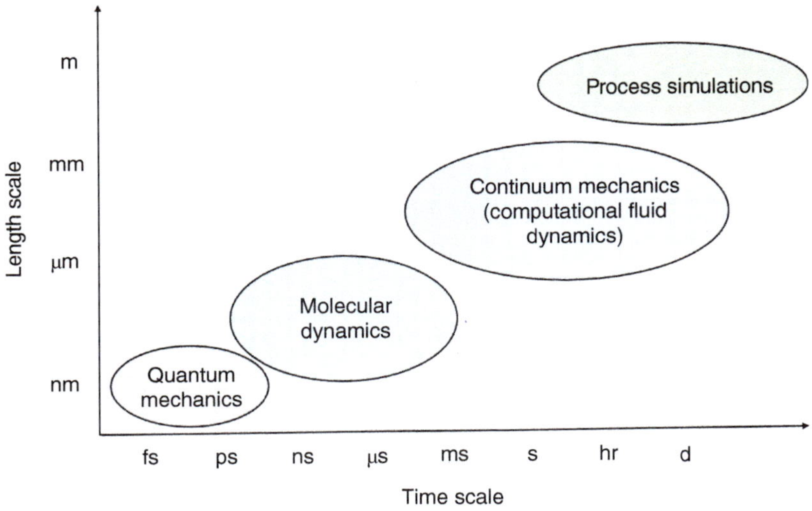

Figure 2.1 Time and length scales of the simulation methods available for CCSU processes from quantum mechanics to molecular dynamics, continuum mechanics, and process engineering simulations.

commented briefly on the capabilities of quantum mechanics and molecular dynamics, the next step toward greater length and time scales is given by the application of continuum mechanics, which assumes that matter is a continuum instead of being formed by discrete particles. There is thus a boundary in terms of length scale between the suitability of the use of molecular dynamics or continuum mechanics for a given problem. Generally speaking, continuum mechanics must be applied instead of molecular dynamics when the representative length scale of the phenomenon to be studied is greater than the mean free path of the particles; that is, when the length scale contains a large number of particles. The Knudsen number $Kn = {}^{\lambda}/_{L}$, where λ is the mean free path and L the length scale of the problem, gives a rule of thumb as to when to apply either methodology, with continuum mechanics being applicable when $Kn \ll 1$. Continuum mechanics is of interest for the study of the unit operations within a CCSU process. For instance, it is useful when the modeler needs to perform a preliminary proof of concept of a packed bed column or an oxy-fuel burner and needs to represent its entire geometry. In some applications, however, it is unfeasible to perform a continuum mechanics simulation of the entire unit process, and thus, a multi-scale approach might be necessary owing to computational capacity limitations. This becomes clear for the case of packed bed columns, where the intricate geometry requires the use of different scales depending on the variable to be studied.

The application of continuum mechanics to fluid flows gives way to the Navier–Stokes equations. As of today, the analytical solution of these equations is known only upon some restrictive assumptions, and therefore, numerical methods must be applied in the majority of cases of interest. The study of the numerical solving of the Navier–Stokes equations conforms the discipline known as computational fluid dynamics (CFD). Although historically the term CFD has referred to the numerical solving of the Navier–Stokes equations, it has been expanded recently to include the study and development of Lattice Boltzmann methods. The development of CFD kicked off within the field of aeronautics but has rapidly entered into other fields such as chemical engineering with the advent of increased computing power and refined algorithms capable of describing multiphase flows, chemical reactions, interface tracking, diffusion, etc.

The Navier–Stokes equations result from a mass and momentum balance on an infinitesimal control volume within the fluid being studied. The mass conservation equation reads

$$\frac{\partial \rho}{\partial t} + \nabla \cdot (\rho \vec{v}) = S_{\text{mass}} \tag{2.1}$$

where ρ is the density of the fluid, t denotes the time, $\vec{v}$ is the velocity vector associated with the infinitesimal control volume, and S_{mass} is a mass source term. Mass source terms are crucial to the description of reactive flows, as it occurs in CO_2 chemisorption in packed columns and in CO_2 utilization by chemical conversion, for instance. On the other hand, the momentum equation is the second Newton's law applied to the infinitesimal control volume, whereby the sum of forces equals its acceleration. The momentum balance in one of its simplest forms is expressed as

$$\rho \frac{D\vec{v}}{Dt} = -\nabla p + \nabla \cdot \tau + \rho \vec{g} + S_{\text{mom}} \tag{2.2}$$

where flow incompressibility is assumed and p denotes pressure, τ is the viscous stress tensor, $\vec{g}$ is the acceleration of gravity, and D denotes the material derivative. Momentum source terms S_{mom} are helpful when the flow in a porous media is to be considered, for instance, as will be described in Section 2.2.

Finally, on the opposite end of the time- and length-scale spectrum, process engineering simulations perform mass balances coupled with the equations that define the functioning of each device within an entire plant or part of it. The body of literature dealing with the application of process simulations in the field of CCSU is considerable, and as of today, process simulations remain the most widely used simulation approach in the field of CCSU. Process simulations are used to test new plant configurations, for instance, and give a general overview of the techno-economic viability of CCSU processes, although the information lacks detail on the physical phenomena taking place within each unit process.

Simulations can therefore give valuable insight into CCSU technologies, which would be otherwise rather expensive to obtain if only experimental methods were to be applied. This chapter aims at providing the reader with a summary of the capabilities of CFD simulation techniques across the field of CCSU, from carbon capture to sequestration and utilization. Particular emphasis is to be put on those specifics that have already undergone comprehensive validation, with the aim of letting the reader know the applications where these models can be applied most confidently.

2.2 Multi-scale Approach for CFD Simulation of Amine Scrubbers

As of today, amine scrubbing in packed columns remains the most advanced method for CO_2 capture, having been deployed commercially for some years [5]. Accordingly, it has also received the greatest attention in terms of CFD modeling among all of the CCSU methods available. High-fidelity simulations of the entire absorber describing the reactive multiphase flow within are, however, far beyond computational capacity. For this reason, the CFD modeling of amine scrubbers must be undertaken using a multi-scale perspective. Although at present, the boundaries between scales are very clear in every CFD study on structured packings, these boundaries will be blurred in the future with the upcoming of more powerful computational resources.

An interesting multi-scale approach for amine-based absorbers was proposed by Raynal and Royon-Lebeaud [6], and reviewed by Raynal et al. [7], who divided the modeling in the three scales illustrated in Figure 2.2. First off, two-dimensional (2-D) simulations were used at the small scale to represent the gravity-driven liquid flow falling down the corrugations of the packing by using an interface tracking numerical technique named Volume-of-Fluid (VOF) method. The latter introduces an additional conservation equation for a colour function f, the volume fraction, which varies within the closed interval from 0 through 1. Cells within the mesh with values at both ends of that range correspond to either phase, whereas values in between correspond to the interface. The VOF method has proved a useful tool for the description of a number of free surface interface problems across many engineering fields. It ideally requires however the use of adaptive grids or in the case of fixed grids at least a greater resolution in those cells that are expected to contain the interface [8]. This requirement comes motivated by the discontinuous character of gas–liquid interfaces in Nature, whereas an unnatural transition area between phases appears instead of a sharp interface when using the VOF algorithm. The

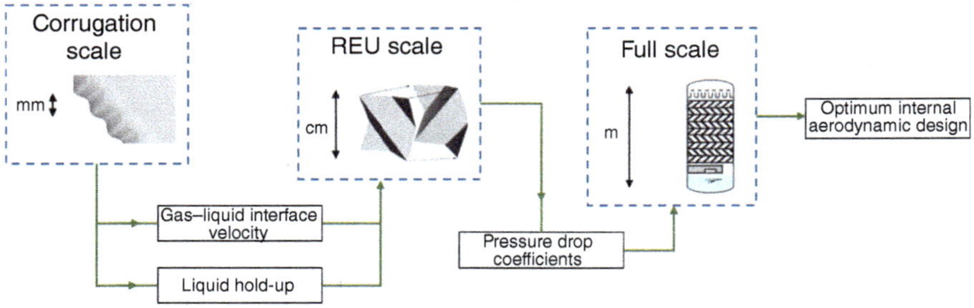

Figure 2.2 This schematic illustration depicts the flow of information between the three scales proposed in the modeling strategy presented by Raynal and Royon-Lebeaud. Source: Adapted from Raynal and Royon-Lebeaud [6].

results can therefore be affected by substantial errors if the mesh is not treated conveniently in the vicinity of the interface. Moreover, the VOF method is characterized by the assumption that the density and viscosity assigned to the interface cells are volume fraction averaged. There is therefore some variability in the density along the computational domain that results in the appearance of additional mass and momentum source terms in the conservation equations. The implications of these additional source terms are yet to be assessed and are minimized when extra mesh refinement is introduced [9]. The average film thickness and the gas–liquid interface velocity were the direct output from the calculations at the small scale in Raynal and Royon-Lebeaud's work. Then, the authors estimated indirectly the total liquid hold-up as the product between the average liquid film thickness from the simulation and the specific surface of the packing given by the manufacturer.

At the intermediate scale, one representative elementary unit (REU, the repeating unit that forms the geometry of the packing) was represented because the pressure drop per unit length in a limited set of REUs matches that of the entire column [10]. Raynal and Royon-Lebeaud [6] proposed a single-phase flow approach (with only gas phase) at the REU level, although in their case, information from the small scale would be fed into their REU set-up in order to account for the effect of the presence of liquid. To do so, two modifications at the REU scale setup were introduced. On the one hand, instead of a non-slip boundary condition on the packing walls, i.e. velocity equal zero, the authors established a fixed velocity (the gas–liquid interface velocity obtained at the small scale). On the other hand, and in order to compare their data with a set of experiments including gas and liquid flow, the authors corrected the F-factor by the liquid hold-up obtained from the average liquid film thickness of the small-scale simulations. The F-factor F is a measure of the kinetic energy of the gas phase that enters the packed bed and comes determined by the product between the superficial velocity and the square root of the gas-phase density. The corrected value of the F-factor F' is obtained by dividing it over the volume occupied by the gas phase per unit volume of packing, that is, the complementary value of the liquid hold-up. A greater F-factor then, than that of a dry simulation, is obtained for the same value of the gas superficial velocity. This is in accordance with the narrowing effect that the presence of the liquid phase has on the channels through which the gas flows, and therefore, greater gas pressure drops are obtained upon wet conditions. These intermediate scale calculations allowed the authors to obtain the pressure drop coefficients K in the three coordinate directions as the ratio per unit length between the pressure drop ΔP and the superficial dynamic pressure of the gas phase.

Finally, the entirety of the absorber is considered at the full scale. The pressure coefficients calculated at the REU scale are the link between the latter and the full scale and are therefore introduced in the full-scale simulations as an additional momentum sink in Eq. (2.2). The final outcome from this multi-scale approach was the velocity field within the absorber, with the objective of avoiding the possibility of flooding and selecting the most appropriate gas distributor so as to minimize strong velocity gradients.

Other multi-scale strategies have followed in order to model amine scrubbers, such as that presented by Sun et al. [11], who proposed a two-scale approach partially based on CFD. At the REU-scale three-dimensional (3-D) multiphase (gas and liquid), CFD simulations of a limited set of REUs of the commercial packing Mellapak 350X were used to obtain the stream split fraction coefficients and the effective wetted area ratio as a function of the liquid flow rate. The authors distinguished between four REU types, namely, inlet of the column, outlet, wall, and interior. From that stage, they mapped the entire set of REUs within the column, assigning an identifier based on their location and subsequently, introduced the parameters from the small-scale calculations mentioned above into a mechanistic model. The final output was the liquid hold-up distribution across the entire column, which is necessary to obtain the wet pressure drop by means of experimental correlations. Although not entirely based on CFD, the scale-based approach reported by Sun et al. [11] proved that multiphase simulations with gas–liquid interface tracking via the VOF method are possible with the advent of computational capabilities with increased power. The use of the computationally costly VOF method was not restricted to 2-D simulations at the corrugation level or 3-D simulations over a simple inclined plate, but at the REU scale.

A number of other models falling under the umbrella of one of the three scales presented by Raynal and Royon-Lebeaud [6] can be found in the literature, showcasing interesting characteristics of the flow within an amine scrubber. At the corrugation scale 2- and 3-D simulations featuring the VOF algorithm have been used to study the fundamental hydrodynamics of rivulets and droplets. Some attempts to introduce the chemistry between carbon dioxide and the amine have also been reported in 2-D by Haroun et al. [12] and in 3-D by Sebastia-Saez et al. [13]. At the REU scale, most of the studies have focused on obtaining the dry pressure drop of the packing, where an accurate selection of the appropriate turbulence model is crucial [14]. Finally, at the full scale, a big step forward has been taken by introducing the multiphase reactive flow on the porous Euler-Euler approach. To accomplish this, Pham et al. [15] introduced mass and momentum source terms added to the Navier–Stokes equations. These are the terms marked as S_{mass} and S_{mom} in Eqs. (2.1) and (2.2), respectively, to account for the anisotropy of the packing. To include these source terms in the calculation, one can make use of a user-defined function, like those used in commercial software such as ANSYS Fluent, or directly code their defining equations in an open-source CFD software such as OpenFOAM or an in-house code. The momentum source terms were accounted for the loss of momentum on both the liquid and the gas phase caused by the porous resistance, momentum exchange between both phases and liquid dispersion. For a complete description of the momentum source terms, the reader is referred to the work of Pham et al. [15]. Table 2.1 gathers information on CFD models related to amine scrubbers in terms of the scale used and the conclusions reached.

Table 2.1 Summary of published CFD studies concerning amine scrubbers.

Authors	Simulation scale	Aspect studied	Short comment on conclusions
van Baten et al. [16]	REU (intermediate scale)	Quantification of axial and radial liquid dispersion on KATAPAK-S commercial structured packing	Enhanced axial dispersion in the case of KATAPAK-S relative to common packings
Petre et al. [17]	REU (intermediate scale)	Description and quantification of the main contributions to dry pressure drop	The most significant causes of pressure loss are elbow effect and jet splitting at packed bed entrance, elbow effect at column walls, elbow effect at layer transitions, and collision at crisscross junctions
Raynal et al. [18]	REU (intermediate scale)	Multi-scale REU + corrugation scale study	Combining dry pressure drop from 3-D REU-scale and liquid hold-up from 2-D corrugation scale is a successful strategy to match wet pressure drop data obtained experimentally
Fernandes et al. [19, 20]	REU (intermediate scale)	Study of dry and wet pressure drop in a considerable number of REUs (whole column section)	The geometry used encompasses one of the largest sets of REUs found in the literature along with that represented by Isoz and Haidl [21]. Wet pressure drop obtained assuming fully developed film flow over the entire packing surface
Armstrong et al. [22]	REU (intermediate scale)	Effect of corrugation angle and F-factor on dry pressure drop caused by friction	Adding the effect of corrugation angle in correlations improves dry pressure drop predictions
Owens et al. [23]	REU (intermediate scale)	Optimization of computational resources to simulate more effectively at the REU scale. Use of X-ray computer tomography to generate geometries	Number of REU considerably increased to include half an element of Mellapak N250Y

Table 2.1 (Continued)

Authors	Simulation scale	Aspect studied	Short comment on conclusions
Haroun et al. [24]	REU (intermediate scale)	Gas–liquid interface tracking in four REU geometries. Their approach constitutes a step forward toward integrating the corrugation and REU scales	Not only is the REU scale valid for representing the dry pressure drop per unit length but also the interface area versus liquid flow rate
Lautenschleger et al. [25]	REU (intermediate scale)	Single-phase (gas) flow characteristics of novel packing PD-10	Optimization of packing geometry to reduce pressure drop without jeopardizing mass transfer
Sebastia-Saez et al. [26]	REU (intermediate scale)	Characteristics of multiphase gas–liquid flow on a limited set of REUs	Dependence between liquid flow rate and both gas–liquid interface area and liquid hold-up successfully represented in a set of four REUs
Li et al. [27]	REU (intermediate scale)	Transition between corrugation and REU scale	Wet pressure drop and liquid hold-up complemented with data obtained at the corrugation scale
Isoz and Haidl [21]	REU (intermediate scale)	Parametric study on the effect of slope of packing channels among others on the dry pressure drop. The authors used the imaging software Blender to obtain accurate representations of the column inner intricacies	Largest set of REUs along with that presented by Fernandes et al. [19, 20]
Yang et al. [28]	REU (intermediate scale)	Multiphase VOF gas–liquid flow within three piled REUs of Mellapak 250 Y. The work focuses on the effect of liquid flow patterns on the formation of dead zones	Portion of dead zones increased at high values of the Weber No. Wetted area highly affected by liquid flow rate but not by gas flow rate. The latter affects film stability and droplet formation
Asendrych et al. [29]	Full scale	Operating parameters including profiles of velocity of both phases, pressure liquid hold-up, and species concentration by using the Euler–Euler multiphase method + mass source terms for chemical reactions	The model was validated in terms of CO_2 capture efficiency by comparison to own experimental data

Table 2.1 (Continued)

Authors	Simulation scale	Aspect studied	Short comment on conclusions
Niegodajew and Asendrych [30]	Full scale	Parametric 2-D study of the entire column using the full-scale porous medium approach	There is a mild effect of the carbon dioxide concentration at inlet conditions on the capture efficiency, which highlights the suitability of amine scrubbing on diverse scenarios (change of fuel, etc.)
Kim et al. [31]	Full scale	Effect of four modification factors in the Ergun equation related to the liquid hold-up and the liquid load in order to determine the pressure drop	Good match between simulation and experimental results in terms of carbon dioxide conversion, wet pressure drop, and hold-up using the new factors in the Ergun equation
Gu et al. [32]	Corrugation scale (small scale)	2-D study of the effect of corrugation geometry on the hydrodynamics of liquid films	Research on surfactants and treatment of packing surface are crucial to obtain desired film flow patterns
Valluri et al. [33]	Corrugation scale (small scale)	2-D multiphase study of gravity-driven liquid films flowing down a corrugated textured solid surface (Mellapak) at low Reynolds numbers	The ratio between gas–liquid interface and specific packing area diminishes with increased surface roughness
Ataki and Bart [34]	Corrugation scale (small scale)	Morphology of rivulets in Rombopak 4M	Simulations used to develop or modify the degree of wetting, liquid hold-up, and effective area
Haroun et al. [35]	Corrugation scale (small scale)	2-D direct multiphase simulation of the formation of liquid recirculation areas in packing corrugation and its effect on mass transfer	Recirculation areas substantially affect the liquid hold-up (liquid dead zones). Mass transfer not affected by the formation of recirculation areas
Iso et al. [36]	Corrugation scale (small scale)	Study of the effect of packing texture patterns on the gas–liquid interface area	To place ridges perpendicularly to the flow direction results in enhanced gas–liquid interface area
Sebastia-Saez et al. [37]	Corrugation scale (small scale)	Study of the hydrodynamics of rivulet instabilities (braiding)	Braiding caused by interplay between surface tension and inertia and results in reduced interface area. Packing texture must be designed to hamper the effect of surface tension

2.3 Eulerian, Eulerian–Lagrangian, and Discrete Element Methods for the Simulation of Calcium Looping, Mineral Carbonation, and Adsorption in Other Solid Particulate Materials

Alongside amine scrubbing, calcium looping (CaL) has also proved a promising technology to be implemented commercially [38]. A CaL facility consists of two reactors: the carbonator and the calciner. In the carbonator, CO_2 is chemically trapped in CaO particles to form limestone ($CaCO_3$); the regeneration of the sorbent and subsequent release of CO_2 takes place in the calciner. CFD offers the possibility of studying the interaction between a particulate solid phase and a gaseous phase in a fluidized bed by means of the Eulerian multiphase method. Phase tracking using the Eulerian method is (similarly to the VOF method) accomplished by solving the volume fraction field. The difference between the VOF and the Eulerian method is that a single set of conservation equations is solved in the VOF method for all of the phases involved, whereas in the Eulerian method, a different set of conservation equations is solved for each phase in the system, thus enabling the description of granular flows. The Eulerian method has been widely used in the literature to model the flow within fluidized beds, while the VOF method has traditionally been reserved to tracking free interface problems such as the gas–liquid flow within a structured packing column. The work published by Atsonios et al. [39] is a good example of the capabilities of the Eulerian model to describe the fluid-granular flow within a CaL plant. They developed a CFD Eulerian model of both the carbonator and the calciner and linked it to advanced thermodynamic models developed using Aspen Plus to account for the reaction kinetics. The CFD setup included the two-fluid model and the energy minimization multi-scale (EMMS) model for calculating the drag forces between the particulate and the gas phase. Figure 2.3 shows visually the details of the mesh and the results obtained including maps of solids volume fraction and species mass fraction. In linking both CFD and process simulations, the authors bridged the gap between continuum mechanics and process modeling. An alternative to feed the reaction kinetics from external sources is to implement them directly on the CFD software by way of mass source terms linked to the mass conservation equation (S_{mass} term in Eq. (2.1)). This however can result in unstable convergence or even divergence, and therefore, using the stiff chemistry solver in the CFD setup might be necessary.

The work of Atsonios et al. [39] proved that a CFD strategy based on the Eulerian method is a powerful tool to obtain values of the relevant variables in any spot within the calciner. Another approach to solve the multiphase fluid–particle flow in some carbon capture technologies is the discrete element method (DEM) and the dense discrete phase model (DDPM), which can also be combined with reaction kinetics. DEM is based on the application of Newton's laws of motion to a set of particles. Maps of particle size distributions can be obtained by using DEM (instead of the less descriptive maps of volume fraction that the Eulerian method delivers). DDPM

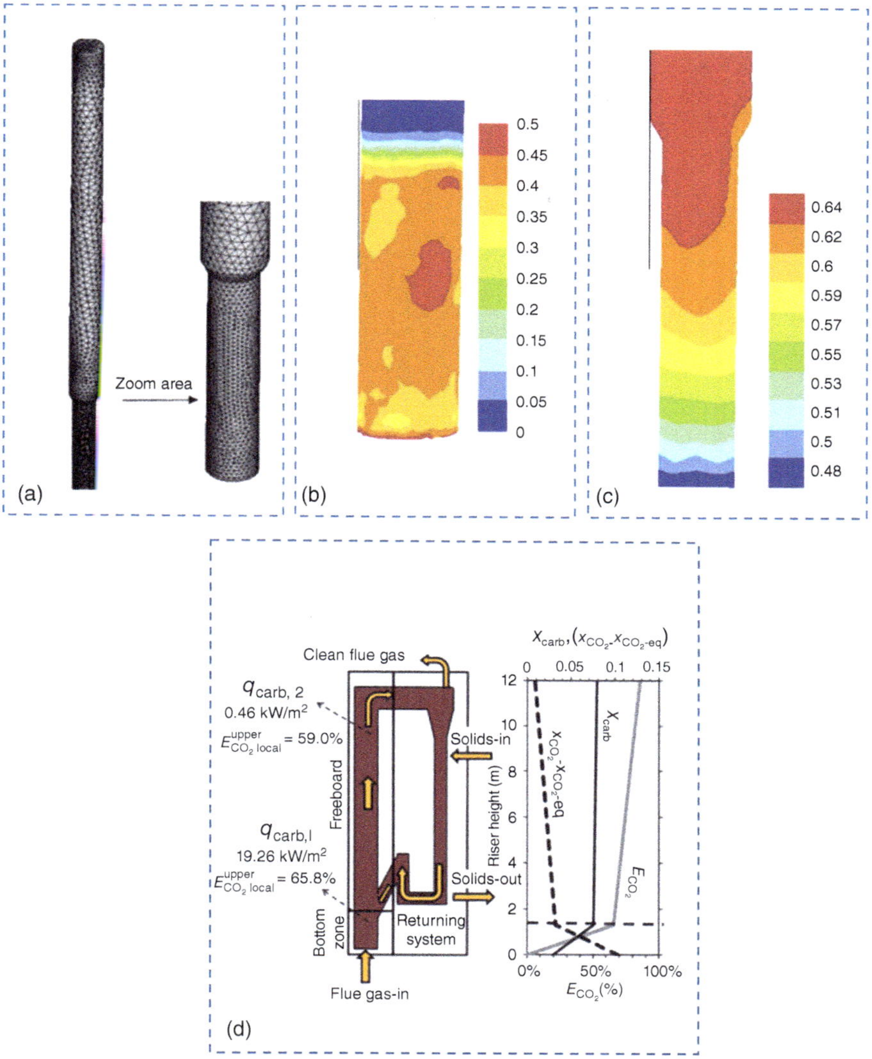

Figure 2.3 (a) Details of the space discretization of the calciner, (b) map of time-averaged solids volume fraction, (c) map of time-averaged CO_2 mass fraction, and (d) schematic of the carbonator and mass fractions of calcium carbonate and carbon dioxide versus riser height. Source: Atsonios et al. [39]. © Elsevier.

on the other hand is based on a hybrid Eulerian–Lagrangian approach, where the solid phase is tracked thanks to the application of Lagrangian mechanics, while the continuous phase is described by the Eulerian approach. Comments on research articles dealing with carbon capture technologies based on particulate solids using CFD or DEM models are gathered in Table 2.2.

Table 2.2 Summary of published CFD studies concerning carbon capture technologies based on adsorption into a particulate material.

Authors	Numerical technique	Capture technology modeled and short comment on findings
Abbasi et al. [40]	CFD Euler–Euler + PBM (population balance model)	MgO solid sorbent. Detailed maps of carbon dioxide concentration, reaction rates, and solid volume fraction
Ryan et al. [41]	CFD (Euler–Euler and Eulerian–Lagrangian)	Comparison between numerical techniques (ANSYS fluent Eulerian–Lagrangian results in unstable calculations)
Barelli et al. [42]	CFD (Euler–Euler)	CaO solid sorbent. Proof of concept of a novel reactor configuration. Maps of chemical species involved
Sornumpol et al. [43]	CFD (Euler–Euler)	Chemical looping. Maps of chemical species conversion
Kim et al. [44]	CFD (Euler–Euler). The lumped element model resulted in economy of computational resources	Mineral carbonation ($Ca(OH)_2$ solution) in a bubble column. Maps of different species hold-up and mass transfer rate were obtained
Chen et al. [45]	CFD (Euler–Euler)	Investigated pressure swing adsorption (PSA). Transient calculations of CO_2 adsorption
Ghadirian et al. [46]	CFD (Euler–Euler)	Circulating fluidized bed (CFB) reacting loop. Contours of solid volume fraction. Transient carbon dioxide conversion rates were obtained
Wang et al. [47]	Dense discrete phase model (DDPM)	Potassium-based solid sorbent. Maps of particle size distribution were obtained

2.4 CFD for Oxy-fuel Combustion Technologies: The Application of Single-Phase Reactive Flows and Particle Tracking Algorithms

Oxy-fuel combustion is another technology that has also received a great share of attention in terms of CFD modeling, given its future potential as an economically viable carbon capture technique [48]. In burning fuel in an (almost pure) oxygen atmosphere instead of air, the products of the reaction are mainly water vapor and carbon dioxide. This results in extraordinary ease of separation of CO_2 from the exhaust gas stream. CFD models of oxy-combustion systems do not have a different setup from other combustion models, which consist mainly in a single-phase, multispecies setup where the relevant reaction kinetics need to be specified. It is worth mentioning however that radiation models need to be incorporated because of the high temperatures attained within the burning chamber. Exceptions to the common single-phase approach are the study of oxy-fuel combustion of solid particulate fuels such as the work presented by Wu et al. [49], where the Eulerian approach discussed earlier was implemented in order to track the movement of the solid phase, and the work of Bhuiyan and Naser [50], who applied the Eulerian–Lagrangian method. Table 2.3 summarizes some recent CFD studies in oxy-fuel technologies. Also, a look at the literature shows that some reported studies on oxy-fuel CFD simulations have been combined with process simulations in co-simulation strategies. Such is the work published by Edge et al. [54] and Fei et al. [55]. Co-simulation is the object of the Section 2.8, where it will be discussed further because it can give way to enhanced numerical predictions with implications also in control engineering.

2.5 CFD for Carbon Storage and Enhanced Oil Recovery (EOR): The Link Between Advanced Imaging Techniques and CFD

In terms of carbon storage and enhanced oil recovery, CFD simulations can be applied at various scales in a way similar to the methodology discussed earlier in this chapter for amine scrubbers. The most interesting scale appears to be (according to the number of articles published) the small scale though, where it is possible to utilize the VOF method to analyze the flow across a small portion of a solid porous medium representing the geometry of the pores. In this direction, He et al. [56] studied the two-phase flow between supercritical CO_2 and water in the walls of a saline aquifer. The porous rock was assumed to be formed by detached spheres in a body-centered cubic (BCC) arrangement. Their simulation setup allowed visualization of the displacement process (the porous medium was initially filled with water). The effect of wettability (i.e. contact angle), surface tension, and viscosity ratio was also assessed, obtaining the permeability saturation curves. The latter certainly constitutes essential information in terms of carbon sequestration, but the approximation of considering the porous medium as a network of perfect

Table 2.3 Summary of published CFD studies concerning oxy-fuel technologies.

Authors	Aspect studied	Short comment on findings
Gharebaghi et al. [51]	Single-phase combustion simulation for a test facility	Comparison between turbulence modeling strategies, i.e. large eddy simulation (LES) and Reynolds averaged Navier–Stokes (RANS), and experimental data
Mayr et al. [52]	3-D steady-state simulation of a natural gas furnace including radiation models and the eddy dissipation concept (EDC) model. Effect of O_2/N_2 ratios on furnace efficiency	To increase the O_2-to-N_2 ratio resulted in better furnace efficiency. Good matching between simulated and experimental results
Bhuiyan and Naser [50]	Co-firing biomass + coal using the Eulerian–Lagrangian approach	The authors included factors describing the irregular shape of the biomass particles. The effect of changing the fuel ratio combustion atmosphere in the performance parameters of the furnace
Carrasco-Maldonado et al. [53]	Single-phase approach to simulate the effect of integrating oxy-fuel technologies in a cement production plant	Validation against experimental data accomplished. The k–ω turbulence model gave way to the best results
Wu et al. [49]	Study of oxy-fuel combustion in a circulating fluidized bed (CFB). The model uses the Eulerian approach and thus this is a multiphase case that deviates from common oxy-fuel CFD studies in the literature	Detailed profiles of temperature and hydrodynamic variables are obtained, which match the experimental results. Gas hold-up is also studied, resulting in identification of gas accumulation spots

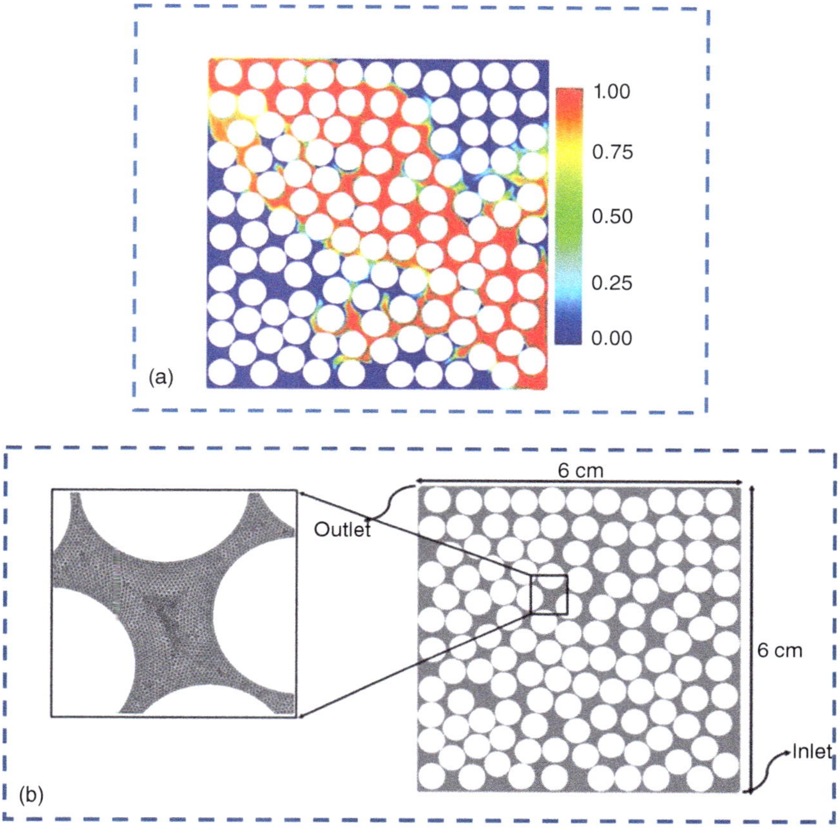

Figure 2.4 (a) Details of the volume fraction map describing the liquid flow within the porous medium. Source: Dezfully et al. [57]. © Trans Tech Publication. (b) Details of the pore geometry considered. Source: Gharibshahi et al. [58]. © Elsevier.

spheres with an ideal BCC arrangement might lead to undesired errors. Similar examples are the simulations reported by Dezfully et al. [57] and Gharibshahi et al. [58] (Figure 2.4), which also considered spheres but in a random arrangement.

The main hurdle to get accurate simulations of CO_2 trapping in porous rocks appears to be the correct representation of the intricate geometry of the pore. To overcome this, a recent trend in the simulation of the flow through porous media microchannel networks consists in applying imaging techniques, which are subsequently exported into computer-aided design (CAD) files and spatially discretized. This approach has been applied in other fields where CFD simulations are of great help to gain insight into the flow within such geometries. For instance, Sznitman et al. point out the possibility of using micro-computed tomography (μCT) or scanning transmission X-ray microscopy (STXM) to reproduce the alveoli network within the lungs. Other examples are the use that Owens et al. [23] did of X-ray computed tomography (CT) to obtain a geometric representation of a structured packing, or the work of Isoz and Haidl [21] also applied to structured packing columns. The use of such imaging techniques to be coupled with VOF-CFD simulations remains rather unexplored and will surely constitute an important research avenue in the future.

2.6 CFD for Carbon Utilization with Chemical Conversion: The Importance of Numerical Techniques on the Study of New Catalysts

Utilization of carbon dioxide with chemical conversion entails its use as a feedstock in order to transform it into other valuable products such as polymers, fuels, methanol, pharmaceuticals, urea, etc., which otherwise would need to be manufactured by using petrochemicals.

A reaction of particular importance in carbon dioxide utilization with chemical conversion is the Sabatier reaction, whereby CO_x is converted to methane by hydrogenation and subsequently introduced into the gas grid:

$$CO_2 + 4H_2 \rightarrow CH_4 + 2H_2O \tag{2.3}$$

CFD simulations of the Sabatier reaction are single phase, which is an advantage from the perspective of the computational resources needed, although they require a multispecies approach and a careful selection of the turbulence model in those cases where the Reynolds number is high. CFD can play an important role in research oriented towards the implementation of different catalysts to accelerate the production of methane. The general approach followed in that case is to obtain experimental data on the reaction kinetics first and introduce them subsequently into the simulation set-up.

A recent example of the modeling of the Sabatier reaction combined with the water gas shift reaction in a microchannel reactor was presented by Engelbrecht et al. [59], who used the commercial software COMSOL Multiphysics to perform their simulations. They assumed equally distributed flow within the microreactor, and therefore, only one microchannel was considered. To successfully model the process, the computational domain was divided in two zones: a free flow region, where the reaction kinetics without the catalyst was defined, and a porous layer, where the modified reaction kinetics was introduced to accommodate the effect of the catalyst. Figure 2.5a–c shows three parity plots between their numerical and experimental results, with fair agreement between them. Figure 2.5d shows a velocity contour plot and indicates the two areas mentioned above (the porous layer marked as a catalyst layer in the plot and the free flow zone at the center), whereas Figure 2.5e displays the computational mesh used. The reaction kinetics published earlier by Ohya et al. [60] were introduced in the software. In their experiments, the catalyst layer was formed using 8.5 wt% Ru–Cs/Al_2O_3. The model was successful in predicting the CO_2 conversion percentage as a function of working temperature and pressure, which were compared to the data obtained experimentally.

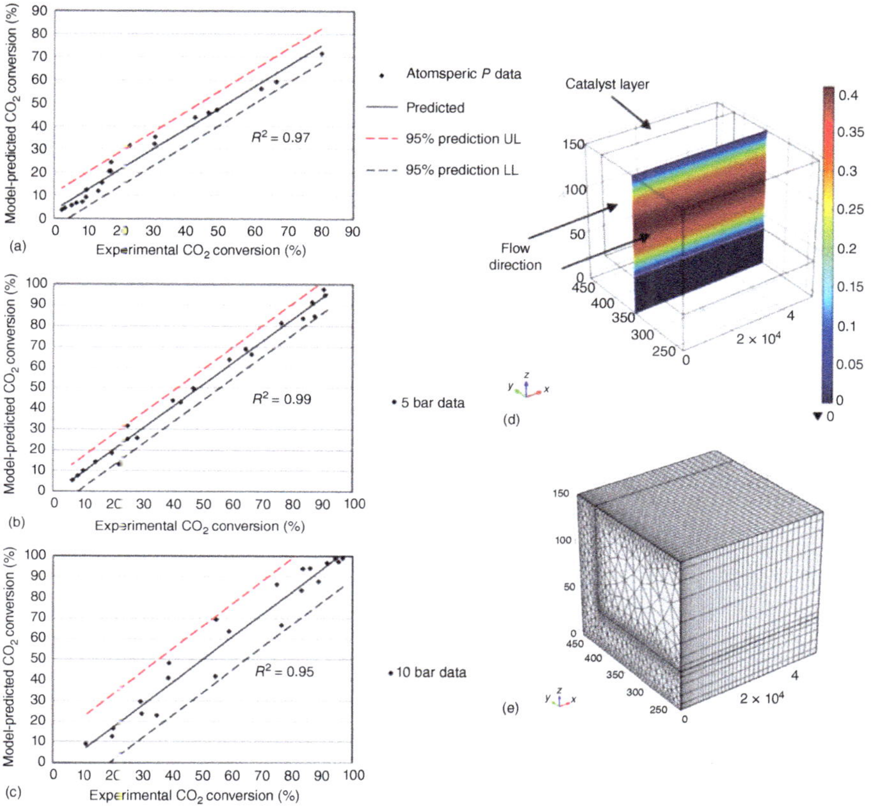

Figure 2.5 (a) Parity plot showing the match between the numerical and the experimental results at atmospheric pressure, (b) at 5 bar, and (c) at 10 bar, (d) contours of axial velocity (legend in $m\,s^{-1}$) at 250 °C and atmospheric pressure, and (e) image of the computational mesh used to obtain the numerical data. Source: Engelbrecht et al. [59]. © Elsevier.

Engelbrecht et al. [59] thus used already published data for the reaction kinetics and introduced them on their numerical model, which was subsequently validated by direct comparison with their own experiments in terms of carbon dioxide conversion percentage. The authors not only confirmed their numerical setup as a useful tool, but also confirmed the data of Ohya et al. [60] as an adequate way of describing the reaction kinetics of the methanation reaction in the presence of 8.5 wt% Ru–Cs/Al_2O_3 catalyst. The added value of their simulation was therefore to spare the experimental load needed otherwise to obtain the reaction kinetics and on the other hand the insight provided in terms of variable contours and quick parametric sweep possibilities.

A similar methodology was developed by Alarcón et al. [61], who used their own experimental setup to obtain the reaction kinetics and introduced it into the simulation. They developed their model using the commercial software ANSYS Fluent and tested the effect of a catalyst formed with 15 wt% Ni as the active component and ceria as the promoter. Similarly, they only simulated one channel and defined both a free flow and a porous zone where the reaction kinetics were modified in order to account for the presence of the catalyst.

In both instances, the effect of the catalyst was thus indirectly considered by measuring the reaction kinetics and introducing their defining parameters in the simulation set-up. The approximation of introducing the modified reaction kinetics in a porous bed is thus a well-established trend in the literature, which can yield valuable insight into how a particular catalyst affects the development of a carbon conversion reaction.

Other numerical techniques might be useful for the numerical study of the methanation reaction in other types of reactors. In those cases where the methanation reaction occurs in a fluidized bed of catalytic powder, the Euler–Euler method and the more computationally expensive CFD-DEM approach can be applied. Liu et al. for instance implemented the methanation reaction of carbon monoxide plus the water shift reaction in an Euler–Euler model of a fluidized bed using the solver *twoPhaseEulerFoam*, within the open-source CFD library OpenFOAM. The main code was modified to accommodate the kinetic model proposed by Kopyscinski et al. [62]. The effect of feed composition and the catalyst inventory on the concentration profiles across the fluidized bed were obtained and compared to the experimental data of Kopyscinski et al. [63] showing fair agreement. Another instance of the application of the Euler–Euler method to the study of the methanation reaction was presented by Sun et al. [64], who used the specific purpose software MFIX to study the methanation reaction of carbon monoxide in a fluidized bed using the Euler–Euler approach. The effect of operating parameters and catalyst inventory are investigated and the reactor will get eventually optimized. An example of the use of the coupled CFD-DEM model to analyze the hydrodynamics within a methanation fluidized bed reactor was presented by Wu and Tian [65]. DEM provides further insight as a result of the assessment of the behavior of single solid particles within the bed at the expense of a substantially increased computational time.

2.7 CFD for Biological Utilization: Microalgae Cultivation

As for the application of CFD to biological utilization of carbon dioxide for photosynthetic production of microalgae, the number of published studies is rather limited. One of the most recent and representative studies available is the work reported by Chatterjee [66]. This article is concerned with the internal hydrodynamics of a photo-bioreactor (PBR) but ignores any other aspects such as the chemistry of the system or the distribution of light across the vessel, which is crucial to the efficiency of the process and depends on the shape of the PBR. Another assumption in this model is that the mass and weight of the microalgae formed in the process were neglected.

Chatterjee [66] carried out a comparative study between a bubble column reactor and a serpentine PBR configuration using ANSYS CFX-14.0 (Euler–Euler multiphase approach) with time steps ranging from 0.1 to 2 seconds and grids with a number of nodes varying between 75k and 109k. The modeler needs to have special care with the development of the mesh, which must have an appropriate resolution relative to the size of the gas bubbles formed within the PBR. The other crucial aspect is the choice of the turbulence model. Given that the scope of this work is to assess the mixing parameters, the turbulent model must be selected carefully. Both the standard k–ε model and the Reynolds stress model (RSM) were tested. The latter option resulted in a better match between the CFD and the experimental data in terms of gas hold-up and gas velocities throughout the reactor. The set-up gave way to valuable maps of turbulence kinetic energy, velocity swirling strength, and gas hold-up within the domain. The results show that the serpentine configuration gives way to more intense turbulence, which in turn should theoretically result in better microalgae growth rate.

Perhaps the most innovative CFD work concerning PBRs is, however, the article presented by Zhang et al. (2019), who carried out a CFD study of a bionic fractal-inspired branch-like PBR. Fractal shapes are the solution provided by Nature for those applications where a high area-to-volume ratio is required. They used a similar set-up to that presented by Chatterjee [66], based on the Euler–Euler method for gas–liquid interface tracking combined with the k–ε turbulence model in order to compare the performance of fractal-inspired branched geometries to that of common configurations such as multi-tubular and serpentine. As for the spatial discretization, finer meshes were employed because of the tiny spaces formed in the successive branching generations of the fractal shape (5th branching generation with diameters in the order of the millimeter). A grid independence study was conducted using three non-structured meshes (a coarse mesh with 1.30 m nodes, a medium mesh with 2.48 m nodes, and a fine one composed of 3.5 m nodes). The prospective user can thus have an approximation of the degree of grid refinement required in these simulations given the diameter of the channels and the number of nodes, which need substantial computational resources. The results of their model suggested that better mixing occurs when using the fractal shape relative to the multi-tubular and serpentine configurations, together with small pressure drops (Figure 2.6).

The literature confirms thus, CFD modeling as a valuable tool to carry out preliminary proof of concept of innovative PBR shapes. An interesting proposal for further work would be, however, to use a multispecies approach including mass source terms to account for the consumption of carbon dioxide and the growth of microalgae.

2.8 What Does the Future Hold?

As presented previously, CFD modeling has proved a beneficial tool for the modeling of CCSU technologies, thus helping to get insight into problems where experimental measurements cannot be obtained or analytical solutions are impossible.

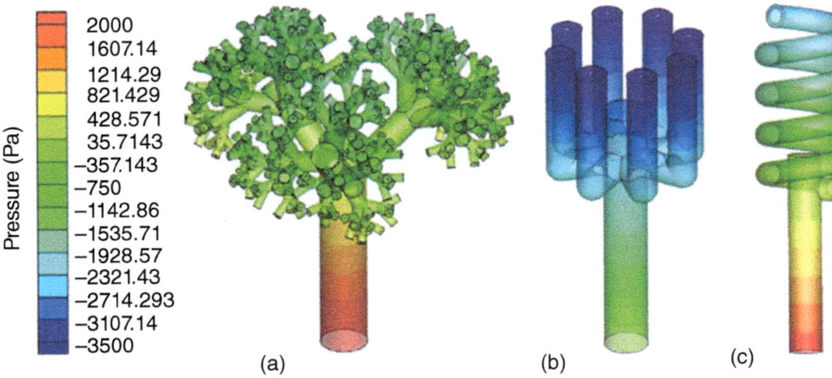

Figure 2.6 Pressure maps within PBR reactors obtained using CFD simulations to shed light on the pressure distribution of three geometric configurations: (a) fractal-inspired shape, (b) multi-tubular, and (c) serpentine. Source: Adapted from Tao et al. (2019).

Process simulations on the other hand are used in order to provide information for the design and operation of entire plants. Both CFD and process simulations have their advantages and disadvantages when applied separately, but the combination of both techniques to run in parallel and live feedback each other would offer new opportunities to analyze and optimize the overall plant performance.

There have been several efforts in different fields of engineering to integrate CFD and process simulations. Proof of this is the recent development a co-simulation software framework at the DOE National Energy Technology Laboratory in the United States, which was applied to fossil energy systems with carbon capture. This tool is called APECS (advanced process engineering co-simulator) and allows the design and optimization of the overall plant performance based on detailed high-fidelity fluid dynamic models (CFD). Other instances of co-simulation strategies applied to CCSU technologies are the work of Zitney [67], where an integrated gasification combined cycle (IGCC) power and hydrogen coproduction plant with carbon capture was analyzed by feeding data from CFD models into Aspen Plus. The results of the integration showed that the overall plant performance is affected by complex thermal and fluid flow phenomena that can only be analyzed at the CFD level; otherwise, process simulations miss those details. Another example of the intertwining between CFD and process simulations can be found in the work of Fei et al. [55], where the link between CFD and process simulations was accomplished by developing reduced order models (ROM) with the CFD data and introducing them into the process model. Edge et al. [54] on the other hand obtained temperature contours, velocities, and mole fraction maps of different species involved by using CFD and introduced the data into the process simulation tool gPROMS to assess the retrofitting of a coal-fired power plant into an oxy-fuel plant. Their approach resulted in guidelines as for the conditions where the system results in the same efficiency as air-firing. Also, similar to Fei et al. [55], Lang et al. [68] presented a co-simulation approach for an IGCC by developing an ROM via CFD in order to reduce the computational time and then optimize the plant by using process simulations. The efficiency of the process was increased by 7%, compared to the conventional configuration.

The above examples show the benefits of the co-simulation approach, in allowing the detailed interactions between fluid mechanics, heat transfer, reaction, and control strategy to be examined, and give valuable outputs to the design and operational model. The aforementioned examples also show that there is a lot to be done, given the scarcity of CFD process co-simulation studies published in the literature. For instance, and to the best of the author's knowledge, no co-simulation study has been reported regarding carbon utilization. As previously mentioned however, the combination of CFD and process simulations will certainly lead to significant research outcomes, especially in cases with CO_2 utilization where new catalysts (CFD) need to be tested in a reactor (part of a bigger process simulation) in which steady-state performance, dynamics, and control strategy depend on mixing and fluid flow behavior. More specifically, in the area of methanation, there are two different aspects that need to be combined: the methanation reactor configuration and the catalysts. Not only is the reactor design clearly influenced by the catalyst applied, its activity, and selectivity, but also are up- and downstream processes [69]. A tight interfacing between CFD calculations for the performance assessment of a given catalyst and process simulation tools for the reactor design will open the possibility for process modeling on a detailed and optimized approach.

It is evident from the aforementioned examples that the combination of process simulations and CFD will lead to a future with improved and optimized CCSU technologies. Also, the combination and implementation of different control strategies shall also provide an extra benefit.

References

1 Shaikh, A.R., Karchanechi, H., Kamio, E. et al. (2016). Quantum mechanical and molecular dynamics simulations of dual-amino-acid ionic liquids for CO_2 capture. *J. Phys. Chem. C* 120 (49): 27734–27745.

2 Cygan, R.T., Romanov, V.N., and Myshakin, E.M. (2012). Molecular simulation of carbon dioxide capture by montmorillonite using an accurate and flexible force field. *J Phys. Chem. C* 116 (24): 13079–13091.

3 Tao, M., Xu, N., Gao, J. et al. (2019). Phase-change mechanism for capturing CO_2 into an environmentally benign nonaqueous solution: a combined NMR and molecular dynamics simulation study. *Energy Fuels* 33 (1): 474–483.

4 Trinh, T.T., Tran, K.Q., Bach, Q.V., and Trinh, D.Q. (2016). A molecular dynamics simulation study on separation selectivity of CO_2/CH_4 mixture in mesoporous carbons. *Energy Procedia* 86: 144–149.

5 Wilcox, J. (2012). *Carbon Capture*. New York, NY: Springer Science + Business Media

6 Raynal, L. and Royon-Lebeaud, A. (2007). A multi-scale approach for CFD calculations of gas-liquid flow within large size column equipped with structured packing. *Chem. Eng. Sci.* 62 (24): 7196–7204.

7 Raynal, L., Ben, R.F., and Royon-Lebeaud, A. (2009). Use of CFD for CO_2 absorbers optimum design: from local scale to large industrial scale. *Energy Procedia* 1: 917–924.

8 Cooke, J.J., Armstrong, L.M., Luo, K.H., and Gu, S. (2014). Adaptive mesh refinement of gas-liquid flow on an inclined plane. *Comput. Chem. Eng.* 60: 297–306.

9 Sebastia-Saez, D., Gu, S., and Ramaioli, M. (2018). Effect of the contact angle on the morphology, residence time distribution and mass transfer into liquid rivulets: a CFD study. *Chem. Eng. Sci.* 176: 356–366.

10 Larachi, F., Petre, C.F., Iliuta, I., and Grandjean, B. (2003). Tailoring the pressure drop of structured packings through CFD simulations. *Chem. Eng. Process.* 42 (7): 535–541.

11 Sun, B., He, L., Liu, B.T. et al. (2013). A new multi-scale model based on CFD and macroscopic calculation for corrugated structured packing column. *AIChE J.* 59 (8): 3119–3130.

12 Haroun, Y., Legendre, D., and Raynal, L. (2010). Direct numerical simulation of reactive absorption in gas-liquid flow on structured packing using interface capturing method. *Chem. Eng. Sci.* 65 (1): 351–356.

13 Sebastia-Saez, D., Gu, S., Ranganathan, P., and Papadikis, K. (2015). Micro-scale CFD modeling of reactive mass transfer in falling liquid films within structured packing materials. *Int. J. Greenhouse Gas Control* 33: 40–50.

14 Said, W., Nemer, M., and Clodic, D. (2011). Modeling of dry pressure drop for fully developed gas flow in structured packing using CFD simulations. *Chem. Eng. Sci.* 66 (10): 2107–2117.

15 Pham, D.A., Lim, Y.I., Jee, H. et al. (2015). Porous media Eulerian computational fluid dynamics (CFD) model of amine absorber with structured-packing for CO_2 removal. *Chem. Eng. Sci.* 132: 259–270.

16 van Baten, J.M., Ellenberger, J., and Krishna, R. (2001). Radial and axial dispersion of the liquid phase within a KATAPAK-S® structure: experiments vs. CFD simulations. *Chem. Eng. Sci.* 56 (3): 813–821.

17 Petre, C.F., Larachi, F., Iliuta, I., and Grandjean, B.P.A. (2003). Pressure drop through structured packings: breakdown into the contributing mechanisms by CFD modeling. *Chem. Eng. Sci.* 58: 163–177.

18 Raynal, L., Boyer, C., and Ballaguet, J.P. (2004). Liquid holdup and pressure drop determination in structured packing with CFD simulations. *Can. J. Chem. Eng.* 82: 871–879.

19 Fernandes, J., Simões, P.C., Mota, J.P.B., and Saatdjian, E. (2008). Applications of CFD in the study of supercritical fluid extraction with structured packing: dry pressure drop calculations. *J. Supercrit. Fluids* 47: 17–24.

20 Fernandes, J., Lisboa, P.F., Simões, P.C. et al. (2009). Application of CFD in the study of supercritical fluid extraction with structured packing: wet pressure drop calculations. *J. Supercrit. Fluids* 50: 61–68.

21 Isoz, M. and Haidl, J. (2018). Computational-fluid-dynamics analysis of gas flow through corrugated-sheet-structured packing: effects of packing geometry. *Ind. Eng. Chem. Res.* 57: 11785–11796.

22 Armstrong, L.M., Gu, S., and Luo, K.H. (2013). Dry pressure drop prediction within Montz-pak B1-250.45 packing with varying inclination angles and geometries. *Ind. Eng. Chem. Res.* 52 (11): 4372–4378.

23 Owens, S.A., Perkins, M.R., and Eldridge, R.B. (2013). Computational fluid dynamics simulation of structured packing. *Ind. Eng. Chem. Res.* 52 (5): 2032–2045.

24 Haroun, Y., Raynal, L., and Alix, P. (2014). Prediction of effective area and liquid hold-up in structured packings by CFD. *Chem. Eng. Res. Des.* 92: 2247–2254.

25 Lautenschleger, A., Olenberg, A., and Kenig, E.Y. (2015). A systematic CFD-based method to investigate and optimise novel structured packings. *Chem. Eng. Sci.* 122: 452–464.

26 Sebastia-Saez, D., Gu, S., Ranganathan, P., and Papadikis, K. (2015). Meso-scale CFD study of the pressure drop, liquid hold-up, interfacial area and mass transfer in structured packing materials. *Int. J. Greenhouse Gas Control* 42: 388–399.

27 Li, Q., Wang, T., Dai, C., and Lei, Z. (2016). Hydrodynamics of novel structured packings: an experimental and multi-scale CFD study. *Chem. Eng. Sci.* 143: 23–35.

28 Yang, L., Liu, F., Saito, K., and Liu, K. (2018). CFD modeling on hydrodynamic characteristics of multiphase counter-current flow in a structured packed bed for post-combustion CO_2 capture. *Energies* 11 (11): 3103.

29 Asendrych, D., Niegodajew, P., and Drobniak, S. (2013). CFD modelling of CO_2 capture in a packed bed by chemical absorption. *Chem. Process Eng.* 34 (2): 269–282.

30 Niegodajew, P. and Asendrych, D. (2016). Amine based CO_2 capture – CFD simulation of absorber performance. *Appl. Math. Modell.* 40: 10222–10237.

31 Kim, I., Pham, D.A., and Lim, Y.I. (2016). Gas-liquid multiphase computational fluid dynamics (CFD) of amine absorption column with structured packing for CO_2 capture. *Comput. Chem. Eng.* 88: 39–49.

32 Gu, F., Liu, C.J., Yuan, X.G., and Yu, G.C. (2004). CFD simulation of liquid film flow on inclined plates. *Chem. Eng. Technol.* 27: 1099–1104.

33 Valluri, P., Matar, O.M., Hewitt, G.F., and Mendes, M.A. (2005). Thin film flow over structured packings at moderate Reynolds numbers. *Chem. Eng. Sci.* 60 (7): 1965–1975.

34 Ataki, A. and Bart, H.J. (2006). Experimental and CFD simulation study for the wetting of a structured packing element. *Chem. Eng. Technol.* 29 (3): 336–347.

35 Haroun, Y., Raynal, L., and Legendre, D. (2012). Mass transfer and liquid hold-up determination in structured packing by CFD. *Chem. Eng. Sci.* 75: 342–348.

36 Iso, Y., Huang, J., Kato, M. et al. (2013). Numerical and experimental study on liquid film flows on packing elements in absorbers for post-combustion CO_2 capture. *Energy Procedia* 37: 860–868.

37 Sebastia-Saez, D., Reina, T.R., and Arellano-Garcia, H. (2017). Numerical modelling of braiding and meandering instabilities in gravity-driven liquid rivulets. *Chem. Ing. Tech.* 89 (11): 1515–1522.

38 Sun, H., Wu, C., Shen, B. et al. (2018). Progress in the development and application of CaO-based adsorbents for CO_2 capture – a review. *Mater. Today Sustainability* 1–2: 1–27.

39 Atsonios, K., Zeneli, M., Nikolopoulos, A. et al. (2015). Calcium looping process simulation based on an advanced thermodynamic model combined with CFD analysis. *Fuel* 153: 370.

40 Abbasi, E., Abbasian, J., and Arastoopour, H. (2015). CFD-PBE numerical simulation of CO_2 capture using MgO-based sorbent. *Powder Technol.* 286: 616–628.

41 Ryan, E.M., DeCroix, D., Breault, R. et al. (2013). Multi-phase CFD modeling of solid sorbent carbon capture system. *Powder Technol.* 242: 117–134.

42 Barelli, L., Bidini, G., and Gallorini, F. (2016). CO_2 capture with solid sorbent: CFD modelling of an innovative reactor concept. *Appl. Energy* 162: 58–67.

43 Sornumpol, R., Uraisakul, W., Kuchonthara, P. et al. (2017). CFD simulation of fuel reactor in chemical looping combustion. *Energy Procedia* 138: 979–984.

44 Kim, M., Na, J., Park, S. et al. (2018). Modeling and validation of a pilot-scale aqueous mineral carbonation reactor for carbon capture using computational fluid dynamics. *Chem. Eng. Sci.* 177: 301–312.

45 Chen, Q., Rosner, F., Rao, A. et al. (2019). Simulation of elevated temperature solid sorbent CO_2 capture for pre-combustion applications using computational fluid dynamics. *Appl. Energy* 237: 314–325.

46 Ghadirian, E., Abbasian, J., and Arastoopour, H. (2019). CFD simulation of gas and particle flow and a carbon capture process using a circulating fluidized bed (CFB) reacting loop. *Powder Technol.* 344: 27–35.

47 Wang, S., Hu, B., Jin, C. et al. (2019). Dense discrete phase model simulations of CO_2 capture process in a fluidized bed absorber with potassium-based solid sorbent. *Powder Technol.* 345: 260–266.

48 Wu, F., Argyle, M.D., Dellenback, P.A., and Fan, M. (2018). Progress in O_2 separation for oxy-fuel combustion–a promising way for cost-effective CO_2 capture: a review. *Prog. Energy Combust. Sci.* 67: 188–205.

49 Wu, Y., Liu, D., Duan, L. et al. (2018). Three-dimensional CFD simulation of oxy-fuel combustion in a circulating fluidized bed with warm flue gas recycle. *Fuel* 216: 596–611.

50 Bhuiyan, A.A. and Naser, J. (2015). CFD modelling of co-firing of biomass with coal under oxy-fuel combustion in a large scale power plant. *Fuel* 159: 150–168.

51 Gharebaghi, M., Irons, M.R.A., Ma, L. et al. (2011). Large eddy simulation of oxy-coal combustion in an industrial combustion test facility. *Int. J. Greenhouse Gas Control* 5S1: S100–S110.

52 Mayr, B., Prieler, R., Demuth, M. et al. (2015). CFD and experimental analysis of a 115 kW natural gas fired lab-scale furnace under oxy-fuel and air-fuel conditions. *Fuel* 159: 864–875.

53 Carrasco-Maldonado, F., Bakken, J., Ditaranto, M. et al. (2017). Oxy-fuel burner investigations for CO_2 capture in cement plants. *Energy Procedia* 120: 120–125.

54 Edge, P.J., Heggs, P.J., Pourkashanian, M., and Stephenson, P.L. (2013). Integrated fluid dynamics-process modelling of a coal-fired power plant with carbon capture. *Appl. Therm. Eng.* 60: 456–464.

55 Fei, Y., Black, S., Szuhánszki, J. et al. (2015). Evaluation of the potential of retrofitting a coal power plant to oxi-firing using CFD and process co-simulation. *Fuel Process. Technol.* 131: 45–58.

56 He, D., Jiang, P., Lun, Z. et al. (2018). Pore scale CFD simulation of supercritical carbon dioxide drainage process in porous media saturated with water. *Energy Sources Part A* https://doi.org/10.1080/15567036.2018.1549155.

57 Dezfully, M.G., Jafari, A., and Gharibshahi, R. (2015). CFD simulation of enhanced oil recovery using nanosilica/supercritical CO_2. *Adv. Mater. Res.* 1104: 81–86.

58 Gharibshahi, R., Jafari, A., and Ahmadi, H. (2019). CFD investigation of enhanced extra-heavy oil recovery using metallic nanoparticles/steam injection in a micromodel with random pore distribution. *J. Pet. Sci. Eng.* 174: 374–383.

59 Engelbrecht, N., Chiuta, S., Everson, R.C. et al. (2017). Experimentation and CFD modelling of a microchannel reactor for carbon dioxide methanation. *Chem. Eng. J.* 313: 847–857.

60 Ohya, H., Fun, J., Kawamura, H. et al. (1997). Methanation of carbon dioxide by using membrane reactor integrated with water vapor permselective membrane and its analysis. *J. Membr. Sci.* 131: 237–247.

61 Alarcón, A., Guilera, J., and Andreu, T. (2018). CO_2 conversion to synthetic natural gas: reactor design over Ni–Ce/Al_2O_3 catalyst. *Chem. Eng. Res. Des.* 140: 155–165.

62 Kopyscinski, J., Schildhauer, T.J., Vogel, F. et al. (2010). Applying spatially resolved concentration and temperature measurements in a catalytic plate reactor for the kinetic study of CO methanation. *J. Catal.* 271: 262–279.

63 Kopyscinski, J., Schildhauer, T.J., and Biollaz, S.M.A. (2011). Fluidized-bed methanation: interaction between kinetics and mass transfer. *Ind. Eng. Chem. Res.* 50: 2781–2790.

64 Sun, L., Luo, K., and Fan, J. (2017). Numerical simulation of CO methanation for the production of synthetic natural gas in a fluidized bed reactor. *Energy Fuels* 31: 10267–10273.

65 Wu, C. and Tian, D. (2010). CFD-DEM simulation of syngas to methane process in a fluidized bed reactor. *The 13th International Conference on Fluidization – New Paradigm in Fluidization Engineering*, 16–21 May 2010 – Hotel Hyundai, Gyeong-ju, Korea.

66 Chatterjee, S. (2017). CFD analysis on hydrodynamic conditions of a designed spiral column photobioreactor for cultivation of microalgae. *World J. Eng.* 14 (5): 443–450.

67 Zitney, S.E. (2010). Process/equipment co-simulation for design and analysis of advanced energy systems. *Comput. Chem. Eng.* 34: 1532–1542.

68 Lang, Y., Zitney, S.E., and Biegler, L.T. (2011). Optimization of IGCC processes with reduced order CFD models. *Comput. Chem. Eng.* 35: 1705–1717.

69 Rönsch, S., Schneider, J., Matthischke, S. et al. (2016). Review on methanation – from fundamentals to current projects. *Fuel* 166: 276–296.

3

Membranes Technologies for Efficient CO_2 Capture–Conversion

Sonia Remiro-Buenamañana, Laura Navarrete, Julio García-Fayos, Sara Escorihuela, Sonia Escolástico, and José M. Serra

Instituto de Tecnología Química (Universitat Politècnica de València-Consejo Superior de Investigaciones Científicas), Av. de los Naranjos s/n, Valencia, E-46022, Spain

3.1 Introduction

The growing interest of the scientific community toward global warming and its consequences over environment has led to investigate new energy resources in order to decrease the dependence on fossil fuels. Combustion processes, mainly represented by power generation and industry sectors, account for more than 50% of CO_2 emissions, increasing at a rate of 2.5 per year with a worldwide amount of 33.1 $GtCO_2$ during 2018 [1]. Therefore, these processes are the main source of the total CO_2 emitted to the atmosphere. As part of the greenhouse effect mitigation, efforts have been made toward decreasing these atmospheric CO_2 emissions. The most considered actions are to capture the CO_2 from point source emissions [2] and to use it as a feedstock to valuable chemicals and fuels [3].

These proposed actions for enabling the CO_2 capture and conversion can be implemented by applying the so-called membrane technology. Membranes are materials acting as semipermeable barriers between two different phases, and because of their intrinsic properties, they are able to selectively transport a particular component (Figure 3.1). The different characteristics between retentate and permeate will work as the driving force and the membrane properties will set the separation degree. The transport mechanism of the membrane will be determined by the membrane composition, structure (homogeneous or heterogeneous), morphology, origin (natural or synthetic), and operation conditions [4]. Most of the membranes are based on the pore dimension principle, where the pore size will act as the cutoff of the compounds that are able to pass through the membrane, blocking in the retentate larger species [5]. Hence, as a general classification, membranes can be divided into microporous membranes and dense membranes. In addition, selective membranes without electrostatic interaction make up a significant proportion of membrane technology. A selective membrane only allows certain molecules to be transported along the membrane to the permeate side. Finally, for ion-selective membranes, the transport of ions is ascribed to the compensation of charges in the system by charge-carrying

Engineering Solutions for CO_2 Conversion, First Edition.
Edited by Tomas R. Reina, José A. Odriozola, and Harvey Arellano-Garcia.

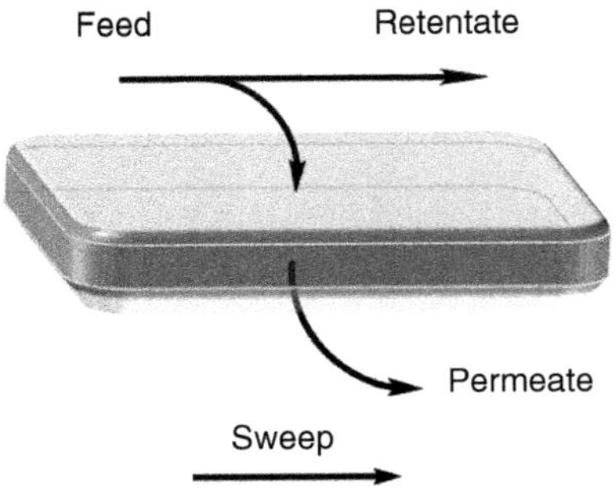

Figure 3.1 Gas transport through a membrane.

species [6]. Furthermore, membranes possess some advantages such as low energy consumption, mild reaction conditions, ability to continuously perform separation processes, generally facile scaling up, and among others [4].

This chapter will take the reader along the most relevant membranes for the aim of CO_2 carbon/capture applications, and these can be divided into polymeric, ionic, protonic, and electrochemical membranes. Metallic membranes are not included in this chapter because of several reasons, such as their production costs and CO poisoning. Some studies report the retarding effects of CO on H_2 permeation through Pd-based membranes. This detriment on H_2 diffusion through the active sites of Pd became more significant at temperatures around 250 °C and/or higher CO concentration [7–9]. In the majority of industrial applications for CO_2 capture or storage, CO may appear as a subproduct. Hence, the use of Pd membranes for H_2/CO_2 separation processes may be limited.

Regarding the different applications, membrane reactors have demonstrated to be promising candidates to tackle climate change, decreasing the levels of CO_2 by using it in capture–conversion technologies to obtain valuable chemicals [10, 11]. The present chapter describes a range of approaches toward CO_2 capture–conversion in the context of catalytic membranes. Because of the extension of the field, this chapter will briefly summarize different types of gas separation and gas absorption membranes that are currently being investigated for CO_2 applications.

3.2 Polymer Membranes

Polymer materials for gas separation membranes have been widely used for different applications in the past three decades. Although the first recorded description of a semipermeable membrane was in 1748 [12], followed by the observation of the permeation of H_2 through balloons in 1831 [13], it was not until the late 1970s when several experiments demonstrated the great commercial potential of polymeric gas separation membranes [14].

Polymer membranes are dense (nonporous) membranes, which follow the solution–diffusion model as a transport mechanism [15]. In this model, gas is transported through a dense polymer membrane in three steps: (i) dissolving into the face of the membrane that is exposed to high gas pressure, (ii) diffusion through the polymer, and finally (iii) desorbing from the face of the membrane that is exposed to low pressure. The rate-limiting step for gas separation polymer membranes is the second step, diffusion of gas through the polymer material [15]. As a consequence, permeability can be expressed as the product of diffusion coefficient and solubility coefficient.

$$\rho = D \cdot S \tag{3.1}$$

Diffusion coefficient is related to the kinetic terms, and it reflects the mobility of the individual molecules in the membrane material. In other words, it depends on the molecular size of the target gas (step ii). On the other hand, solubility coefficient links the concentration of a component in the fluid phase with its concentration in the membrane polymer phase and reflects the number of molecules dissolved in the membrane material (step i and iii). It depends on molecular interaction; hence, it is an equilibrium term [15].

Regarding the selectivity or separation factor of polymer membranes, it is defined as a product of mobility selectivity, (D_i/D_j) the fraction of the diffusion coefficients of the two gases, the solubility coefficient (S_i/S_j), and the ratio of the solubility coefficients of the two gases.

$$\alpha_{ij} = \alpha_D \cdot \alpha_S = \left[\frac{D_i}{D_j}\right] \cdot \left[\frac{S_i}{S_j}\right] \tag{3.2}$$

Based on these previous explanations, membrane material is more related to diffusion coefficient, being more affected by the membrane changes, than solubility coefficient. In order to explain the concept in a more visual way, as an example, diffusion and solubility coefficients of four different gases in a group of several related polyimides are plotted against each other. Both coefficients are reasonably well grouped for each gas as it can be shown in Figure 3.2. Thus, for any gas, the difference in diffusion coefficient from the highest value is 100 times bigger than the lowest value, while the deviation in solubility coefficients is only 2–4 times. Changes in polymer chemistry affect both diffusion and solubility coefficients, but the effect on the diffusion coefficient is more meaningful [16]. Diffusion coefficient will differ depending on the polymer nature (rubber or glass). There is a considerable difference between the motion of polymer segments in a malleable rubbery polymer and in a stiff glassy polymer. Consequently, diffusion coefficients of glassy polymer are usually lower than the diffusion coefficients of rubbery polymers, and in addition, diffusion coefficients of glassy polymers decrease faster with increasing permeate molecular size. In other words, the mobility selectivity term for rubbery membranes is smaller than the mobility coefficient of glassy membranes.

Fractional free volume (FFV) parameter is used in order to correlate permeation properties with the structure and chemical properties of the polymers. Permeability factor of gases in glassy polymers are highly dependent on the FFV, which can be defined as the free space that is not occupied by the polymer molecules [18]. FFV is defined by

$$v_f = \frac{v - v_0}{v} \tag{3.3}$$

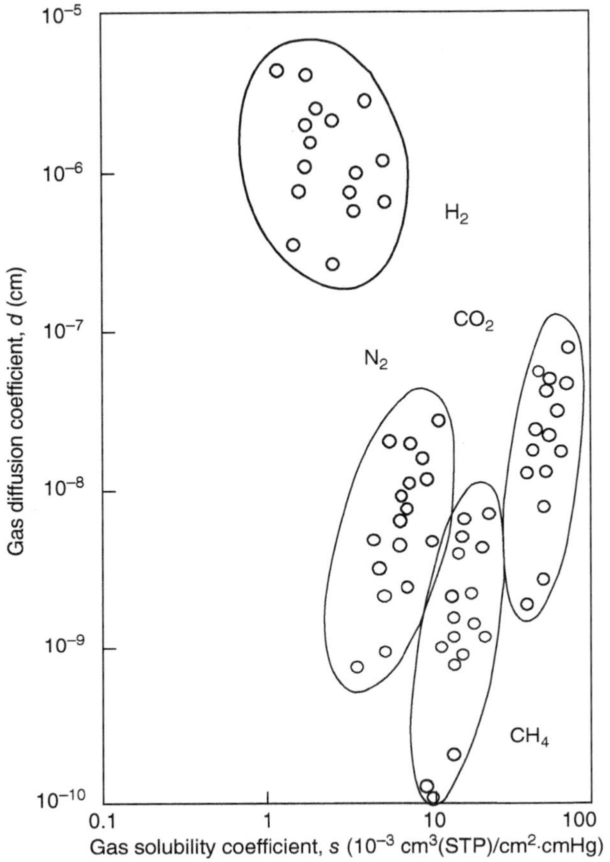

Figure 3.2 Diffusion and solubility coefficient for different polyimides. Source: Adapted from Baker and Tanaka et al. [16, 17].

where v is the specific volume of the polymer (cm^3 g^{-1}), i.e. the correlative of the polymer density, and v_0 is the volume occupied by the polymer molecules (cm^3 g^{-1}). In principle, FFV is the sum of all the spaces between the polymer chains.

Nowadays, polymer membranes for CO_2 capture are being intensively investigated, specifically regarding post-combustion carbon capture (CO_2/N_2), hydrogen purification (CO_2/H_2), and natural gas sweetening (CO_2/CH_4) [19]. In terms of post-combustion carbon capture, competing with chemical absorption, gas separation membranes appear as a more environmentally friendly solution. As an example, a simulation of two potential polymer membranes, Polyactive® and PVAm/polyvinyl alcohol membranes, was performed for a 600 MW (gross) reference power plant [20]. Other materials, such Pebax, were also investigated for similar purposes, regarding CO_2 capture in post-combustion processes [21].

Regarding H_2 purification, the predominant technology is steam methane reforming (SMR), where H_2 is obtained together with, mainly, CO_2 (15–20%) among other gases [22]. Polymeric membranes not only show advantageous properties, such as ease of processing or low cost, but also would be more accessible for large-scale applications over other types of membranes. Consequently, and regarding high-temperature processes, different polyimides such as 6FDA–6FpDA and Matrimid, with outstanding H_2/CO_2 separation properties at temperatures up to 270 °C, have been recently reported [23, 24].

With regard to natural gas sweetening, or in other words, CO_2 removal from natural gas, membrane systems show greater potential than conventional chemical absorption because they have main advantages such as low cost, environmentally friendly technologies, and process flexibility [25]. Several commercial membranes are available for CO_2 elimination from natural gas. The most common representative materials for CO_2 removal are cellulose acetate/triacetate and polyimide [25].

In order to study the properties of different membrane materials, and to be able to compare different polymer materials, Robeson published several charts, the first time in 1991 and the latest in 2008, about membrane permeability as a function of selectivity. These charts are called upper bound correlation [26, 27]. These experimental data are obtained from single gas tests at 30 °C and 1 bar. This indicates that permeabilities were measured using pure gas tests, and selectivities were obtained from the ratio of the pure gas permeabilities. This gives ideal selectivities, despite the fact that industrial processes of gas separation membranes are commonly performed with gas mixtures. Nevertheless, it is possible to extrapolate to industrial applications, this is because, in principle, if gases do not interact with the membrane material, the difference between single gas selectivity and gas mixture selectivity will be small. Consequently, in gas mixtures, molecules that possess high-solubility coefficients will be sorbed enough by the membrane material to affect the other gas permeabilities.

Hence, upper bound correlations are described for several gas pairs. Regarding CO_2 separation, two different gas pair (CO_2/N_2 and CO_2/CH_4) are studied, CO_2/CH_4 being the second most investigated gas pair for membrane separation. In Figure 3.3a,b, a large number of polymers have been shown, which are mostly placed between a selectivity of 2–100 and a permeability of 0.1–1000. In addition, two black lines are identifying the prior upper bound and the present upper bound. Any polymer that surpasses the present upper bound can be considered as the most promising polymer for CO_2 separation.

As an example, PIM-1 with a permeability of 2300 Barrer and a selectivity of 18.4, or PIM-7 with a permeability of 1100 Barrer and a selectivity of 17.7 [28], is considered as a ladder polymer for the gas pair CO_2/CH_4, which surpasses the upper bound [29]. Thermally rearranged polymers have been recently published, which exhibit outstanding improved separation characteristics, although the present upper bound is determined without these data [30], see Figure 3.3a.

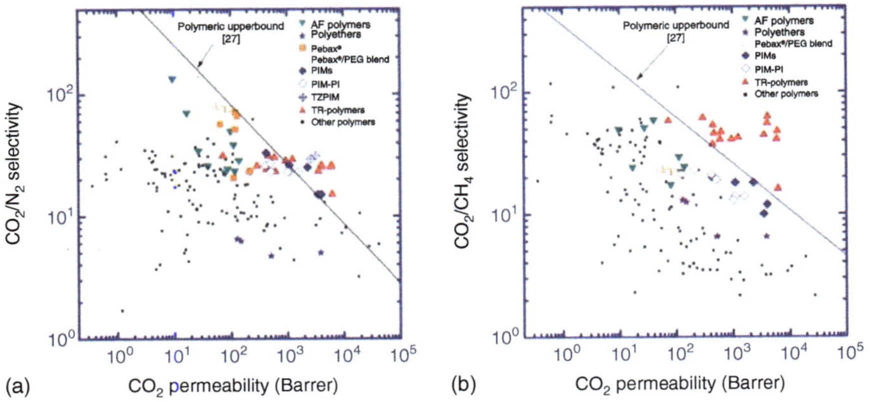

Figure 3.3 Upper bound correlation for (a) CO_2/CH_4 separation and (b) CO_2/N_2 separation in 2008. Source: Kim and Lee [28].

3.3 Oxygen Transport Membranes for CO_2 Valorization

The application of strategies for capturing the CO_2 from combustion processes is of great importance for effectively implementing CO_2 conversion approaches and thus reducing the impact of industrial activity regarding CO_2 emissions. Oxyfuel technology is one of the most considered options for conducting the implementation of this capture and sequestration of CO_2 from exhaust gases, as well as for improving the efficiency of the aforementioned industrial processes [31–33]. The oxyfuel approach applied to combustion processes consists of burning the fuel with a pure or O_2-rich stream, thus obtaining mainly CO_2 and H_2O as products. Consequently, CO_2 sequestration can be easily carried out with a simple separation step.

Nevertheless, oxyfuel technology is still far from being implemented in the most of targeted industrial processes. The main reason is related to the O_2 supply, which is not economically feasible for those applications. Currently, almost all the O_2 is produced by cryogenic distillation of air. This process, which is operated at very low temperatures and high pressures, requires from high energy and large production plants that avert its consideration for O_2 on-site production in small- and medium-scale installations. An appealing solution for implementing an O_2 supply system in such applications can be performed by considering oxygen transport membranes (OTMs). OTMs are ceramic membranes consisting of metallic oxides with the ability of diffusing O^{2-} ions through the oxygen vacancies present in their crystal lattice. This is due to the mixed ionic–electronic conductivity (MIEC) of these materials at high temperatures (>600 °C), thus allowing O_2 separation with 100% selectivity from a high pO_2 feed stream (pressurized air) to a low pO_2 sweep stream (vacuum or recirculated flue gas stream). Furthermore, waste heat streams generated in high-temperature processes in several industries requiring from O_2 supply for conducting combustions (e.g. cement plants, ceramic and glass production, and power plants) can be used for heating the OTM modules up to their operating temperature. This thermal integration can then result in a significant increase in plant efficiency and in a reduction in O_2 production costs, which can be lowered up to 35% with respect to conventional cryogenic distillation [34].

3.3.1 Oxygen Transport Membrane Fundamentals

The ceramic materials exhibiting MIEC properties that are considered for constituting OTMs because of their good O_2 permeation properties are known since the 1970s [35–37]. OTMs typically consist of dense layer(s) of multimetallic oxides comprising alkali, alkali earth, rare earth, or transition metals together in the same crystalline structure. The O_2 permeation through a ceramic membrane comprises several steps, which are depicted in Figure 3.4a,b.

As can be seen in Figure 3.4a,b, O_2 permeation depends on two main processes: oxygen bulk diffusion (step 5) and oxygen surface reactions (steps 2–4 and 6–8). Figure 3.4c depicts O_2-bulk diffusion through oxygen vacancies in the material's cristal lattice. Typically, O_2 permeation is described by using Wagner's equation (Eq. (3.4))

$$J\left(O_2\right) = \frac{RT}{16F^2L}\int_{pO_2'}^{pO_2''}\frac{\sigma_{el}\sigma_{ion}}{\sigma_{el}+\sigma_{ion}}d\ln pO_2 \tag{3.4}$$

where $J(O_2)$ is the O_2 flux through the membrane, R is the gas constant, T is the temperature, F is Faraday's constant, L is the membrane thickness, pO_2'' and pO_2' stand for the O_2 partial pressures at feed and permeate sides, respectively, and σ_{el} and

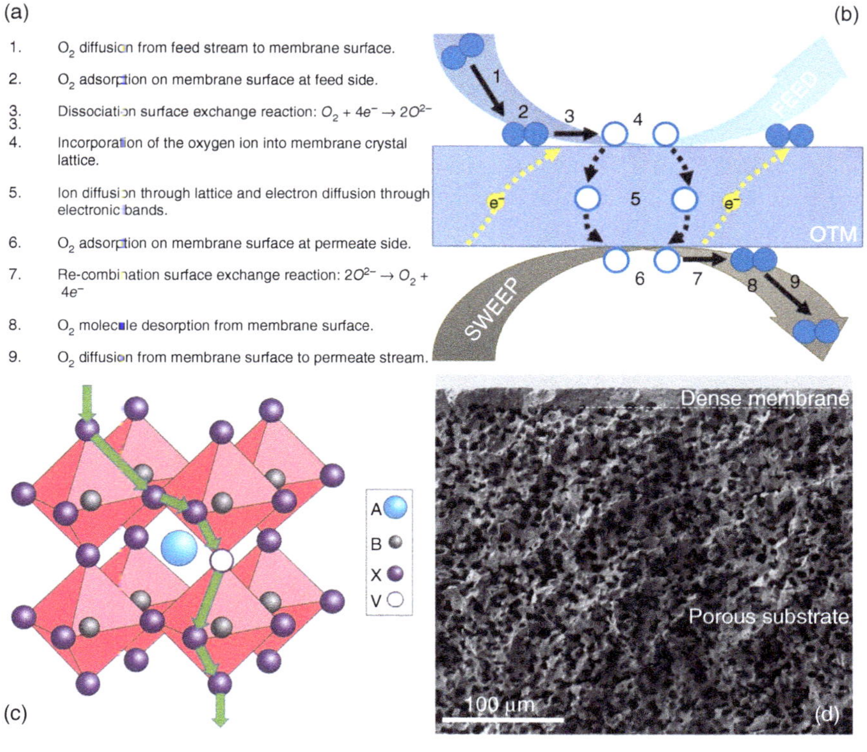

Figure 3.4 (a, b) Steps involved in the oxygen permeation through an oxygen transport membrane and (c) representation of the oxygen anion diffusion through the oxygen vacancies present in a perovskite's crystal lattice and (d) fracture cross. Source: (c) Bouwmeester and Burggraaf [38].

σ_{ion} are the electronic and ionic conductivities of the considered membrane material, respectively. Wagner's equation describes the O_2 permeation through the material bulk (step 5 in Figure 3.4a,b), when the membrane is thicker than the value known as the characteristic thickness L_c [38], which is distinctive of every material.

For thinner membranes and for temperatures typically lower than 800 °C, $J(O_2)$ does not depend on O^{2-} bulk diffusion but on oxygen surface exchange kinetics. These surface reactions involve a number of steps that may include O_2 adsorption, O_2 dissociation into O^{2-} and electrons, charge transfer, surface diffusion of intermediate species, and finally O^{2-} incorporation into the material crystal lattice [39, 40]. A general expression for describing $J(O_2)$ can be obtained by assuming linear kinetics of the relevant rate laws when both surface exchange reactions and bulk diffusion are governing the O_2 permeation and then $J(O_2)$ [41] can be expressed by means of the following expression:

$$J\left(O_2\right) = \frac{1}{1 + 2L_c/L}\frac{1}{16F^2}\frac{\sigma_{el}\sigma_{ion}}{\sigma_{el} + \sigma_{ion}}\frac{\Delta\mu_{O_2}^{total}}{L} \tag{3.5}$$

where $\Delta\mu_{O_2}^{total}$ is the total O_2 chemical potential difference across the membrane.

Therefore, the O_2 permeation of an OTM depends on temperature, membrane thickness, pO_2 gradient between the two membrane sides, and the MIEC properties of the considered material. Then, higher O_2 fluxes can be achieved by operating at higher temperatures and larger pO_2 gradients (use of pressurized feeding or high vacuum or sweeping flows), by using thinner membranes (material brittleness will require the use of porous supports for ensuring mechanical stability as depicted in Figure 3.4d where a thin supported membrane is shown) and by selecting materials with high ionic and electronic conductivities.

The most considered materials for being used as OTMs are those with perovskite and fluorite crystal structures [38]. In addition to these materials, other compounds that also exhibit interesting properties are pyrochlore ($A_2B_2O_7$), brownmillerite ($A_2B_2O_5$), Ruddlesden–Popper series ($A_{n+1}B_nO_{3n+1}$), orthorhombic K_2NiF_4-type structure materials, and $Sr_4Fe_{6-x}Co_xO_{13}$ [42–45]; nevertheless, the performance of this materials is very low in comparison with fluorites and perovskites. Among the mentioned structures, the perovskite with composition $Ba_{0.5}Sr_{0.5}Co_{0.8}Fe_{0.2}O_{3-\delta}$ (BSCF) is that presenting the highest O_2 permeation up to date, achieving an O_2 production of up to 67.7 ml min^{-1} cm^{-2} at 1000 °C for a 70 μm thick membrane [46]. Nevertheless, the unpractical stability under certain environments (especially when exposed to CO_2 atmospheres) makes BSCF-based membranes unsuitable for most of the industrial applications if a direct contact between flue gases and membranes is considered. Fluorites, on the contrary, exhibit outstanding chemical and mechanical stability when exposed to oxyfuel and reaction atmospheres, but their low electronic conductivity averts them for being considered for practical applications because of the low O_2 permeation performance. One solution is then the use of dual-phase structures consisting of a mixture of ionic-conductive and electronic conductive materials. These materials with good O_2 permeation and stability when subjected to harsh environments have attracted a lot of interest within the past years in studies focused on oxyfuel applications [47–51], achieving interesting O_2 fluxes of c. 3 ml min^{-1} cm^{-2} at 925 °C under full CO_2 environments [51].

3.3.2 Application Concepts of OTMs for Carbon Capture and Storage (CCS)

As already stated for the application of CCS strategies, OTMs are considered as O_2 supply units for the conduction of combustion processes. In that way, an OTM module is integrated within an industrial process following approaches that use waste heat streams for heating up the module to its operation temperature (typically 800–900 °C). Therefore, the operation and integration can be done according to two different configurations, known as 3-end and 4-end modes (Figure 3.5a,b). These modes differ mainly on the number of streams connected to the OTM module. As can be seen in Figure 3.5a, the 4-end mode uses a fraction of the flue gas (mainly consisting of CO_2 and H_2O) from the burner for heating the module and for performing O_2 separation as a sweeping agent. As already pointed out, the recirculation of flue gases increases the plant efficiency, also reducing O_2 production costs due to the highly synergetic thermal integration of OTM modules in the process. Differently, in the 3-end mode, any contact between flue gases and membrane materials is avoided. In this configuration, O_2 is extracted using a vacuum pump at the permeate side, thus obtaining a pure O_2 stream that is used for combusting the fuel. The thermal integration is done by heating the air feed stream with heat exchangers where the heat is transferred from the exiting flue gases. This operation mode is considered for the cases where membrane material cannot operate under flue gas atmospheres. The selection of one or another configuration will influence parameters such as module design (especially the required membrane surface area) and plant layout, also determining the overall plant efficiency [52, 53]. Nevertheless, there is no preferred mode for conducting the OTM module integration. This can be observed in the OTM developments carried out by Praxair [54] and Air Products and Chemicals Inc. [55] (the most advanced up to date), which consider 4-end and 3-end operation modes, respectively.

3.3.3 Existing Developments

Currently, the R&D efforts are mainly focused on the development of new materials and membrane architectures with better performance and stability, as well on the search of new applications. All these developments are being carried out at laboratory scale; nevertheless, there are companies and institutions that are advancing in the upscaling of OTM concepts, with significant progresses in integrating ceramic membrane modules in industrial environments. Among all, the most advanced progresses have been achieved by Praxair and Air Products [56, 57], with the construction of demonstrative plants with OTM technology at high Technology Readiness Levels (TRL). These companies have worked for more than 20 years in the development of industrial-scale OTM modules and integrated gasification combined cycle (IGCC) systems with ceramic membranes for O_2 separation.

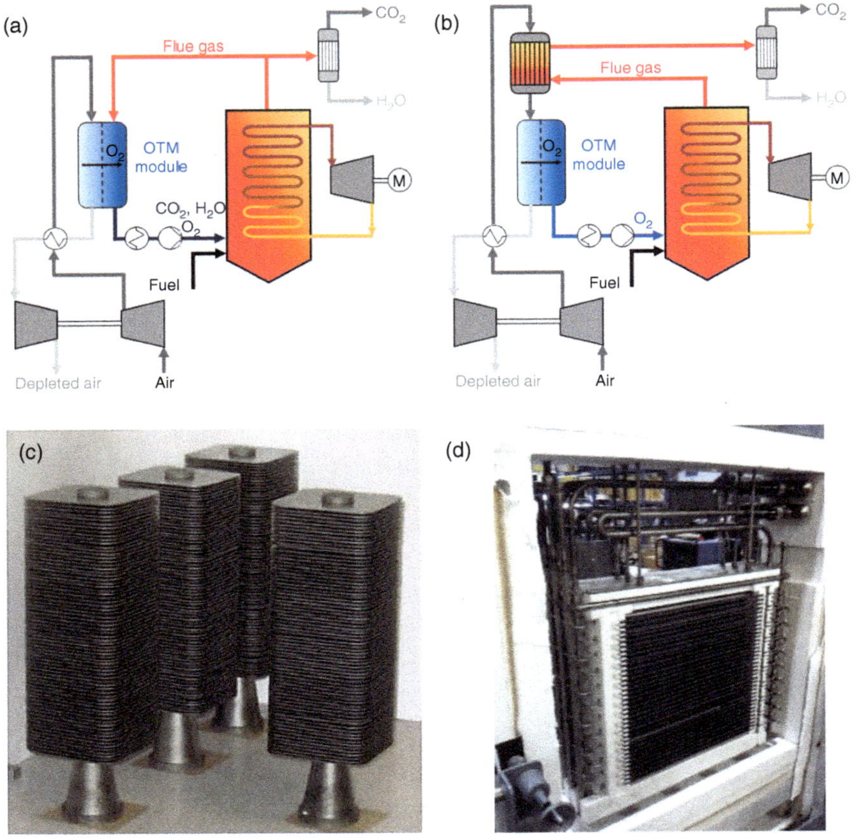

Figure 3.5 Simplified process layouts for oxygen permeating membrane modules integrated in oxyfuel power plants following (a) 4-end and (b) 3-end mode approaches, (c) Air Products' planar stacks, and (d) combined system steam reformer-OTM-ATR developed by Praxair. Source: Linde.

With regard to Air Products, the most advanced developments consisted of an intermediate scale testing with a capacity of 100 temperature programme desorption (TPD) O_2 (corresponding to an IGCC output of 12 MW) [58] and a membrane vessel consisting of several 1 TPD O_2 OTM modules (as those shown in Figure 3.5c), with a total production of 2000 TPD O_2. Despite that Air Products developments are the most advanced in terms of integration and demonstration, performance, and TRLs, they have been apparently abandoned since 2015 because of a company structure reorganization.

Praxair's developments present a tubular geometry where OTM tubes and CH_4 reforming tubes are combined in systems for oxy-combustion and syngas production applications by using advanced boilers and heaters in combustion processes [59–61]. As it can be observed in Figure 3.5d, Praxair's OTM concept consists of a multi-panel tubular reactor system where natural gas steam reforming, O_2

separation, and autothermal reforming is carried out by using integrated U-shaped reformer and OTM tubes.

Research centers such as RWTH-Aachen and the Fraunhofer Institute for Ceramics Technologies and Systems (IKTS) – both located in Germany – are conducting other of the most advanced developments in the OTM field. RWTH-Aachen designed, fabricated, and tested in a realistic environment an OTM module within the OXYCOAL-AC Project [62, 63]. The main aim of this development was to demonstrate a zero-CO_2 emission proof of concept for coal-fired power plants using an OTM module as an O_2 supply unit for conducting and oxy-combustion [64]. For that, an OTM module was developed consisting of BSCF tubular membranes (15 m^2 membrane area with 570 tubes) with a production capability of 0.6 TPD O_2, generating up to 120 kW by combusting pulverized coal. With regard to IKTS, which are specialized in the manufacturing and testing of 3-end OTM module systems considering BSCF tubes, they constructed the first stand-alone O_2 production unit in 2009 producing 2.7 l min^{-1} O_2 at 850 °C [65], being later improved achieving up to 2 kg O_2 h^{-1} (23.3 l min^{-1}).

3.4 Protonic Membranes

As previously mentioned, the impact of the CO_2 emissions on Earth is triggering the energetic transition from fossil fuels to environmentally friendly energy sources. H_2 is a promising energy carrier allowing the storage of chemical energy; nowadays, its main use is as a reactant for the synthesis of NH_3 and CH_3OH, in the refining and other industrial applications. H_2 can be used in fuel cell cars, as feed into the natural gas network, and in H_2/O_2 fuel cells among others [66–69].

Therefore, H_2 separation is an important process and its utility has been demonstrated over the past years. Pressure gradient is the main driving force for H_2 separation in these type of membranes, giving a large H_2 partial pressure, hydrogen will migrate across the membrane. These membranes operate at a wide range of temperatures, and they can be divided into six different types depending on their properties, temperature ranges, and H_2 permeation performance. Table 3.1 shows a comparison between H_2-selective membranes.

Among these membranes, mixed protonic–electronic conductor (MPEC)-based membranes are the most appropriate candidates for application to high-temperature H_2 (>500 °C) separation-based processes. These membranes allow to separate H_2 because of their ambipolar conductivity (electronic and protonic) when a hydrogen partial pressure difference is applied across the membrane [72–76]. This technology is the focus of many research groups worldwide because it offers the advantage of process intensification by shifting the thermodynamic equilibrium of a reaction [77–79].

3.4.1 Proton Defects in Oxide Ceramics

In an environment containing hydrogen or water, protons are dissolved in the oxide lattice forming positively charged defects following Eq. (3.6) written in Kröger–Vink

Table 3.1 Hydrogen-selective membrane types [70, 71].

	Dense polymer	Microporous ceramic	Dense metallic	Porous carbon	Dense ceramic
Temperature range (°C)	<100	200–600	300–600	500–900	600–900
H_2 selectivity	Low	5–139	>1000	4–20	>1000
H_2 flux ($\times 10^{-3}$ mol m^{-2} s^{-1}) DP = 100 kPa	Low	60–300	60–300	10–200	6–80
Stability issues	Swelling, compaction, mechanical strength	Stability in H_2O	Phase transition	Brittle, oxidizing	Stability in CO_2
Poisoning issues	HCl, SO_2, CO_2		H_2S, HCl, CO	Strong adsorbing vapors, organics	H_2S
Materials	Polymers	Silica, alumina, zirconia, titania, zeolites	Pd alloy	Carbon	Proton conducting ceramics (mainly $SrCeO_3$, $BaCeO_3$)
Transport mechanism	Solution/diffusion	Molecular sieving	Solution/diffusion	Surface diffusion; molecular sieving	Solution/diffusion (proton conduction)
Development status	Commercial by air products, Linde, BOC, and Air Liquide	Prototype tubular silica membranes available up to 90 cm. Other materials only small samples (cm^2)	Commercial by Johnson Matthey; prototype membrane tubes available up to 60 cm	Small membrane modules commercial, mostly small samples (cm^2) available for testing	Small samples available for testing

Source: Adapted from Kluiters [70] and Al-Mufachi et al. [71].

notation [78]. $V_o^{\cdot\cdot}$ is a positively charge (2+) oxygen vacancy, O_o^x represents an oxygen atom with a neutral charge placed on its original place in the crystal lattice, and $OH_o^{\cdot}$ is a proton defect (+1 charged).

$$H_2O + V_o^{\cdot\cdot} + O_o^x \leftrightarrow 2OH_o^{\cdot} \tag{3.6}$$

The equilibrium constant of proton defect formation reaction in oxide ceramic materials ($K_{OH^{\cdot}}$) is depicted in Eq. (3.7), where $[OH_O^{\cdot}]$ represents the proton defect concentration, $[V_o^{\cdot\cdot}]$ is the oxygen vacancy concentration, $[O_o^x]$ represents the concentration of oxygen atoms with a neutral charge placed on its original place in the crystal lattice, and p_{H2O} is the water vapor pressure.

$$K_{OH^{\cdot}} = \frac{[OH_O^{\cdot}]^2}{\left([V_o^{\cdot\cdot}]\,[O_o^x]\,p_{H_2O}\right)} \tag{3.7}$$

For large bandgap oxide materials (e.g. Ce, Ti, and Zr), the formation of proton defects at moderate temperatures takes places through the dissociative absorption of water [80]. Water dissociates into a hydroxide ion and a proton, the hydrogen ion then occupies an oxide ion vacancy, and the proton forms a covalent bond with a lattice oxygen. The formation of proton defects implies a significant weight gain; hence, the concentration of such defects can be measured by thermogravimetric analysis (TGA) as a function of temperature and water partial pressure.

3.4.2 Proton Transport Membrane Fundamentals

Understanding the mechanism of proton conduction is of utmost importance for the development of novel materials. It is generally accepted that proton diffusion in protonic conductors occur via the Grotthuss-type mechanism assisted by water molecules [81–83]. Moreover, hydrogen separation is driven by the hydrogen partial pressure difference across the membrane.

Proton conductivity has been observed in different types of materials. Perovskite-type oxide ceramics are known to be proton conductors since the early 1980s. In general, perovskite structure with a general formula $A^{2+}B^{4+}O_3$ (type II–IV), where A is Ba and B is Zr, Tb, Ce, or Th, exhibits the best proton conductivities being higher than $10^{-2}\,S\,cm^{-1}$, the lowest activation energies for proton transport, and high negative hydration enthalpies [78]. In particular, ceramic materials such $SrCeO_3$, $BaCeO_3$, or $SrZrO_3$ are the most widely studied high-temperature proton-conducting perovskite-type materials. Zirconate-based materials are more interesting than cerates regarding their application in CO_2 environments because of their higher stability under reducing atmospheres; however, they present an important grain boundary resistance and a high sintering temperature is needed to produce dense samples. In order to overcome the disadvantage of both families of materials, solid solutions of doped $BaCeO_3$ and $BaZrO_3$ have been developed by different research groups [84–87].

The reader is referred to other literature sources to dwell into other examples of proton conducting materials such as rare earth oxides, rare earth ortho-niobates and tantalates, rare earth tungstates, phosphates, and pyrochlores [75, 78, 88, 89].

Because of the low electronic conductivity, the abovementioned materials do not present a sufficiently high hydrogen permeation flux for their integration in practical applications. Composite membranes composed by two phases, an electronic and a protonic phase, have been developed in the past years to overcome this problem, obtaining promising permeation values [90–93].

3.4.3 Application Concepts of Proton Conducting Membranes

The applications of CCS using proton conducting membrane technologies has emerged as a hot topic to provide an industrial solution for the mitigation of the greenhouse effect. H_2-related membranes can operate at intermediate and high temperatures (400–900 °C), and the processes in which they have been integrated to can be divided into (i) CO_2 reduction into valuable chemicals such as methane or methanol (catalytic membrane reactors [CMR]), (ii) conversion of chemicals into electrical energy (fuel cells), and (iii) generation of H_2 as a fuel.

SMR has proven to be one of the most energy-efficient way to produce hydrogen from methane from an industrial point of view. SMR converts methane and water into syngas in an endothermic reaction (Eq. (3.8)); this step can be subsequently combined with water gas shift reaction (WGSR) to produce extra H_2 together with CO_2 (see Eq. (3.9)).

$$CH_4 + H_2O \rightarrow CO + 3H_2, \qquad \Delta H_{298K} = +206\ \text{kJ mol}^{-1} \tag{3.8}$$

$$CO + H_2O \rightarrow CO_2 + H_2, \qquad \Delta H_{298K} = -41\ \text{kJ mol}^{-1} \tag{3.9}$$

Using hydrogen-selective membranes, a pure H_2 stream is obtained, resulting in the shift of the thermodynamic equilibrium and hence in process intensification. The majority of the reported membranes are metal-based membranes, i.e. Pd or Pd/Ag alloys [71]. However, these membranes do not achieve full CH_4 conversion nor H_2 permeation [94–96].

WGSR at temperatures ranging from 700 to 900 °C have been performed using MPEC-based membranes, $SrCe_{0.9}Eu_{0.1}O_{3-\delta}$- and $SrCe_{0.7}Zr_{0.2}Eu_{0.1}O_{3-\delta}$-supported tubular membranes, yielding interesting results [97, 98]. By using the $SrCe_{0.7}Zr_{0.2}Eu_{0.1}O_{3-\delta}$-membrane, an increase of 77% in the CH_4 conversion as compared with thermodynamic conversion was obtained at 900 °C (H_2O/CO ratio = 1/1).

The selective conversion of natural gas to higher hydrocarbons and aromatics remains an important industrial challenge. Non-oxidative coupling of methane to produce olefins and aromatics (see Eq. (3.10)) are reactions limited thermodynamically. The selective extraction of H_2 will allow the shift of the thermodynamic equilibrium, i.e. the conversion toward the product side, giving rise to a significant improvement in the reaction yield. However, the H_2 extraction accelerates coking and catalyst deactivation.

$$6CH_4 \rightarrow C_6H_6 + 9H_2, \qquad \Delta H = 88.7\ \text{kJ mol}^{-1} \tag{3.10}$$

Methane dehydroaromatization (MDA) reaction was performed by Li and coworkers [99] using a 2 μm dense membrane made of $SrCe_{0.95}Yb_{0.05}O_{3-\delta}$ and 4 wt%

Mo/HZSM-5 catalyst. A modest increase of the CH_4 conversion as compared with the conventional reactor was obtained at 720 °C. However, a slightly higher catalyst deactivation was also observed.

Caro and coworkers studied the MDA reaction by using a U-shape $La_{5.5}W_{0.6}Mo_{0.4}O_{11.25-\delta}$ [100] membrane and 6 wt% Mo/HZSM-5 as a catalyst at 700 °C. Higher aromatics yield than that without membrane was obtained during the first five hours on the stream because of the important H_2 extraction, reaching 40–60% of the H_2 produced in the reaction. In this case, a catalyst deactivation was also observed, giving rise to aromatic yield lower than that without H_2 extraction after 10 hours on the stream.

3.5 Membranes for Electrochemical Applications

As previously mentioned, hydrogen has been identified as a potential alternative fuel source and a key energy carrier for the near future energy supply with a low CO_2 footprint. Hydrogen can be used directly to produce energy (fuel cells) or be easily transformed into other forms of energy for different end applications. Nevertheless, pure hydrogen is not abundant in the atmosphere. Its production is generally accomplished by steam reforming, coal gasification, or the partial oxidation of some heavy hydrocarbons. On the other hand, CO_2-neutral processes are also studied and are based on reforming, pyrolyzing, and fermentation of biomass, but efficiencies are not good enough for decreasing time-to-market. H_2 produced from water is an efficient, green, and commercial technology, where the common techniques include solar thermochemical or photocatalytic water splitting and electrical water electrolysis. One advantage of such water electrolysis systems is the possibility to perform CO_2 electrolysis or H_2O/CO_2 co-electrolysis to produce syngas [101] (H_2 and CO). The opportunity to conduct H_2O and CO_2 co-electrolysis is a very interesting way for the production of clean synthetic fuels: power to gas and power to fuel technologies. Therefore, these technologies have been pointed out as promising alternatives to store energy from renewable intermittent energy sources.

3.5.1 Electrolysis and Co-electrolysis Processes

The equivalent energy required for the water and carbon dioxide split reaction (ΔH) is determined from the free energy and the entropy as follows:

$$\Delta H = \Delta G + T\Delta S \tag{3.11}$$

where the electrical energy needed for the electrolysis will be defined by the free energy term (ΔG), whereas the thermal energy is defined by the entropy term (ΔS). As can be seen in Figure 3.6, the enthalpy of the water splitting reaction drops sharply at 100 °C because of the phase transition from liquid water to steam. In addition, the electrical energy needed for the electroreduction of water and carbon monoxide decreases as the temperature is increased. Furthermore, as temperature increases, the enthalpy of the CO_2 electrolysis decreases. Therefore, by increasing

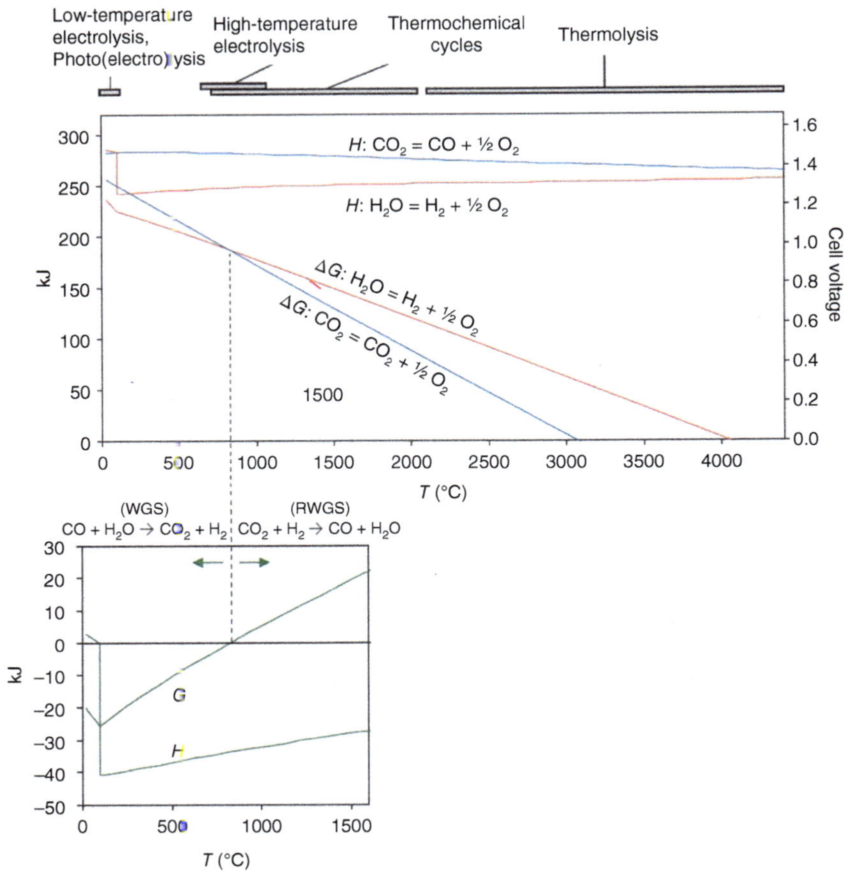

Figure 3.6 Enthalpy and free energy of CO_2 and H_2O reduction reactions and water gas shift reaction (WGSR). The present figure has been taken from the Grave's review. Source: Graves et al. [102].

temperature and operating at higher temperatures, the system will work more efficiently. Reaction kinetics are also enhanced at higher temperatures, which indicates higher H_2 or/and CO production at the same operation voltage.

If co-electrolysis is conducted, the WGSR has to be taken into account, and the system becomes more complicated. At temperatures above the zero free energy, reverse water gas shift reaction (RWGSR) is favored. As results, co-electrolysis is usually run by systems that work at high temperature (solid oxide electrolyzers).

3.5.1.1 Water Electrolysis

The electrolyzer can be depicted as an electrolyte placed between two electrodes (anode and cathode) connected electrically by an external circuit. In most electrolyzers, the electrolyte is an ion-selective membrane and works due to the electrochemical gradient generated in the cell. The principle of the transport is ascribed to the compensation of charges in the system by charge-carrying species. For polymeric membranes, a functional group is introduced in the polymer structure to induce the ionic conductivity, whereas in solid oxide materials, defects are generated into the crystalline structure to enhance the O^{2-} or H^+ transport.

Electrolyzers can be classified according to the materials used for the cell construction, operation temperature, and charge carriers. Considering the charge carrier in the electrolyte, the main electrolyzers can be divided into three groups:

- Hydroxide anion (OH^-).
  - *Alkaline electrolysis cell (AEC):* Water is decomposed into hydrogen and HO^- at the cathode. The hydroxide anion migrates to the anode, through the electrolyte, generating oxygen. The electrolyte solution consists of a mixture of water and NaOH or KOH. Among the electrolysis technologies, the alkaline electrolysis of water is the most mature technology and a long-time commercial technology.
  - *Polymer alkaline electrolyzer cell (PAEC)*: The main difference with the previous electrolyzer lies in the electrolyte; instead of using a liquid electrolyte, a polymeric membrane with OH^- ions conductivity is employed.
- Proton (H^+).
  - *Proton exchange membrane electrolysis cell (PEMEC)*: The two half cells are separated by a polymeric membrane. Protons are generated in the anode and then they pass by the membrane, generating hydrogen in the cathode. Because of the low temperature, expensive noble metals are used generally on both electrodes, the most common is platinum, and high external voltages are required to overcome the reaction kinetics.
  - *Proton conductor solid oxide electrolyzer cell (PC-SOEC)*: All components in the cell are solids and high operation temperatures (600–1000 °C) are required for the cell operation to favor the electronic and/or ionic conductivity of materials. As in the PEM electrolyzers, the protons are generated in the cathode and are recombined in the anode to produce hydrogen. Membrane is based on a proton conductor material.
- Oxygen ions (O^{2-}).
  - *Solid oxide electrolyzer cell (SOEC)*: Oxygen ions generated in the cathode are conducted toward the anode through an oxygen ion conductor membrane. All components of the cell are solids and work at high operation temperature.

The main components and most common materials for the cell construction of each type of electrolyzer are summarized in Table 3.2.

Despite the fact that SOEC technology is a more mature technology, proton conductor electrolysis cell (PCEC) technology presents the advantage of producing directly dry pressurized H_2 and subsequently the need of less separation steps. In addition operation temperatures are lower (500–700 °C), allowing the reduction of the material cost.

Recently, a study about the first fully operational PCEC has been published [106]. A tubular supported electrolyte made of $BaZr_{0.7}Ce_{0.2}Y_{0.1}O_{2.95}$ and $Ba_{1-x}Gd_{0.8}La_{0.2+x}Co_2O_{6-\delta}$ as a steam anode was employed and a hydrogen

Table 3.2 Summary of main characteristics of electrolyzers [103–105].

		AEC	PAEC	PEMEC	PC-SOEC	SOEC
Anode	Reaction	$4OH^- \rightarrow 2H_2O + O_2 + 4e^-$	$4OH^- \rightarrow 2H_2O + O_2 + 4e^-$	$2H_2O \rightarrow 4H^+ + O_2 + 4e^-$	$2H_2O \rightarrow 4H^+ + O_2 + 4e^-$	
$O^{2-} \rightarrow {}^1/_2O_2 + 2e^-$	Materials	Ni–Co–Fe, Ni_2CoO_4, La–Sr–CoO_3, Co_3O_4	Ni-based	Ir, Ru oxide	BCZY, SCZY	$La_xSr_{1-x}MnO_3$ + Y-stabilized ZrO_2 (YSZ)
Electrolyte	Charge carrier	OH^-	OH^-	H^+	H^+	O^{2-}
	Materials	Liquid: 25–30 wt% $(KOH)_{aq}$	Solid: polymeric	Solid: polymeric	Solid: BCZY, BZY, BCY, proton conductors	Solid: Y_2O_3–ZrO_2, Sc_2O_3–ZrO_2, MgO–ZrO_2, CaO–ZrO_2
Cathode	Reaction	$2H_2O + 4e^- \rightarrow 4OH^- + 2H_2$	$2H_2O + 4e^- \rightarrow 4OH^- + 2H_2$	$4H^+ + 4e^- \rightarrow 2H_2$	$4H^+ + 4e^- \rightarrow 2H_2$	$H_2O + 2e^- \rightarrow O^{2-} + H_2$
	Materials	Nickel foam/Ni alloys; Ni–Mo/ZrO_2–TiO_2	Ni, Ni–Fe, $NiFe_2O_4$	Pt/C, MoS_2	Ni cermets	Ni-YSZ
Operation temperature		20–80 °C	20–200 °C	20–200 °C	600–1000 °C	600–1000 °C

Source: Adapted from Millet et al. [103], Laguna-Bercero [104], and Sakai et al. [105].

Table 3.3 Summary of main features of CO electrolyzers or co-electrolyzers.

		MCEC	SOEC	PC-SOEC
Anode	Reaction	$CO_3^{2-} \rightarrow 1/2O_2 + CO_2 + 2e^-$	$O^{2-} \rightarrow 1/2O_2 + 2e^-$	$2H_2O \rightarrow 4H^+ + O_2 + 4e^-$
	Materials	Nickel based	$La_xSr_{1-x}MnO_3$ + Y-stabilized ZrO_2 (YSZ)	BCZY, SCZY
Electrolyte	Charge carrier	CO_3^{2-}	O^{2-}	H^+
	Materials	Molten carbonates	Solid: Y_2O_3–ZrO_2, Sc_2O_3–ZrO_2, MgO–ZrO_2, CaO–ZrO_2	Solid: BCZY, BZY, BCY, proton conductors
Cathode	Reaction	$H_2O + CO_2 + 2e^- \rightarrow H_2 + CO_3^{2-}$	$H_2O + 2e^- \rightarrow O^{2-} + H_2$	$4H^+ + 4e^- \rightarrow 2H_2$
		$2CO_2 + 2e^- \rightarrow CO + CO_3^{2-}$	$CO_2 + 2e^-$ rarr; $CO + O^{2-}$	$CO_2 + 2H^+ + 2e^- \rightarrow CO + H_2O$
	Materials	Nickel-based	Ni-YSZ	Ni cermets
Operation temperature		600–800 °C	600–1000 °C	600–1000 °C

production above 15 N ml min^{-1} was obtained. In addition, Faradaic efficiencies close to 100% at high steam pressures were observed.

3.5.1.2 CO_2 Co-electrolysis

Electrochemical CO_2 reduction has gained importance in the field of energy storage and conversion, and the catalysts and electrolytes influence not only the catalytic activity and selectivity of the reaction, but also on the CO_2 reduction mechanism to different species [107].

High temperature is desired for the CO_2 electrolysis (Figure 3.6), and among the electrolysis systems, molten carbonate electrolyzer cells (MCECs), SOEC, and PC-SOEC are favored for the CO_2 direct valorization by electrolysis (Table 3.3).

Because CO_2, H_2O, and H_2 are involved in the reactions, the system is further complicated when the co-electrolysis takes places, a scheme with all the species and reactions involved is represented in Figure 3.7.

Among the Co-electrolyzers at high temperatures, solid oxide cells (PC-SOEC and SOEC) present two main advantages. First, all components in the cells are solids and the risk of liquid leakage is avoided. Secondly, high temperature facilitates the electrolysis because the kinetics and thermodynamics are favored.

Different materials and configurations can be used for the electrolyte and electrodes in solid oxide electrolyzers. The selection of the most suitable material will depend on the operation conditions, such as temperature or gas atmosphere. The type of electrolyte selected (protonic or ionic conductor) will set the reactions in each electrode. All electrolytes must possess some characteristics to ensure a good

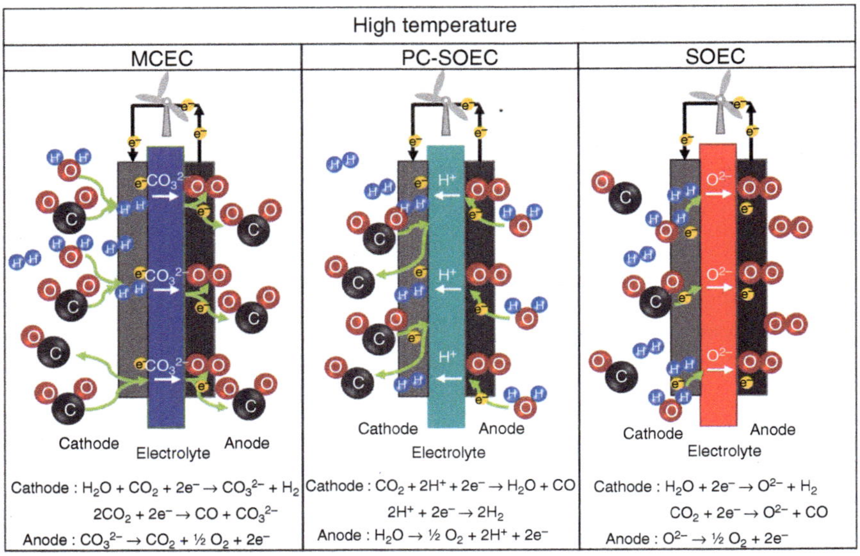

Figure 3.7 Schematic representation of the CO_2 and Co-electrolyzer systems.

performance [108]. The electrolyte must be chemically, morphologically, and dimensionally stable in both atmosphere of the cell (oxidizing and reducing) and for all range of operation conditions. In addition, in order to minimize the ohmic losses in the cell performance, the electrolyte should have a good ionic or protonic conductivity in the cell operation conditions. The electronic conductivity should be as small as possible to avoid electron leakage in the electrolyte and the resulting low Faraday efficiency. Likewise, the porosity in the electrolyte has to be negligible to avoid gas leakage in the cell and consequently low performance. Finally, thermal expansion coefficient (TEC) should match with the adjacent components of the cell to avoid problems such as cracks and delamination. Moreover, TEC should be unalterable with oxygen partial pressure and temperature changes.

Regarding electrodes, some features have to be accomplished as well. Both, the oxidant and fuel electrodes have to be chemically, morphologically, and dimensionally stable in the working atmospheres and temperatures. To improve cell performance, the electronic conductivity has to be as large as possible. Additionally, oxygen ion or proton conductivity is required to extend the triple phase boundary (the point where electrons, gases, and ions are in contact and the electrochemical reaction takes place) along the whole electrode surface. Electrodes should have enough porosity to allow fast gas transport from/to the active reaction sites. TEC of electrodes should match the electrolyte and adjacent components along the operation conditions. Furthermore, electrodes have to match with other components in the operation and fabrication conditions. Finally, electrodes must exhibit enough catalytic activity (low polarization) for the different reactions that take place in the active sites, such as oxygen reduction reaction, hydrogen oxidation reaction, water splitting reaction, CO_2 reduction, etc.

Other characteristics are also desirable for all components of the cell as the low cost, low toxicity, and manufacturability.

In the past years, several studies about PCEC have been published. Steam and CO_2 co-electrolysis was performed by Ruiz-Trejo and Irvine using $BaCe_{0.5}Zr_{0.3}Y_{0.16}Zn_{0.04}O_{3-\delta}$ around 500 °C, obtaining promising results [109, 110]. Recently, Bausá et. al. reported co-electrolysis at 700 °C using a $BaCe_{0.2}Zr_{0.7}Y_{0.1}O_{3-\delta}$ electrolyte [111].

3.5.2 Synthesis Gas Chemistry

Electrolysis consists in the dissociation of H_2O and/or CO_2 by means of electricity. One of the main advantages of the co-electrolysis is the production of H_2 and CO simultaneously (synthesis gas or syngas). Syngas is widely used to produce a wide range of chemicals [112, 113]. The Fischer–Tropsch process is one of the main processes where syngas is involved. The Fischer–Tropsch process consists of the H_2 and CO transformation into hydrocarbons of longer chains than methane. Hydrocarbons are the basis for the production of gasoline, diesel, and chemicals such as olefins and waxes. The type of catalyst selected (usually Fe and Co), design of the reactor, and the process conditions will shape the product selectivity in the Fischer–Tropsch process. Usually, Fischer–Tropsch synthesis takes place at temperatures between 200 and 300 °C and pressures comprised between 1 and 6 MPa. Among all the reactions that occur during Fischer–Tropsch synthesis, the main desired reactions can be addressed with the production of alkanes or paraffins (see Eq. (3.12)), alkenes or olefins (Eq. (3.13)), and alcohols (Eq. (1.14)) [114].

$$nCO + (2n+1)\,H_2 \rightarrow C_nH_{2n+2} + nH_2O \tag{3.12}$$

$$nCO + 2nH_2 \rightarrow C_nH_{2n} + nH_2O \tag{3.13}$$

$$nCO + 2nH_2 \rightarrow C_nH_{2n+1}OH + (n-1)\,H_2O \tag{3.14}$$

Apart from the reactions described above, some undesired reactions might also occur in the process as WGSR, carbonaceous materials, Boudouard reaction, and bulk carbide formation.

The synthesis of methanol and methane can also be derived from the syngas [115]. Methanation and methanol can be performed directly from the syngas or by CO_2 hydrogenation. The main reactions for both syntheses are described below. For both processes, WGSR plays an important role and affects the selectivity to the final products.

Power to methane.

(a) $CO_2 + 4H_2 \rightarrow CH_4 + 2H_2O$
(b) $CO + 3H_2 \rightarrow CH_4 + H_2O$
(c) $CO + H_2O \rightarrow CO_2 + H_2$

Power to methanol.

(a) $CO_2 + 3H_2 \rightarrow CH_3OH + H_2O$
(b) $CO + 2H_2 \rightarrow CH_3OH$
(c) $CO + H_2O \rightarrow CO_2 + H_2$

Furthermore, products obtained from the Fischer–Tropsch reaction and methanol can also be recombined with syngas, CO, or H_2 to obtain different products highly valued in the chemical industry. Some examples for different products are, for instance, ethanol to gasoline reaction by means of zeolites [116, 117]; hydroformylation reaction of olefins with syngas to aldehydes, and alcohols with cobalt and rhodium catalysts [118]; carbonylation reactions [119], where CO from the electrolysis can be employed; and methanol plus syngas or carbon monoxide, for the synthesis of value-added chemicals [117].

The important role of electrolysis as a bridge between renewable energies, energy storage technology, and value-added products (chemicals, fuels, etc.) is therefore obvious. The combination in the same step of both technologies, CO_2 co-electrolysis to produce syngas and the production of hydrocarbons by Fischer–Tropsch process, will give rise to a more efficient, compact, and environmental friendly technology. The eCOCO$_2$ project [120] (sponsored by the EU commission via the H2020 program) based on CO_2 conversion focuses on the development of co-ionic electrochemical cells, which enable both: the electrolysis of water and the hydrocarbon synthesis in the same step. The CO_2 converter consists of electrochemical cell constructed with a co-ionic electrolyte that allows the injection of protons to the reaction cell, and the simultaneous extraction of oxygen ions. In addition, a multifunctional catalyst will be integrated in the electrochemical cell for the hydrocarbon generation. The final objective of this project is the production of more than 250 g of jet fuel per day.

3.5.3 Other Applications

3.5.3.1 Methane Steam Reforming

The feasibility of performing SMR using the proton-conducting material $BaZr_{0.7}Ce_{0.2}Y_{0.1}O_{2.9}$ (BZCY72) was successfully examined at low temperatures (450–650 °C) under atmospheric pressure by M. Stoukides and coworkers [121]. The system exhibited strong dependence on gas concentration, temperature, and applied voltage, as well as excellent chemical stability.

In 2017, J.M. Serra, C. Kjølseth, and colleagues reported an outstanding breakthrough regarding the production of pure hydrogen from methane. A protonic

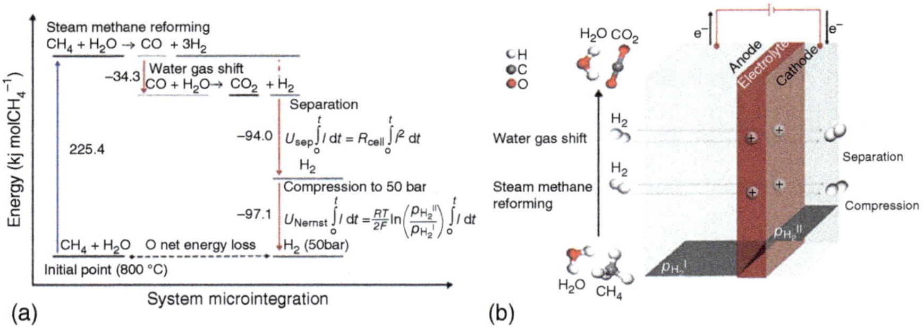

Figure 3.8 (a) Representation of the four chemical steps. (b) Protonic membrane reformer. Source: Malerød-Fjeld et al. [122].

membrane reformer (PMR) electrochemically driven capable of realizing four process steps simultaneously, achieving near-zero energy loss (see Figure 3.8a). The reformer is made out of a dense proton-conducting layer of $BaZr_{0.8-x-y}Ce_xY_yO_{3-\delta}$ (BZCY), sandwiched between two porous electrodes (BZCY and Ni) [122]. Figure 3.8b represents the PMR concept, where methane is reformed with steam (H_2O) to produce hydrogen that is subsequently transported through the membrane and finally compressed as a consequence of the applied voltage. The study reported full methane conversion permeating 99% of the produced hydrogen at 800 °C. The permeated hydrogen was electrochemically compressed up to 50 bar.

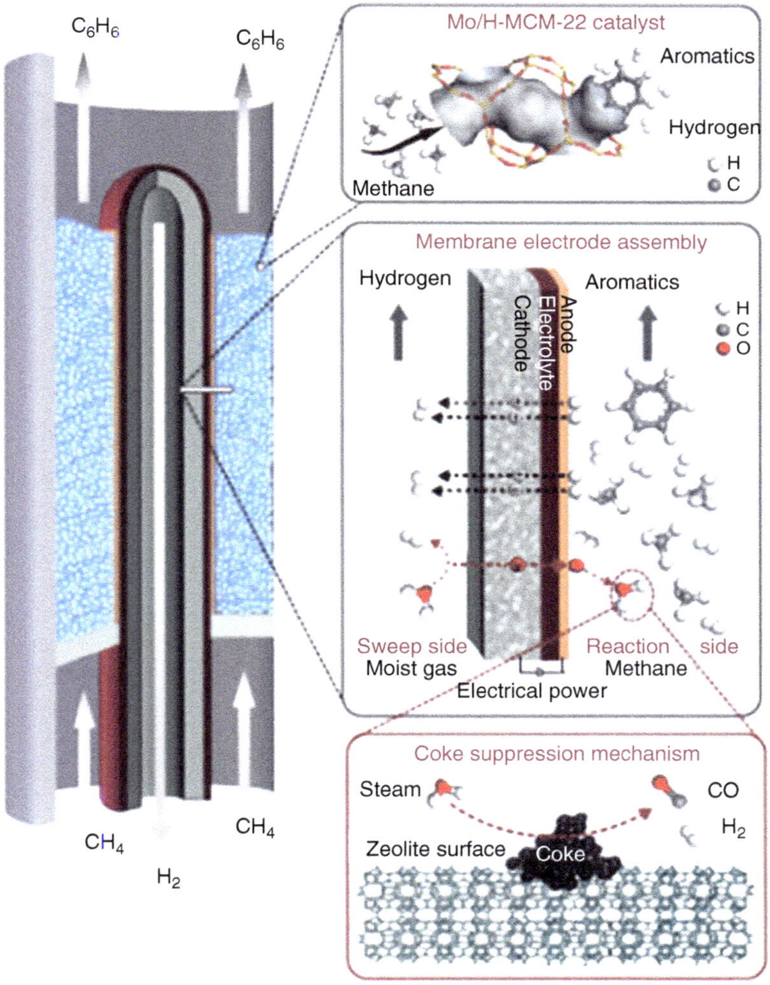

Figure 3.9 Co-ionic catalytic membrane reactor. Source: Morejudo et al. [125].

3.5.3.2 Methane Dehydroaromatization

A clear example of the contribution of membranes toward chemical production and process intensification is the direct conversion of methane-containing sources, i.e. natural gas or biogas, into valuable petrochemicals using MDA reaction [123]. This reaction, normally carried out at 700 °C, suffers from some drawbacks, and rapid catalyst deactivation occurs because of the accumulation of polyaromatic-type coke, impeding the access to the catalyst-active sites and the limitation imposed by thermodynamics of the per-pass conversion to aromatics [124]. In order to overcome the abovementioned disadvantages, a recent publication reports a CMR for MDA process intensification [125]. The system is composed of an electrochemical $BaZrO_3$-based tubular membrane that exhibits proton and ion conductivity. Figure 3.9 depicts a representation of the MDA reaction and the different components/materials that form the CMR.

In this case, methane is converted to benzene and hydrogen using a Mo/H-MCM-22 catalyst. Protons are transported to the sweep side, whereas oxide ions are transported to the reaction chamber and react with H_2 to form water that will subsequently react with coke to form CO and H_2, enhancing catalyst stability. The authors observed that the electrochemically driven simultaneous extraction and injection of proton and oxide ions, respectively, allows obtaining high aromatic yields while drastically reducing the catalyst deactivation rate by coking.

3.6 Conclusions and Final Remarks

Membrane technologies are striking as strong candidates to tackle climate change by presenting potential and effective solutions for capturing CO_2 and also for performing its conversion toward highly valuable chemicals. In order to set the different available options within the membrane field, different types of membranes have been summarized in this chapter, together with the outline of the main strategies and novel solutions toward the CO_2 valorization.

Therefore, the main issues concerning the most promising concepts, their particularities and properties, as well as their advantages and drawbacks when considering their industrial application have been presented. Also, a discussion on how these options can be integrated into the existing engineering layouts has been performed. This led to the presentation of some of the most advanced developments up to date, with an extended description of their operating concepts, and how their implementation can help to the effective valorization of CO_2 while reducing the emission problem.

It is clear that many research groups are working to develop new materials to enhance these processes; however, more efforts are necessary from the research and exploitation point of view of this technology in order to promote membranes in an industrial way.

References

1 IEA (2020). *CO_2 Emissions from Fuel Combustion: Overview*. Paris: IEA. https://www.iea.org/reports/co2-emissions-from-fuel-combustion-overview.

2 Yu, X., Yang, J., Yan, J., and Tu, S. (2015). Membrane technologies for CO_2 capture. In: *Handbook of Clean Energy Systems* (ed. J. Yan), 1–13. Wiley.

3 Remiro-Buenamañana, S. and García, H. (2019). *ChemCatChem* 11 (1): 342–356.

4 Mulder, J. (2013). *Basic Principles of Membrane Technology*. Netherlands: Springer.

5 Mohanty, K. and Purkait, M.K. (2011). *Membrane Technologies and Applications*. CRC Press.

6 Paidar, M., Fateev, V., and Bouzek, K. (2016). *Electrochim. Acta* 209: 737–756.

7 Li, H., Caravella, A., and Xu, H.Y. (2016). *J. Mater. Chem. A* 4 (37): 14069–14094.

8 Amano, M., Nishimura, C., and Komaki, M. (1990). Effects of high concentration CO and CO_2 on hydrogen permeation through the palladium membrane. *Mater. Trans. JIM* 31: 404–408.

9 O'Brien, C.P. and Lee, I.C. (2017). *J. Phys. Chem. C* 121 (31): 16864–16871.

10 Brunetti, A. and Fontananova, E. (2019). *J. Nanosci. Nanotechnol.* 19 (6): 3124–3134.

11 Pomilla, F.R., Brunetti, A., Marcì, G. et al. (2018). *ACS Sustainable Chem. Eng.* 6 (7): 8743–8753.

12 Mason, E.A. (1991). *J. Membr. Sci.* 60 (2): 125–145.

13 Mitchell, J.K. (1995). *J. Membr. Sci.* 100 (1): 11–16.

14 Koros, W.J. and Fleming, G.K. (1993). *J. Membr. Sci.* 83 (1): 1–80.

15 Wijmans, J.G. and Baker, R.W. (1995). *J. Membr. Sci.* 107 (1): 1–21.

16 Baker, R.W. (2012). *Membrane Technology and Applications*, 3e. Wiley.

17 Tanaka, K., Kita, H., Okano, M., and Okamoto, K.-i. (1992). *Polymer* 33 (3): 585–592.

18 White, R.P. and Lipson, J.E.G. (2016). *Macromolecules* 49 (11): 3987–4007.

19 Han, Y. and Ho, W.S.W. (2018). *Chin. J. Chem. Eng.* 26 (11): 2238–2254.

20 Zhao, L., Weber, M., and Stolten, D. (2013). *Energy Procedia* 37: 1125–1134.

21 Yave, W. and Car, A. (2011). Chapter 6: Polymeric membranes for post-combustion carbon dioxide (CO_2) capture. In: *Advanced Membrane Science and Technology for Sustainable Energy and Environmental Applications* (eds. A. Basile and S.P. Nunes), 160–183. Woodhead Publishing.

22 Ockwig, N.W. and Nenoff, T.M. (2007). *Chem. Rev.* 107 (10): 4078–4110.

23 Weigelt, F., Escorihuela, S., Descalzo, A. et al. (2019). *Membranes* 9 (4): 51.

24 Escorihuela, S., Tena, A., Shishatskiy, S. et al. (2018). *Membranes* 8 (1): 16.

25 He, X. (2016). Membranes for natural gas sweetening. In: *Encyclopedia of Membranes* (eds. E. Drioli and L. Giorno), 1266–1267. Berlin, Heidelberg: Springer Berlin Heidelberg.

26 Robeson, L.M. (1991). *J. Membr. Sci.* 62 (2): 165–185.

27 Robeson, L.M. (2008). *J. Membr. Sci.* 320 (1): 390–400.

28 Kim, S. and Lee, Y.M. (2013). *Curr. Opin. Chem. Eng.* 2 (2): 238–244.

29 Budd, P.M., Msayib, K.J., Tattershall, C.E. et al. (2005). *J. Membr. Sci.* 251 (1): 263–269.

30 Park, H.B., Jung, C.H., Lee, Y.M. et al. (2007). *Science* 318 (5848): 254.

31 Perrin, N., Dubettier, R., Lockwood, F. et al. (2015). *Appl. Therm. Eng.* 74: 75–82.

32 Lupion, M., Alvarez, I., Otero, P. et al. (2013). *Ghgt-11* 37: 6179–6188.

33 Monne, J. and Prinet, C. (2013). *Ghgt-11* 37: 6444–6457.

34 Smart, S., Lin, C.X.C., Ding, L. et al. (2010). *Energy Environ. Sci.* 3 (3): 268–278.

35 Takahashi, T., Esaka, T., and Iwahara, H. (1976). *J. Solid State Chem.* 16 (3–4): 317–323.

36 Cales, B. and Baumard, J.F. (1982). *J. Mater. Sci.* 17 (11): 3243–3248.

37 Cales, B. and Baumard, J.F. (1984). *J. Electrochem. Soc.* 131 (10): 2407–2413.

38 Bouwmeester, H. and Burggraaf, A. (1997). Chapter 14: Dense ceramic membranes for oxygen separation. In: *CRC Handbook of Solid State Electrochemistry* (eds. P.J. Gellings and H. Bouwmeester), 435–528. Boca Raton, FL: CRC Press.

39 Verkerk, M.J., Hammink, M.W.J., and Burggraaf, A.J. (1983). *J. Electrochem. Soc.* 130 (1): 70–78.

40 Kuklja, M.M., Kotomin, E.A., Merkle, R. et al. (2013). *Phys. Chem. Chem. Phys.* 15 (15): 5443–5471.

41 Bouwmeester, H.J.M., Kruidhof, H., and Burggraaf, A.J. (1994). *Solid State Ionics* 72, Part 2: 185–194.

42 Kharton, V.V., Tsipis, E.V., Yaremchenko, A.A. et al. (2003). *J. Solid State Electrochem.* 7 (8): 468–476.

43 Shaula, A.L., Kolotygin, V.A., Naumovich, E.N. et al. (2013). Oxygen ionic transport in Brownmillerite-type $Ca_2Fe_2O_5$-delta and calcium ferrite-based composite membranes. In: *Oxide Materials for Electronic Engineering – Fabrication, Properties and Applications*, vol. 200 (eds. S. Ubizskii, L. Vasylechko and Y. Zhydachevskii), 286–292. Trans Tech Publications Ltd.

44 Xue, J., Liao, Q., Chen, W. et al. (2015). *J. Mater. Chem. A* 3 (37): 19107–19114.

45 Bochkov, D.M., Kharton, V.V., Kovalevsky, A.V. et al. (1999). *Solid State Ionics* 120 (1): 281–288.

46 Baumann, S., Serra, J.M., Lobera, M.P. et al. (2011). *J. Membr. Sci.* 377 (1–2): 198–205.

47 Luo, H., Jiang, H., Klande, T. et al. (2012). *Chem. Mater.* 24 (11): 2148–2154.

48 Luo, H., Efimov, K., Jiang, H. et al. (2011). *Angew. Chem. Int. Ed.* 50 (3): 759–763.

49 Balaguer, M., Garcia-Fayos, J., Solis, C., and Serra, J.M. (2013). *Chem. Mater.* 25 (24): 4986–4993.

50 Garcia-Fayos, J., Balaguer, M., and Serra, J.M. (2015). *ChemSusChem* 8 (24): 4242–4249.

51 Gaudillere, C., Garcia-Fayos, J., Balaguer, M., and Serra, J.M. (2014). *ChemSusChem* 7 (9): 2554–2561.

52 Engels, S., Beggel, F., Modigell, M., and Stadler, H. (2010). *J. Membr. Sci.* 359 (1–2): 93–101.

53 Stadler, H., Beggel, F., Habermehl, M. et al. (2011). *Int. J. Greenhouse Gas Control* 5 (1): 7–15.

54 Kelly, S.M., Kromer, B.R., Litwin, M.M., et al. (2013). Synthesis gas method and apparatus. US Patent 8,349,214, filed 08 July 2011 and issued 08 January 2012.

55 John Repasky, D.M., Armstrong, P., and Carolan, M. (2014). ITM technology for carbon capture on natural gas and hybrid power systems. In *Workshop on Technology Pathways Forward for Carbon Capture & Storage on Natural Gas Power Systems*, Washington DC (22 April 2014).

56 Miller, C.F., Chen, J., Carolan, M.F., and Foster, E.P. (2014). *Catal. Today* 228: 152–157.

57 Repasky, J.M., Waldron, W.E., and Miller, C.F. (2010). ITM syngas: ceramic membrane technology for lower cost conversion of natural gas. In *10th Topical Conference on Gas Utilization 2010 – Topical Conference at the 2010 AIChE Spring Meeting and 6th Global Congress on Process Safety*, San Antonio, TX (21–25 March 2010). AIChE (pp. 48–59).

58 Anderson, L.L., Armstrong, P.A., Broekhuis, R.R. et al. (2016). *Solid State Ionics* 288: 331–337.

59 Rosen, L., Degenstein, N., Shah, M. et al. (2011). *10th International Conference on Greenhouse Gas Control Technologies*, vol. 4 (eds. J. Gale, C. Hendriks and W. Turkenberg), 750–755, ISSN 1876-6102.

60 Kromer, B.R., Litwin, M.M., and Kelly, S.M. (2014). Oxygen transport membrane based advanced power cycle with low pressure synthesis gas slip stream. US Patent 7,856,829. US20140183866 A1, filed 23 December 2013 and issued 27 September 2016.

61 Shah, M.M., Jamal, A., Dmevich, R.F., et al. (2012). Electrical power generation apparatus. US Patent 8,196,387, filed 17 November 2010 and issued 12 June 2012.

62 Kneer, R., Toporov, D., Förster, M. et al. (2010). *Energy Environ. Sci.* 3 (2): 198–207.

63 Gutiérrez-Guerra, N., Valverde, J.L., Romero, A. et al. (2017). *Electrochem. Commun.* 81: 128–131.

64 Pfaff, E.M., Kaletsch, A., and Broeckmann, C. (2012). *Chem. Eng. Technol.* 35 (3): 455–463.

65 Middelkoop, V. and Michielsen, B. (2015). Oxy fuel combustion power production using high temperature O_2 membranes. In: *Process Intensification for Sustainable Energy Conversion* (eds. F. Gallucci and M.V.S. Annaland), 321–385. Wiley.

66 Gröger, O., Gasteiger, H.A., and Suchsland, J.-P. (2015). *J. Electrochem. Soc.* 162 (14): A2605–A2622.

67 Büchi, F.N., Hofer, M., Peter, C. et al. (2014). *RSC Adv.* 4 (99): 56139–56146.

68 Lewandowska-Bernat, A. and Desideri, U. (2017). *Energy Procedia* 105: 4569–4574.

69 Götz, M., Lefebvre, J., Mörs, F. et al. (2016). *Renewable Energy* 85: 1371–1390.

70 Kluiters, S.C.A. (2004). Status Review on Membrane Systems for Hydrogen Separation. *Intermediate Report EU project MIGREYD NNE5-2001*. p. 670.

71 Al-Mufachi, N.A., Rees, N.V., and Steinberger-Wilkens, R. (2015). *Renewable Sustainable Energy Rev.* 47: 540–551.

72 Escolástico, S., Schroeder, M., and Serra, J.M. (2014). *J. Mater. Chem. A* 2 (18): 6616–6630.

73 Escolástico, S., Solís, C., Scherb, T. et al. (2013). *J. Membr. Sci.* 444: 276–284.

74 Haugsrud, R. and Kjølseth, C. (2008). *J. Phys. Chem. Solids* 69 (7): 1758–1765.

75 Haugsrud, R. and Norby, T. (2006). *Nat. Mater.* 5 (3): 193–196.

76 Magrasó, A. and Haugsrud, R. (2014). *J. Mater. Chem. A* 2 (32): 12630–12641.

77 Serra, J.M. (2019). *Nat. Energy* 4 (3): 178–179.

78 Meulenberg, W.A., Ivanova, M.E., Serra, J.M., and Roitsch, S. (2011). Chapter 17: Proton-conducting ceramic membranes for solid oxide fuel cells and hydrogen (H_2) processing. In: *Advanced Membrane Science and Technology for Sustainable Energy and Environmental Applications* (eds. A. Basile and S.P. Nunes), 541–567. Woodhead Publishing.

79 Bareiß, K., de la Rua, C., Möckl, M., and Hamacher, T. (2019). *Appl. Energy* 237: 862–872.

80 Kreuer, K.D. (2003). *Annu. Rev. Mater. Res.* 33 (1): 333–359.

81 de Grotthuss, C. (1806). *Philos. Mag.* 25: 330–339.

82 Marx, D. (2006). *ChemPhysChem* 7 (9): 1848–1870.

83 Kreuer, K.-D. (1996). *Chem. Mater.* 8 (3): 610–641.

84 Ricote, S., Bonanos, N., Manerbino, A., and Coors, W.G. (2012). *Int. J. Hydrogen Energy* 37 (9): 7954–7961.

85 Ricote, S., Bonanos, N., Marco de Lucas, M.C., and Caboche, G. (2009). *J. Power Sources* 193 (1): 189–193.

86 Yang, L., Wang, S., Blinn, K. et al. (2009). *Science* 326 (5949): 126.

87 Choi, S., Kucharczyk, C.J., Liang, Y. et al. (2018). *Nat. Energy* 3 (3): 202–210.

88 Escolástico, S., Somacescu, S., and Serra, J.M. (2014). *Chem. Mater.* 26 (2): 982–992.

89 Escolástico, S., Somacescu, S., and Serra, J.M. (2015). *J. Mater. Chem. A* 3 (2): 719–731.

90 Escolástico, S., Solis, C., Kjolseth, C., and Serra, J.M. (2014). *Energy Environ. Sci.* 7 (11): 3736–3746.

91 Ivanova, M.E., Escolástico, S., Balaguer, M. et al. (2016). *Sci. Rep.* 6: 34773.

92 Mortalò, C., Rebollo, E., Escolástico, S. et al. (2018). *J. Membr. Sci.* 564: 123–132.

93 Rebollo, E., Mortalo, C., Escolástico, S. et al. (2015). *Energy Environ. Sci.* 8 (12): 3675–3686.

94 Iulianelli, A., Liguori, S., Wilcox, J., and Basile, A. (2016). *Catal. Rev.* 58 (1): 1–35.

95 Chang, H.-F., Pai, W.-J., Chen, Y.-J., and Lin, W.-H. (2010). *Int. J. Hydrogen Energy* 35 (23): 12986–12992.

96 Iulianelli, A., Ribeirinha, P., Mendes, A., and Basile, A. (2014). *Renewable Sustainable Energy Rev.* 29: 355–368.

97 Li, J., Yoon, H., Oh, T.-K., and Wachsman, E.D. (2009). *Appl. Catal., B* 92 (3–4): 234–239.

98 Li, J., Yoon, H., Oh, T.-K., and Wachsman, E.D. (2012). *Int. J. Hydrogen Energy* 37 (21): 16006–16012.

99 Liu, Z., Li, L., and Iglesia, E. *Catal. Lett.* 82 (3): 175–180.

100 Xue, J., Chen, Y., Wei, Y. et al. (2016). *ACS Catal.* 6 (4): 2448–2451.

101 Wender, I. (1996). *Fuel Process. Technol.* 48 (3): 189–297.

102 Graves, C., Ebbesen, S.D., Mogensen, M., and Lackner, K.S. (2011). *Renewable Sustainable Energy Rev.* 15 (1): 1–23.

103 Millet, P. and Grigoriev, S. (2013). Chapter 2: Water electrolysis technologies. In: *Renewable Hydrogen Technologies* (eds. L.M. Gandía, G. Arzamendi and P.M. Diéguez), 19–41. Amsterdam: Elsevier.

104 Laguna-Bercero, M.A. (2012). *J. Power Sources* 203: 4–16.

105 Sakai, T., Arakawa, K., Ogushi, M. et al. (2015). *J. Solid State Electrochem.* 19 (6): 1793–1798.

106 Vøllestad, E., Strandbakke, R., Tarach, M. et al. (2019). *Nat. Mater.* 18 (7): 752–759.

107 Kortlever, R., Shen, J., Schouten, K.J.P. et al. (2015). *J. Phys. Chem. Lett.* 6 (20): 4073–4082.

108 Minh, N.Q. and Takahashi, T. (1995). *Science and Technology of Ceramic Fuel Cells.* Elsevier Science.

109 Ruiz-Trejo, E. and Irvine, J.T.S. (2013). *Solid State Ionics* 252: 157–164.

110 Ruiz-Trejo, E. and Irvine, J.T.S. (2012). *Solid State Ionics* 216: 36–40.

111 Bausá, N., Escolástico, S., and Serra, J.M. (2019). *J. CO_2 Util.* 34: 231–238.

112 Speight, J.G. (2019). Chapter 3: Unconventional gas. In: *Natural Gas*, 2e (ed. J.G. Speight), 59–98. Boston, MA: Gulf Professional Publishing.

113 Rostrup-Nielsen, J.R. (2000). *Catal. Today* 63 (2): 159–164.

114 Dalai, A.K. and Davis, B.H. (2008). *Appl. Catal., A* 348 (1): 1–15.

115 Wang, L., Chen, M., Küngas, R. et al. (2019). *Renewable Sustainable Energy Rev.* 110: 174–187.

116 Choi, M., Na, K., Kim, J. et al. (2009). *Nature* 461: 246.

117 Stöcker, M. (1999). *Microporous Mesoporous Mater.* 29 (1–2): 3–48.

118 Franke, R., Selent, D., and Börner, A. (2012). *Chem. Rev.* 112 (11): 5675–5732.

119 Beller, M. (2006). *Catalytic Carbonylation Reactions.* Springer.

120 Haworth, P.F., Smart, S., Serra, J.M., and Diniz da Costa, J.C. (2012). *Phys. Chem. Chem. Phys.* 14 (25): 9104–9111.

121 Kyriakou, V., Garagounis, I., Vourros, A. et al. (2016). *Appl. Catal., B* 186: 1–9.

122 Malerød-Fjeld, H., Clark, D., Yuste-Tirados, I. et al. (2017). *Nat. Energy* 2 (12): 923–931.

123 Xu, Y., Bao, X., and Lin, L. (2003). *J. Catal.* 216 (1): 386–395.

124 Tempelman, C.H.L. and Hensen, E.J.M. (2015). *Appl. Catal., B* 176–177: 731–739.

125 Morejudo, S.H., Zanón, R., Escolástico, S. et al. (2016). *Science* 353 (6299): 563.

4

Computational Modeling of Carbon Dioxide Catalytic Conversion

Javier Amaya Suárez, Elena R. Remesal, Jose J. Plata, Antonio M. Márquez, and Javier Fernández Sanz

Universidad de Seville, Department of Physical Chemistry, Sevilla 41012, Spain

4.1 Introduction

Because of the demonstrated impact on the present and future life on the Earth, control of carbon dioxide (CO_2) emissions is not enough nowadays to mitigate its undesired effects, and steps toward CO_2 reduction in the atmosphere must be undertaken. Among a wide range of possibilities, chemical transformation of CO_2 into valuable chemicals (Figure 4.1) has attracted great interest for long (see [2] for a recent review). However, CO_2 is the last product in the carbon compound oxidation reaction, which is actually one of the most common chemical processes that takes place on the nature. Because of its stability and chemical inactivity, transformation of CO_2 generally involves endothermic processes that require stringent physical conditions and the use of catalysts. Beyond the traditional approach to develop catalysts, mostly based on trial-and-error procedures, theoretical tools have been proven to reliably help both in the discovery and optimization of catalytic processes.

Catalysis is a complex process involving physicochemical phenomena taking place at different length and time scales. Because of the multiscale nature of catalysis, a variety of theoretical approaches can be used to address its study, from continuum theories and finite differences in the macroscale range to molecular dynamics (MD) simulations for meso- and microscale. The molecular-level processes that take place at the active sites and in their surroundings ultimately determine the performance of a catalyst. These processes take place in the smallest time and length scales, which is the scale of individual atoms and bonds ($\sim 10^{-10}$ m and $\sim 10^{-15}$ seconds). In order to study these phenomena, *first-principles* electronic structure methods are mandatory.

4.2 General Methods for Theoretical Catalysis Research

Within the so-called *first-principles* approach, two different categories of electronic structure methodologies can be found: wave function-based and density functional

Engineering Solutions for CO_2 Conversion, First Edition.
Edited by Tomas R. Reina, José A. Odriozola, and Harvey Arellano-Garcia.

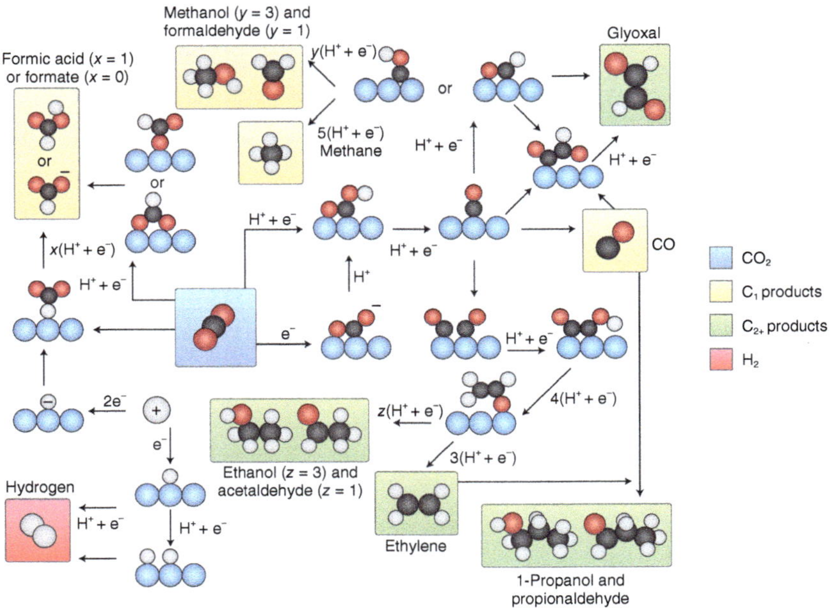

Figure 4.1 Overview of reaction pathways for CO_2 toward different products. Black spheres, carbon; red spheres, oxygen; white spheres, hydrogen; and blue spheres, (metal) catalyst. The arrows indicate whether proton, electron, or concerted proton–electron transfers (CPETs) take place. Source: Birdja et al. [1].

theory (DFT) methods. The simplest and cheapest wave function-based calculation is the Hartree–Fock (HF) method, where the wave function is approximated by a single Slater determinant of one-electron functions called spin orbitals. In this approximation, the exchange energy is calculated exactly while Coulomb correlation effects are completely dismissed, which significantly influence the energetics and structure of the system. For this reason, post-HF methods, such as the second-order perturbation Møller–Plesset (MP2) method, reintroduce the Coulomb correlation at the expense of a considerable increase in the computational cost with the system size [3].

DFT is an attractive alternative to wave function-based methods. It approximately includes the effects of correlation and exchange at a computational cost as good as HF calculations in the case of local density approximation (LDA) and generalized gradient approximation (GGA), while hybrid functionals cause a significant increase in the calculation time. DFT yields structural and energetic results in the range of the so-called "chemical accuracy," reproducing bond lengths within ±0.02 Å, vibrational frequencies within 5–10%, and adsorption energies in the range of tenths of eV [4]. Because of the computational cost in combination with the results' accuracy, DFT has become the main first-principles theoretical tool to investigate catalytic reactions on solid surfaces.

The role of the catalyst is to decrease the activation barrier present in the uncatalyzed reaction by altering the reaction mechanism providing new binding sites on the catalyst, which stabilize intermediates and transition states (TS). The transition state theory (TST) is based on the idea that the potential energy surface (PES) of an elementary reaction can be divided into reactant and product regions. The minimum energy points of these regions are the initial state and the final state. The border between the two regions is the TS, corresponding to a saddle point in the minimum energy path (MEP) connecting the initial and final states [4]. One of the most frequently used computational techniques to determine the TS in heterogeneous catalysis is the nudged elastic band (NEB) method [5]. Here, a given number of intermediate images in proximity to the MEP between the initial and final states are optimized in such a way that the energy of all images is minimized, while equal spacing to neighboring images is maintained. In a variant of the NEB method, the highest energy image is driven up to the saddle point by inverting the forces along the tangent, going through the point on the MEP. This method is called climbing-image nudged elastic band (CI-NEB) method [6]. The energy barrier is then estimated as the energy difference between the TS and the initial state.

4.3 Characterizing the Catalyst and Its Interaction with CO_2 Using DFT Calculations

In the past years, because of the ever-increasing computational capability, there has been an increment in the use of DFT methods to study or even predict certain properties before the experiment. For example, Montoya et al. [7] presented a first-principles theoretical study of carbon–carbon coupling in CO_2 electroreduction on the Cu(211) surface. In this work, it was demonstrated that reaction barriers decrease with an increasing grade of hydrogenation of the adsorbates, which can be easily tuned by the applied potential. The study carried out by Yoo et al. [8] is another example of a pure theoretical work in this field. In this case, they explored 27 different metal surfaces to get an insight into the trends that guide the electrochemical reduction of carbon dioxide to formic acid, which competes with another two reactions, the electroreduction of CO_2 to CO and the hydrogen evolution reaction. Lead and silver surfaces were identified as the most promising catalysts. First-principles calculations can also be used to help into the interpretation of experimental data, unveiling the reaction mechanism and studying the intermediate species. Jiang et al. [9] and von Wolff et al. [10] used the Gaussian code and a hybrid DFT functional for supporting their studies in the catalytic activation of CO_2 by frustrated Lewis pairs.

For the electrocatalytic reduction of CO_2, Bourrez et al. [11] combined DFT calculations with sophisticated spectroscopic methods such as UV–vis absorption and pulsed electron paramagnetic resonance (EPR) techniques to characterize a key intermediate in the electroreduction of CO_2 to CO catalyzed by a manganese tris(carbonyl) complex. Kim et al. [12] studied the efficiency and selectivity of the CO_2 electroreduction using immobilized silver nanoparticles (NPs) with cysteamine as an anchoring agent. DFT calculations showed that the specific interactions developed between Ag nanoparticle and the anchoring agent leads the unpaired electron to be localized at the surface of the nanoparticle, which helps to selectively stabilize

COOH over CO. Luo et al. [13] employed DFT calculations combined with a plane wave basis set using the VASP code to model the CO_2 electroreduction to CO at the copper/indium interface. DFT calculations showed that the interface of the two metals significantly decreases the free energy barrier for the formation of the key reaction intermediate (*COOH).

Among all transition metals, copper was identified first as a promising electrode for the electroconversion of CO_2 to hydrocarbons and alcohols [14]. Since then, copper and its alloys have been extensively studied [15]. Peterson et al. [16] used DFT calculations to explain copper's unique ability to convert CO_2 into hydrocarbons, concluding that the key enabling step in their formation from CO_2 is the protonation of adsorbed CO to form adsorbed CHO. Wang et al. [17] used both cluster and periodic DFT calculations to study the adsorption and activation of CO_2 in different Cu surfaces, finding that CO_2 adsorbs and reacts following the order Cu(110) > Cu(100) > Cu(111). Durand et al. [18] showed that the key intermediates in the CO_2 electroreduction on copper are stabilized by the surfaces in the order Cu(211) > Cu(100) > Cu(111).

Methanol is one of the most desired compounds in which CO_2 can be turned into. Mixed metal and metal oxides catalysts are used to obtain methanol through CO_2 by hydrogenation, given the importance of this molecule on the chemical industry. One of the first works that theoretically examined the reaction mechanism and kinetics of the CH_3OH synthesis from CO_2 on industrial Cu/ZnO/Al_2O_3 catalysts was reported by Yang et al. [19] This work focused on the effects of different dopants on two possible pathways. On the pure Cu(111) surface, the first pathway (via a formate intermediate) is restricted by the hydrogenation of HCOO and H_2COO, while the second (via a reverse water gas shift, RWGS: $CO_2 + H_2 \rightarrow CO + H_2O$) is limited by the hydrogenation of CO and HCO species. Using DFT calculations, the authors determined the intermediates for the two reaction mechanisms on Ni-, Pd-, Pt-, Rh-, and Au-doped Cu(111). Later, kinetic Monte Carlo simulations over a network of 15 elementary steps were performed. From their results, the authors conclude that all dopants but Au promote the methanol synthesis. The also showed that on pure Cu(111) and on Au-doped Cu(111), the formate pathway dominates, while on all the other systems, the RWGS mechanism is more relevant, although both pathways are significant in the global production of methanol. In a related study, Graciani et al. [20] (Figure 4.2) developed highly active copper–ceria and copper–ceria–titania catalysts for methanol synthesis from CO_2, where DFT calculations were used to characterize the reaction path and intermediate species. Using a model constituted of a ceria nanoparticle supported on a Cu(111) surface, these authors showed that CO_2 activation takes place at the metal/metal oxide interface (Figure 4.2a,b). After reduction to CO, subsequent hydrogenation steps lead to formation of methoxy group (Figure 4.2e–h). This study illustrates the substantial benefits obtained by properly tuning the properties of a metal/metal oxide interface in catalysts for methanol synthesis.

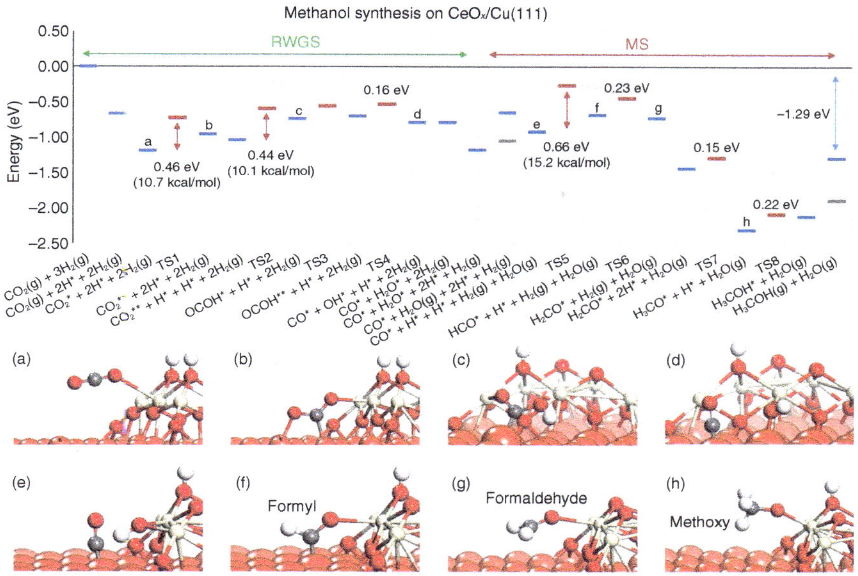

Figure 4.2 Reaction path for methanol synthesis (MS) by CO_2 hydrogenation on the CeO_x/Cu(111) system. Step colors are as follows: local minima states (blue), transition states (TS) (red), and energies including the entropy contribution (gray). Colors: O (red), C (gray), H (white), cerium (light-beige), and Cu (dark pink). (a) Linear CO_2 adsorption on a Ce^{3+} center. Two near O atoms are bonded to H atoms, (b) bent CO_2 adsorption on both Ce^{3+} and Cu, (c) CO_2 hydrogenation produces an adsorbed OCOH carboxyl species, (d) carboxyl group separates to yield a CO molecule and an OH group, both interacting with Cu and Ce^{3+}, (e) CO molecule separates from Ce^{3+}, adsorbed only on Cu. OH group interacting with two Ce^{3+} centers, (f) hydrogen atom from the OH group is bonded to the CO to form a formyl HCO group. Oxygen atom from OH group remains bonded to two Ce^{3+}, (g) hydrogen atom from the pre-hydrogenated oxygen atoms is bonded to the formyl group, yielding a formaldehyde H_2CO group, (h) hydrogen atom from the pre-hydrogenated oxygen atoms is bonded to the formaldehyde group, yielding a methoxy H_3CO group. Source: Graciani et al. [20].

4.4 Microkinetic Modeling in Heterogeneous Catalysis

The development of new, more active, and selective catalytic materials, optimizing the reaction conditions, is the primary goal of heterogeneous catalysis. To reach this objective, the understanding of the elementary reaction steps that takes place at the catalyst surface is fundamental [21]. This includes the knowledge of the thermodynamic and kinetic parameters of the individual steps that constitute the full reaction mechanism. When these steps and their main kinetic parameters are known, it is convenient to build a model to connect the microscopic description of the reaction to the macroscopic experimental data, including concentration effects [22]. This task can be tackled using microkinetic models of the reaction. In a microkinetic model, a reaction network including all relevant elementary steps and their reaction rates is modeled [4, 23, 24]. This model is expressed as a system of differential equations that can be solved for given initial conditions (concentrations of reactants/products). The model can then be used to predict information about surface coverages and reaction behavior at given experimental conditions and establish whether a rate-determining step exists. Although experimental data may allow initial guesses for some kinetic parameters, a microkinetic model constructed using parameters rigorously computed from DFT (adsorption energies, vibrational frequencies, and reaction barriers) and a statistical thermodynamics treatment has true predictive power, obtaining a microscopic understanding of the reaction mechanism [25].

CO_2 conversion by dry reforming of methane (DRM) has been found to be an interesting solution to transform greenhouse gases into valuable fuels via the production of syngas. Industrially, Ni-based catalysts have shown to be effective for this reaction because of their high activity and low cost. However, excessive coking is a severe problem under DRM conditions, leading to catalyst deactivation. This has motivated many studies [26, 27] (Figure 4.3), exploring by theoretical methods the reforming mechanism. However, as there are many elementary reaction steps

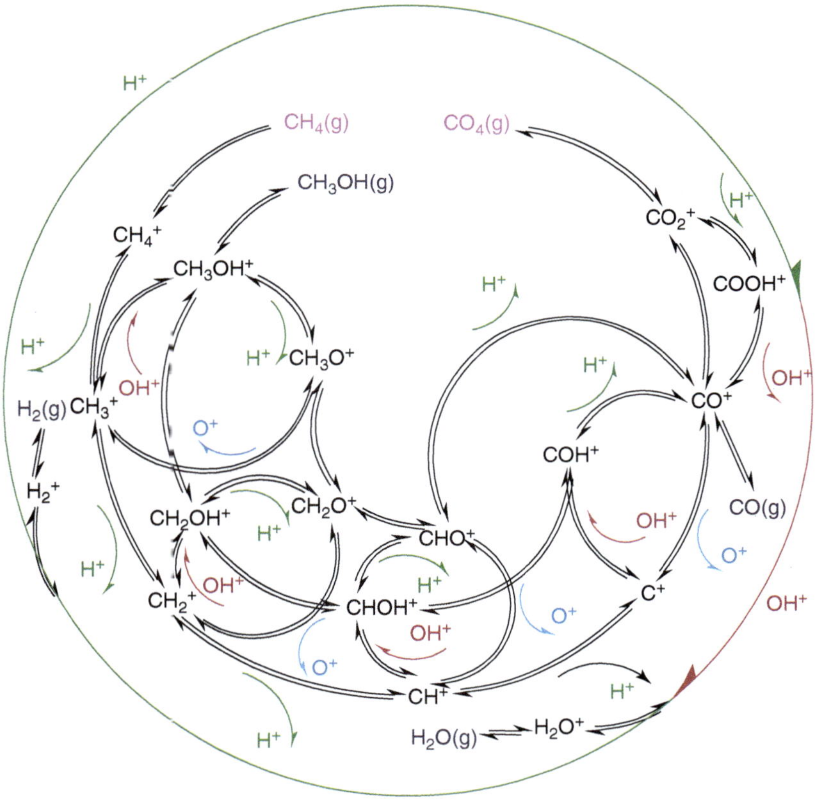

Figure 4.3 Proposed mechanism for dry reforming of methane on Ni catalyst. Source: Zhu et al. [26].

involved, the detailed kinetics is rather complicated. In one of these studies, Fan et al. [27] have developed a detailed microkinetic model to explore CO_2 conversion via DRM on Ni-based catalysts. They included contributions from the different kinds of surfaces that can be present: the close-packed Ni(111) surface, the open Ni(100) surface, and the stepped Ni(211) surface. DFT calculations based on a GGA functional were used to compute the adsorption energies of reactants, intermediates, and products. It was found that the adsorption of reaction intermediates was stronger on the low-coordinated faces, Ni(100) and Ni(211), and that adsorption energies were sensitive to the adsorption configuration. The microkinetic analysis identifies three leading reaction pathways and establishes the oxidation of adsorbed C or CH intermediates by surface O adatom or OH as the rate-determining steps. The rapid deactivation of Ni-based catalysts due to the carbon deposition has motivated different studies for the development of a thermally stable catalyst that will resist deactivation. Noble metals have been generally found to be highly active and, at the same time, more resistant to coke formation. In order to minimize the catalyst deactivation, it is necessary to understand the elementary steps involved in the DRM over the chosen catalysts. In this regard, Niu et al. [28] have used DFT calculations based on the B3LYP functional and cluster models to create a kinetic model for the DRM reaction on the Pt(111) surface. The authors conclude that H-assisted activation is the dominant reaction pathway in CO_2 dissociation and, thus, CH_4 must firstly dissociate to produce the atomic H required for CO_2 activation. Although some small amount of C may be deposited on the Pt surface, the authors find two different paths for C oxidation that can eliminate any residual C. Rh-substituted lanthanum zirconate pyrochlores have been found to be catalytically active and stable for DRM conversion, which has motivated different studies to understand the reaction mechanism. In a couple of recent theoretical works, Polo-Garzon et al. [29, 30] have used DFT + U-based calculations and a microkinetic model that includes 62 elementary reaction steps to model the complete catalytic system. The model showed a general agreement with experiment regarding trends in reactant conversion and H_2/CO ratio with temperature. Moreover, the model predicted the formation of low quantities of carbon, in agreement with the experiment. Quite recently and on the same footing (using other than a supported pure transition metal as a catalyst), Guharoy et al. [31] have explored the potential application of Ni_2P in CO_2 conversion via DRM. They have employed DFT calculations based on the Perdew–Burke–Ernzerhof functional on Ni_2P (0001) surface models and a microkinetic reaction network that includes 17 elementary steps. CO_2 was found to chemisorb on the surface, with C and O atoms strongly bonded to the Ni_2P surface. However, methane was found to physisorb on the surface with the CH_4 activation step ($CH_4(g) \rightarrow CH_3^* + H^*$) and CH cracking steps ($CH^* \rightarrow C^* + H^*$), requiring high activation energies. The microkinetic calculations showed that the O coverage is higher than the C coverage, resulting in the inhibition of C nucleation and an easy propagation of C oxidation on the surface. Thus, the authors concluded that Ni_2P present the active characteristics of Ni surfaces toward DRM but with a reduced rate of carbon formation at the relevant experimental conditions. The activation of CH_4 and CO_2, in this case through direct C–C coupling, has also been examined recently by Zhao et al. [32] using different oxides supported on In_2O_3. The $(ZnO)_3/In_2O_3$ system shows the best activity as CH_4 can dissociate both on the supported $(ZnO)_3$ cluster and on the In_2O_3 support. Based on their microkinetic analysis, the authors show that CH_4 activation plays a key role on the formation of acetic acid.

Chemical fixation of CO_2 with hydrogen to produce fuels and other chemicals of added value is an alternative that is being explored to reduce global warming. Combining DFT + U calculations with a microkinetic model, Tang et al. [33] have examined the interplay of components of the Cu/ZnO catalysts. It was found that the ZnO support does not take part directly in the methanol synthesis because there are few undercoordinated surface O atoms at the interface with Cu. However, a large amount of H_2 dissociates at the interface that acts as a reservoir of H atoms that are spilled onto the Cu(111) surface. The H diffusion provides the atomic hydrogen required for the hydrogenation steps that take place on the Cu surface promoting the whole CO_2 hydrogenation reaction. Indium oxide, In_2O_3, has been found recently to be an excellent catalyst for the methanol synthesis reaction. It combines high methanol selectivity that results in high methanol formation rates with exceptional durability. In a recent paper, Frei et al. [34] have examined the kinetic details of the CO_2 hydrogenation reaction on In_2O_3, both experimentally and using DFT calculations and microkinetic simulations. The DFT calculations showed that the oxygen vacancies created under reaction conditions activate CO_2 and heterolytically dissociate H_2. The microkinetic simulations predict values for the apparent activation energies for CO_2 hydrogenation and the competing RWGS reaction that closely match the experimental data being slightly lower for the CO_2 hydrogenation reaction, explaining the selectivity of the process.

4.5 New Trends: High-Throughput Screening, Volcano Plots, and Machine Learning

4.5.1 High-Throughput Screening

DFT fundamental concepts were developed in the 1960s [35, 36], but it took around 20 years to have the first practical applications and efficient DFT codes and algorithms [37–39]. Since then, DFT packages have become one of the main tools to model materials and chemical systems [40]. The steady growth of computational power and the development of more efficient DFT algorithms have drastically reduced the required time to perform DFT calculations, opening the door to the study of larger systems and more computationally demanding properties. Simultaneously, frameworks and infrastructures have been created to manage automatically tens of hundreds or thousands of calculations. The so-called high-throughput (HT) DFT calculations facilitate the systematic predictions of material or chemical properties and their development [41]. This strategy has been successfully applied for the prediction of crystal structures [42], optoelectronic devices design [43], or catalyst optimization [44].

4.5.2 Volcano Plots and Scaling Relations

Although HT frameworks have automatized DFT calculations and the availability of methodologies such as microkinetic models to predict the activity of a catalyst, the combination of both would become extremely computationally demanding if hundreds or thousands of catalysts needed to be explored. One of the most used methods to overcome this computational bottleneck are volcano plots [45, 46], which connect a computable descriptor with the kinetics or thermodynamics of the catalytic cycle. This method stems from Sabatier's principle, identifying as best candidates those catalysts with intermediates that are not adsorbed on the surface too strongly or two weakly [47]. If an intermediate strongly binds to the surface, high barriers are expected. However, weakly adsorbed intermediates can desorb and stop the reaction at that step [48]. For some cases, multidimensional volcano plots can be built when more than one descriptor is required, for instance, the adsorption energy of two different reactants or competing reactions and pathways [49]. In most cases, the number of variables can be reduced to one or two using scaling relations [50], which are linear correlations between set of adsorption energies.

One of the first works that exploited this approach looked for the right dopant for an Ag-based catalyst for the CO_2 electrochemical conversion [51] (Figure 4.4a). Lim et al. predicted the theoretical reduction potential of CO_2 to CO as a function of the free energy binding energy of CO and COOH. They found that S-doped or As-doped Ag(111) surfaces reduce the overpotential up to 0.5 V. Similarly, Cheng et al. studied single-atom bimetallic alloys using as a host Au and Ag (111) and (100) surfaces [54]. These catalysts follow a two-pot tandem approach where Ag and Au are known for their ability to reduce CO_2 to CO and the single-atom dopant plays the role of active site for the further reduction of CO to hydrocarbons. Approximately half of the alloys that were explored (28) present a preference for CO reduction instead of proton reduction, suppressing the competitive evolution of hydrogen. Most of these alloys contained Co, Rh, and Ir as dopants. If selectivity to different products, beyond CO, wants to be considered, the adsorption energy of four non-coupled intermediates should be considered: H*, COOH*, CO*, and CH_3O* [55]. Volcano plots were also used for the screening of almost 200 near-surface alloys (NSAs) as candidates for CO_2 reduction considering HCOOH, CO, CH_4, CH_3OH, and C_2H_4 as products [52] (Figure 4.4b). Only 26 of them favor HCOO* formation over hydrogen evolution reaction and 20 of them are predicted to be selective for HCOOH formation, highlighting Pd/W, Au/Hf, and Au/Zr. On the other hand, Ag/Ta and Ag/Nb are found as the most promising candidates for CH_3OH and C_2H_4 synthesis. This approach is not exclusive of heterogeneous catalysts. Wodrich et al. have predicted turnover frequency (TOF) with volcano plots and scaling relations for CO_2 reduction to formate using homogeneous catalysts based on transition metals and tridentate pincer ligands [53] (Figure 4.4c).

4.5.3 DFT and Machine Learning

Advances in machine learning (ML) has revolutionized our society in many ways, and its impact on catalysis is a prime example of this. In the previous sections, we

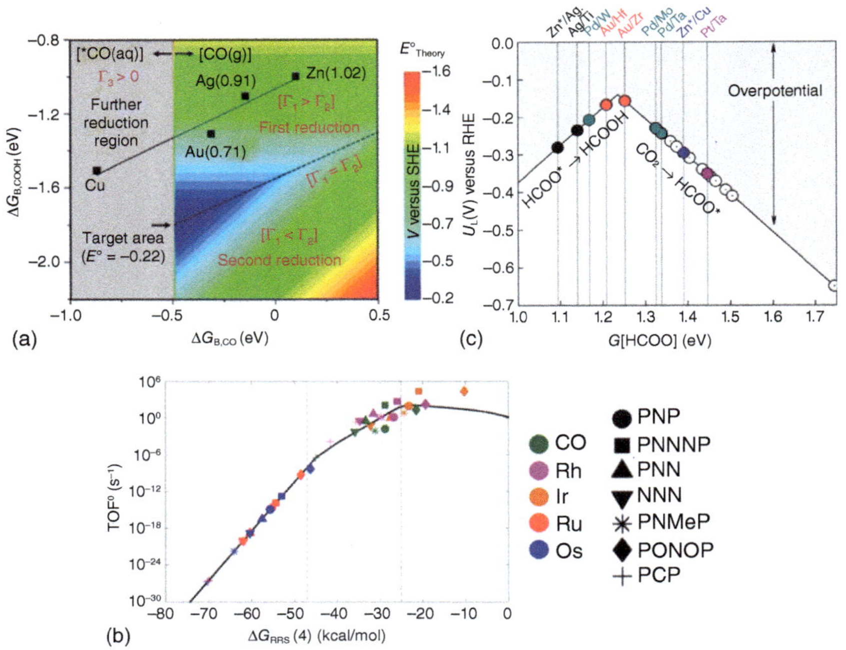

Figure 4.4 (a) 2D contour map of the theoretical reduction potential in terms of the solvated binding free energies of COOH and CO on the catalyst surface. The theoretical minimum bound of the reduction potential is −0.22 V versus SHE, which is shifted from the linear correlation for pure metal catalysts. Source: (a) Lim et al. [51]. (b) Overpotential volcano plot for formic acid production as a function of G[HCOO]. Source: Zhao and Lu [52]. (c) Represent changes in the nature of the turnover determining transition state and turnover determining intermediate within the context of the energy span model. Source: Wodrich et al. [53].

have relied on physical principles, for instance, the use of density functional density, to model and describe surfaces. However, ML builds and trains predictive models based on previous data [56]. Although similar approaches have been used before in catalysis using experimental data [57, 58], ML has rapidly grown in the chemistry [59] and materials science [60, 61] communities because of: (i) the availability of open-source ML tools such as Scikit-learn [62] or TensorFlow [63] that have democratized his use; (ii) the significant growth of computational power; and (iii) the progress on new high-throughput frameworks (PyMatgen [64], ASE [65], and Aflow [66]) and databases (Materials Project [67], Citrination [68], and ioChem-BD [69]).

All the earlier examples such as active site characterization, structure–activity relations, or volcano plots have proved that computational modeling based on first principles can accelerate catalyst screening and discovery. Despite all this progress, the large multiconfigurational space that represents catalyst optimization is practically impossible to explore just relying on the state-of-the-art quantum mechanical methods because of their high computational cost. That is why, ML tools stand as the perfect complement to DFT calculations. ML models are extremely efficient exploring large spaces when previous data are available. In this section, we discuss different approaches and examples in combining DFT and ML for CO_2 capture and catalysis.

4.5.3.1 Machine-Learned Potentials

Despite the rapid growth of computational power and the development of more efficient and scalable DFT codes, the number of variables and the size of a typical catalytic system make impossible to create exhaustive models under reaction conditions using DFT. For instance, ab initio molecular dynamics (AIMD) simulations require thousand or hundreds of thousands of DFT calculations, which rarely can be afforded for realistic catalytic models. State-of-the-art AIMD simulations can manage few hundred of atoms over tens of picoseconds. Classical potentials have traditionally overcome this problem, offering an inexpensive solution that scales more favorably with size. Molecular dynamics (MD) using empirical force fields can simulate microsecond-long trajectories with millions of atoms. However, developing interatomic potentials that accurately captures the heterogeneous chemical environment exhibited at reaction conditions remains as a challenging task. In order to combine the accuracy of DFT calculations and the lower cost of force fields, machine-learned interatomic potentials (MLIP) trained with DFT data have been proposed recently [70], neural networks (NNs) [71], Gaussian process [72], or kernel ridge regression (KRR) [73] are some of the methodologies that have been used to develop open-source codes [74, 75] for generating these potentials. Since then, catalyst dynamics have been increasingly studied using MLIPs because they have opened the door to the accurate study of phenomena that could not be tackled with AIMD: surface dynamics [76], reaction trajectories [77, 78], or liquid/solid interactions [79, 80]. For instance, MLIPs have been used to study the structure of bimetallic Au/Cu nanoparticles (NPs) in aqueous solution, which is a common CO_2 reduction catalyst [79, 80]. Artrith and Kolpak combined first-principles calculations and Monte Carlo simulations with accurate NN potentials [80]. They observed that the NP morphology and composition strongly depend on the size and solvation. While Au/Cu NPs present a Cu-core/Au-shell morphology at small sizes, Wulf-like morphology is observed when the NP size is in the 4–6 nm range in vacuum. However, mixed Au–Cu surfaces are preferred in aqueous solution [79]. These environment-dependent surface reconstructions completely change CO_2 reduction activity and can be used to design synthesis conditions for alternative catalysts with improved properties.

4.5.3.2 Descriptors to Predict Catalytic Properties

ML algorithms can excel describing the behavior of properties, such as the catalytic activity, in terms of some descriptors. That is why, the selection of these descriptors or features stands as the critical step to obtain accurate models [81]. One of the most recurrent descriptors in heterogeneous catalysis for metals is the d-band theory that connects the d-band center, with respect to the Fermi level, with the absorption energies of adsorbates [47, 82–84]. Most importantly, features need to be physically meaningful properties that are correlated with the activity, regardless of whether or not they have been obtained from DFT calculations (band filling level, band width, and atomic charge) or not (geometry, coordination, and ionic potential). For CO_2 reduction, CO adsorption energy is the most common predicted property using ML and descriptors [85]. Recent studies of CO_2 electroreduction have suggested that the limiting steps to C_1 and C_2 species is directly related with the CO adsorption energy. While CO* dimerization is the critical step for C_2 pathways, CO* protonation is the rate-limiting step for C_1 pathways [86, 87]. In both cases, weak CO binding energies reduce the barrier. However, catalyst with a positive CO adsorption energy would desorb CO as a reaction product. Ma et al. trained an artificial NN model using DFT calculations to predict CO binding energies for hundreds of transition bimetallic alloys at their (100) surface [88]. This model predicts CO adsorption energies with a root mean square error (RMSE) of 0.16 eV with respect to the DFT calculations and identified potential Cu-based bimetallic alloys with promising activity and selectivity for CO_2 electroreduction to C_2 species. This model was recently revised using only ab initio free input features such as electronegativity, electron affinity, or coordination, obtaining an RMSE of 0.1 eV [89]. Other authors have explored different surfaces using similar methodologies. NN and KKR methods were used to predict CO adsorption energies on (100), (111), and (211) facets of more than 250 alloys with an RMSE of 0.05–0.18 eV [90]. Also, high-throughput calculations were combined with NN to explore CO_2 reduction on thousands of active sites at 40 different facets for Ni_xGa_y alloys [91] (Figure 4.5). This framework led to the discovery of previously unconsidered but very active sites that consist on Ni atoms surrounded of Ga atoms.

4.5.3.3 Future Challenges in HT-DFT Applied to Catalysis

The combination of high-throughput screening based on DFT calculations, HT-DFT, together with ML has open the door to new possibilities in catalysis. The number of applications and size of the systems that are explored grow day by day, and this is not different for CO_2 reduction reaction [92]. This reaction has been used for testing new frameworks, which automatize data generation, analysis, descriptor, and fingerprint design without losing chemical insight [93]. Databases also facilitate the implementation of these methodologies providing the required data for their training and validation. For instance, hundreds of thousands of metal–organic frameworks were explored to screened using ML recognition to find best candidates for CO_2 capture [94]. Despite all this progress in the field, there are still new methodologies that have not been used for CO_2 catalysis. The study of reaction networks using combining DFT and ML has been already tested for syngas reaction on Rh(111) and could be applied to CO_2 where many different pathways need to be considered [77]. Also, while CO adsorption energy has been correlated with catalytic activity, ultimately activation energies are what govern the kinetic behavior of the catalysts. Reaction barriers for dehydrogenations, N_2 and O_2 dissociation on metal surfaces, have been already predicted using ML models [95].

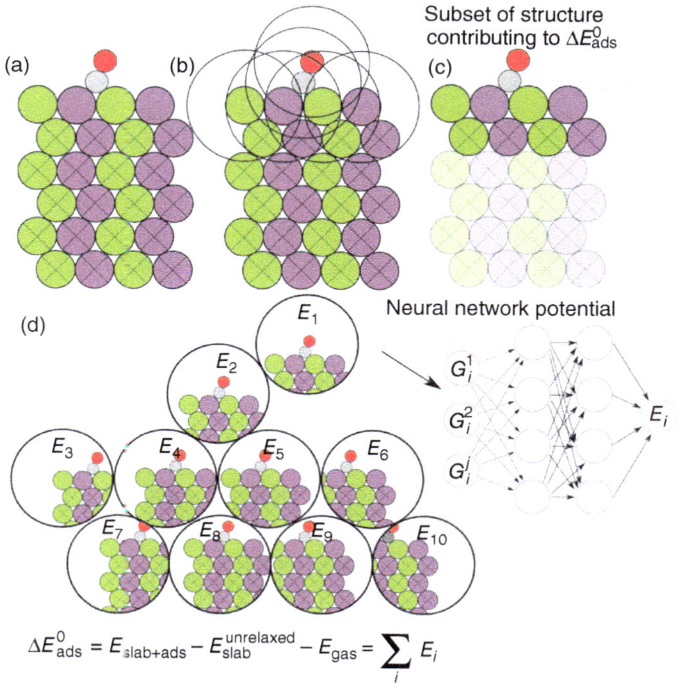

Figure 4.5 Cartoon of the neural network potential used to directly relax and predict adsorption energies for small molecules. (a) Atomistic representation of the bietallic surface and a CO adsorbed on it. (b) Identification of the atoms that will be include in the reduced representation. (c) Subset of the structure used to predict the adsorption energy. (d) The local region around each atom is used to generate a geometric fingerprint, which is fed through a neural network to provide an atomic contribution to the adsorption energy. The predicted adsorption energy is a summation over these atomic contributions. Source: From Ulissi et al. [91].

References

1 Birdja, Y.Y., Pérez-Gallent, E., Figueiredo, M.C. et al. (2019). Advances and challenges in understanding the electrocatalytic conversion of carbon dioxide to fuels. *Nat. Energy* 4 (9): 732–745.

2 Burkart, M.D., Hazari, N., Tway, C.L., and Zeitler, E.L. (2019). Opportunities and challenges for catalysis in carbon dioxide utilization. *ACS Catal.* 9 (9): 7937–7956.

3 Koch, W. and Holthausen, M.C. (2001). *A Chemist's Guide to Density Functional Theory*. Wiley-VCH.

4 Nørskov, J.K., Studt, F., Abild-Pedersen, F., and Bligaard, T. (2014). *Fundamental Concepts in Heterogeneous Catalysis*. Hoboken, NJ: Wiley.

5 Mills, G. and Jónsson, H. (1994). Quantum and thermal effects in H_2 dissociative adsorption: evaluation of free energy barriers in multidimensional quantum systems. *Phys. Rev. Lett.* 72 (7): 1124–1127.

6 Henkelman, G., Uberuaga, B.P., and Jónsson, H. (2000). A climbing image nudged elastic band method for finding saddle points and minimum energy paths. *J. Chem. Phys.* 113 (22): 9901–9904.

7 Montoya, J.H., Peterson, A.A., and Nørskov, J.K. (2013). Insights into C–C coupling in CO_2 electroreduction on copper electrodes. *ChemCatChem* 5 (3): 737–742.

8 Yoo, J.S., Christensen, R., Vegge, T. et al. (2016). Theoretical insight into the trends that guide the electrochemical reduction of carbon dioxide to formic acid. *ChemSusChem* 9 (4): 358–363.

9 Jiang, Y., Blacque, O., Fox, T., and Berke, H. (2013). Catalytic CO_2 activation assisted by rhenium hydride/$B(C_6F_5)_3$ frustrated Lewis pairs – metal hydrides functioning as FLP bases. *J. Am. Chem. Soc.* 135 (20): 7751–7760.

10 von Wolff, N., Lefèvre, G., Berthet, J.-C. et al. (2016). Implications of CO_2 activation by frustrated Lewis pairs in the catalytic hydroboration of CO_2: a view using N/Si^+ frustrated Lewis pairs. *ACS Catal.* 6 (7): 4526–4535.

11 Bourrez, M., Orio, M., Molton, F. et al. (2014). Pulsed-EPR evidence of a manganese(II) hydroxycarbonyl intermediate in the electrocatalytic reduction of carbon dioxide by a manganese bipyridyl derivative. *Angew. Chem. Int. Ed.* 53 (1): 240–243

12 Kim, C., Jeon, H.S., Eom, T. et al. (2015). Achieving selective and efficient electrocatalytic activity for CO_2 reduction using immobilized silver nanoparticles. *J. Am. Chem. Soc.* 137 (43): 13844–13850.

13 Luo, W., Xie, W., Mutschler, R. et al. (2018). Selective and stable electroreduction of CO_2 to CO at the copper/indium interface. *ACS Catal.* 8 (7): 6571–6581.

14 Hori, Y., Murata, A., Takahashi, R., and Suzuki, S. (1987). Electroreduction of carbon monoxide to methane and ethylene at a copper electrode in aqueous solutions at ambient temperature and pressure. *J. Am. Chem. Soc.* 109 (16): 5022–5023.

15 Rendón-Calle, A., Builes, S., and Calle-Vallejo, F. (2018). A brief review of the computational modeling of CO_2 electroreduction on Cu electrodes. *Curr. Opin. Electrochem.* 9: 158–165.

16 Peterson, A.A., Abild-Pedersen, F., Studt, F. et al. (2010). How copper catalyzes the electroreduction of carbon dioxide into hydrocarbon fuels. *Energy Environ. Sci.* 3 (9): 1311–1315.

17 Wang, G.-C., Jiang, L., Morikawa, Y. et al. (2004). Cluster and periodic DFT calculations of adsorption and activation of CO_2 on the Cu(hkl) surfaces. *Surf. Sci.* 570 (3): 205–217.

18 Durand, W.J., Peterson, A.A., Studt, F. et al. (2011). Structure effects on the energetics of the electrochemical reduction of CO_2 by copper surfaces. *Surf. Sci.* 605 (14–15): 1354–1359.

19 Yang, Y., White, M.G., and Liu, P. (2012). Theoretical study of methanol synthesis from CO_2 hydrogenation on metal-doped Cu(111) surfaces. *J. Phys. Chem. C* 116 (1): 248–256.

20 Graciani, J., Mudiyanselage, K., Xu, F. et al. (2014). Highly active copper-ceria and copper-ceria-titania catalysts for methanol synthesis from CO_2. *Science* 345 (6196): 546–550.

21 Chorkendorff, I. and Niemantsverdriet, J.W. (2017). *Concepts of Modern Catalysis and Kinetics*. Wiley-VCH.

22 Stoltze, P. and Nørskov, J.K. (1985). Bridging the pressure gap between ultrahigh-vacuum surface physics and high-pressure catalysis. *Phys. Rev. Lett.* 55 (22): 2502–2505.

23 Dumesic, J.A. (1993). *The Microkinetics of Heterogeneous Catalysis*. American Chemical Society.

24 Cortright, R.D. and Dumesic, J.A. (2001). Kinetics of heterogeneous catalytic reactions: analysis of reaction schemes. *Adv. Catal.* 46: 161–264.

25 Gokhale, A.A., Kandoi, S., Greeley, J.P. et al. (2004). Molecular-level descriptions of surface chemistry in kinetic models using density functional theory. *Chem. Eng. Sci.* 59 (22–23): 4679–4691.

26 Zhu, Y.-A., Chen, D., Zhou, X.-G., and Yuan, W.-K. (2009). DFT studies of dry reforming of methane on Ni catalyst. *Catal. Today* 148 (3–4): 260–267.

27 Fan, C. Zhu, Y.-A., Yang, M.-L. et al. (2015). Density functional theory-assisted microkinetic analysis of methane dry reforming on Ni catalyst. *Ind. Eng. Chem. Res.* 54 (22): 5901–5913.

28 Niu, J., Du, X., Ran, J., and Wang, R. (2016). Dry (CO_2) reforming of methane over Pt catalysts studied by DFT and kinetic modeling. *Appl. Surf. Sci.* 376: 79–90.

29 Polo-Garzon, F., He, M., and Bruce, D.A. (2016). Ab initio derived reaction mechanism for the dry reforming of methane on Rh doped pyrochlore catalysts. *J. Catal.* 333: 59–70.

30 Polo-Garzon, F., Scott, J.K., and Bruce, D.A. (2016). Microkinetic model for the dry reforming of methane on Rh doped pyrochlore catalysts. *J. Catal.* 340: 196–204.

31 Guharoy, U., Ramirez Reina, T., Olsson, E. et al. (2019). Theoretical insights of Ni_2P (0001) surface toward its potential applicability in CO_2 conversion via dry reforming of methane. *ACS Catal.* 9 (4): 3487–3497.

32 Zhao, Y., Wang, H., Han, J. et al. (2019). Simultaneous activation of CH_4 and CO_2 for concerted C–C coupling at oxide–oxide interfaces. *ACS Catal.* 9 (4): 3187–3197.

33 Tang, Q.-L., Zou, W.-T., Huang, R.-K. et al. (2015). Effect of the components' interface on the synthesis of methanol over Cu/ZnO from CO_2/H_2: a microkinetic analysis based on DFT + U calculations. *Phys. Chem. Chem. Phys.* 17 (11): 7317–7333.

34 Frei, M.S., Capdevila-Cortada, M., García-Muelas, R. et al. (2018). Mechanism and microkinetics of methanol synthesis via CO_2 hydrogenation on indium oxide. *J. Catal.* 361: 313–321.

35 Hohenberg, P. and Kohn, W. (1964). Inhomogeneous electron gas. *Phys. Rev.* 136 (3B): B864–B871.

36 Kohn, W. and Sham, L.J. (1965). Self-consistent equations including exchange and correlation effects. *Phys. Rev.* 140 (4A): A1133–A1138.

37 Perdew, J.P. and Zunger, A. (1981). Self-interaction correction to density-functional approximations for many-electron systems. *Phys. Rev. B: Condens. Matter* 23 (10): 5048–5079.

38 Ceperley, D.M. and Alder, B.J. (1980). Ground state of the electron gas by a stochastic method. *Phys. Rev. Lett.* 45 (7): 566–569.

39 Ihm, J., Yin, M.T., and Cohen, M.L. (1981). Quantum mechanical force calculations in solids: the phonon spectrum of Si. *Solid State Commun.* 37 (6): 491–494.

40 Hafner, J., Wolverton, C., and Ceder, G. (2006). Toward computational materials design: the impact of density functional theory on materials research. *MRS Bull.* 31 (9): 659–668.

41 Curtarolo, S., Hart, G.L.W., Nardelli, M.B. et al. (2013). The high-throughput highway to computational materials design. *Nat. Mater.* 12: 191–201.

42 Fischer, C.C., Tibbetts, K.J., Morgan, D., and Ceder, G. (2006). Predicting crystal structure by merging data mining with quantum mechanics. *Nat. Mater.* 5 (8): 641–646.

43 Li, Y. and Yang, K. (2019). High-throughput computational design of organic–inorganic hybrid halide semiconductors beyond perovskites for optoelectronics. *Energy Environ. Sci.* 12 (7): 2233–2243.

44 Andersson, M.P., Bligaard, T., Kustov, A. et al. (2006). Toward computational screening in heterogeneous catalysis: pareto-optimal methanation catalysts. *J. Catal.* 239 (2): 501–506.

45 Gerischer, H. (1958). Mechanismus der Elektrolytischen Wasserstoffabscheidung und Adsorptionsenergie von Atomarem Wasserstoff. *Bull. Soc. Chim. Belg.* 67 (7–8): 506–527.

46 Parsons, R. (1958). The rate of electrolytic hydrogen evolution and the heat of adsorption of hydrogen. *Trans. Faraday Soc.* 54: 1053–1063.

47 Nørskov, J.K., Bligaard, T., Rossmeisl, J., and Christensen, C.H. (2009). Towards the computational design of solid catalysts. *Nat. Chem.* 1: 37–46.

48 Ananikov, V.P. and Barone, V. (2015). *Understanding Organometallic Reaction Mechanisms and Catalysis: Computational and Experimental Tools*. Weinheim, Germany: Wiley-VCH Verlag GmbH & Co. KGaA.

49 Calle-Vallejo, F. and Koper, M.T.M. (2017). Accounting for bifurcating pathways in the screening for CO_2 reduction catalysts. *ACS Catal.* 7 (10): 7346–7351.

50 Abild-Pedersen, F., Greeley, J., Studt, F. et al. (2007). Scaling properties of adsorption energies for hydrogen-containing molecules on transition-metal surfaces. *Phys. Rev. Lett.* 99 (1): 016105.

51 Lim, H.-K., Shin, H., Goddard, W.A. et al. (2014). Embedding covalency into metal catalysts for efficient electrochemical conversion of CO_2. *J. Am. Chem. Soc.* 136 (32): 11355–11361.

52 Zhao, Z. and Lu, G. (2018). Computational screening of near-surface alloys for CO_2 electroreduction. *ACS Catal.* 8 (5): 3885–3894.

53 Wodrich, M.D., Sawatlon, B., Solel, E. et al. (2019). Activity-based screening of homogeneous catalysts through the rapid assessment of theoretically derived turnover frequencies. *ACS Catal.* 9 (6): 5716–5725.

54 Cheng, M.-J., Clark, E.L., Pham, H.H. et al. (2016). Quantum mechanical screening of single-atom bimetallic alloys for the selective reduction of CO_2 to C_1 hydrocarbons. *ACS Catal.* 6 (11): 7769–7777.

55 Bagger, A., Ju, W., Varela, A.S. et al. (2017). Electrochemical CO_2 reduction: a classification problem. *ChemPhysChem* 18 (22): 3266–3273.

56 Hastie, T., Tibshirani, R., and Friedman, J.H. (2009). *The Elements of Statistical Learning: Data Mining, Inference, and Prediction*. Springer.

57 Klanner, C., Farrusseng, D., Baumes, L. et al. (2004). The development of descriptors for solids: teaching "catalytic intuition" to a computer. *Angew. Chem. Int. Ed.* 43 (40): 5347–5349.

58 Corma, A., Serra, J.M., Serna, P., and Moliner, M. (2005). Integrating high-throughput characterization into combinatorial heterogeneous catalysis: unsupervised construction of quantitative structure/property relationship models. *J. Catal.* 232 (2): 335–341.

59 Kitchin, J.R. (2018). Machine learning in catalysis. *Nat. Catal.* 1 (4): 230–232.

60 Ramprasad, R., Batra, R., Pilania, G. et al. (2017). Machine learning in materials informatics: recent applications and prospects. *npj Comput. Mater.* 3 (1): 54.

61 Tabor, D.P., Roch, L.M., Saikin, S.K. et al. (2018). Accelerating the discovery of materials for clean energy in the era of smart automation. *Nat. Rev. Mater.* 3 (5): 5–20.

62 Pedregosa, F., Varoquaux, G., Gramfort, A. et al. (2011). Scikit-learn: machine learning in Python. *J. Mach. Learn. Res.* 12: 2825–2830.

63 Baylor, D., Breck, E., Cheng, H.-T. et al. (2017). TFX: a TensorFlow-based production-scale machine learning platform. In: *Proceedings of the 23rd ACM SIGKDD International Conference on Knowledge Discovery and Data Mining*, Part F1296, 1387–1395. https://doi.org/10.1145/3097983.3098021.

64 Ong, S.P., Richards, W.D., Jain, A. et al. (2013). Python materials genomics (pymatgen): a robust, open-source python library for materials analysis. *Comput. Mater. Sci.* 68: 314–319.

65 Hjorth Larsen, A., JØrgen Mortensen, J., Blomqvist, J. et al. (2017). The atomic simulation environment – a Python library for working with atoms. *J. Phys. Condens. Matter* 29 (27): 273002.

66 Curtarolo, S., Setyawan, W., Hart, G.L.W. et al. (2012). AFLOW: an automatic framework for high-throughput materials discovery. *Comput. Mater. Sci.* 58: 218–226.

67 Jain, A., Ong, S.P., Hautier, G. et al. (2013). Commentary: the materials project: a materials genome approach to accelerating materials innovation. *APL Mater.* 1 (1): 011002.

68 O'Mara, J., Meredig, B., and Michel, K. (2016). Materials data infrastructure: a case study of the citrination platform to examine data import, storage, and access. *JOM* 68 (8): 2031–2034.

69 Álvarez-Moreno, M., De Graaf, C., López, N. et al. (2015). Managing the computational chemistry big data problem: the ioChem-BD platform. *J. Chem. Inf. Model.* 55 (1): 95–103.

70 Botu, V., Batra, R., Chapman, J., and Ramprasad, R. (2017). Machine learning force fields: construction, validation, and outlook. *J. Phys. Chem. C* 121 (1): 511–522.

71 Behler, J. (2015). Constructing high-dimensional neural network potentials: a tutorial review. *Int. J. Quantum Chem.* 115 (16): 1032–1050.

72 Bartók, A.P., Payne, M.C., Kondor, R., and Csányi, G. (2010). Gaussian approximation potentials: the accuracy of quantum mechanics, without the electrons. *Phys. Rev. Lett.* 104 (13): 136403.

73 Huan, T.D., Batra, R., Chapman, J. et al. (2017). A universal strategy for the creation of machine learning-based atomistic force fields. *npj Comput. Mater.* 3 (1): 37.

74 Khorshidi, A. and Peterson, A.A. (2016). Amp: a modular approach to machine learning in atomistic simulations. *Comput. Phys. Commun.* 207: 310–324.

75 Kolb, B., Lentz, L.C., and Kolpak, A.M. (2017). Discovering charge density functionals and structure-property relationships with PROPhet: a general framework for coupling machine learning and first-principles methods. *Sci. Rep.* 7 (1): 1192.

76 Boes, J.R. and Kitchin, J.R. (2017). Neural network predictions of oxygen interactions on a dynamic Pd surface. *Mol. Simul.* 43 (5–6): 346–354.

77 Shakouri, K., Behler, J., Meyer, J., and Kroes, G.-J. (2017). Accurate neural network description of surface phonons in reactive gas–surface dynamics: N_2 + Ru(0001). *J. Phys. Chem. Lett.* 8 (10): 2131–2136.

78 Gerrits, N., Shakouri, K., Behler, J., and Kroes, G.-J. (2019). Accurate probabilities for highly activated reaction of polyatomic molecules on surfaces using a high-dimensional neural network potential: CHD_3 + Cu(111). *J. Phys. Chem. Lett.* 10 (8): 1763–1768.

79 Artrith, N. and Kolpak, A.M. (2014). Understanding the composition and activity of electrocatalytic nanoalloys in aqueous solvents: a combination of DFT and accurate neural network potentials. *Nano Lett.* 14 (5): 2670–2676.

80 Artrith, N. and Kolpak, A.M. (2015). Grand canonical molecular dynamics simulations of Cu–Au nanoalloys in thermal equilibrium using reactive ANN potentials. *Comput. Mater. Sci.* 110: 20–28.

81 Ghiringhelli, L.M., Vybiral, J., Levchenko, S.V. et al. (2015). Big data of materials science: critical role of the descriptor. *Phys. Rev. Lett.* 114 (10): 105503.

82 Hammer, B., Morikawa, Y., and Nørskov, J.K. (1996). CO chemisorption at metal surfaces and overlayers. *Phys. Rev. Lett.* 76 (12): 2141–2144.

83 Hammer, B. and Nørskov, J.K. (1995). Electronic factors determining the reactivity of metal surfaces. *Surf. Sci.* 343 (3): 211–220.

84 Xin, H., Vojvodic, A., Voss, J. et al. (2014). Effects of d-band shape on the surface reactivity of transition-metal alloys. *Phys. Rev. B: Condens. Matter* 89 (11): 115114.

85 Schouten, K.J.P., Qin, Z., Pérez Gallent, E., and Koper, M.T.M. (2012). Two pathways for the formation of ethylene in CO reduction on single-crystal copper electrodes. *J. Am. Chem. Soc.* 134 (24): 9864–9867.

86 Calle-Vallejo, F. and Koper, M.T.M. (2013). Theoretical considerations on the electroreduction of CO to C_2 species on Cu(100) electrodes. *Angew. Chem. Int. Ed.* 52 (28): 7282–7285.

87 Montoya, J.H., Shi, C., Chan, K., and Nørskov, J.K. (2015). Theoretical insights into a CO dimerization mechanism in CO_2 electroreduction. *J. Phys. Chem. Lett.* 6 (11): 2032–2037.

88 Ma, X., Li, Z., Achenie, L.E.K., and Xin, H. (2015). Machine-learning-augmented chemisorption model for CO_2 electroreduction catalyst screening. *J. Phys. Chem. Lett.* 6 (18): 3528–3533.

89 Li, Z., Ma, X., and Xin, H. (2017). Feature engineering of machine-learning chemisorption models for catalyst design. *Catal. Today* 280: 232–238.

90 Noh, J., Back, S., Kim, J., and Jung, Y. (2018). Active learning with non-*ab initio* input features toward efficient CO_2 reduction catalysts. *Chem. Sci.* 9 (23): 5152–5159.

91 Ulissi, Z.W., Tang, M.T., Xiao, J. et al. (2017). Machine-learning methods enable exhaustive searches for active bimetallic facets and reveal active site motifs for CO_2 reduction. *ACS Catal.* 7 (10): 6600–6608.

92 Artrith, N. (2019). Machine learning for the modeling of interfaces in energy storage and conversion materials. *J. Phys. Energy* 1 (3): 032002.

93 Pankajakshan, P., Sanyal, S., De Noord, O.E. et al. (2017). Machine learning and statistical analysis for materials science: stability and transferability of fingerprint descriptors and chemical insights. *Chem. Mater.* 29 (10): 4190–4201.

94 Fernandez, M., Boyd, P.G., Daff, T.D. et al. (2014). Rapid and accurate machine learning recognition of high performing metal organic frameworks for CO_2 capture *J. Phys. Chem. Lett.* 5 (17): 3056–3060.

95 Singh, A.R., Rohr, B.A., Gauthier, J.A., and Nørskov, J.K. (2019). Predicting chemical reaction barriers with a machine learning model. *Catal. Lett.* 149 (9): 2347–2354.

5

An Overview of the Transition to a Carbon-Neutral Steel Industry

Juan C. Navarro[1], *Pablo Navarro*[2], *Oscar H. Laguna*[3], *Miguel A. Centeno*[1], *and José A. Odriozola*[1,4]

[1]*Universidad de Sevilla, Instituto de Ciencia de Materiales de Sevilla, Centro Mixto CSIC, Departamento de Química Inorgánica, Av. Américo Vespucio 49, Seville 41092, Spain*
[2]*Acerinox Europa S.A.U., R+D Department, Polígono Industrial Palmones s/n, Los Barrios ES-11319, Spain*
[3]*Medioambiental y de los Materiales. Escuela Politécnica Superior de Linares – Universidad de Jaén, Departamento de Ingeniería Química, Av. de la Universidad s/n, Linares (Jaén) 23700, Spain*
[4]*University of Surrey, Department of Chemical & Process Engineering, Guilford GU27HX, UK*

5.1 Introduction

The steel industry is one of the main axes of industrial development. Therefore, it has been essential for the growth of the most important economies and will be the key in the case of emerging countries. Furthermore, the application of steel in many fields makes this material one of the most relevant worldwide. However, obtaining steel requires large energy investments and generates huge amounts of greenhouse gas emissions. Therefore, given the current situation, in which humanity is aware of the environmental problem of global warming and the need to act to reverse this damage to the planet, the steel industry has been identified as one of the major contributors to CO_2 emissions. In fact, steel, cement, and the chemical industry are the main emitters of CO_2 in the world and are known as hard-to-decarbonize sectors, thus requiring a comprehensive rethinking that allows them to become environmentally friendly industries, without losing profitability. This is a challenging task that requires the contribution of different actors, not only the responsible of the steel's production but also scientists, engineers, and consumers and politicians.

Therefore, considering the relevance of the steel sector in the worldwide economy and the transition to a carbon-neutral production that has to carry out, in the present chapter, we first present an overview of the relevance of this industry worldwide. Then, we will show some of the political strategies drawn by leaders in the production of steel in the world, followed by clues about the challenge of the cited transition to a carbon-neutral production. Subsequently, the case study of the methanation of CO_2 will be presented as a particularly interesting pathway for the valorization of CO_2 emitted during the production of steel. Finally, the most outstanding information of the projects (most of them in initial phases) will be presented in which technologies are integrated to reduce the carbon footprint in steel production.

Engineering Solutions for CO_2 Conversion, First Edition.
Edited by Tomas R. Reina, José A. Odriozola, and Harvey Arellano-Garcia.

5.2 Global Relevance of the Steel Industry

Steel is a product whose consumption has been increasing every year since 1950 to become the most common material in the current modern society as can be observed in Figure 5.1 [1–3]. Consequently, the steel industry makes an important contribution to the global economy growth around US$ 2.5 trillion in 2017 based on Oxford Economics data [2, 3]. This document shows that at least 6 million of employees work in the sector in the same year, and only the production process benefit was estimated in US$ 500 billion. Additionally, the global importance of the stainless steel industry is also observable in the fact that for every 2 jobs generated, the other 13 jobs are indirectly created; because of the acquisition of raw materials to the distribution of processed steel, indirect impacts are taking place as a result of raw material extraction, heavy machinery, electronic devices, energy consumption, and transportation.

The close relation between steel industry and development of society produces trends such as that presented in Figure 5.2, which illustrates the evolution of percentage of regional share of steel production from 2000 to 2018 [1]. For instance, the high steelmaking capacity of China is in good agreement with the rapid economic growth of this country in the past two decades. Consequently, the impact in the market is clear. In 2000, the Chinese production represents a quarter of the total market and becomes at the same level of Europe in 2002. Despite the drop in 2015, nowadays, China maintains the supremacy in the steel sector. The increasing trend of China in the steel market has made this region the largest manufacturer in the

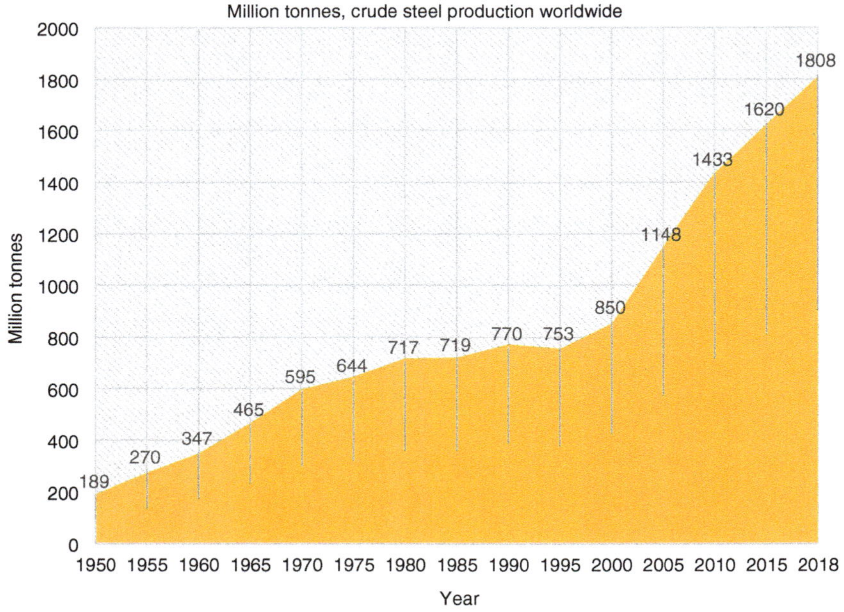

Figure 5.1 World crude steel production in million tonnes reported by the World Steel Association. Source: Adapted from Ref. [1].

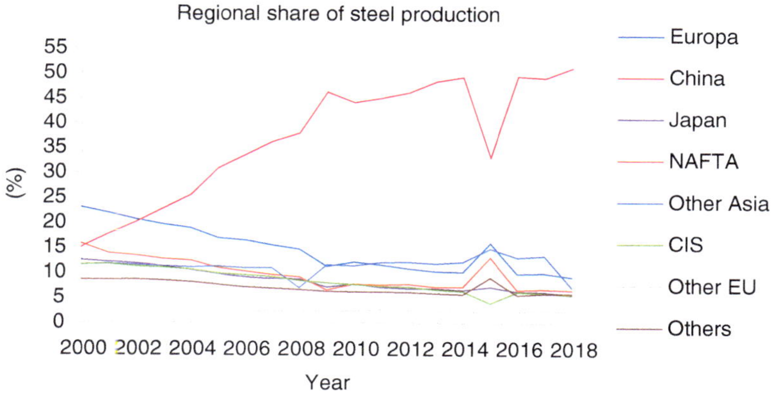

Figure 5.2 Regional share of steel production in 2018.

steel industry since 2003, having 51% of the total market in the present day. This has motivated other countries to exercise aggressive strategies to counteract it, as in the case of the US Government under the President Trump's administration with the imposition of 25% tariff to the distribution of Chinese steel in the United States [4].

5.2.1 Features that Make Steel a Special Material

Steel is commonly alloyed with chromium and other elements. The alloying of steel improves the corrosion resistance behavior and also increases the mechanical properties such as strength, hardness, toughness, and ductility [5]. Steel alloys are named stainless steels and are classified in four families attending their crystal structure: austenitic, ferritic, martensitic, and duplex. The long life service coupled with the mechanical properties provides a wide range of applications that represents an apparent use of steel at 90% of capacity production [6], which indicates the large demand for such material worldwide for further manufacturing. The World Steel Association has carried out a market analysis of the use of steel in 2017, which is summarized in the scheme presented in Figure 5.3. The analysis reports that over half of the produced alloys go to building sector [8], and because of its good strain–stress features and ductility as well, stainless steel is broadly used in architecture, civil engineering infrastructures, and construction materials [8]. Its efficiency provides solutions such as reinforced bars, pipelines, external cladding, roofs, walls, and panels. Even infrastructures are made of stainless steel-like stairs, bridges, and tunnels among others.

Within the different industries that use steel, a very important one that has to be remarked is the automotive industry, which is a key factor in the global economy, considering that it is estimated in 70.5 million the passenger cars production in 2018 according to the International Organization of Motor Vehicle Manufacturers production statistics [9]. In the new designs, which seek to improve the performance of vehicles, making them safer, smarter, and more sustainable, steel is still the main material used in its manufacturing [10]. The formability and high strength

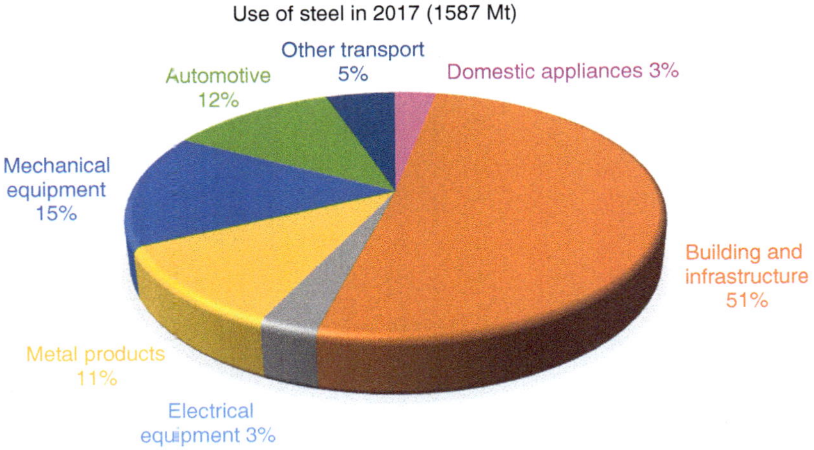

Figure 5.3 Stainless steel market diagram adapted from that published by the World Steel Association. Source: Adapted from Ref. [7].

properties make stainless steel a good candidate for security elements such as doors, trunk, suspension, fuel tank, and break systems. In addition, the corrosion resistance of this material provides a solution in the exhaust system to aggressive atmospheres produced by combustion gases. Furthermore, the long life and low cost of maintenance and cleaning of stainless steel make this material suitable for the production of massive public transportation vehicles and facilities.

Other major consumer of steel is the energy sector, and according to International Energy Statistic report, the world energy demand has been increasing by around twice in 2015 compared with that needed in 1973 [11]. Stainless steel is a key part of the technology for the energy supply and particularly for the renewable energy industry. For instance, stainless steel is the material used for manufacturing turbines in wind farms [12]. Although the blades are made of lighter materials, the principal elements such as the base, the tower, the nacelle, and the rotor are manufactured out of stainless steel [13]. Moreover, water power turbines are manufactured with stainless steel alloys with high saline corrosion resistance, making this material ideal for marine environments. Additionally, this material is a good choice for solar power plants, providing a solid structure and atmospheric corrosion resistance to the photovoltaic panels. Finally, stainless steels become determinant within the scenario of nonrenewable energy industries, principally the oil-based ones, because they usually operate in aggressive and dangerous process conditions such as those required in industrial extracting in offshore and land-field oil platforms and refineries. For these applications, special stainless steel alloys are required with different grades that avoid the carburization produced by corrosive environments. Additionally, since 1970s, nuclear power plants use stainless steel alloys for the coating of the nuclear fuel reactors [9].

According to what has been mentioned so far, the strength of the steel industry and its great projection are undeniable. However, the large volumes of production and the procedures used to obtain the different steels generate a considerable impact on the environment. This represents a great challenge at different levels for the steel industry (as will be shown in Section 5.3) because in the current world scenario, it is looking to rethink all industrial processes to make them more sustainable and friendly to the environment [11, 14].

5.3 Current Trends in Emission Policies in the World's Leading Countries in Steel Industry

Recently, we presented a comprehensive overview of policies and motivations for the CO_2 valorization through the Sabatier reaction using structured reactors [15] and the most relevant legal initiatives were cited and described, including the Paris Agreement [16] and the consequent EU Climate Policy or some of the Chinese plans [17]. From the evaluation of the current emission levels of different sectors considered within the different plans, the steel sector is rapidly highlighted as one of the bigger emitters, thus requiring greater efforts to achieve a real and effective transition to a zero or at least neutral emission technology for the production of steel.

In particular, the Paris Agreement is the cornerstone of the new international climate policy and allowed to set the main goals for facing the global warming in the next years. However, the agreement suffered a severe blow with the resignation of the United States by the decision of the administration of President Trump. The consequences are not only negative from the economic point of view, taking into account that the United States was one of the main contributors to the agreement, but also because this country is one of the technological leaders worldwide. Despite this, in October 2017, some US nonfederal actors (sometimes referred to internationally as "non-state" or "subnational" actors) such as states, cities, universities, and businesses have assumed the American leadership on climate change, as well as the stepping up with commitments to reduce greenhouse gas emissions. Therefore, under the motto "We are still in," this initiative, known as the *America's Pledge* and led by relevant politicians such as the New York City Major or the California Governor, has launched a contingency plan for assuming the commitments of the Paris Agreement and has recently presented a diagnosis [18] showing all the actors involved, the commitments to be acquired, and the possible strategies to also generate new market opportunities. In this diagnosis, the big emitter character of the steel sector is also inferable, so strategies such as the recycling of structural steel, the formulation of federal laws for promoting the use of steel produced with less CO_2 emissions, or the transition to better production technologies were featured [18].

In the case of China, which is the other great technological leader worldwide, the opposite occurs. They not only remain in the 2015 agreement but also show an important interest in leading the initiative [19]. In addition, regarding its particular case, they have established a plan so that their economic growth does not continue to generate a negative impact on the environment, according to the *National Strategy on Energy Production and Consumption Revolution (2016–2030)* [19], and the preliminary evaluations of the launched actions show that they will fulfill the objectives set out in the medium term ahead of schedule. As for the steel sector, China is the world leader and already has a reserve that allows it to slow down the production. With this, they seek to stop the ecological impact of the industrial sector and give time to the development of new technologies that allow greater efficiency in production. In the same way, they develop strategies to incorporate the energy supply in the said sector, using renewable energies [19].

On the other hand, European Union (EU), which is the most important group of countries in the world, presented its Climate Policy in 2016 [17], where a plan for the reduction of emissions was set for 2050 with a relevant check point by 2030. Most recently (November 2018), the 2050 long-term strategy was launched [20] with more detailed plans for different sectors, and, in the case of steel (also included within the great CO_2 emitters), an intensive use of renewable hydrogen was established as a first approach for decarbonizing the sector. Other relevant strategy highlighted was the adaptation of power-to-X technologies to the process of production of steel, aiming to reduce the emission of wastes. Therefore, the reuse of wastes (such as CO_2) for the production of valuable by-products arises as a suitable way for the transition to a circular economy model in the steel sector.

5.4 Transition to a Carbon-Neutral Production. A Big Challenge for the Steel Industry

As mentioned above, the massive production of steel has a considerable impact on the environment, and one of the most important effects is the enormous generation of CO_2. Nowadays, energy-related emissions from the steel and iron industries are around 2.8 Gt of CO_2 per year, and the projections suggest that these would grow to 3.1 Gt by 2040, representing 7.5% of global emissions and 34% of industry emissions [15, 21, 22]. This is why the steel sector is included in the "hard-to-abate" sectors, along with the heavy-duty transport-trucking, shipping, aviation, and plastics industries, as reported by the Energy Transitions Commission (ETC) [22], considering that these sectors that are likely to be harder to decarbonize represent 40% of carbon emissions from the current energy system. Consequently, new approaches are sought that are able to fully decarbonize them, thanks also to the accelerated action from key policy, industry, and finance players [22]. In this sense, it results remarkable that an organization such as the ETC is confident on a complete decarbonization of the steelmaking industry by mid-century. For that, it recommends specific actions in the fields of R&D, the industry/business, and public policy, as can be observed in Figure 5.4.

A first approach to tackle the issue of emissions in the steel sector could be to promote the reduction of the total demand or to fulfill the demand with more recycled secondary steel and less primary production. For the latter option, the involved technologies have to be considered because most primary steel (95%) is produced in blast furnaces (BFs) by means of basic oxygen furnace (BOF) process, using coking coal as a reductant agent as well as source of heat, and this emits lots of CO_2 especially in the blast furnace, which accounts for the 70% CO_2 emissions of the whole

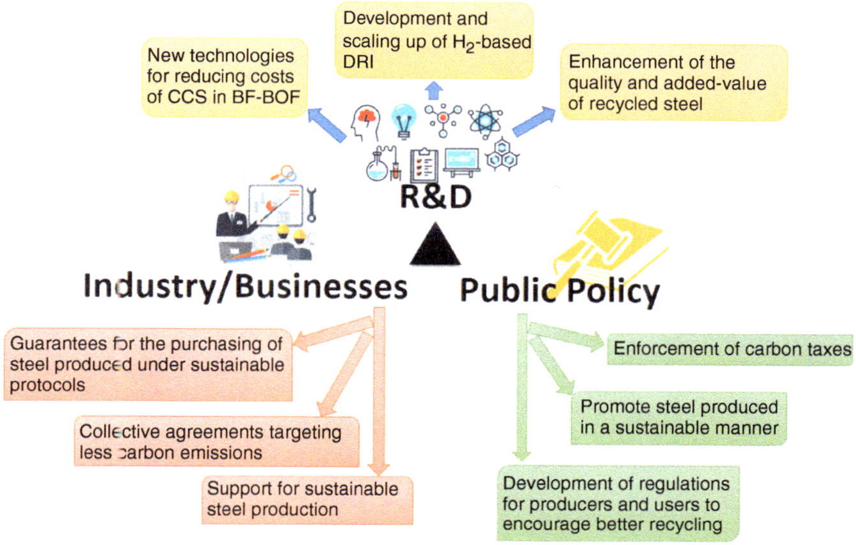

Figure 5.4 Main actors and actions for the transition toward a zero-emission steel industry. Source: Adapted from Ref. [22].

process [23]. The rest of new steel is produced via direct reduction iron (DRI) combined with electric arc furnaces (EAFs). In the case of the DRI-EAF technology, this uses syngas ($CO + H_2$) obtained by means of reforming of methane, for carrying out the reduction process. Moreover, secondary steel recycling is often achieved in EAFs.

The transition to a greater response to demand with secondary steel instead of primary steel is a complex process in which there are different factors that affect the main actors (steelmakers, governments, consumers, etc.). About this regard, the International Energy Agency [21] shows projections from today to 2050, and in the case of the main producer (China), a fall in demand from 800 Mt to 550 Mt, along with the shift from BF-BOF primary production to EAF steel recycling, is expected [22]. These would represent a 60% fall of coal-based steel in such country. However, in the cases of Africa and India, a rapid growth is expected and the coal-based primary production would be applied. As for Europe, the transition to a heavily EAF-based technology is feasible for fulfilling the local demand, although primary production would still be required to fulfill exports [22].

The promotion of the recycling of steel results in capital for the achievement of the cited transition. In fact, it is estimated that 83% of steel is already recycled at end-of-life in several countries around the world. Nevertheless, steel recycling has to overcome serious drawbacks such as the losses of steel that are not recycled resulted from end-of-life structures, which are inaccessible or too corroded, old craps simply lost, losses in remelting process, and new scraps lost in fabrication but not collected. Another drawback is the losses of quality and value in the recycled steel because of the presence of pollutants. Particularly important is the case of copper because this element makes difficult the production of some alloy categories, being a key one for the steel's downcycling [22].

According to that described above, a superior use of secondary steel instead of primary one is not an easy task in most developing regions (except in China where a deceleration has been observable of the economic growth), where the primary production is still required. Therefore, a complementary alternative is to develop low or zero-carbon production routes for primary steel [22]. However, there is not a only pathway for that task because it will depend, among others, on local electricity prices, local feasibility, and cost of the new technologies implied.

The ETC remarks four pathways for the decarbonization of primary steel production [22]: **(i) *Bioenergy***: The use of charcoal obtained from residual biomass as fuel in BF-BOF plants. This is a mature technology whose application is remarkable in the case of Brazil. However, the success of such technology strongly depends on the real renewable character of the biomass feedstocks that can be variable in different places of the world. Biogas (methane obtained by means of the anaerobic digestion of biomass) is other suitable alternative to be included in the production of steel processes, although both generation and use of biogas in technologies such as DRI production require further developments aiming superior efficiency and profitability [22]; **(ii) *H_2 as the reduction agent***: H_2 is already involved in processes such as DRI because the initial CH_4 input is transformed into syngas, which acts as the reduction agent. Therefore, a gradual transition to pure hydrogen instead of methane or syngas can be carried out in the existing DRI facilities. Additionally, the BF-BOF plants could be replaced by new pure hydrogen-based DRI ones. There are some remarkable initiatives such as those of the steel producers Salzgitter [24] and SSAB [25] (from Germany and Sweden, respectively) that have launched a transition toward pure hydrogen-based technologies aiming to achieve important reduction of emissions by 2040. However, these advances strongly depend on the renewability of the generation of the required hydrogen, which is directly associated with a clean production of electricity; **(iii) *Electrolysis***: the possible production of clean electricity also motivates the theoretically viable iron ore reduction via direct electrolysis. However, in the case of steel, unlike aluminum, this technology is under development and requires further advances since still in the laboratory scale [22]; and finally **(iv) *Carbon capture and storage/use (CCS/U)***. The adaptation of carbon capture and utilization (CCU) strategies to the steel industry, on one hand would result in the depleting of CO_2 emissions, and on the other hand, it would help to turn carbon dioxide from waste into a renewable source of carbon for the production of fuels and value-added products. The value of such chemicals can compensate the high energy requirements and economy investments of CCU technologies [15, 26, 27]. This implies the emergence of a CO_2-based economy, which is in good agreement with the principles of the green economy and the circular economy models [15, 28, 29]. In the steel sector, the ETC remarks that CCS/U could be retrofitted on existing BF-BOF production without relevant changes of the current technology. Nowadays, some initiatives of adapting CCS/U to steel manufacturing are being launched in Europe, and these will be presented with more detail in Section 5.6.

Nevertheless, before the present examples of the inclusion of CCU within the steel production process, some general features of such technology and possible routes of CO_2 valorization will be presented for a better understanding of the potential benefits that it can provide. Firstly, there are two types of utilizations: direct and non-direct. Direct utilization is applied, for example, in food, drink, and pharmaceutical industries [30, 31]. Nevertheless, for these kinds of applications, the purification of the used CO_2 is required, which is a procedure that requires energy and economy investment [31–33]. In the case of steel industry, Perez-Fortes et al. [34] have studied the possibility of direct utilization through the recycling of the gas produced in the BF that can act as a reducing agent for the iron process [35, 36], and the results demonstrated that this approach can be successfully adapted to the steel industry.

Regarding the non-direct utilization, the transformation of CO_2 into fuels and/or value-added chemicals is implied and there are different pathways for carrying out this through mineralization, biological, or chemical processes. Firstly, mineralization or mineral carbonation is a chemical process where CO_2 reacts with metal oxides (calcium, magnesium, or iron) to generate carbonates, such as calcite ($CaCO_3$), dolomite ($Ca_{0.5}MgO{\cdot}5CO_3$), magnesite ($MgCO_3$), or siderite ($FeCO_3$) [37]. The principal advantage of this technology is the formation of stable carbonates with the capacity of capture CO_2 capturing for long periods [38], but the problem is the low development for large scale and the elevated cost.

Secondly, biofuel production via biological conversion of CO_2 is an interesting alternative that takes advantage of the natural ability of living species to process such compound. For instance, several species of algae are used for the transformation of CO_2 using sunlight and water into biofuels [39]. Nevertheless, the efficiency of this procedure is limited and requires large areas for cultivation and the careful control of the process, making it expensive and difficult to be applied in large CO_2 emitter sectors [40–42].

As for the chemical conversion procedures, these can be reached through carboxylation reactions, being CO_2 a precursor for organic compounds such as carbonates, acrylates, and polymers, or reduction reactions, breaking the C=O bonds to produce methane, methanol, syngas, urea, dimethyl ether, or formic acid [31, 32, 38, 43]. Some of the different products that can be obtained from CO_2 are presented below because their involvement in the steel industry is highly feasible mainly because of the involved raw materials.

5.4.1 Urea

Large amounts of CO_2 can be used for the production of urea with ammonia that industrially proceeds with an important energy investment (Eq. (5.1)) [44]. What is interesting of this reaction is that its assembling to the stainless steel production is feasible because besides the high production of CO_2, considerable amounts of NH_3 are used during the bright annealing step. Therefore, a certain portion of NH_3 could be used for the transformation of the CO_2 generated.

$$2NH_3 + CO_2 \rightarrow CON_2H_4 + H_2O \quad \Delta H^0_{298K} = 37\ \text{kJ mol}^{-1} \tag{5.1}$$

5.4.2 Methanol and Formic Acid

Both products are generated via hydrogenation of CO_2 (Eqs. (5.2, 5.3)) and their production has been widely studied [44]. In these reactions and others that imply the reduction with hydrogen, not only the supply of such compound but also its renewable character will be determinant for the coupling of these processes of CO_2 valorization in the steel industry. This means that within the technology for the hydrogen supply, renewable energy sources such as solar or wind have to be implemented [45].

Methanol has a great potential because of the capacity to storage hydrogen and can be used as a fuel or as a feedstock for the synthesis of olefins [46], whereas formic acid is applied in textiles, pharmaceuticals, and food chemicals [47]. Moreover, formic acid can also be used as a hydrogen carrier, which is attractive to solve the problem of pure hydrogen transportation.

$$CO_2 + 3H_2 \leftrightarrow CH_3OH + H_2O \quad \Delta H^0_{298K} = -49.5\ \text{kJ mol}^{-1} \tag{5.2}$$

$$CO_2 + H_2 \rightarrow HCOOH \quad \Delta H^0_{298K} = -31.2\ \text{kJ mol}^{-1} \tag{5.3}$$

5.4.3 Carbon Monoxide

The conversion of CO_2 via reverse water gas shift (R-WGS) for the production of CO is another feasible route for its valorization (Eq. (5.4)) [48, 49]. One of the remarkable features of this process is that it commonly presents higher conversions at high temperatures (550–750 °C), so considering that the CO_2 in the steel industry comes mainly in hot flue gas, a direct feed of an R-WGS unit for the CO_2 valorization would avoid important heat losses, which would make the process more energy efficient.

$$CO_2 + H_2 \leftrightarrow CO + H_2O \quad \Delta H^0_{298K} = 41\ \text{kJ mol}^{-1} \tag{5.4}$$

5.4.4 Methane

The CO_2 conversion into CH_4 through the Sabatier reaction (Eq. (5.5)) is an attractive reaction that allows treating large amounts of CO_2 in short times [44, 50]. It implies a hydrogenation process, so the supply of renewable hydrogen is also mandatory and presents a series of advantages that deserve to be exposed in a more detailed way in Section 5.5.

$$CO_2 + 4H_2 \leftrightarrow CH_4 + 2H_2O \quad \Delta H^0_{298K} = -165\ \text{kJ mol}^{-1} \tag{5.5}$$

5.5 CO_2 Methanation: An Interesting Opportunity for the Valorization of the Steel Industry Emissions

Sabatier reaction was discovered by Paul Sabatier and Jean-Baptiste Senderens in 1902 [51, 52], so this is an old process that has been used within the ammonia synthesis for the previous purification of the required H_2, aiming to remove CO_2

[53, 54] and resulting in a substitute of natural gas also known as synthetic natural gas (SNG) [55]. The direct transformation of CO_2 in CH_4 is a highly exothermic reaction (Eq. (5.6)) [56].

$$CO + 3H_2 \rightarrow CH_4 + H_2O \quad \Delta H^0_{298K} = -206\ \text{kJ mol}^{-1} \tag{5.6}$$

One of the most accepted mechanism of this reaction is the combination of R-WGS, and the hydrogenation of the C—O bond [57, 58]. Regarding the catalysts, the most studied have been ruthenium, rhodium, or cobalt, supported over different oxides such as TiO_2, SiO_2, Al_2O_3, CeO_2, or ZrO_2 [59, 60]. Despite this, nickel-based catalysts have been the most widely applied because of their relatively low cost and high activity.

The renewed interest on the Sabatier reaction relies on the current need to promote alternative sources of energy. For this purpose, H_2 may be generated by means of procedures such as electrolysis fed with electricity obtained from wind or solar power. Therefore, this clean H_2 combined with CO_2 (preferably captured from the atmosphere) through the Sabatier reaction is a pathway for transforming this renewable energy (solar or wind) into a valuable gaseous compound. This is known as the power-to-gas (P2G) technology, and it is called to play an important role in the future energy system [61] because the produced methane (SNG) can be injected in the already existing natural gas distribution grid of many cities. In Figure 5.5, an overview of the P2G within the power-to-X concepts recently proposed by Wulf et al. [62] is presented. It results interestingly the possibility of recirculating CO_2 to generate value-added products, which shows that this compound can be the basis of a business model with circular economy concepts.

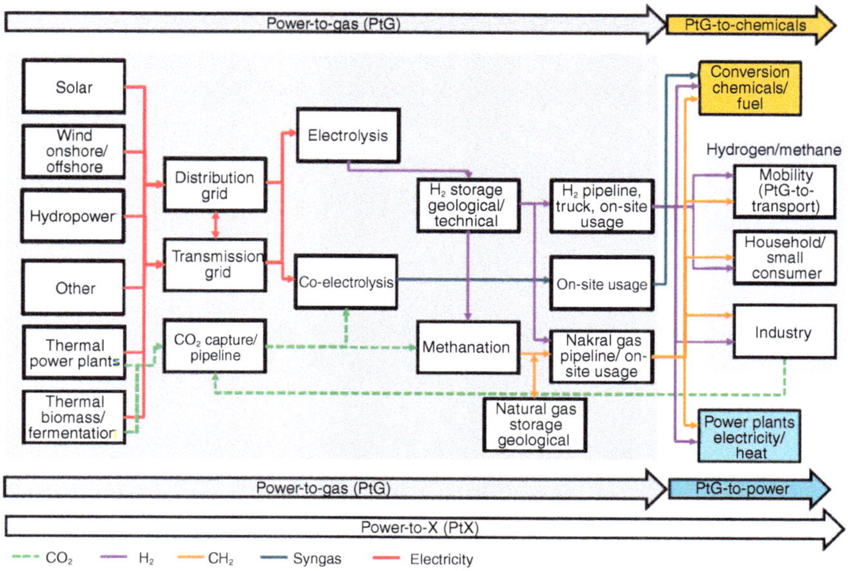

Figure 5.5 Power-to-X concepts overview. Source: Wulf et al. [62].

The P2G technology may also generate oxygen as a by-product during the electrolysis, which could also be commercialized. With regard to methane, this is a valuable compound especially attractive to be used as a fuel because of its high heat capacity and that its combustion may proceed cleanly, generating just H_2O and CO_2 that can be recaptured and valorized again. As an example of the use of P2G, Ancona et al. [63] investigated a P2G as a long-term storage for the production of CH_4. Because the electric energy market of renewable energy sources is continuously increasing, storage systems become essential. To carry out the methanation reaction, the hydrogen is produced via water electrolysis. The results of this study showed that the most influencing parameters on the production of natural gas are the position of the compression section and the water inlet temperature for the hydrolysis. Manipulating these parameters, the process could be more efficient [63].

In Reno (Nevada – USA), the Desert Research Institute (DRI) demonstrates the feasibility of reducing CO_2 with renewably produced H_2 to obtain SNG and water through Sabatier reaction [64]. Moreover, Wulf et al. [62] reviewed 128 P2G research and demonstration projects realized or already finished in Europe by 2018, which demonstrate the great interest on the development of this technology by different governments, scientific institutions, and companies. In most cases, H_2 is produced by water electrolysis, being the electrolyzers powdered by renewable energy sources. Additionally, almost pure CO_2 is used after separating it from biogas using amine scrubbers. Among them, it is remarkably the pioneer Audi e-gas plant that is operating since 2013, located in Wertle (Germany) [65].

For the case of the steel industry, so far, only a few projects that will be presented in Section 5.6 have been already launched for the valorization of CO_2. However, considering its large CO_2 emissions and the imperative need to reverse this, together with the enormous energy demand to keep the blast furnaces active during the production of steel, the possibility of reusing one of the main wastes, making it a useful fuel for the activity of the plant, is a very attractive idea and even more attractive for stainless steel producers that have bright annealing stages for polishing purposes in which ammonia is used. Ammonia decomposes and generates large amounts of H_2 that could be used to supplement the H_2 supply generated (for instance) via electrolysis for a P2G unit, thus reducing the emission of waste of the entire process and increasing its profitability.

5.6 Relevant Projects Already Launched for the Valorization of the CO_2 Emitted by the Steel Industry

European initiatives in the steel sector are the most visible and the most relevant and were recently cited in a review by Hensmann et al. in the EU Industrial Day (22 February 2018, Brussels [66]). Firstly, attending an approach of process integration with reduced use of carbon, the following projects were highlighted:

- *HIsarna* [67]: Led by the company Tata Steel. This project (2017), founded by the EU and the Dutch government, aims to develop a new approach for producing steel, removing a series of preprocessing steps and requiring less-stringent conditions on the quality of the raw materials used. This will represent an enhancement of the efficiency through the use of less energy, with less CO_2 emissions (by 20% less) and less emissions of fine particles, sulfur dioxide, and nitrogen oxide (60–80%).

In the information provided by the company, the operation of HIsarna's technology is described as follows: "...*HIsarna consists of a reactor with temperatures above the melting point of iron throughout the vessel, so that the injected iron ore instantly melts and converts into liquid iron. In the HIsarna furnace temperatures are above the melting point of iron throughout the reactor. The process gasses in the melting vessel have a high temperature. At the top of the reactor (in the cyclone) the temperature is increased further by adding pure oxygen, which reacts with the carbon monoxide present. Because of the turbulence in the cyclone there is enough contact time for the hot gas to melt the iron ore (which is injected at the top). The iron ore immediately melts and drips into the bottom of the vessel. That is where the powder coal is injected, causing the oxygen from the iron ore (= iron oxide) to bind with the carbon, thus creating pure liquid iron, which can then be tapped*..." [67].

- *TGR-BF-Plasma (Injection de Gaz Réformé – IGAR)* [68]: This project, launched in 2018, aims to reformate emitted steel plant gases and then to inject the obtained reformate gases into a blast furnace (BF) to reduce coal/coke consumption. For this purpose, a plasma torch and a reactor to heat and reform gases are used. Additionally, potential CO_2 savings between 0.1 and 0.3 tonnes CO_2 per tonne of crude steel are expected, and the validation of this technology was projected for 2020.
- *STEPWISE* [69]: This project is also founded by the H2020 program of the EU and was launched in 2017. Its principal aim is to scale up the sorption enhanced water gas shift (SEWGS) technology [70] for the CO_2 capture from BF gas. The SEWGS is an already known technology for CO_2 capture from fuel gases in combination with water gas shift and acid gas removal. The success of STEPWISE will also be determined by the achievement of three objectives [69]: (i) high carbon capture rate compared with traditional technologies, (ii) high energy efficiency due to a lower energy consumption for CO_2 capture, and (iii) better economy through the reduction of the CO_2 avoided costs, which indicates that the reduction of CO_2 emission is less expensive.

Moreover, within the approach of using CO/CO_2 from steel mill as a raw material for the production of valuable products, which implies the chemical conversion of CO/CO_2, the following projects are remarked.

- *Steelanol* [71]: This recent project (2019), led by the company ArcelorMittal, aims to produce ethanol at the steel mill in Ghent (Belgium) using a previously developed technology by LanzaTech, whereby gases produced during the chemistry of steel production are treated in a biological procedure with microbes to generate ethanol. This plant will provide the first data for carrying out a life cycle assessment for evaluating the real environmental impact of this technology. However, the production of ethanol from the waste generated during the steel production will represent a clear advance in the CCS and CCU capabilities of ArcelorMittal, in good agreement with the circular economy model, because the obtained ethanol can be incorporated into the liquid fuel supply network (http://www.esrl.noaa.gov/gmd/ccgg/trends/global.html).

- *Carbon2Chem* [72]: Since 2015, this project sponsored by different companies and by the German Federal Ministry of Education and Research aims to transform smelting gases from the steel industry into valuable chemicals such as methanol, thanks to the support provided by partners such as the company Clariant. Different subprojects are integrated in the development of Carbon2Chem: water electrolysis (because surplus electricity from renewable energies is used), gas cleaning, production of higher alcohols and polyalcohols, and production of polymers. Currently, the obtained chemicals are being used as a source of power in the same steel power plant. The commercial implementation of Carbon2Chem will be launched in 2023.
- *FReSME* [73]: According to the information of the project's website, Carbon Recycling International (CRI) and a consortium of European industrial firms and research institutions have been awarded a €11 million grant under the EU's Horizon 2020 program to implement CRI's Emissions-to-Liquids technology in a Swedish steel manufacturing plant, demonstrating how residual blast furnace gases can be turned into liquid fuel (methanol). The project, entitled FreSMe will be implemented in the Swerea MEFOS facility in Luleå, Sweden. The FReSMe project will leverage infrastructure from the Stepwise research project, which separates CO_2 from blast furnace gas and from the MefCO2 project, which demonstrates how CRI's technology can utilize intermittent renewable electricity sources [73].

  Finally, the following projects that follow the strategy of using renewable electricity in basic steelmaking, e.g. production of H_2 to replace carbon, can be remarked.
- *SALCOS* [24]: This project has been formulated by the Salzgitter AG, which is a pan-European steel and technology group. The project proposes that iron ore could initially be reduced to iron with the aid of natural gas and a higher volume of hydrogen in a direct reduction reactor. The reaction would take place at 950 °C, and sponge iron would be produced. Based on this method, a reduction of iron of up to 85% can be achieved. Moreover, the facilities in question involve an integrated process. Gas is introduced in a circular pattern and, after separation of the water produced by the reduction, cleansed of any remaining CO_2 and reused. The challenge inherent in direct reduction consists of integrating the new facilities into the existing steelworks, through the gradual implementation of a reactor of this kind, and CO_2 savings up to 50% are theoretically possible if, in the future, switching the entire production to a direct reduction plant is possible [24].

  As a remarkable feature of this project, it has to be highlighted that the leaders of the idea are aware about the fact that a direct implementation of the project is not technically possible [24]. However, this demonstrates the compromise of steelmakers to launch initiatives that can be viable and profitable in the long term.
- *GrInHy* [24]: Salzgitter AG also launched in 2016 the parallel project GrInHy attending the need of integration of new facilities into the existing infrastructures. In this case, the project is focused in the adaptation of renewable production of

hydrogen for the valorization of waste gases emitted by the steel industry. The description of GrInHy in the website summarizes the following information: "...*GrInHy (= Green Industrial Hydrogen via reversible high-temperature electrolysis). The facilities in Salzgitter also feature the world's currently most powerful reversible high-temperature electrolyser.*
The technology of high-temperature electrolysis enables the reversible operation of the trial facilities. While, during electrolysis operation with the highest electrical efficiency levels, the technology turns industrially generated steam from waste heat into hydrogen, in the reverse case of fuel cell operation, it produces electricity and heat from hydrogen or natural gas.
The electrolytically generated hydrogen can already be used today as a shield gas enabling annealing processes to be used for steel production and replaces fossil-based hydrogen. In the future, naturally with significantly larger capacities, use as a substitute for carbon for the direct reduction of iron oxide is feasible. Along with the proof of electrical efficiencies of more than 80% (in terms of calorific value), the test operation of the techno-economic assessment of the technology also serves as proof in the context of competitive hydrogen production" [24].

- *SIDERWIN* [74]: This European project (led by ArcelorMittal, the world's leading steel and mining company) for the development of new methodologies for industrial CO_2-free steel industry production by electrowining is funded by the European Commission. It aims to develop an innovative electrochemical process to transform iron oxide into steel metal plates, without direct CO_2 emissions. According to the information posted on the project's web site, "...*in this operation, electrical energy and iron oxide are converted into chemical energy consisting of separated iron metal from oxygen gas,*" thus avoiding the use of carbon fossil fuels. In addition, a remarkable feature of this project is that it proposes important technological advances that reduce the investment requirements as well as the risk for industrial participation according to techno-economic studies, which indicates that a comprehensive approach has been initially proposed in order to make this project profitable.

5.7 Concluding Remarks

Considering everything described so far, it can be said that there is a need for a rethinking of the steel industry to reduce its CO_2 emission levels without losing profitability and adapting to a sustainable model in accordance with the circular economy. In addition, taking into account that the steel sector is one of the main emitters in the world, the transition toward a more environmentally friendly production model is a very important technological challenge.

This technological challenge must be assumed as soon as possible and an appropriate political scenario is already being established for this. Likewise, the first initiatives are already being launched with exploration projects involving leaders from the steel sector and two aspects that will be decisive in all future developments can now be identified: first, the incorporation of hydrogen obtained cleanly, both in stages of the same production of steel and in complementary processes for the valuation of the generated emissions, and secondly, strategies for the recovery of waste from the steel industry, especially CO_2, converting it into a raw material for obtaining fuels and/or value-added chemical products.

The road to a carbon-neutral production of steel is long, but important technological advances are expected soon, given that we are immersed in what is known as the fourth industrial revolution (Industry 4.0). Therefore, the rethinking of current production technologies will have new tools to develop increasingly efficient processes and fully adaptable to the particularities of different environments. However, to ensure the success of technological advances, changes must take place at the social, political, and cultural levels that allow for a new way of interacting with the planet and taking advantage of its resources, avoiding their destruction.

References

1 Worldsteel Association (2019). World steel in figures. https://www.worldsteel.org/steel-by-topic/statistics/World-Steel-in-Figures.html (accessed 06 May 2019).

2 International Iron and Steel Institute (1978). *A Handbook of World Steel Statistics*. International Iron and Steel Institute.

3 Worldsteel Association (2014). World steel production 1980–2014. https://www.worldsteel.org/en/dam/jcr:e1f8ca82-b51f-4b10-9edf-5498780a9059/World%2520Steel%2520in%2520Figures%25202020%2520infographic.pdf (accessed 06 May 2019).

4 Cerasa, A. and Buscaglia, D. (2019). A hedonic model of import steel prices: is the EU market integrated? *Resour. Policy* 61: 241–249.

5 Cunat, P.-J. (2002). *The Euro Inox Handbook of Stainless Steel*, Materials and Applications Series, vol. 1. Euro Inox.

6 Chalabyan, A., Mori, L., and Vercammen, S. (2018). The Current Capacity Shake-Up in Steel and How the Industry is Adapting. Metals and Mining Practise, McKinsey&Company.

7 Steel Markets (2019). Steel markets and durability. https://www.worldsteel.org/steel-by-topic/steel-markets.html (accessed 05 June 2019).

8 Steel (2016). The permanent material in the circular economy. https://www.worldsteel.org/media-centre/press-releases/2016/steel-permanent-material-circular-economy.html (accessed 06 May 2019).

9 Barcellini, C., Dumbill, S., and Jimenez-Melero, E. (2018). Isothermal annealing behaviour of nuclear grade 20Cr–25Ni austenitic stainless steel. *Mater. Charact.* 145: 303–311.

10 Hilditch, T.B., de Souza, T., and Hodgson, P.D. (2015). Properties and automotive applications of advanced high-strength steels (AHSS). In: *Welding and Joining of Advanced High Strength Steel* (eds. M. Shome and M. Tumuluru), 9–28. Elsevier Ltd.

11 IEA (2017). *Key World Energy Statistics*. International Energy Statistic.

12 Jaen-Sola, P., McDonald, A.S., and Oterkus, E. (2019). Lightweight design of direct-drive wind turbine electrical generators: a comparison between steel and composite material structures. *Ocean Eng.* 181: 330–341.

13 Greening, B. and Azapagic, A. (2013). Environmental impacts of micro-wind turbines and their potential to contribute to UK climate change targets. *Energy* 59: 454–466.

14 W.S. Association (2017). *Steel's Contribution to a Low Carbon Future and Climate Resilient Societies.* World Steel Association.

15 Navarro J.C., Centeno, M.A., Laguna, O.H., and Odriozola, J.A. (2018). Policies and motivations for the CO_2 valorization through the Sabatier reaction using structured catalysts. A review of the most recent advances. *Catalysts* 8 (12): 578.

16 United Nations (2015). Paris Agreement – United Naitons 2015.

17 Delbeke J. and Vis, P. (2016). *EU Climate Policy Explained.* European Union.

18 America's Pledge (2017). America's Pledge Phase 1 Report: States, Cities, and Businesses in the United States Are Stepping Up on Climate Action.

19 Qi, Y., Stern, N., He, J. et al. (2018). *China's Peaking Emissions and the Future of Global Climate Policy.* Brookings-Tsighua Center for Public Policy.

20 European Commission (2018). Communication from the Commission of the European Parliament, the European Council, the European Economic and Social Committee of the Regions and the European Investment Bank: A Clean Planet for all A European strategic long-term vision for a prosperous, modern, competitive and climate neutral economy. COM/2018/773 final.

21 IEA (2016). Energy Technology Perspectives (Towards Sustainable Urban Energy Systems), International Energy Agency, Paris.

22 Energy Transitions Commission (2018). Reaching zero carbon emissions from steel.

23 Brown, T., Gambhir, A., Florin, N., and Fennell, P. (2012). Reducing CO_2 Emissions from Heavy Industry: A Review of Technologies and Considerations for Policy Makers. Imperial College London, London. p. Paper 7.

24 S. AG (2019). SALCOS: Steelmaking. Reinvented. https://salcos.salzgitter-ag.com/en/index.html?no_cache=1 (accessed 10 July 2019).

25 HYBRIT (2019). Toward fossil-free steel. https://www.ssab.com/company/sustainability/sustainable-operations/hybrit (accessed 11 July 2019).

26 Mac Dowell, N., Fennell, P.S., Shah, N., and Maitland, G.C. (2017). The role of CO_2 capture and utilization in mitigating climate change. *Nat. Clim. Change* 7: 243.

27 Onarheim, K., Arasto, A., Hannula, I., et al. (2016). Carbon capture and utilization (The role of carbon capture and utilization in transitioning to a low-carbon future), VTT (Discussion Paper). pp. 1–21.

28 Ghisellini, P., Cialani, C., and Ulgiati, S. (2016). A review on circular economy: the expected transition to a balanced interplay of environmental and economic systems *J. Cleaner Prod.* 114: 11–32.

29 Ganesh, I. (2014). Conversion of carbon dioxide into methanol – a potential liquid fuel: fundamental challenges and opportunities (a review). *Renewable Sustainable Energy Rev.* 31: 221–257.

30 NETL (2019). Commercial carbon dioxide uses: carbon dioxide enhanced oil recovery. https://netl.doe.gov/research/coal/energy-systems/gasification/gasifipedia/eor (accessed 08 October 2020).

31 Yu, K.M., Curcic, I., Gabriel, J., and Tsang, S.C. (2010). Recent advances in CO_2 capture and utilization. *ChemSusChem* 11 (1): 893–899.

32 Markewitz, P., Kuckshinrichs, W., Leitner, W. et al. (2012). Worldwide innovations in the development of carbon capture technologies and the utilization of CO_2. *Energy Environ. Sci.* 5 (6): 7281–7305.

33 Metz, B., Davidson, O., de Coninck, H., et al. (2005). IPCC Special Report on Carbon Dioxide Capture and Storage. Intergovernmental Panel on Climate Change. https://www.ipcc.ch/report/carbon-dioxide-capture-and-storage/ (accessed 14 October 2020).

34 Pérez-Fortes, M., Moya, J.A., Vatopoulos, K., and Tzimas, E. (2014). CO_2 capture and utilization in cement and iron and steel industries. *Energy Procedia* 63: 6534–6543.

35 van der Stel, J., Hattink, M., Sert, D., et al. (2011). Developments of the ULCOS low CO_2 blast furnace process at the LKAB experimental BF in Lulea. METEC in Steel Con 2011. Volume: Lean CO_2 ironmaking, Session 5. Project: Experimental BF, Düsseldorf, Germany.

36 van der Stel, J., Hirsch, A., Janhsen, U., et al. (2014). ULCOS Top Gas Recycling Blast Furnace Process (ULCOS TGRBF). Research fund for coal and steel (EC), EUR 26414.

37 Olajire, A.A. (2013). A review of mineral carbonation technology in sequestration of CO_2. *J. Petrol. Sci. Eng.* 109: 364–392.

38 Styring, P., Jansen, D., Coninck, H.D., et al. (2011). Carbon capture and utilisation in the green economy. Centre for Low Carbon Futures. http://co2chem.co.uk/wp-content/uploads/2012/06/CCU%20in%20the%20green%20economy%20report.pdf (accessed 05 June 2019).

39 Rodionova, M.V., Poudyal, R.S., Tiwari, I. et al. (2017). Biofuel production: challenges and opportunities. *Int. J. Hydrogen Energy* 42: 8450–8461.

40 Cuéllar-Franca, R.M. and Azapagic, A. (2015). Carbon capture, storage and utilisation technologies: a critical analysis and comparison of their life cycle environmental impacts. *J. CO_2 Util.* 9: 82–102.

41 Brennan, L. and Owende, P. (2010). Biofuels from microalgae – a review of technologies for production, processing and extractions of biofuels and co-products. *Renewable Sustainable Energy Rev.* 14: 557–577.

42 Li, Y., Horsman, M., Wu, N. et al. (2008). Biofuels from microalgae. *Biotechnol. Progr.* 24 (4): 815–820.

43 von der Assen, N., Jung, J., and Bardow, A. (2013). Life-cycle assessment of carbon dioxide capture and utilization: avoiding the pitfalls. *Energy Environ. Sci.* 6: 2721–2734.

44 Aresta, M., Dibenedetto, A., and Angelini, A. (2014). Catalysis for the valorization of exhaust carbon: from CO_2 to chemicals, materials, and fuels. Technological use of CO_2. *Chem. Rev.* 114 (3): 1709–1742.

45 Trovarelli, A., de Leitenburg, C., and Dolcetti, G. (1997). Design better cerium-based oxidation catalysts. *Chem. Tech.* 27 (6): 32–37.
46 Olah, G. (2005). Beyond oil and gas: the methanol economy. *Angew. Chem. Int. Ed.* 44: 2636–2639.
47 Pérez-Fortes, M., Schöneberger, J.C., Boulamanti, A. et al. (2016). Formic acid synthesis using CO_2 as raw material: techno-economic and environmental evaluation and market potential. *Int. J. Hydrogen Energy* 41: 16444–16462.
48 Zhang, Q., Pastor-Perez, L., Jin, W. et al. (2019). Understanding the promoter effect of Cu and Cs over highly effective β-Mo_2C catalysts for the reverse water-gas shift reaction. *Appl. Catal., B* 244: 889–898.
49 Pastor-Perez, L., Baibars, F., Le Sache, E. et al. (2017). CO_2 valorisation via reverse water-gas shift reaction using advanced Cs doped Fe–Cu/Al_2O_3 catalysts. *J. CO_2 Util.* 21: 423–428.
50 Davis, W. and Martin, M. (2014). Optimal year-round operation for methane production from CO_2 and water using wind and/or solar energy. *J. Cleaner Prod.* 80: 252–261.
51 Senderens, J.-B. and Sabatier, P. (1902). *Comptes Rendus Des Seances De L'Academie Des Sciences Section VI - Chimie.* Paris: Imprimerie Gauthier-Villars.
52 Sabatier, P. (1912). The method of direct hydrogenation by catalysis. https://www.nobelprize.org/prizes/chemistry/1912/sabatier/lecture/ (accessed 06 June 2019).
53 Araki, V.P.M. (1976). Methanation of carbon monoxide on nickel and nickel–copper alloys. *J. Catal.* 44: 439–448.
54 Martin, G.A., Primet, M., and Dalmon, J.A. (1978). Reactions of CO and CO_2 on Ni/SiO_2 above 373K as studied by infrared spectroscopic and magnetic methods. *J. Catal.* 53: 321–330.
55 Garbarino, G., Riani, P., Magistri, L., and Busca, G. (2014). A study of the methanation of carbon dioxide on Ni/Al_2O_3 catalysts at atmospheric pressure. *Int. J. Hydrogen Energy* 39: 11557–11565.
56 Bobadilla, L.F., Muñoz-Murillo, A., Laguna, O.H. et al. (2019). Does shaping catalysts modify active phase sites? A comprehensive in situ FTIR spectroscopic study on the performance of a model Ru/Al_2O_3 catalyst for the CO methanation. *Chem. Eng. J.* 1: 248–257.
57 Jürgensen, L., Ehimen, E.A., Born, J., and Holm-Nielsen, J.B. (2015). Dynamic biogas upgrading based on Sabatier process: thermodynamic and dynamic process simulation. *Bioresour. Technol.* 178: 323–329.
58 Navarro-Jaén, S., Navarro, J.C., Bobadilla, L.F. et al. (2019). Size-tailored Ru nanoparticles deposited over γ-Al_2O_3 for the CO_2 methanation reaction. *Appl. Surf. Sci.* 483: 750–761.
59 Martinez Tejada, A.M.L.M., Centeno, M.A., and Odriozola, J.A. (2016). In situ Raman spectroscopy study of Ru/TiO_2 catalyst in the selective methanation of CO. *J. Raman Spectrosc.* 47: 189–197.
60 Razzaq, C.S.L.R., Usman, M., Suzuki, K., and Zhang, S.J. (2015). A highly active and stable Co_4N/γ-Al_2O_3 catalyst for CO and CO_2 methanation to produce synthetic natural gas (SNG). *Chem. Eng. J.* 262: 1090–1098.

61 Götz, M., Lefebvre, J., Mörs, F. et al. (2016). Renewable power-to-gas: a technological and economic review. *Renewable Energy* 85: 1371–1390.

62 Wulf, C., Linßen, J., and Zapp, P. (2018). Review of power-to-gas projects in Europe. *Energy Procedia* 155: 367–378.

63 Ancona, M.A., Antonioni, G., Branchini, L. et al. (2016). Renewable energy storage system based on a power-to-gas conversion process. *Energy Procedia* 101: 854–861.

64 Hoekman, S.K., Broch, A., Robbins, C., and Purcell, R. (2010). CO_2 recycling by reaction with renewably-generated hydrogen. *Int. J. Greenhouse Gas Control* 4 (1): 44–50.

65 Köbler, J. (2012). *Audi Future Lab: Mobility*. Ingolstadt: Audi.

66 Hensmann, M., Meijer, K., and Oles, M. (2018). *Smart Carbon Usage, Process Integration and Carbon Capture and Usage*. EU Industry Day, European Commission.

67 Tata Steel (2019). Tata Steel Factsheet Dec 2017: HIsarna: Game changer in the steel industry. https://www.ademe.fr/igar (accessed 14 October 2020).

68 IGAR (2018). *Validation pré-industrielle de l'injection de gaz réducteur dans un haut-fourneau sidérurgique*. Agence de l'Environment et de la Maitrise de l'Energie (ADEME).

69 STEPWISE (2019). The STEPWISE project. https://www.stepwise.eu/ (accessed 11 July 2019).

70 Gazzani, M., Romano, M.C., and Manzolini, G. (2015). CO_2 capture in integrated steelworks by commercial-ready technologies and SEWGS process. *Int. J. Greenhouse Gas Control* 41: 249–267.

71 Kennedylaan, J. (2019). Steelanol: fueling a sustainable future. http://www.steelanol.eu/en (accessed 11 July 2019).

72 Bouwens, T. (2016). Clariant supports Carbon2Chem project for the reduction of industrial CO_2 emissions. *Focus Catal.* 2016 (8): 3.

73 FReSMe Project (2019). From residual steel gases to methanol. https://www.carbonrecycling.is/projects#projects-fresmehttps://www.carbonrecycling.is/projects#projects-fresme (accessed 11 July 2019).

74 SIDERWIN (2019). Development of new methodologies for industrial CO_2-free steel production by electrowining. https://www.siderwin-spire.eu/content/home (accessed 10 July 2019).

6

Potential Processes for Simultaneous Biogas Upgrading and Carbon Dioxide Utilization

Francisco M. Baena-Moreno[1,2], *Mónica Rodríguez-Galán*[1], *Fernando Vega*[1], *Isabel Malico*[2], *and Benito Navarrete*[1]

[1] *University of Seville, Technical School of Engineering, Chemical and Environmental Engineering Department, C/Camino de los Descubrimientos s/n, Sevilla 41092, Spain*
[2] *University of Évora, Department of Physics, R. Romão Ramalho, 59, 7000-671, Évora, Portugal*

6.1 Introduction

Climate change and global warming caused by human activity is one of the most challenging environmental problems that need to be solved nowadays. Among the greenhouse gases responsible for this problem, carbon dioxide (CO_2) stands out because the emissions of this contaminant have considerably increased from 2013 to 2018 in agreement with the Scripps Institution of Oceanography [1], as represented in Figure 6.1. CO_2 production mainly comes from energy production through fossil fuels, and its reduction seems infeasible in the short term [2]. For this reason, many countries have adopted environmentally friendly energy policies based on renewable energy. Nevertheless, these green energy sources are questionable because their use presents some barriers [3].

The most challenging barriers for the majority of renewable energy development are siting and transmission, market entry, reliability, and cost [4, 5]. The latter is not a consequence of operation but is caused as building the necessary facilities to obtain energy [6]. For this reason, the future affordability of renewable energy production relies on obtaining not only energy but also potential subproducts that could be launched into the market [7].

Among the several renewable energies, biogas has stood out recently because the barriers previously explained can be overcome [8]. Indeed, biogas production plants have progressively increased during the past years as shown in Figure 6.2. Biogas is a reliable source of bioenergy because it does not face the dependence of weather patterns as solar or wind do [10]. Generally, biogas comes from the anaerobic digestion of biomass from landfills or wastewaters; hence, its utilization reduces the amount of waste [11]. Unlike other renewable energies, biogas requires low capital investment and production plants can be developed in rural areas that can also help to balance the energy dependence of developing countries [12].

Engineering Solutions for CO_2 Conversion, First Edition.
Edited by Tomas R. Reina, José A. Odriozola, and Harvey Arellano-Garcia.

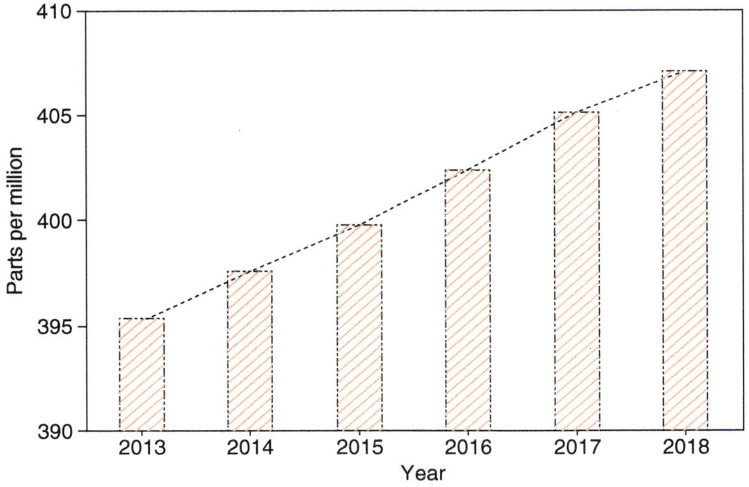

Figure 6.1 Annual record of CO_2 concentration in the atmosphere from 2013 to 2018. Source: Modified from Scripps Institution of Oceanography [1].

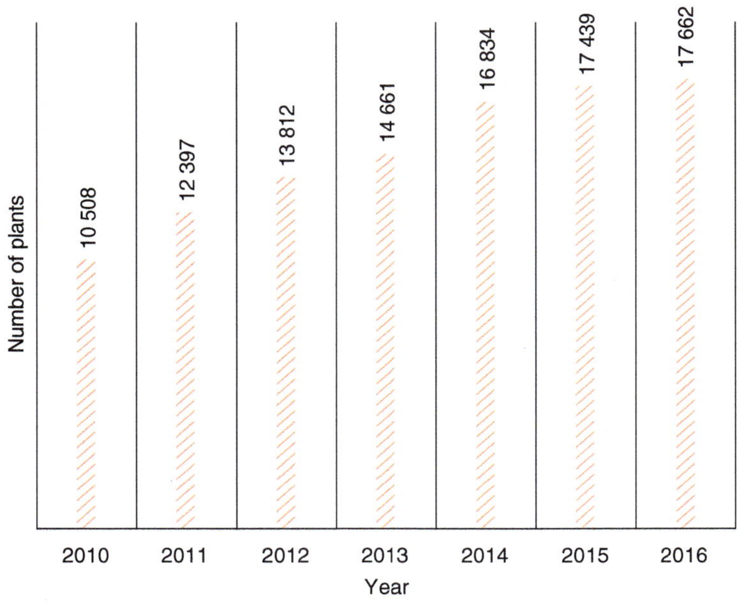

Figure 6.2 Evolution of biogas production plants from 2010 to 2016. Source: Adapted from Deremince and Königsberger [9].

Biogas is mainly composed by methane (CH_4) and CO_2 in a 60–40%, respectively [13]. Thus, its direct utilization without CO_2 removal will end in a non 100% clean energy use. Consequently, a previous stage to remove CO_2 and other minor impurities is required to obtain clean bio-methane product [14]. This stage is well known as biogas upgrading process, and multiple different configurations are commercially available [15, 16]. The necessity of a previous conditioning/CO_2 removal step to obtain bio-methane makes the overall process more expensive.

To face this problem, many researchers have focused their efforts in minimizing the economic output of biogas upgrading. One of the outstanding paths is the production of value-added by-products from CO_2 removal, which could be sold [17]. This chapter serves as a guide for those interested in bio-methane production and CO_2 utilization. To this end, the chapter is organized as follows: firstly, an overview of biogas general characteristics and upgrading technologies to produce bio-methane is presented. Secondly, carbon capture and utilization (CCU) technologies are briefly explained. Then, the central topic of this chapter is covered by an exhaustive examination of the most suitable processes to synergize biogas upgrading and CCU. Finally, a conclusion section, which also includes some future perspectives, is presented.

6.2 Overview of Biogas General Characteristics and Upgrading Technologies to Bio-methane Production

6.2.1 Biogas Composition and Applications

As addressed before, biogas is obtained as the main product from the anaerobic digestion of multitude of wastes and other organic materials. Therefore, its composition strongly depends on the feedstock employed to obtain the final biogas. Table 6.1 provides a range for biogas compositions depending on different sources. After removing contaminants from biogas, CO_2 can also be scrubbed to obtain clean bio-methane that can be used in several applications. Some of these applications that are relevant for the chemical and transport sectors are summarized in Figure 6.3.

Direct combustion of biogas to produce heat and steam is the most common industrial application so far considered. Bio-methane injection into the natural gas grid has the advantage of using already existing facilities in the majority of the countries, which makes its implementation easier for practical and legal reasons. Regarding bio-methane utilization in light vehicles as a gasoline substitutive, high benefits are obtained in terms of emission reductions [23–25].

6.2.2 Biogas Upgrading Processes

Removal of biogas contaminants is not only an environmental advantage but also there are multiple benefits such as the expansion of the life span of the

Table 6.1 Biogas composition from different sources [13, 15, 18–22].

Compound	Biogas from sewage	Biogas from landfill	Biogas from wastewater
CH_4 (%)	60–70	35–65	55–58
CO_2 (%)	34–38	30–45	32–50
H_2O (%)	1–7	1–5	1–5
NH_3 (ppm)	50–100	0–5	0–100
H_2 (%)	Traces	0–5	Traces
O_2 (%)	Traces	0–1	Traces
N_2 (%)	0–2	5–15	Traces
H_2S (ppm)	0–4000	0–100	0–4000
Siloxanes (%)	0–0.2	0–0.2	0–0.5

Source: Adapted from Baena-Moreno et al. [15], Weiland [22], Baciocchi et al. [18], Lombardi et al. [20], Bekkering et al. [13], Rasi et al. [21], Kerroum et al. [19].

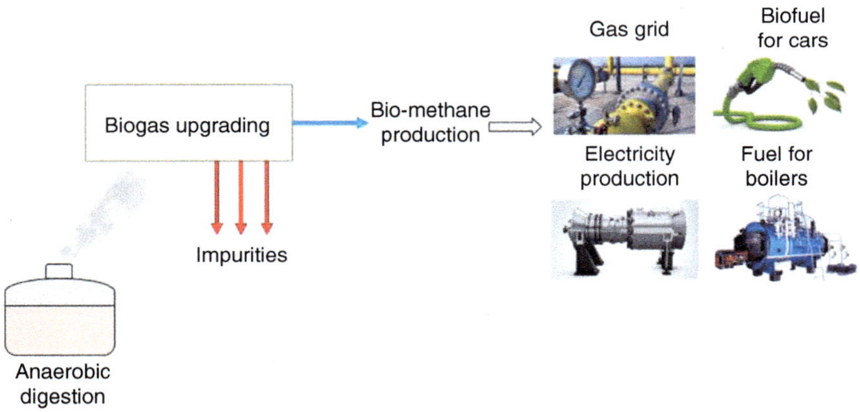

Figure 6.3 Some options for bio-methane use after upgrading process. Sources: DarioEgidi/Getty Images; maksym yemelyanov/123RF.

process equipment, the increase of the calorific value of biogas from 20–25 to 35–40 MJ m^{-3}, and the overall improvement of the economic balance of the process [26, 27]. The economic difference is even higher because biogas can be sold at 0.89–2.97 pence kWh^{-1} and bio-methane at 1.49–3.30 pence kWh^{-1} [28, 29].

Many biogas upgrading technologies are available both at lab scale and commercially. The most important biogas upgrading technologies are water scrubbing, pressure swing adsorption, physical separation, chemical absorption, membrane processes, and cryogenic upgrading. A brief description of each process can be found in Sections 6.2.2.1–6.2.2.6.

6.2.2.1 Water Scrubbing

This technology is based on the different solubility of CO_2 and CH_4 in water. Therefore, absorption of CO_2 in water is carried out physically in a packed tower filled with random packing to increase the contact surface. This process is usually carried out at high pressures of around 6–10 bar and can also be employed to selective removal of H_2S [30]. After the removal stage, water can be regenerated by a stripping air process; hence, it can be reused to decrease the water fresh feed.

6.2.2.2 Pressure Swing Adsorption

Pressure swing adsorption is one of the most employed technologies for biogas upgrading. Indeed several companies such as Carbotech or Xebec Inc. have developed commercial modules of this technology for small–medium-scale plants. Biogas passes through an absorption column where CO_2 is trapped in the adsorbent while CH_4 is recovered at the top of the column. Afterward, to regenerate the adsorbent, pressure is reduced to vacuum. The working pressure for this technology is between 4 and 10 bars for the adsorption column, and the most employed adsorbents are carbon molecular sieves, activated carbons, and zeolites [31].

6.2.2.3 Chemical Scrubbing

Chemical absorption by means of traditional solvents such as monoethanolamine (MEA) or piperazine have demonstrated to be an effective alternative for biogas upgrading because of the high efficiencies obtained. In this process, CO_2 is first absorbed in a packed tower similarly to water scrubbing. Subsequently, the regeneration of the solvent is carried out by increasing the working temperature in a second tower, hence boosting the energy penalty. During the past decade, some studies have tested caustic solvents as an alternative to the traditional solvents because the regeneration stage is relatively simple and can be accomplished by means of chemical reaction [32–34].

6.2.2.4 Organic Physical Scrubbing

This technology is very similar to water scrubbing. The main difference is that CO_2 solubility is higher in a prepared organic solvent than in water. This results in a lower demand of fresh solvent feed. One of the solvents typically employed is Selexol, which has proved high CO_2 removal efficiencies. As in water scrubbing biogas upgrading, H_2S can also be adsorbed physically in the solvent [35]. On the other hand, organic solvents are much more expensive than water; hence, the economic balance can be negatively affected in some cases.

6.2.2.5 Membrane Separation

Gas cleaning by means of membrane separation is a common technology employed in chemical industries. For biogas upgrading, hollow fiber membranes made of different polymers are employed to obtain a stream with high CH_4 concentration and another one with CO_2, water, and NH_3. In a previous stage, biogas should be compressed to 16 bars. Despite the simplicity of the technology, not high recovery efficiencies can be obtained with only one module operation, so a membrane cascade is necessary in those cases where high CH_4 purity is needed [36].

6.2.2.6 Cryogenic Separation

Cryogenic separation to obtain liquid bio-methane from biogas is typically carried out in three different stages. First, a pretreatment for H_2S, water, and other contaminant removal is needed. This stage works first at −40 °C for water removal and then at −75 °C for other contaminant elimination. Second, carbon dioxide is removed from biogas in a −120 °C column and recovered in liquid phase for its use in many applications. Finally, the liquefaction of clean bio-methane is carried out in a cryogenic column at 15 bars and −120 °C. Then, the final product is stored in a cryogenic vessel. Because of the low temperatures required, the operational costs of this technology are quite elevated particularly in those places where cold resources are not available [37, 38].

Comparing biogas upgrading technologies in terms of product result, removal capacity and CH_4 losses have been the purpose of several studies. The technical features of each technology are summarized in Table 6.2. As it can be seen, water scrubbing and organic physical scrubbing result in the highest amount of impurities as well as the highest CH_4 losses. Also membrane technology gives high losses when a recirculation stage is not implemented. On the other hand, chemical scrubbing and cryogenic separation provide the best CH_4 purity in the product stream.

Nowadays, many research teams are focusing their efforts on improving the overall efficiency of the process as well as on reducing energy consumption and hence operational costs. Table 6.3 summarizes the energy consumption and average efficiency of biogas upgrading technologies. According to these data, the average efficiency is higher in chemical and water scrubbing, whereas membrane and cryogenic separation employ more energy per mass of CO_2 removed.

Table 6.2 Technical features of biogas upgrading technologies.

Technology	CH_4 (vol%)	CH_4 loss (%)	Major impurities (%)
Water scrubbing	96–98	2–6	CO_2 (0.5–3), N_2 (0.5–1), O_2 (0–0.7)
Cryogenic separation	98–99	0.65–1	CO_2 (<3)
Organic physical scrubbing	93–97	4–10	CO_2 (3–5)
Chemical absorption	98–99	0.1–0.5	CO_2 (0–1)
Pressure swing adsorption	95–98	2–5	CO_2 (1–4)
Membrane technology	90–98	2–15	CO_2 (2–9)

Source: Adapted from Sun et al. [16].

Table 6.3 Energy consumption and efficiency range of biogas upgrading technologies.

Technology	Energy consumption (electricity) (MJ/ton CO_2)	Efficiency range (%)
Water scrubbing	770	94.4
Cryogenic separation	1275	90.8
Organic physical scrubbing	1069	92.8
Chemical absorption	433	93.1
Pressure swing adsorption	915	89.2
Membrane technology	1264	90.2

Source: Adapted from Sun et al. [16].

6.3 CCU Main Technologies

CCU technologies have emerged as a potential solution not only to reduce emissions from industry but also to make them more and more competitive through the manufacture of by-products. In this sense, it seems clear that also new applications for CO_2 derived products, which substitute other raw materials, are desired by researchers [39]. Inasmuch these methods cannot achieve by themselves a zero-emission strategy, its potential combination with other techniques and processes is vital for the proposed scope.

Typically, CCU technologies have been classified into four different groups, which include CO_2 employment as a solvent, chemicals obtained from CO_2 as a raw material, fuels obtained from CO_2, and enhanced oil recovery (EOR) [40]. Additionally, some authors include mineral carbonation and bio-processes involving algae as alternative options [41]. Figure 6.4 depicts all these options. At this point, we briefly present the main characteristics of each division. According to the final use for CO_2, it can be distinguished among technological, biological, or chemical characteristics [2]. Technological use is commonly bounded to those industrial processes such as extraction with supercritical solvents, water treatment, and food industry. On the contrary, biological use refers, for instance, to direct fixation of CO_2 as a feed-stream for microalgae production, whereas chemical use refers to the production of chemical value-added products.

6.3.1 Supercritical CO_2 as a Solvent

The use of organic solvents is quite frequent in industrial operations such as extraction or separation. These organic solvents inevitably cause environmental problems such as toxicity and waste disposal. Moreover, organic solvents are characterized for their high cost; hence, the operational costs are incremented. For these very reasons, supercritical CO_2 is very interesting for chemical industry. Its attractiveness is not only because CO_2 is a cheap solvent but also because CO_2 is not considered a volatile organic compound, so it is subject to fewer legal restrictions [42, 43]. Additionally,

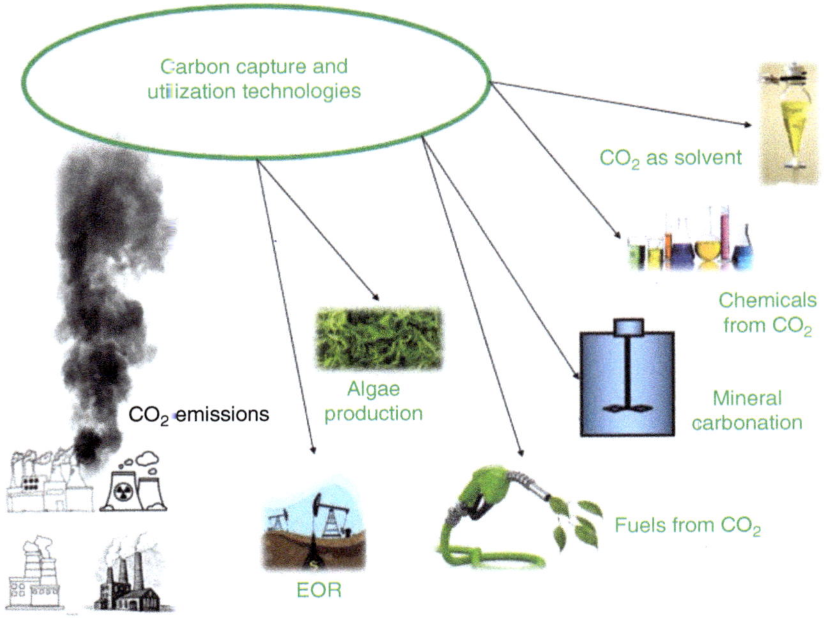

Figure 6.4 Main CCU technologies. Sources: Gino's Premium Images/Alamy Stock Photo; moonoi172/Shutterstock.com; maksym yemelyanov/123RF.

CO_2 critical conditions are not very hard ($T = 32.1\,°C$; $P = 73.8\,bar$) in comparison with other organic compounds (e.g. isopropanol $T = 235.2\,°C$; $P = 47.6\,bar$).

6.3.2 Chemicals from CO_2

During the past years, emerging technologies for CO_2 conversion into chemicals have been developed for a multitude of products. Classically, methanol, dimethyl ether, formic acid, ethylene, ethanol, dimethyl carbonate, diphenyl carbonate, carboxylic acid, and even polymers production have been the most studied chemicals to obtain from CO_2 [44]. Nevertheless, these production processes need to be optimized in terms of energy efficiency because they have proved to be highly energy intensive. Furthermore, these chemicals have limited storage periods for captured CO_2 because of their short life span. Therefore, CO_2 is again released to the atmosphere. Some researchers have focused on mineral carbonates, which can be used in construction industry with larger life span, a solution that has encouraged many works to focus on this option [2, 39].

6.3.3 Mineral Carbonation

Mineral carbonation consists of the permanent fixation of CO_2 into alkaline oxides, typically MgO and CaO. This option occurs naturally in the earth in rocks such as serpentine [45]. Even though this option is very appealing from a storage point of view, economical analyses shows that CO_2 sequestration using a large-scale infrastructure would have a cost of 120 \$/tCO_2 [46]. Therefore, the key to make this process affordable is implementing mineral carbonation for small–medium CO_2 producers. A potential process to optimize mineral carbonation is biogas upgrading, as will be seen in Section 6.4 of this chapter.

6.3.4 Fuels from CO_2

Fuels pose a fundamental dilemma to the society both in the present and in the future. Their shortage in the next 100 years drives the research community to develop new fuels that can substitute the traditional fossil fuels. The transformation of CO_2 into CH_4 or other hydrocarbons that can act as green fuels is very attractive to solve this fossil fuel problem [47]. Nevertheless, the technical limit of this route falls on the Gibbs free energy, which is not favorable when converting CO_2 to other molecules [48]. Some authors affirm that gas hydrates are another option for fuel production from CO_2 [49], whereas other researchers try to use CO_2 for algae production in order to obtain biofuels [50].

6.3.5 Algae Production

Algae poses a high potential both as a biomass source and a CO_2-negative emission technology. The cultivation of biomass from algae can provide biofuel feedstock (in the form of biomass) and has a favorable environmental impact because of CO_2 consumption on the process. Furthermore, this biomass can be transformed into biogas by means of anaerobic digestion as explained at the beginning of this chapter [51]. Additionally, this option does not compete with food feedstocks or resources suitable for conventional agriculture. Algae production from CO_2 can also be coupled with other techniques, for example, for wastewater treatment. Some challenges need to be overcome such as the full technology for commercialization of microalgae to biofuel and dewatering the final biomass before its employment [50].

6.3.6 Enhanced Oil Recovery (EOR)

This route is commonly known as one of the direct utilizations of CO_2, along with CO_2 for food or drink industries. EOR intends to extract crude oil from unrecoverable reserves through the injection of high-pressure gases. In this sense, CO_2 is a cheap and suitable gas for compression and injection into reservoirs for oil extraction [52]. EOR can extract about 30–60% more crude than initially extracted [39]. The main disadvantage of this route for CO_2 utilization is that most of CO_2 returns to the surface with the extracted oil and inevitably to the atmosphere [53].

6.4 Potential Processes for Biogas Upgrading and Carbon Utilization

In agreement with what was presented above, in this section, some potential processes for biogas upgrading to bio-methane, which could admit CCU as a value-added stage will be explained. In fact, some of these processes have been

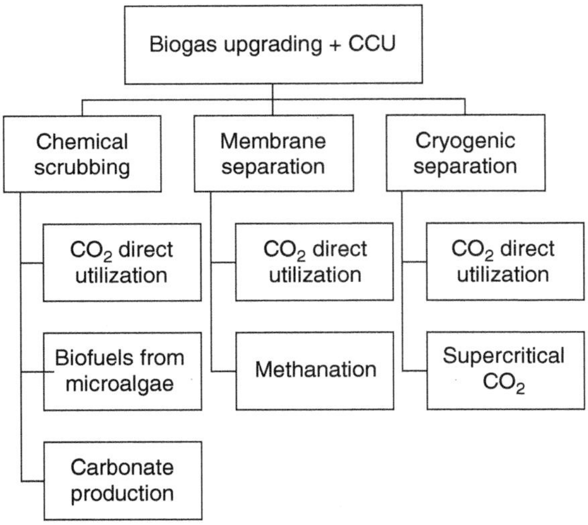

Figure 6.5 Options for coupling biogas upgrading and CCU.

reported in the literature with potential advantages and promising results, whereas others represent new ideas that could be further studied in future research. From the data available in the literature, the processes chosen for a detailed description are direct utilization of CO_2, carbonate production from biogas CO_2, and biomass obtained from microalgae and biogas upgrading.

Several biogas upgrading techniques are suitable for the processes that were selected, but not all of them are the most appropriate both because of final CO_2 content and CH_4 composition in the final stream. Therefore, taking into account these limitations and the data from the literature, we have explained ideas for potential solutions by coupling biogas upgrading and CCU technologies. These potential solutions for biogas upgrading and CCU application are represented in the scheme shown in Figure 6.5. Nevertheless, apart from the options shown in Figure 6.5, more connections between biogas upgrading and CCU are possible.

6.4.1 Chemical Scrubbing Coupled with CCU

Chemical scrubbing can be performed with traditional solvents (e.g. MEA or piperazine) and with caustic solvents (NaOH and KOH) [54]. The first ones have proved to achieve high capture efficiencies (>90%) and solvent regeneration is possible by means of physical methods such as temperature increase. Indeed, some recent works are directed to reduce the high energy consumption of the regeneration stage [55–58]. Results show that high-purity CO_2 (99%) can be obtained and easily used in direct utilization in the food or drink industry, or in microalgae cultivation for biofuels [59]. Figure 6.6 shows a possible configuration for this proposal.

Besides its advantages, this process still has yet elevated energy consumption (3–8 MJ kg^{-1} CO_2) [57, 60]. To face this high-energy penalty, many researchers have investigated alternative routes based on caustic solvents that can be regenerated or transformed into other products through chemical reaction [28, 61, 62]. This

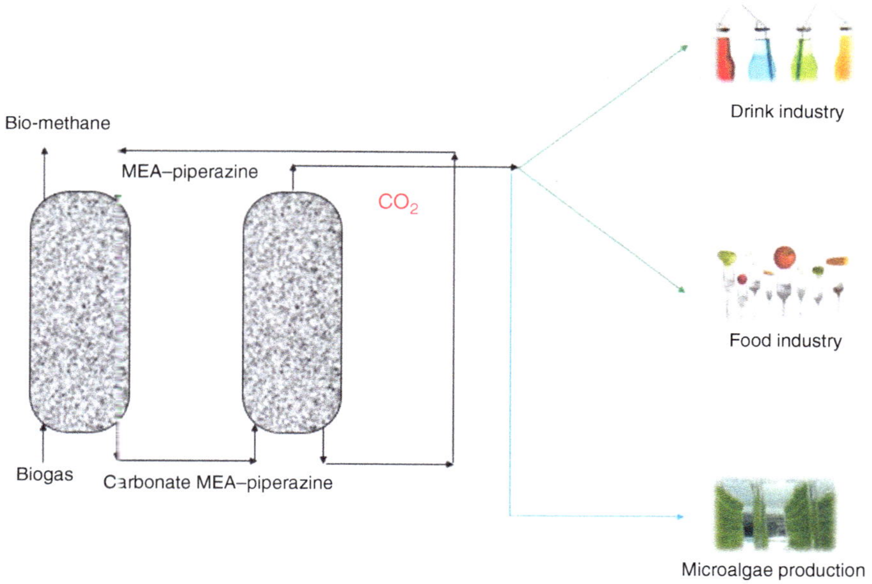

Figure 6.6 CO_2 capture from biogas and utilization in drink–food industry or microalgae production. Sources: pinkomelet/Getty Images; gerenme/Getty Images.

process is shown in Figure 6.7. In this alternative, CO_2 from biogas is first chemically absorbed by NaOH or KOH. Afterward, the regeneration step is carried out by chemical reaction between the carbonated solvent and Ca^+–Mg^+-rich agents to precipitate calcium–magnesium carbonate. Regenerated NaOH or KOH can be obtained as subproducts of this reaction, along with some other products that can be employed in industrial applications can be a good result (e.g. NaCl).

The precipitating agent employed by several researchers varies from calcium–magnesium hydroxides–chlorides to calcium–magnesium-rich wastes, and valuable products to be sold in the market can be obtained. Table 6.4 reflects some results obtained from these studies. As shown in Table 6.4, precipitation efficiencies oscillate approximately from 15 to 80. In case of wastes, the efficiency is considerably lower because of the influence of other undesired compounds that can interfere the precipitation process.

6.4.2 Membrane Separation Coupled with CCU

Membranes have been investigated for CO_2 separation from biogas by several authors, but only a few proposed to employ the separated CO_2 in CCU applications [17, 68, 69]. The reason is that not many applications would be acceptable because of the conditions of the high CO_2 stream in case of only one membrane module. Indeed, Esposito et al. [68] had to build a three-stage membrane module to obtain a food-grade CO_2 for direct application in food industry. The CO_2 purity obtained was 99%. The idea developed in this study is shown in Figure 6.8.

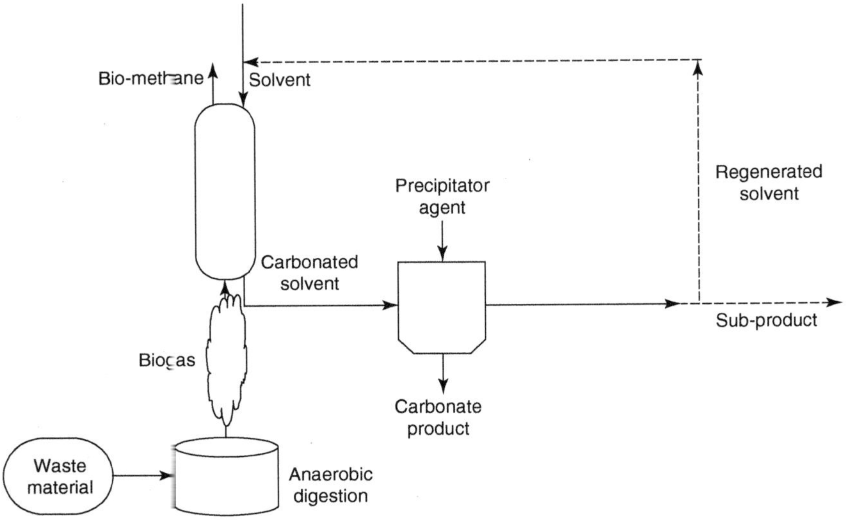

Figure 6.7 CO_2 capture from biogas and precipitation with $Ca^{2+}-Mg^{2+}$-rich compounds.

Table 6.4 Precipitation with $Ca^{+}-Mg^{+}$-rich compound studies.

Material	Efficiency (%)	References
$Ca(OH)_2$	82	Baena-Moreno et al. [2]
$MgCl_2$	76	Chen et al. [63]
$CaCl_2$	78	Baena-Moreno et al. [28]
Stainless steel slag	13	Baciocchi et al. [64]
Blended hydraulic cement slag	28	Chang et al. [65]
Air pollution control residues	25	Baciocchi et al. [66]
Oil shale ash	17	Uibu and Kuusik [67]

In the study of Pfister et al. [69], a fixed site carried membrane is used to obtain in a single-membrane module 95% CO_2. In this case and in agreement with the required grade for food industry, it is not possible to use CO_2 for this end. Nevertheless, methanation reaction could be employed as a CCU pathway for this stream to obtain additional CH_4. Figure 6.9 presents this idea proposed by some authors [70, 71]. The results clearly reflect the feasibility of this process because the quality for gas grid injection was reached for the produced synthetic natural gas (SNG) (>96% v/v CH_4).

6.4.3 Cryogenic Separation Coupled with CCU

The utilization of the potential CO_2 recovered from biogas is completely necessary if the upgrading technology chosen is cryogenic separation. The high amount of

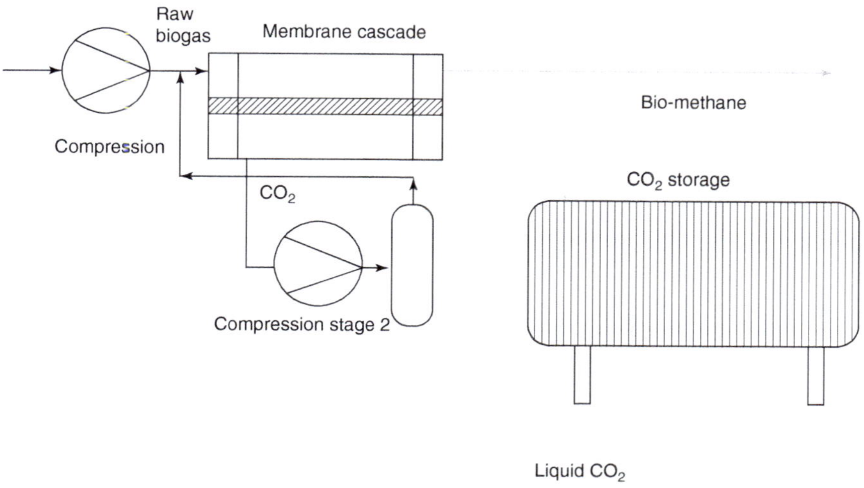

Figure 6.8 Membrane plus CO_2 storage for food application.

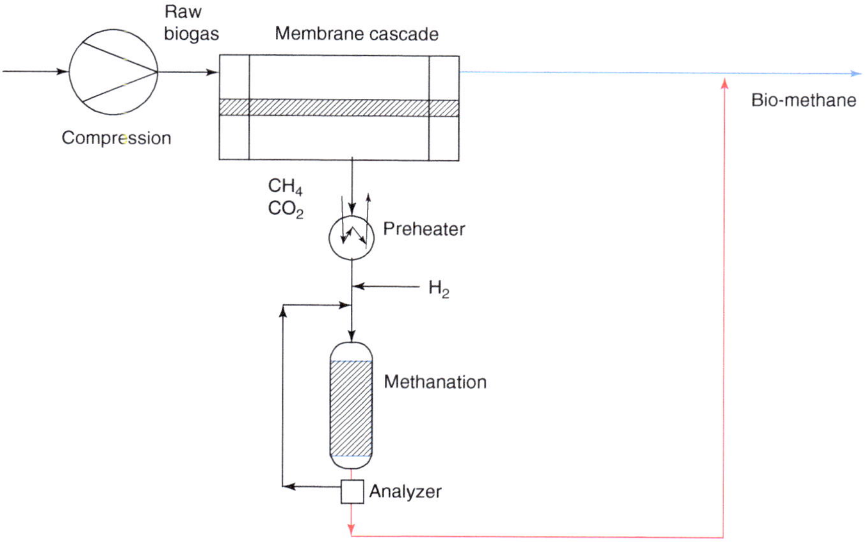

Figure 6.9 Biogas upgrading through membrane separation and methanation.

energy consumed by cryogenic techniques makes CO_2 utilization needed to balance the overall economic of the process. In this case, CO_2 can be directly utilized as a cold medium because of the low temperature of the CO_2 source or it can also be used in the food and drink industries where a high-pressure CO_2 is needed. Moreover, this CO_2 can also be employed as dry ice, a novel application that has been recently described [72]. This process is described in Figure 6.10. Two options are available for selling the CO_2 produced from biogas at cryogenic temperatures: direct sale to the end user of CO_2 or indirect sale via traders dealing with dry ice. The amount of

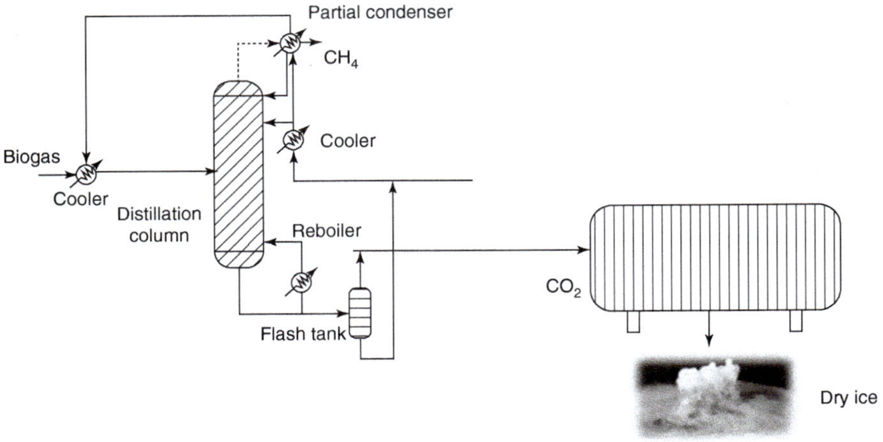

Figure 6.10 Biogas upgrading via cryogenic methods and dry ice production.

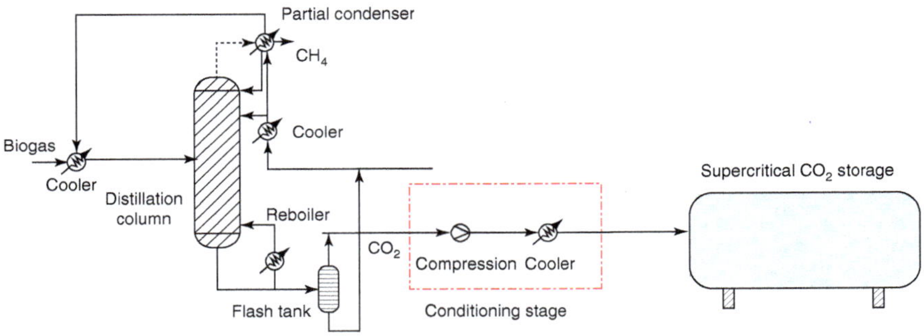

Figure 6.11 Biogas upgrading via cryogenic methods and supercritical CO_2 production.

dry ice consumed by food, chemical, or pharmaceutical industry fits in the range of cryogenic CO_2 from biogas and its price is about 0.25 € kg^{-1} [72]. Nevertheless, if higher quantities of cryogenic CO_2 are available as dry ice, the price may decrease.

Additionally, cryogenic biogas upgrading provides a high-quality and high-pressure CO_2 that could be used as a solvent because of its near conditions to supercritical CO_2 ($T = 32.1\,°C$; $P = 73.8\,bar$). Depending on the cryogenic process chosen, CO_2 can be obtained in a range of pressures, from 10 bars [73] up to 80 bars [14]. Thus, in upgrading processes with the highest pressures, CO_2 could be stored at supercritical conditions for solvent utilization. Figure 6.11 reflects graphically this idea.

6.5 Conclusions

This study shows that technologies for biogas upgrading to obtain bio-methane and for CCU are still under development and have not reached sufficient level of maturity to be coupled. Biogas is available not only in developed societies but also in developing areas because it comes from the anaerobic digestion of organic matter. Thus, its utilization could also reduce the energy dependency of developing areas. Several biogas upgrading technologies are available to obtain a high-purity bio-methane as a clean energy source. The upgrading technologies briefly explained in this chapter are water scrubbing, pressure swing adsorption, chemical scrubbing, organic physical scrubbing, membrane separation, and cryogenic separation. Regarding CCU technologies, supercritical CO_2 as a solvent, chemicals obtained from CO_2, mineral carbonation – carbonates production, fuels from CO_2, algae production, and EOR have been typically employed as end used for CO_2.

Water scrubbing and organic physical scrubbing employment results in higher impurities in the final bio-methane stream as well as in the highest CH_4 losses. On the other hand, chemical scrubbing and cryogenic separation provide the best CH_4 purity in the product stream. According to the analyzed data, the average efficiency is higher for chemical and water scrubbing, whereas membrane and cryogenic separation employ the most energy per ton of CO_2 removed. This inevitability invites people to think that carbon utilization technologies should be implemented in those biogas upgrading technologies that provide highest CH_4 purity or are less energy intensive.

Considering the reasons previously presented, this chapter proposes CCU technologies as a coupling process for biogas upgrading in chemical scrubbing, membrane separation, and cryogenic separation. Further research should be focused on improving the processes and technologies presented, as well as to scale-up the processes proposed in the literature. Among the different alternatives, chemical scrubbing with carbonate production stands out because the economic balance could be potentially interesting. Membrane separation to obtain bio-methane and CO_2 for direct utilization in food industry has been tested at pilot scale with promising results. Cryogenic separation to provide bio-methane and dry ice CO_2 can be an interesting option to further pursue the research.

References

1 Scripps Institution of Oceanography (2018). The keeling curve. UC San Diego. https://scripps.ucsd.edu/programs/keelingcurve/ (accessed 08 October 2020).

2 Baena-Moreno, F.M., Rodríguez-galán, M., Vega, F. et al. (2018). Carbon capture and utilization technologies: a literature review and recent advances. *Energy Sources Part A* . Taylor & Francis: 1–31. https://doi.org/10.1080/15567036.2018.1548518.

3 Sen, S. and Ganguly, S. (2017). Opportunities, barriers and issues with renewable energy development – a discussion. *Renewable Sustainable Energy Rev.* https://doi.org/10.1016/j.rser.2016.09.137.

4 Ellabban, O., Abu-Rub, H., and Blaabjerg, F. (2014). Renewable energy resources: current status, future prospects and their enabling technology. *Renewable Sustainable Energy Rev.* https://doi.org/10.1016/j.rser.2014.07.113.

5 IRENA (2017). *Renewable Energy and Jobs: Annual Review 2017*. International Renewable Energy Agency (IRENA). https://www.irena.org/-/media/Files/IRENA/Agency/Publication/2018/Jan/IRENA_2017_Power_Costs_2018_summary.pdf?la=en&hash=6A74B8D3F7931DEF00AB88BD3B339CAE180D11C3#:~:text=1%20Some%20projects%20undercut%20fossil,Energy%20Agency%20(IRENA)%20shows.&text=was%20USD%200.05%20per%20kilowatt,for%20bioenergy%20and%20geothermal%20projects.

6 International Renewable Energy Agency (IRENA) (2018). *Renewable Power Generation Costs in 2017*. International Renewable Energy Agency https://doi.org/10.1007/SpringerReference_7300.

7 Sinha, S. and Chandel, S.S. (2014). Review of software tools for hybrid renewable energy systems. *Renewable Sustainable Energy Rev.* https://doi.org/10.1016/j.rser.2014.01.035.

8 Angelidaki, I., Treu, L., Tsapekos, P. et al. (2018). Biogas upgrading and utilization: current status and perspectives. *Biotechnol. Adv.* https://doi.org/10.1016/j.biotechadv.2018.01.011.

9 Deremince, B. and Königsberger, S. (2017). Statistical Report of the European Biogas Association 2017. 20.

10 Johansson, N. (2008). Production of liquid biogas, LBG, with cryogenic and conventional upgrading technology – description of systems and evaluations of energy balances. http://lup.lub.lu.se/student-papers/record/4468178/file/4469242.pdf (accessed 08 October 2020).

11 Saboor, A., Khan, S., Shah, A.A. et al. (2017). Enhancement of biomethane production from cattle manure with codigestion of dilute acid pretreated lignocellulosic biomass. *Int. J. Green Energy* https://doi.org/10.1080/15435075.2017.1313740.

12 Zhou, K., Chaemchuen, S., and Verpoort, F. (2017). Alternative materials in technologies for biogas upgrading via CO_2 capture. *Renewable Sustainable Energy Rev.* 79: 1414–1441. https://doi.org/10.1016/J.RSER.2017.05.198.

13 Bekkering, J., Broekhuis, A.A., and van Gemert, W.J.T. (2010). Optimisation of a green gas supply chain – a review. *Bioresour. Technol.* https://doi.org/10.1016/j.biortech.2009.08.106.

14 Awe, O.W., Zhao, Y., Nzihou, A. et al. (2017). A review of biogas utilisation, purification and upgrading technologies. *Waste Biomass Valorization* https://doi.org/10.1007/s12649-016-9826-4.

15 Baena-Moreno, F.M., Rodríguez-Galán, M., Vega, F. et al. (2019). Review: recent advances in biogas purifying technologies. *Int. J. Green Energy* . Taylor & Francis: 1–12. https://doi.org/10.1080/15435075.2019.1572610.

16 Sun, Q., Li, H., Yan, J. et al. (2015). Selection of appropriate biogas upgrading technology – a review of biogas cleaning, upgrading and utilisation. *Renewable Sustainable Energy Rev.* 51: 521–532. https://doi.org/10.1016/j.rser.2015.06.029.

17 Esposito, E., Jansen, J.C., Dellamuzia, L. et al. (2017). CO_2 recovery from biogas and successful upgrading to food-grade quality: a case study. *WASET*.

18 Baciocchi, R., Carnevale, E., Costa, G. et al. (2013). Performance of a biogas upgrading process based on alkali absorption with regeneration using air

pollution control residues. *Waste Manage. (Oxford)* https://doi.org/10.1016/j.wasman.2013.08.022.

19 Kerroum, D., Mossaab, B.L., and Hassen, M.A. (2014). Production of bio-energy from organic waste: effect of temperature and substrate composition. *Int. J. Energy Res.* https://doi.org/10.1002/er.3044.

20 Lombardi, L., Corti, A., Carnevale, E. et al. (2011). Carbon dioxide removal and capture for landfill gas up-grading. *Energy Procedia* 4: 465–472. https://doi.org/10.1016/j.egypro.2011.01.076.

21 Rasi, S., Läntelä, J., and Rintala, J. (2011). Trace compounds affecting biogas energy utilisation – a review. *Energy Convers. Manage.* https://doi.org/10.1016/j.enconman.2011.07.005.

22 Weiland, P. (2010). Biogas production: current state and perspectives. *Appl. Microbiol. Biotechnol.* https://doi.org/10.1007/s00253-009-2246-7.

23 Barbera, E., Menegon, S., Banzato, D. et al. (2019). From biogas to biomethane: a process simulation-based techno-economic comparison of different upgrading technologies in the Italian context. *Renewable Energy* https://doi.org/10.1016/j.renene.2018.12.052.

24 Paolini, V., Petracchini, F., Segreto, M. et al. (2018). Environmental impact of biogas: a short review of current knowledge. *J. Environ. Sci. Health. Part A Toxic/Hazard. Subst. Environ. Eng.* https://doi.org/10.1080/10934529.2018.1459076.

25 Sheets, J.P. and Shah, A. (2018). Techno-economic comparison of biogas cleaning for grid injection, compressed natural gas, and biogas-to-methanol conversion technologies. *Biofuels, Bioprod. Biorefin.* https://doi.org/10.1002/bbb.1848.

26 Pipatmanomai, S., Kaewluan, S., and Vitidsant, T. (2009). Economic assessment of biogas-to-electricity generation system with H_2S removal by activated carbon in small pig farm. *Appl. Energy* https://doi.org/10.1016/j.apenergy.2008.07.007.

27 Khan, I.U., Imran, M.H.D.O., Hashim, H. et al. (2017). Biogas as a renewable energy fuel – a review of biogas upgrading, utilisation and storage. *Energy Convers. Manage.* https://doi.org/10.1016/j.enconman.2017.08.035.

28 Baena-Moreno, F.M., Rodríguez-Galán, M., Fernando, V. et al. (2019). Synergizing carbon capture storage and utilization in a biogas upgrading lab-scale plant based on calcium chloride: influence of precipitation parameters. *Sci. Total Environ.* https://doi.org/10.1016/j.scitotenv.2019.03.204.

29 Hoo, P.Y., Hashim, H., and Ho, W.S. (2018). Opportunities and challenges: landfill gas to biomethane injection into natural gas distribution grid through pipeline. *J. Cleaner Prod.* 175: 409–419. https://doi.org/10.1016/j.jclepro.2017.11.193.

30 Wheeler, P., Holm-Nielsen, J.B., Jaatinen, T. et al. (1999). *Biogas Upgrading and Utilisation*, 3–20. IEA Bioenergy http://www.iea-biogas.net/files/daten-redaktion/download/publi-task37/Biogas%20upgrading.pdf.

31 Grande, C.A. and Rodrigues, A.E. (2007). Biogas to fuel by vacuum pressure swing adsorption I. Behavior of equilibrium and kinetic-based adsorbents. *Ind. Eng. Chem. Res.* 46 (13): 4595–4605. https://doi.org/10.1021/ie061341+.

32 Baciocchi, R., Carnevale, E., Corti, A. et al. (2013). Innovative process for biogas upgrading with CO_2 storage: results from pilot plant operation. *Biomass Bioenergy* https://doi.org/10.1016/j.biombioe.2012.11.016.

33 Baena-Moreno, F.M., Rodríguez-Galán, M., Vega, F. et al. (2018). Regeneration of sodium hydroxide from a biogas upgrading unit through the synthesis of precipitated calcium carbonate: an experimental influence study of reaction parameters. *Processes* 6 (205) https://doi.org/10.3390/pr6110205.

34 Baena-Moreno, F.M., Rodríguez-Galán, M., Vega, F. et al. (2019). Understanding the influence of the alkaline cation K^+ or Na^+ in the regeneration efficiency of a biogas upgrading unit. *Int. J. Energy Res.* (December 2018): 1–8. https://doi.org/10.1002/er.4448.

35 Niesner, J., Jecha, D., and Stehlík, P. (2013). Biogas upgrading technologies: state of art review in European region. *Chem. Eng. Trans.* https://doi.org/10.3303/CET1335086.

36 Ryckebosch, E., Drouillon, M., and Vervaeren, H. (2011). Techniques for transformation of biogas to biomethane. *Biomass Bioenergy* https://doi.org/10.1016/j.biombioe.2011.02.033.

37 Baena-Moreno, F.M., Rodríguez-Galán, M., Vega, F. et al. (2019). Biogas upgrading by cryogenic techniques. *Environ. Chem. Lett.*, no. 0123456789. Springer International Publishing. doi:https://doi.org/10.1007/s10311-019-00872-2.

38 Tuinier, M.J. and Van Sint Annaland, M. (2012). Biogas purification using cryogenic packed-bed technology. *Ind. Eng. Chem. Res.* 51 (15): 5552–5558. https://doi.org/10.1021/ie202606g.

39 Cuéllar-Franca, R.M. and Azapagic, A. (2015). Carbon capture, storage and utilisation technologies: a critical analysis and comparison of their life cycle environmental impacts. *J. CO_2 Util.* https://doi.org/10.1016/j.jcou.2014.12.001.

40 Aresta, M. (2010). *Carbon Dioxide as Chemical Feedstock*. https://doi.org/10.1002/9783527629916.

41 Styring, P., Jansen, D., de Coninck, H. et al. (2011). *Carbon Capture and Utilisation in the Green Economy*. Centre for Low Carbon Futures.

42 Clarke, C.J., Wei Chien, T., Levers, O. et al. (2018). Green and sustainable solvents in chemical processes. *Chem. Rev.* https://doi.org/10.1021/acs.chemrev.7b00571.

43 Nalawade, S.P., Picchioni, F., and Janssen, L.P.B.M. (2006). Supercritical carbon dioxide as a green solvent for processing polymer melts: processing aspects and applications. *Prog. Polym. Sci. (Oxford)* https://doi.org/10.1016/j.progpolymsci.2005.08.002.

44 Centi, G. and Perathoner, S. (2014). *Green Carbon Dioxide: Advances in CO_2 Utilization*. https://doi.org/10.1002/9781118831922.

45 Sanna, A., Uibu, M., Caramanna, G. et al. (2014). A review of mineral carbonation technologies to sequester CO_2. *Chem. Soc. Rev.* https://doi.org/10.1039/c4cs00035h.

46 Rackley, S.A. and Rackley, S.A. (2017). Chapter 10: Mineral carbonation. In: *Carbon Capture and Storage*. https://doi.org/10.1016/B978-0-12-812041-5.00010-6.

47 le Saché, E., Johnson, S., Pastor-Pérez, L. et al. (2019). Biogas upgrading via dry reforming over a Ni-Sn/CeO_2-Al_2O_3 catalyst: influence of the biogas source. *Energies* 12 (6): 1007. https://doi.org/10.3390/en12061007.

48 Jiang, Z., Xiao, T., Kuznetsov, V.L., and Edwards, P.P. (2010). Turning carbon dioxide into fuel. *Philos. Trans. R. Soc. London, Ser. A* 368 (1923): 3343–3364. https://doi.org/10.1098/rsta.2010.0119.

49 Dashti, H., Yew, L.Z., and Lou, X. (2015). Recent advances in gas hydrate-based CO_2 capture. *J. Nat. Gas Sci. Eng.* https://doi.org/10.1016/j.jngse.2015.01.033.

50 Shuba, E.S. and Kifle, D. (2018). Microalgae to biofuels: 'promising' alternative and renewable energy, review. *Renewable Sustainable Energy Rev.* https://doi.org/10.1016/j.rser.2017.08.042.

51 Moreira, D. and Pires, J.C.M. (2016). Atmospheric CO_2 capture by algae: negative carbon dioxide emission path. *Bioresour. Technol.* https://doi.org/10.1016/j.biortech.2016.03.060.

52 Andrei M., De Simoni, M., Delbianco, A., et al. 2010. Enhanced oil recovery with CO_2 capture and sequestration. 21st World Energy Congress (WEC) 2010, Montreal, Quebec (Canada), 12–16 September 2010.

53 Denney, D. (2015). Industry experience with CO_2-enhanced-oil-recovery technology. *J. Pet. Technol.* https://doi.org/10.2118/0110-0039-jpt.

54 Vega, F., Cano, M., Gallego, M. et al. (2017). Evaluation of MEA 5 M performance at different CO_2 concentrations of flue gas tested at a CO_2 capture lab-scale plant. *Energy Procedia* 114 (114): 6222–6228. https://doi.org/10.1016/j.egypro.2017.03.1760.

55 Liang, Z., Idem, R., Tontiwachwuthikul, P. et al. (2016). Experimental study on the solvent regeneration of a CO_2-loaded MEA solution using single and hybrid solid acid catalysts. *AIChE J.* https://doi.org/10.1002/aic.15073.

56 Liu, Y., Fan, W., Wang, K., and Wang, J. (2016). Studies of CO_2 absorption/regeneration performances of novel aqueous monothanlamine (MEA)-based solutions. *J. Cleaner Prod.* https://doi.org/10.1016/j.jclepro.2015.08.116.

57 Oh, S.Y., Binns, M., Cho, H., and Kim, J.K. (2016). Energy minimization of MEA-based CO_2 capture process. *Appl. Energy* https://doi.org/10.1016/j.apenergy.2016.02.046.

58 Zhang, X., Zhang, X., Liu, H. et al. (2017). Reduction of energy requirement of CO_2 desorption from a rich CO_2-loaded MEA solution by using solid acid catalysts. *Appl. Energy* https://doi.org/10.1016/j.apenergy.2017.05.135.

59 Dowell, N.M., Fennell, P.S., Shah, N., and Maitland, G.C. (2017). The role of CO_2 capture and utilization in mitigating climate change. *Nat. Clim. Change* https://doi.org/10.1038/nclimate3231.

60 Øi, L.E. and Kvam, S.H.P. (2014). Comparison of energy consumption for different CO_2 absorption configurations using different simulation tools. *Energy Procedia* https://doi.org/10.1016/j.egypro.2014.11.128.

61 Lombardi, L., Baciocchi, R., Carnevale, E., and Corti, A. (2012). Investigation of an innovative process for biogas up-grading –pilot plant preliminary results. In: *Proceedings of ECOS 2012 – the 25th International Conference on Efficiency,*

Cost, Optimization, Simulation and Environmental Impact of Energy Systems June 26–29, Perugia, Italy, 1–12.

62 Morone, M., Costa, G., Polettini, A. et al. (2014). Valorization of steel slag by a combined carbonation and granulation treatment. *Miner. Eng.* https://doi.org/10.1016/j.mineng.2013.08.009.

63 Chen, G., Song, X., Dong, C. et al. (2016). Mineralizing CO_2 as $MgCO_3 \cdot 3H_2O$ using abandoned $MgCl_2$ based on a coupled reaction-extraction-alcohol precipitation process. *Energy Fuels* https://doi.org/10.1021/acs.energyfuels.6b01297.

64 Baciocchi, R., Costa, G., Polettini, A., and Pomi, R. (2009). Influence of particle size on the carbonation of stainless steel slag for CO_2 storage. *Energy Procedia* https://doi.org/10.1016/j.egypro.2009.02.314.

65 Chang, E.E., Pan, S.Y., Chen, Y.H. et al. (2011). CO_2 sequestration by carbonation of steelmaking slags in an autoclave reactor. *J. Hazard. Mater.* https://doi.org/10.1016/j.jhazmat.2011.08.006.

66 Baciocchi, R., Costa, G., Di Bartolomeo, E. et al. (2009). The effects of accelerated carbonation on CO_2 uptake and metal release from incineration APC residues. *Waste Manage. (Oxford)* https://doi.org/10.1016/j.wasman.2009.07.012.

67 Uibu, M. and Kuusik, R. (2009). Mineral trapping of CO_2 via oil shale ash aqueous carbonation: controlling mechanism of process rate and development of continuous-flow reactor system. *Oil Shale* https://doi.org/10.3176/oil.2009.1.06.

68 Esposito, E., Dellamuzia, L., Moretti, U. et al. (2019). Simultaneous production of biomethane and food grade CO_2 from biogas: an industrial case study. *Energy Environ. Sci.* https://doi.org/10.1039/c8ee02897d.

69 Pfister, M., Belaissaoui, B., and Favre, E. (2017). Membrane gas separation processes from wet postcombustion flue gases for carbon capture and use: a critical reassessment. *Ind. Eng. Chem. Res.* https://doi.org/10.1021/acs.iecr.6b03969.

70 Kirchbacher, F., Miltner, M., Lehner, M. et al. (2016). Demonstration of a biogas methanation combined with membrane based gas upgrading in a promising power-to-gas concept. *Chem. Eng. Trans.* https://doi.org/10.3303/CET1652206.

71 Kirchbacher, F., Biegger, P., Miltner, M. et al. (2018). A new methanation and membrane based power-to-gas process for the direct integration of raw biogas – feasability and comparison. *Energy* https://doi.org/10.1016/j.energy.2017.05.026.

72 Horschig, T., Welfle, A., Billig, E., and Thrän, D. (2019). From Paris agreement to business cases for upgraded biogas: analysis of potential market uptake for biomethane plants in Germany using biogenic carbon capture and utilization technologies. *Biomass Bioenergy* https://doi.org/10.1016/j.biombioe.2018.11.022.

73 Song, C., Liu, Q., Deng, S. et al. (2019). Cryogenic-based CO_2 capture technologies: state-of-the-art developments and current challenges. *Renewable Sustainable Energy Rev.* 101 (July 2018). Elsevier Ltd: 265–278. https://doi.org/10.1016/j.rser.2018.11.018.

7

Biogas Sweetening Technologies

Nikolaos D. Charisiou[1], *Savvas L. Douvartzides*[1,2], *and Maria A. Goula*[1]

[1] *University of Western Macedonia, Department of Chemical Engineering, Laboratory of Alternative Fuels and Environmental Catalysis (LAFEC), GR-50100, Kozani, Greece*
[2] *University of Western Macedonia, Department of Mechanical Engineering, GR-50100, Kozani, Greece*

7.1 Introduction

Biogas is a renewable and environmentally sustainable gaseous fuel produced by the anaerobic digestion of biomass through the degrading action of various microorganisms [1–4]. It is a combustible gas mixture with a typical composition of 45–70% methane (CH_4), 30–55% carbon dioxide (CO_2), 0–2000 ppmv hydrogen sulfide (H_2S), and 0–590 ppmv NH_3 [5–8]. It is usually saturated with water (H_2O), and, depending on the raw biomass matter and the production conditions, it may also contain hydrogen (H_2), carbon monoxide (CO), and various other impurities such as sulfur-containing compounds (e.g. mercaptans), siloxanes, volatile organic compounds (VOCs), nitrogen (N_2), and oxygen (O_2) [5–8].

Biogas sweetening is a term that describes all the available technologies for the purification or the upgrading of biogas before its final use. Purification aims at the cleaning of raw biogas from its undesired impurities without removing CO_2 from the gaseous mixture and is selected for final applications such as the direct combustion of biogas for the production of heat (or the cogeneration of heat and power), in the utilization of biogas for the production of useful liquid fuels and chemicals, where CO_2 is a useful reactant of the synthesis reaction, and in the utilization of biogas for the production of electricity in high-temperature fuel cells such as the molten carbonate fuel cells (MCFCs) or the solid oxide fuel cells (SOFCs) [9–12].

On the other hand, upgrading involves all the technologies that not only clean the biogas but also remove CO_2 from the gas mixture, leading to a gaseous product containing more than 95% CH_4, which is known as biomethane or renewable natural gas (RNG) [13–19]. Biomethane is a product resembling the chemical synthesis of fossil natural gas, and therefore, it has a higher energy density than biogas and the appropriate Wobbe index to be used as an alternative fuel in devices designed for natural gas [20–22]. As a result, biogas upgrading is necessary for final applications in natural gas fuelled burners, turbines, automotive internal combustion engines, and

Engineering Solutions for CO_2 Conversion, First Edition.
Edited by Tomas R. Reina, José A. Odriozola, and Harvey Arellano-Garcia.

fuel cells or when biogas is to be injected in the grid of natural gas for transportation inside the same supplying pipelines [23–25].

7.2 Biogas Purification Technologies

Biogas purification technologies intend to clean the gaseous mixture mainly from $H_2O_{(g)}$, H_2S and sulfur compounds, siloxanes, VOCs, NH_4, N_2, and O_2. Technologies of adsorption, absorption, biofiltration, and/or refrigeration are mainly used, and their ability to clean each of the biogas contaminant is summarized in Table 7.1. The different purification methods have individual characteristics, which indicates that the appropriate technology should be selected by considering the purification efficiency, operational conditions, investment, and maintenance cost [27].

7.2.1 Removal of Water Vapor ($H_2O_{(g)}$)

Water is the most common compound that needs to be cleaned from biogas either during the anaerobic digestion of biomass or through appropriate after treatment [9, 26]. Water vapor condensates under low-temperature and/or high-pressure conditions, forming hydrates that may plug the biogas pipelines and other equipment. At the same time, it tends to react with the acid gases present in the biogas mixture (CO_2 and H_2S) forming carbonic acid (H_2CO_3) and sulfuric acid (H_2SO_4), which are serious contaminants and corrode the metals of the biogas plant [16]. Finally, the presence of water in biogas decreases the heating and economic value of the fuel. For all these reasons, the removal of water vapors from biogas is a common task in the biogas production industry. In almost all biogas applications, it must be cooled, drained, and dried immediately after production.

Water is always present during the anaerobic digestion of biomass, and therefore, some of it evaporates leaving a biogas mixture that is saturated in water vapors. The quantity of the water vapors in the untreated biogas depends on the thermodynamic conditions of temperature and pressure in the digestor, and hence, a possibility to reduce it is by selecting digestion conditions of lower temperature and/or higher pressure, which make the saturation amount lower. Other methods that reduce the

Table 7.1 Contaminant treatability for the major purification technologies [26].

Purification technology	$H_2O_{(g)}$	H_2S	Siloxanes	VOCs	NH_4	N_2	O_2
Adsorption	+3	+3	+3	+3	+2	0	+1
Absorption (water scrubbing)	−1	+3	+3	+3	+3	−1	−1
Biological filters	−1	+3	+1	+3	+1	−1	−1
Refrigeration	+3	+1	+2	+1	+3	0	0

Note: +3: high removal (intended); +2: removal (pre-removal by other purification method is preferred); +1: partial removal; 0: no removal; −1: addition of the contaminant.
Source: Based on Duran et al. [26].

Table 7.2 Advantages and disadvantages of techniques for removal of water [16].

Method	Advantages	Disadvantages
Condensation method (demister, cyclone, moister trap, and water taps)	(i) Higher HC's dust and oil are removed and (ii) simple techniques that are often used as pretreatment	(i) Atmospheric pressure: dew point minimum 1 ° C, (ii) gas at higher pressure to reach lower dew point (minimal −18 ° C) but freezing can occur
Adsorption dryer (silica and alumina)	(i) High removal: dew point −10 to −20 ° C, (ii) low operational costs, and (iii) regeneration possible	(i) More expensive investment: pressure 6–10 bar and (ii) dust needs to be removed in advance
Absorption with glycol	(i) High removal: dew point −5 to 15 °C, (ii) higher HC's and dust are removed, and (iii) nontoxic	More expensive investment: high pressure and 200 °°C for regeneration
Absorption with hygroscopic salts	(i) High removal efficiency and (ii) nontoxic	(i) Higher gas volumes (>500 $m^3 h^{-1}$) to be economical and (ii) not possible to regenerate

Source: Ryckebosch et al. [16]. Elsevier.

concentration of $H_2O_{(g)}$ in the untreated biogas include the removal of the water phase from the digestor [28] and the utilization of chemical inhibitors. On the other hand, water vapors can be removed from the untreated biogas by appropriate after treatment with drying or dehydration techniques such as adsorption dehydration, absorption dehydration, and cooling dehydration. Table 7.2 presents the advantages and disadvantages of common techniques for the removal of water.

Adsorption dehydration takes place into drying agents such as silica gel, activated carbon, alumina, zeolitic molecular sieves, magnesium oxide, or any other material that can effectively adsorb water. Biogas is compressed and fed into a column filled with a packed bed of the drying agent. The packed drying material retains the water molecules and then needs regeneration. For this reason, usually two columns operate in parallel so that one will be regenerated, while the other removes water from biogas. Regeneration may take place by a stream of air or through decompression and/or heating [29].

Absorption dehydration is achieved by the dissolution of the water vapors into liquid triethylene glycol or glycol [30]. After adsorption, this is pumped into a regeneration unit and is regenerated at about 200 °C. Another water adsorption method uses hygroscopic salts that dissolve as they adsorb water from biogas. Because these cannot be regenerated, they are discarded and need replacement by new salts.

In cooling dehydration, the biogas is cooled at −18 to 2 °C, and the water vapors condense into cooling coils and are collected in a trap. This method also removes some of the NH_3 because of its high solubility in water as well as some traces of other impurities. When the cooling temperature is below −70 °C, more than 99.3% of the siloxanes of the biogas are also removed, but the accomplishment of such low temperatures is expensive. A cheap method of cooling dehydration is to bury the biogas line for a long distance into the ground using only a condensate trap.

The low underground temperature will cause some water vapor condensation, but this method does not have the capability of an artificial refrigeration cycle.

7.2.2 Removal of Hydrogen Sulfide (H_2S) and Other Sulfur-Containing Compounds

Hydrogen sulfide (H_2S) is formed as a constituent of biogas together with other sulfur-containing compounds because of the degradation of proteins and other sulfur compounds of the biomass feedstock [31]. During the biogas combustion, it is converted into sulfur oxides (SO_x) and then reacts with H_2O forming H_2SO_4, which is very corrosive. The pure H_2S is highly toxic, odorous, and also strongly corrosive to steel. In concentrations above 300 ppmv, it may also cause other problems such as the poisoning of the catalysts [32, 33]. For these reasons, H_2S is removed from biogas in almost all applications except, maybe, its simple direct combustion for the production of heat.

Typical technologies for the removal of H_2S include in situ precipitation of sulfur in the digestate, adsorption in iron oxides, activated carbon or zeolite molecular sieves, physical adsorption in water (water scrubbing) or organic solvents (organic solvent scrubbing), chemical scrubbing, and biological filters [9, 11, 34]. Membranes are also used for the removal of H_2S as it will be discussed in Section 7.3.5. These H_2S purification technologies also remove all the other sulfur-containing compounds of biogas.

7.2.2.1 In Situ Precipitation of H_2S Through Air/Oxygen Injection

This is the most common method for the removal of H_2S in biogas-producing farm plants and when biogas upgrading is not desired. Air or oxygen is injected directly on the biogas layer, which is formed on the surface of the digester, causing the biological oxidation of the sulfides, which are present inside it by *Thiobacillus* bacteria [16, 35]. This process produces hydrogen and leaves a yellow layer of elemental sulfur and sulfates on the surface of the digestate while effectively reducing the levels of H_2S in the produced biogas by more than 95%, at concentration levels of 20–100 ppm. The air quantity injected is usually 2–6% to biogas ratio and must be carefully controlled because higher amounts can form explosive mixtures and cause serious safety issues.

The injection of air directly into the digestor has the disadvantage that it disturbs the digestion process and impairs the formation of methane. Also the remained layers of sulfur and sulfates can become sources of new H_2S formation with time and cause corrosion problems. These side effects may be avoided by the injection of air downstream of the digestor into a secondary tank or a biofilter carrying the desired bacteria. Finally, the air/oxygen injection method increases the levels of O_2 and N_2 into the produced biogas. The removal of O_2 and N_2 is generally very difficult and expensive, and therefore, the method of air/oxygen injection is not used when biogas upgrading is intended.

7.2.2.2 In Situ Precipitation of H_2S Through Iron Chloride/Oxide Injection

In a method similar to the air/oxygen injection, the levels of H_2S may be decreased inside the digestor by the direct injection of iron chlorides, phosphates, or oxides [16, 36]. The compound injected is most frequently liquid iron chloride ($FeCl_2$), but alternatively, solid iron hydroxide ($Fe(OH)_2$ or $Fe(OH)_3$) or ferrous chloride ($FeCl_3$) may be used. After their injection, these compounds react chemically with H_2S and form iron sulfide salt particles, which remain insoluble on the digestate according to the following reaction scheme:

$$FeCl_2: Fe^{2+} + S^{2-} \rightarrow FeS \downarrow \tag{7.1}$$

$$Fe(OH)_3: 2Fe(OH)_3 + H_2S \rightarrow 2Fe(OH)_2 + S + 2H_2O$$

$$Fe(OH)_2 + H_2S \rightarrow FeS \downarrow + 2H_2O \tag{7.2}$$

$$FeCl_3: 2FeCl_3 + 3H_2S \rightarrow 2FeS \downarrow + S \downarrow + 6HCl \tag{7.3}$$

This method is able to reduce the H_2S levels in the range of 100–200 ppm but must be combined with another H_2S cleaning technology when lower H_2S levels are desired as for biogas applications in automotive engines or for injection in the natural gas grid. The injection of the iron compound can also take place downstream of the digestor in a secondary tank.

7.2.2.3 Adsorption by Activated Carbon

H_2S may be physically adsorbed and catalytically converted in sieves of solid-activated carbon [37–39]. This is the method used most frequently when low H_2S concentrations are required. Activated carbon not only facilitates the physical adsorption of H_2S but also acts as a catalyst for the oxidation of H_2S into elemental sulfur and sulfate [40, 41]. This bifunctional action of activated carbon facilitates the removal of H_2S from the biogas with very high efficiency. The catalytic action of the activated carbon is optimized at temperatures of 50–70 °C and pressures of 7–8 bar and also by the addition of 4–6% air in the biogas to support the partial oxidation of H_2S according to the reaction:

$$2H_2S + O_2 \rightarrow 2S + 2H_2O \tag{7.4}$$

in which the produced elemental sulfur is adsorbed onto the surface of the activated carbon [18].

The activated carbon must have a moisture content of 20–30%, and the required temperatures are easily achieved by the compression of the biogas stream. The carbon filling usually has a lifetime of 4000–8000 hours, and if the untreated biogas has a high H_2S content (above 3000 ppm), frequent regeneration is necessary [29]. Although pure activated carbon is an attracting option for its low cost, the reaction above can be promoted by the impregnation of the activated carbon with suitable chemicals. This means that a cation is added to assist as a catalyst in the adsorption process [42], and thus, the impregnated activated carbon is more efficient in the removal of H_2S. In the role of the cation, various compounds such as sodium hydroxide (NaOH), potassium hydroxide (KOH), sodium bicarbonate ($NaHCO_3$), sodium carbonate (Na_2CO_3), potassium iodide (KI), or potassium permanganate ($KMnO_4$) may be added [34, 43–46]. The impregnation of the activation carbon enhances the removal capacity of the carbon sieve from about 10–20 $kgH_2S\,m^{-3}$ to almost 120–140 $kgH_2S\,m^{-3}$ of carbon [29]. It is noted that the capacity for physical adsorption of activated carbon may be decreased by the impregnation process, as it results in the reduction of the available micropore volume [47]. Moreover, impregnation of the carbon matrix can alter its catalytic properties, and by varying the impregnation method, it is possible to synthesize materials with very different characteristics [47]. Physical adsorption of H_2S dominates under the more usual conditions, such as dry and anaerobic environment and low temperatures. On the contrary, at higher temperatures and with functionalized surfaces, the mechanism of chemical adsorption occurs [48]. However, a significant drawback in the use of activate carbons is that the spent carbons need expensive re-impregnation or landfilling.

7.2.2.4 Zeolite-Based Sieve (Molecular Sieve)

Zeolites are excellent molecular sieves that can selectively remove a target molecule from a fluid mixture and have also found applications in the removal of H_2S from biogas [49–51]. Their separation ability is due to their porous structure and the different adsorption capacity of the various gases [52]. Zeolites can occur naturally (examples include Clinoptilolite, Erionite, Phillipsite, and Ferrierite) or be manufactured [49]. As an example, faujasite (FAU) zeolites have been widely used for acid gas adsorption because of their high specific surface area, pore volume, surface basic counterions present in their structure, and stability at high temperatures [53]. NaY zeolite structures are hydrophilic mainly because of their high surface area and presence of terminal silanol groups (Si–OH). Therefore, degasification can be achieved using an activation procedure before equilibrium adsorption measurements. For example, with a Na-zeolite, the following process is observed [53]:

$$\text{Na} - \text{zeolite} + H_2S \rightarrow \text{H} - \text{zeolite} + \text{NaHS} \tag{7.5}$$

Song et al. [54, 55] developed a nanoporous composite sorbent called "molecular basket" by loading polyethylenimine (PEI) on mesoporous molecular sieves SBA-15 and MCM-41, demonstrating high sorption capacity (87.0 mg H_2S/g-sorbent) at 22 °C. Natural zeolites such as clinoptilolite have high adsorption capacity for H_2S, but this can be enhanced even more by modifications with metals or metals oxides (Zn, Cu, Co, Fe, Ce, Mo, Mn, W, Na, etc.) [49, 56]. Ozerkmekci et al. [49] have recently reviewed the removal of H_2S over more than 30 different zeolite types. According to their study, the highest selectivity was found for a FAU-type zeolite known as 13X and a good choice may also be the zeolite ETS-2 because of its adsorbent properties.

7.2.2.5 Water Scrubbing

Water scrubbing is the processing of biogas with pressurized water that facilitates the separation of its acid gases (CO_2 and H_2S) from the gas stream through physical absorption. Absorption indicates the dissolution of a gas or vapor in a liquid agent. This is a classical and well-established technology based on the fact that both CO_2 and H_2S exhibit higher solubility in water than CH_4. Water scrubbing can also be used to selectively remove H_2S from biogas because the solubility of H_2S is also different from the corresponding solubility of CO_2 [9]. The physical adsorption of the gases is governed by Henry's law, which states that at constant temperature, the amount of any dissolved gas is directly proportional to its partial pressure in the gas stream.

The process of water scrubbing is taking place in a column carrying a fixed bed of commercial packing materials such as pall ring, rasching ring, berl saddle, or Tellerire. These have dimensions between 10 and 80 mm, a void fraction of 63–95%, and a surface area of 64–964 $m^2\ m^{-3}$ [57]. The packed column supports an efficient gas–liquid mass transfer in a countercurrent flow scheme, where compressed biogas at 6–10 bar is fed at the bottom and pumped water is fed at the top. The purification efficiency depends on the packing surface area, on the gas–liquid contact time, and on the solubility of the impurity into the liquid used. The latter improves when the pressure increases and/or the temperature decreases. Then, the used water is regenerated in a separate desorption column and the retained impurity is released. Regeneration is accomplished by reducing the pressure or by stripping with air or steam. The produced biomethane is saturated with water and therefore a drying processing is necessary [58, 59]. Even when both H_2S and CO_2 need to be removed, it is suggested that the cleaning of H_2S will take place separately in a previous stage because the dissolved H_2S is very corrosive and bad smelling can be problematic.

7.2.2.6 Organic Solvent Scrubbing

The solubility of H_2S or CO_2 in some organic solvents such as methanol (CH_3OH), *N*-methyl pyrrolidone (NPM), and commercial mixtures Selexol® and Genosorb® is much higher than in water. Selexol and Genosorb are organic mixtures of dimethyl ethers of polyethylene glycol and are used extensively for these purposes [9, 16, 60]. These organic solvents are used in the same way as water in water scrubbing, and they are capable of simultaneously removing H_2S, CO_2, and $H_2O_{(g)}$ because all these impurities have higher solubility in polyethylene glycol than CH_4. During the regeneration of Selexol with air, the formation of elementary sulfur is very common, and hence, the removal of H_2S usually takes place separately at a former stage from the removal of CO_2. The higher purification capacities of the organic solvents indicates that their required volume will be less than water and that the scrubbing plant will be smaller, more compact, and less expensive [61].

7.2.2.7 Sodium Hydroxide Scrubbing

An aqueous solution of sodium hydroxide (NaOH) has higher scrubbing capability than water because it combines physical absorption with a chemical reaction of H_2S with NaOH [62, 63]. This translates into a smaller and more economic scrubbing apparatus, but because the solvent in this case is not easily regenerated, this method is selected only in cases where small amounts of H_2S are removed. The reactions involved in sodium hydroxide scrubbing are as follows:

$$H_2S + NaOH \rightarrow NaHS + H_2O \tag{7.6}$$

$$NaHS + NaOH \rightarrow Na_2S + H_2O \tag{7.7}$$

These reactions proceed to an extent that depends on the available quantity of NaOH relative to the amount of H_2S. The higher the NaOH quantity is, the higher is the pH of the solvent, and then, the continuation from (7.6, 7.7) is favored. When the biogas also contains CO_2, it complicates the removal of H_2S because it also absorbs easily in the NaOH solution and forms solid carbonate salts, which plug the purification stream lines and equipment. Also, the CO_2 acts competitively and consumes NaOH from the solution according to the general reactions [64, 65]:

$$CO_2 + NaOH \rightarrow NaHCO_3 \tag{7.8}$$

$$NaHCO_3 + NaOH \rightarrow Na_2CO_3 + H_2O \tag{7.9}$$

7.2.2.8 Chemical Adsorption via Iron Oxide or Hydroxide (Iron Sponge)

This method is based on the selective adsorption of H_2S on iron oxides or hydroxides to form iron sulfide [66]. This process is often referred as "iron sponge" because rust-covered steel wool may be used to form the reaction bed. Alternatively, the absorbents are formed by an organic packing material impregnated with iron oxide (Fe_2O_3), hydroxide oxide ($Fe(OH)_3$), or zinc oxide (ZnO), which adsorbs H_2S during biogas circulation and are then appropriately regenerated. This adsorption–regeneration scheme is accomplished according to the following reactions [66]:

$$Fe_2O_3 + 3H_2S \rightarrow Fe_2S_3 + 3H_2O \tag{7.10}$$

$$2Fe(OH)_3 + 3H_2S \rightarrow Fe_2S_3 + 6H_2O \tag{7.11}$$

$$2Fe_2S_3 + 3O_2 \rightarrow 2Fe_2O_3 + 6S \tag{7.12}$$

where the adsorption reactions (7.10) and (7.11) are endothermic, taking place at 25–50 °C while the regeneration (7.12) is highly exothermic.

To facilitate the continuous operation of this purification method, two reaction beds are usually installed in parallel so that one will be regenerated according to reaction (7.11) while the other adsorbs H_2S according to the possible reactions (7.10) or (7.11). The adsorption reactions require residence times of 1–15 minutes and are able to efficiently remove H_2S from biogas by more than 99%. Given that the method is highly efficient and has low operating costs, it is widely used.

7.2.2.9 Biological Filters

H_2S can be removed from biogas by biological filtration in methods that combine water scrubbing and biological conversion [9]. The latter is accomplished by sulfur oxidizing bacteria such as *Thiobacillus*, *Beggiatoa*, and *Paracoccus*, which convert H_2S into elemental sulfur or sulfate [67]. Biological filtration may be applied in three variations as shown in Figure 7.1, using a bioscrubber, a biofilter, or a biotrickling filter. The bioscrubber is an absorption column operating with countercurrent flows similarly to a water scrubber [68, 69]. The untreated biogas is fed on the bottom of the column and H_2S is retained into a liquid fed from the top. This liquid is then routed into a bioreactor where the bacteria cause the biological oxidation of H_2S. The biofilter is a packed bed of an organic material that supports the growth of a community of the oxidizing bacteria in the form of a biofilm [69, 70]. Biogas is forced to pass through the biofilter bed and H_2S is oxidized by the biofilm after natural adsorption and absorption. This is the most common variation of biological filtration

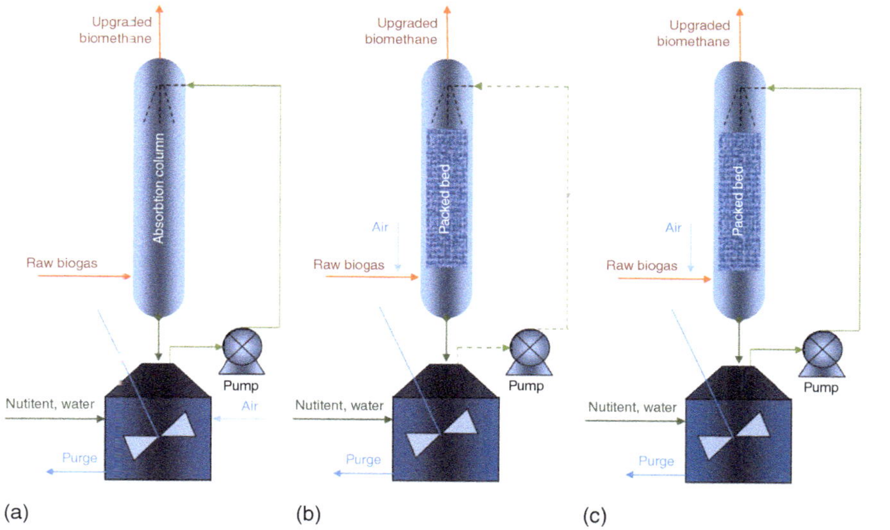

Figure 7.1 Biological filters: (a) bioscrubber, (b) biofilter, and (c) biotrickling filter.

of H_2S, but the continuous operation may increase the pH on the bed disturbing the activity of the bacteria and causing the gradual deactivation of the biofilter, especially when the biogas content in H_2S is high [71].

The biotrickling filter is similar to a biofilter because it is also a packed bed of chemically inert materials, which supports the growth of the biofilm, but in this variation, a liquid flows from the top, providing nutrients to the bacteria and controlling the pH of the bed [72, 73]. Because of this countercurrent flow scheme, a biotrickling filter is essentially a combination of a bioscrubber and a biofilter and solves the problem of the gradual filter deactivation efficiently. The biogas entering the biotrickling bed is premixed with 4–6% air so that the bacteria will have plenty of O_2 to oxidize H_2 into elemental sulfur and H_2S_4O.

The biological filters are able to remove 89–99.9% of the H_2S present in biogas tackling inlet H_2S concentrations of 50–4000 ppm [74]. When the H_2S concentration in biogas is low (at about 200 ppm), these filters can also remove other biogas impurities such as 90–99% of the VOCs and up to 92% of ammonia [67]. At high H_2S concentrations, they achieve the removal of ammonia at about 30% [75]. The optimal conditions for the growth of the bacteria is at about 35 °C with a neutral pH and a moisture content of 20–60%. Because large deviations from these conditions may significantly hinder the growth of the biofilm, these methods are easily affected by the environmental and operation conditions and need careful control. Another disadvantage of the biological filters is the need for large startup periods that range between one and three months.

7.2.3 Removal of Siloxanes

Siloxanes are a class of organometallic compounds characterized by Si–O–Si linkages [76], as shown for some representative cases in Figure 7.2.

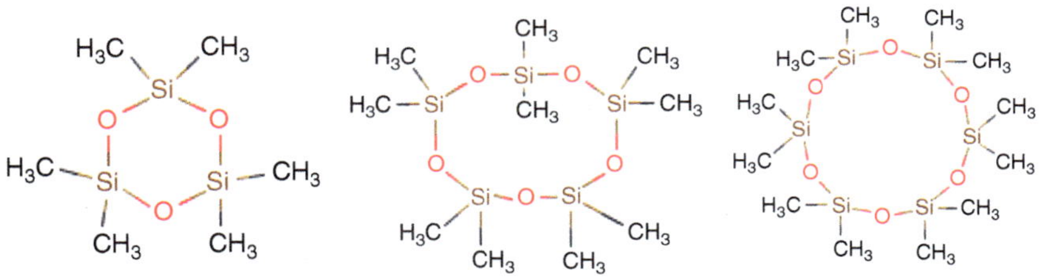

Figure 7.2 The chemical structure of three typical siloxanes.

They are polymeric compounds with numerous useful applications (pharmaceuticals, cosmetics, shampoos, detergents, etc.), but when they are part of biogas, they cause serious problems because combustion transforms them into microcrystalline silica (MCS), and this causes abrasion and fouling inside the equipment of the automotive engines, on the fuel cell anodes and on the catalysts used for the synthesis of chemicals [9–11]. For these reasons, they are considered highly undesired and need removal even from the first stages of any biogas after treatment.

Siloxanes are present in the raw biomass entering the digester and especially in the landfill wastes and composts. During the anaerobic digestion, they do not decompose, but they get volatilized and become part of the produced biogas in concentrations up to 60 mg m^{-3}. Of these, the most common are the volatile methyl siloxanes (VMS) such as L2 ($C_6H_{18}OSi_2$), L3 ($C_8H_{24}O_2Si_3$), L4 ($C_{10}H_{30}O_3Si_4$), L5 ($C_{12}H_{36}O_4Si_5$), D3 ($C_6H_{18}O_3Si_3$), D4 ($C_8H_{24}O_4Si_4$), D5 ($C_{10}H_{30}O_5Si_5$), D6 ($C_{12}H_{36}O_6Si_6$), and TMOH ($C_3H_9O_3SiOH$). The major methods for the cleaning of siloxanes include organic solvent scrubbing, adsorption on activated carbon, molecular sieves and silica gel, membrane separation, biological filtration, and cryogenic removal [77].

7.2.3.1 Organic Solvent Scrubbing

Siloxanes may be removed from biogas in packed column scrubbers through physical absorption into an appropriate liquid. The scrubbers are similar with those described previously for the removal of H_2S. Absorption is accomplished by physical contact in the countercurrent packed column, where the biogas is again fed on the bottom and the liquid absorbent is fed from the top. The most appropriate liquids for the removal of siloxanes are polar organic solvents that leave CH_4 unaffected such as Selexol, which is also used for the removal of H_2S and CO_2. Cold water, tetradecane, and carbohydrite solution are also used, but the use of Selexol was reported to efficiently remove up to 99% of the siloxanes of biogas [77, 78]. The used solvent is regenerated by evaporation or stripping by an inert gas. The absorption is more efficient at low temperatures and frequently biogas is cooled prior the scrubbing stage. Also, scrubbing is frequently followed by activated carbon adsorption to further remove any small concentrations of the siloxanes left.

7.2.3.2 Adsorption on Activated Carbon, Molecular Sieves, and Silica Gel

The removal of siloxanes may take place by physical adsorption in packed columns of appropriate solid materials such as activated carbon, molecular sieves, and silica gel. Given that the cleaning process gradually covers the sorption sites of the adsorbent, its sorption capacity decreases, and after a critical saturation instant, the concentration of the siloxanes inside the biogas start to increase. After saturation, the adsorbent may be thermally regenerated, but this leads to a loss of its sorption ability by about 5–25% in comparison to the initial value [79]. Therefore, the adsorbent usually needs frequent replacement.

Activated carbon has a high surface area of 600–1600 $m^2\,g^{-1}$, high internal porosity of 55–75% with nonuniform pore size distribution and shape, and it is nonpolar, indicating that it preferentially adsorbs nonpolar molecules such as siloxanes [80–82]. Its adsorption capacity varies between 5000 and 15 000 $mg\,kg^{-1}$ of carbon. However, activated carbons tend to interact with CH_4 of biogas accelerating saturation and causing a reduction of CH_4 in the effluent biogas. Also, their performance is hindered significantly by the presence of water in biogas, and for that reason, the removal of siloxanes usually follows a water removal technology such as cooling dehydration. The cost of these combined water and siloxane removal systems is considered acceptable for very low siloxane concentrations below 2 $mg\,m^{-3}$, but at higher siloxane concentrations, it becomes expensive.

Molecular sieves are polar hydrous aluminosilicate minerals with the chemical formula $Na_2O–Al_2O_3–nSiO_2–xH_2O$. Their surface area is about 600–700 $m^2\,g^{-1}$, and their internal porosity is about 40–55% with a mean pore diameter of 3–9 nm [83, 84]. They preferentially adsorb polar molecules such as water, but they effectively remove together water, siloxanes, and the acid gases of biogas even at low concentrations of 1 ppm. Molecular sieves have a regular crystalline porous structure, and depending on their pore volume and geometry, they are able to selectively adsorb a specific biogas impurity without affecting CH_4. Mycock et al. [57] reported that their adsorption forces are highest when the pore size of the sieve is not more than the size of the molecule of the impurity.

Silica gel is an amorphous polar material with the chemical formula $(SiO_2)–nH_2O$ with high affinity to polar molecules such as water. It has a surface area of about 750 $m^2\,g^{-1}$, an internal porosity of 70%, and a mean pore diameter of 2.2 nm. Its ability to retain siloxanes is due to the partial polar behavior of the siloxane molecules because of their —Si—O—Si— bonds. Silica gel is used for the removal of water and siloxanes and their capacity for the removal of siloxanes is about 100 $mg\,g^{-1}$ of silica gel [79]. This capacity is about 10 times higher than the capacity of the activated carbon, and because its thermal regeneration is easier, silica gel is generally considered a better option. The thermal regeneration may be achieved passing a stream of hot air.

7.2.3.3 Membrane Separation

The removal of siloxanes is possible using thin polymeric or inorganic microporous membranes. These have large surface areas, and at the same time, a thickness of only 0.2–2 μm providing a purification system of very low volume and cost. The membranes are permeable to the biogas stream and retain the siloxanes because of their specific porosity and via molecular interactions [85]. High- and low-pressure membrane purification systems are discerned depending on the pressure of the biogas stream. In the high-pressure systems, biogas is forced to pass through the membrane with high pressures of about 35 bar. In the low pressure systems, on the other hand, the pressure of the biogas stream is nearly atmospheric and the siloxanes initially diffuse inside the membrane and then are absorbed by a specific absorbent. After saturation, the membranes are regenerated by the flow of an inert gas or a solvent. The lifetime of a membrane can reach two to three years, but it is important that they operate with a biogas stream, which is already free of H_2S and solid particles. H_2S is harmful to the membranes because its acidity and solid particles can cause their mechanical damage. In the high-pressure systems, the biogas stream must also be free of any oil vapors that originate in the compressors because these can cause the fouling of the membrane. Fouling may also occur by the accumulation of the siloxanes and this effect needs periodic monitoring.

7.2.3.4 Biological Filters

Bioscrubbers, biofilters, and biotrickling filters are common bioreactors in the biogas industry and may also be used for the removal of siloxanes. The biological degradation of siloxanes can take place by specific microorganisms such as *Pseudomonas*, *Agrobacterium*, *Arthrobacter*, and *Fusarium oxysporum*, which were found able to degrade D4 and other silicon organic compound into dimethylsilanediol (DMSD, $C_2H_8O_2Si$), silicic acid, water, and carbon dioxide, which are all soluble in the aqueous phase solution in the reactor bed and cause the cleaning of is sorption sites [77]. The operation principle of the bioscrubbers, biofilters, and biotrickling filters has been described previously for the removal of H_2S.

7.2.3.5 Cryogenic Condensation

As it was mentioned previously, siloxanes may be efficiently cleaned from biogas through dehydration cooling below −70 °C. Despite the high purification efficiency of this method, which achieves the removal of siloxanes by more than 99.3%, the required cryogenic cooling requires high investment and operating costs, which at the moment make the scale up of the technology too expensive. At −25 °C, only 26% of siloxanes condense as liquids [16].

7.2.4 Removal of Volatile Organic Compound (VOCs)

The VOCs that are commonly found in raw biogas are various paraffins, halogenated hydrocarbons, and siloxanes. These are formed in the raw biogas depending on the biomass feedstock. Halogenated hydrocarbons are hydrocarbons also containing chlorine, bromine, or fluorine. During combustion, they form acids such as hydrochloric acids, which are highly corrosive. These are usually removed by activated carbon filtration using two packed beds in parallel so that the one will be regenerated while the other executes the adsorption [16, 86, 87].

7.2.5 Removal of Ammonia (NH_3)

Ammonia becomes a constituent of biogas through the anaerobic digestion of biomass feedstocks that are rich in proteins. Ammonia formation is favored when digestion takes place at high pH and high temperature and so such conditions may be avoided. In most cases, the concentration of ammonia in biogas is very low, below 0.1 mg m^{-3}. In appreciable concentrations, ammonia is a source of nitrogen oxides (NO_x), which are toxic emissions of biogas combustion. Ammonia is usually cleaned when the biogas is dried or upgraded. Therefore, its removal usually does not need a separate processing [8, 88].

7.2.6 Removal of Oxygen (O_2) and Nitrogen (N_2)

Because the production of biogas is carried out in anaerobic conditions, oxygen and nitrogen are generally not constituents of biogas. However, they can be added in biogas through leaks during the production, purification, or upgrading processes. The presence of these impurities decreases the heating value of biogas and causes corrosion on the metal equipment of the biogas plant. Also, in high concentrations (6–12%), oxygen forms an explosive mixture with the methane in biogas which has obvious safety implications. In low quantities, oxygen can be removed from biogas during the removal of H_2S or during the upgrading. In higher quantities, O_2 and N_2 may be removed in activated carbon, molecular sieves, or membranes. The removal of O_2 and N_2 from biogas is generally difficult and expensive and therefore their presence should be avoided [8].

7.3 Biogas Upgrading Technologies

The biogas upgrading technologies also remove CO_2 and are used for the production of biomethane. These technologies usually involve physical absorption (scrubbing), physical and chemical adsorption, biological filtration, mechanical separation in molecular sieves or membranes, and cryogenic treatment.

Figure 7.3 illustrates typical unit operations for some of these technologies such as water scrubbing, chemical scrubbing, pressure swing adsorption (PSA), and polymeric membranes. As shown, depending on the selected CO_2 removal method (methane enrichment), various other processes usually precede or follow aiming at the separate removal of the other contaminants. In that sense, when upgrading is desired the biogas purification and upgrading technologies are almost always in close connection and are used in combination.

7.3.1 Water Scrubbing

Water scrubbing has been explained previously as a physical absorption method for the removal of H_2S. The method is also used for the removal of CO_2 for the upgrading of biogas into biomethane. It is based on the fact that the solubility of CO_2 in water is different from the solubility of CH_4 or H_2S and can produce biomethane with a CH_4 purity of 90–99%. Given that the solubility of a gas in a liquid solvent improves by increasing the pressure or decreasing temperature, CO_2 removal with water scrubbing usually takes place at 6–10 bar and after a relevant cooling of the biogas stream

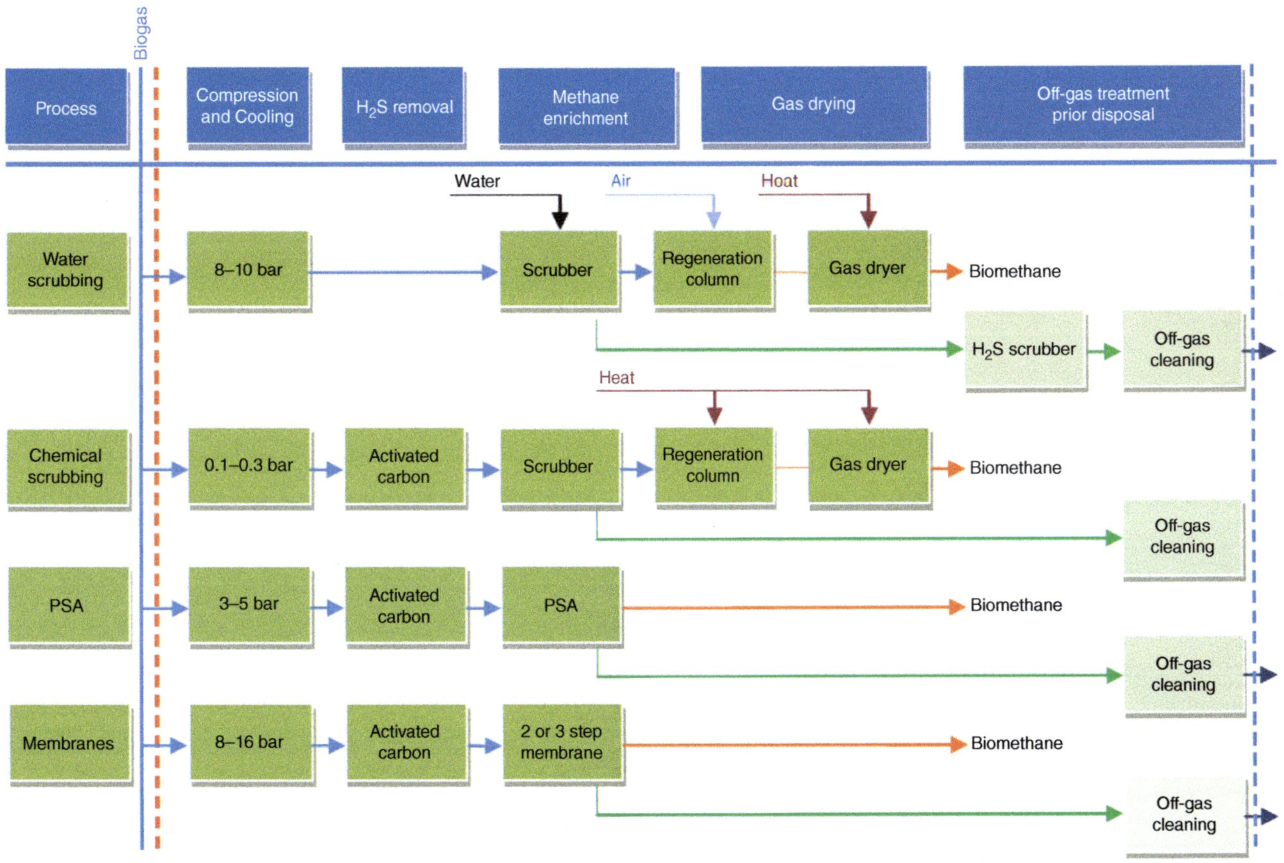

Figure 7.3 Typical unit operations for the upgrading of biogas.

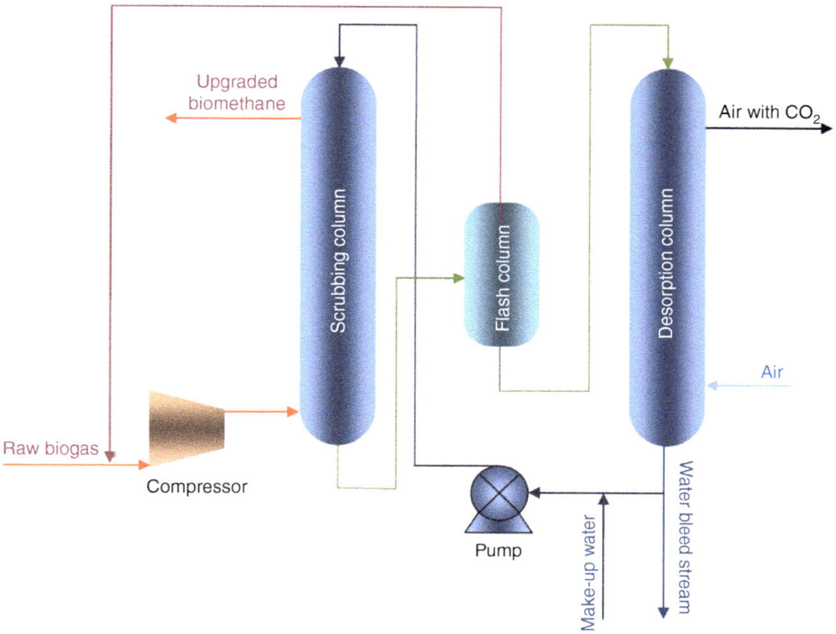

Figure 7.4 Schematic representation of a typical water scrubbing system.

which is heated from the compression [89, 90]. When used for the removal of CO_2, water scrubbing also removes most of the H_2S of the gas. As a result, it simplifies the desulfurization of biogas, but usually, an extra desulfurization technology is also required to clean the remaining H_2S at the desired levels.

A typical CO_2 cleaning water scrubbing plant is shown in Figure 7.4. Water is sprayed at the top of a tall scrubbing column and the biogas is fed upward from the bottom. Inside the column, fixed beds of the packed materials are installed to provide extra transitional surface area and a better contact between the gas and liquid flows. Additionally, multiple intermediate floors are installed in which the water is collected and sprayed again into the lower column sections. The purified biomethane is collected from the top of the scrubbing column while the used water is collected at the bottom. The used water carries the absorbed CO_2 as well as lower quantities of H_2S, NH_4, and CH_4. Initially, the used water is pumped into a flash vessel where the pressure is partially reduced at about 2.5–3.5 bar with a simultaneous discharge of CH_4 and CO_2 [17]. The gaseous mixture produced by this discharge, known as flash gas, is redirected into the scrubbing column minimizing the losses of the plant in CH_4. The used water then flows to the top of the stripping column where it expands at atmospheric pressure inside an upward air flow. Finally, the regenerated water is recirculated to spray against the scrubbing column. Inside the stripping column, most of the captured CO_2 is released and is usually blown off into the atmosphere as an exhaust gas. Water scrubbing almost always causes a small loss of CH_4, which also absorbs into the water and also discharges in the stripping column. Given that CH_4 is a very serious greenhouse gas emission, all pressurized water scrubbers must be equipped with an afterburning equipment to convert the CH_4 emissions into CO_2 and H_2O.

7.3.2 Organic Solvent Scrubbing

Organic solvents such as Selexol are better CO_2 retainers than water and are also used for the upgrading of biogas. The solubility of CO_2 and H_2S in such adsorbents is higher than in water and the higher purification efficiencies result in lower solvent volumes, smaller scrubbing columns, and lower costs. The cost is also lower for the compression of biogas, which in Selexol scrubbers is pressurized at lower pressures, at about 6–8 bar [17].

However, because the adsorption forces in these solvents are higher, their regeneration is more difficult. For that reason, the solvent expansion and stripping also require the heating of the liquid at about 40–80 °C with a heating demand of about 0.1–0.15 kWh m^{-3} of biogas. The heat is usually supplied from the exhaust gas afterburner [91, 92]. Although the removal of H_2S is also possible together with the removal of CO_2, desulfurization is typically carried out before the CO_2 scrubbing process. A typical organic solvent scrubbing plant is shown in Figure 7.5. The biogas is compressed and is then cooled before it will be fed on the bottom scrubbing column. The solvent is also cooled and enters the scrubbing column from the top. The temperature inside the packed bed remains at about 20 °C, and the used solvent is routed to a heat exchanger and then into a flash tank where some of the retained CO_2 and CH_4 are released and redirected to the absorption column.

7.3.3 Chemical Scrubbing

When physical adsorption is combined with the chemical reaction of the CO_2 or H_2S with the liquid adsorbent, chemical scrubbing is accomplished. This is

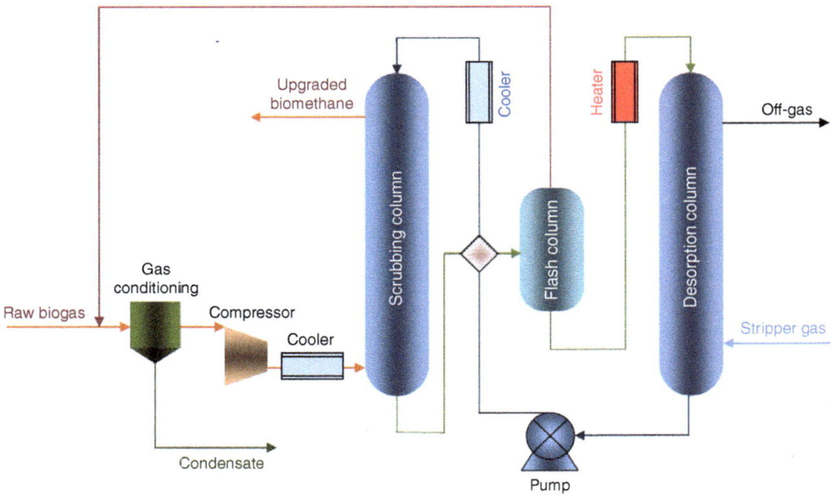

Figure 7.5 Schematic representation of a typical water organic solvent scrubbing system.

possible is aqueous solutions of specific amines such as monoethanolamine (MEA, C_2H_7NO)), diethanolamine (DEA, $C_4H_{11}NO_2$), and methyldiethanolamine (MDEA, $C_5H_{13}NO_2$) [16]. During the past years, a mixture of MDEA and piperazine (PZ) is typically used as such a solvent. This is known as activated methyldiethanolamine (AMDEA) and has a higher adsorption capacity than MDEA, probably because of the presence of primary and secondary amines in the PZ [13]. The adsorption of CO_2 in an amine solution is generally very complex and involves mass transfer and chemical reactions [93]. Initially, CO_2 has to be transported from the gas to the liquid surface, and then it is adsorbed in the liquid solution where it may react chemically with other components. The used amine solution is then regenerated by heating or by pressure reduction. The overall amine scrubbing may be represented by the following general chemical reactions:

$$CO_2\text{adsorption}: RNH_2 + H_2O + CO_2 \rightarrow RNH_3^+ \; HCO_3^- \text{ (under pressure)} \quad (7.13)$$

$$CO_2\text{desorption}: RNH_3^+ \; HCO_3^- \rightarrow RNH_2 + H_2O + CO_2 \text{ (low pressure, some heat)} \quad (7.14)$$

where R represents the remaining organic component of the amine molecule.

Chemical scrubbing is more selective on the cleaning of the desired impurity and has higher purification efficiency than physical scrubbing. Because of the higher selectivity of the method, the solvent does not retain CH_4 and afterburning is not necessary. Also, the scrubbing can take place at nearly atmospheric pressure, and because of the high cleaning capacity, the solvent volume is much lower. However, because the strength of the bonds of the impurity with the washing solution is much higher, regeneration is more complex. The regeneration of the amine solution requires a heating at about 110–160 °C and then it must be cooled at about 40 °C to be able to be used again in the scrubbing column. The heat is usually provided by heat exchangers of the plant. During biogas upgrading, the removal of H_2S typically takes place separately before the removal of CO_2. If H_2S is present in biogas, it will be absorbed in the amine solution and then higher regeneration temperatures will be required for the desorption of H_2S.

A typical amine scrubber system is shown in Figure 7.6. It consists of a scrubbing column where CO_2 is absorbed and a stripping column where CO_2 separates from the waste amine solution through heating at a reduced pressure [94]. In the scrubbing column, the CO_2 reacts chemically with the amine and is absorbed. This process is exothermic and causes the temperature of the absorber to increase from 20 to 40 °C to about 45–65 °C [17, 95]. The increase in temperature reduces the solubility of CO_2 in water, but in amine chemical scrubbing, it enhances the reaction rate between the amine and the CO_2 and causes an overall increase on the absorption of CO_2. The pressure on the scrubbing column is about 1–2 bar [96]. The used amine solution is extracted from the bottom of the scrubbing column, and after it passes a heat exchanger, it is routed on the top of the stripping column. In the stripping column, the CO_2 is released and the amine solution is regenerated through a counterflow of steam. The bottom of the stripping column is equipped with a boiler that boils the amine solution at about 120–150 °C and provides the

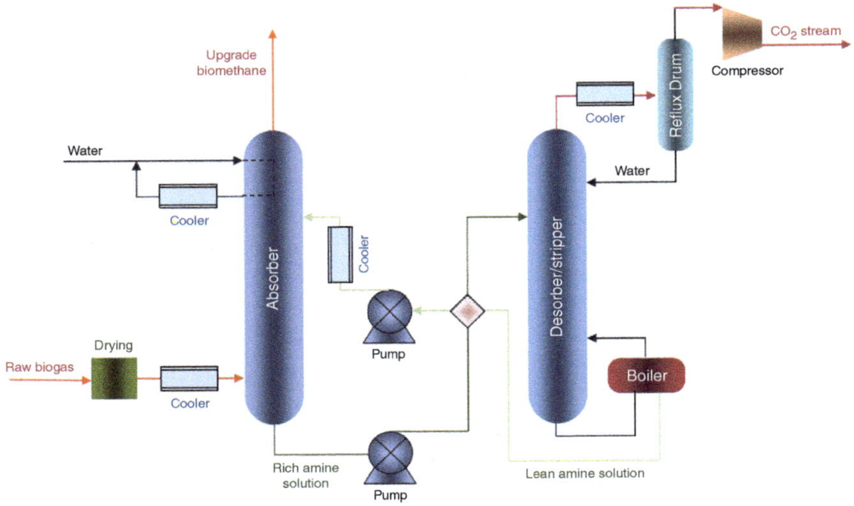

Figure 7.6 Schematic representation of a typical chemical (amine) scrubbing system.

heat for the regeneration and the cleaning of the waste amine solution. The pressure in the stripper is typically 1.5–3 bar [96].

7.3.4 Pressure Swing Adsorption

PSA is a well-established biogas upgrading method carried out in columns with packed beds of activated carbon, zeolitic molecular sieves, or carbon molecular sieves [97, 98]. The gas separation takes place by physical adsorption on the surface of the high surface area packing material and also due to penetration into its microporous structure. After a previous purification treatment for H_2S, the biogas is compressed at 2–7 bar, cooled to about 70 °C, and is directed at the bottom of the packed column. There, CO_2 is retained and CH_4 is left unaffected because of the difference in their molecular size. The top of the packed column is equipped with a valve that opens and provides the formed biomethane. Then, the valve closes and the pressure inside the column is released. This causes desorption of CO_2 from the packed material, forming a CO_2-rich exhaust gas, which is removed at the bottom. Then, the column is filled again with biogas and the process is repeated [98, 99]. PSA is a batch process, and therefore, the continuous supply of biomethane is possible only when a series of columns work in parallel.

Figure 7.7 describes a typical PSA plant with four PSA columns. This plant executes the conventional Skarstrom cycle that is composed of four processes: (i) adsorption of CO_2, (ii) depressurization (blowdown) with CO_2 desorption, (iii) purge with gas recycle, and (iv) pressurization with blowdown gas and feed or recycle gas [100]. As shown in Figure 7.7, at each instant, the four PSA columns execute a different process of the Skarstrom cycle. Usually 4–9 PSA columns are used [8]. Typically, 1.5–2.5% of the CH_4 volume in biogas is lost within the CO_2-rich exhaust stream, and therefore, the utilization of CH_4 afterburner is necessary.

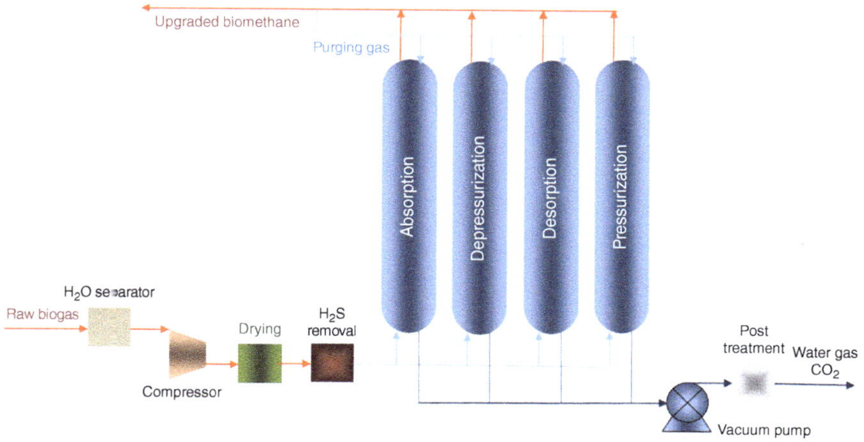

Figure 7.7 Schematic representation of a typical pressure swing adsorption (PSA) system with four PSA columns.

Sometimes, higher CH_4 losses are allowed and the heat provided by the afterburner is supplied to useful demands of the biogas plant. H_2S is undesired in the PSA stage because of its high toxicity and also because its adsorption is normally irreversible. Therefore, the desulfurization of biogas at a previous stage is a prerequisite [101]. On the whole, PSA can provide high-purity biomethane (with 95–99% CH_4) but needs extensive process control, high investments, and high operational costs.

Alternative technologies of PSA are those of thermal swing adsorption (TSA) and electrical swing adsorption (ESA). In the TSA, the regeneration of the packed column is not accomplished by a decrease in pressure but by an isobaric increase of temperature [102]. This can be very advantageous if a cheap source of heat is available. In the ESA, the regeneration of the packed column takes passing electric current that generates heat through the Joule effect [103, 104]. This is a promising technology with lower operational costs compared to PSA and TSA, but it needs the packed bed material to conduct electricity. Various studies focused on the development of appropriate semiconductor adsorbents from activated carbon [104, 105].

7.3.5 Polymeric Membranes

Polymeric membranes are used to separate CO_2, O_2, H_2O, and H_2S from larger molecules such as CH_4 and N_2. Most of them are organic materials such as cellulose acetate, polyimide, polysulfone, polycarbonate, and polydimethyl siloxane [13, 106, 107]. They are usually formed into hollow fiber polymers, which are combined in a tube bundle to provide the highest possible surface per volume ratio [108]. These are then fitted into a tube forming a separation device, which operates continuously, as shown in Figures 7.8 and 7.9.

Because of continuous operation, this technology has the advantage that does not need regeneration. Biogas is typically fed at 7–20 bar, and depending on the different molecular sizes of the impurities, these diffuse through the membranes

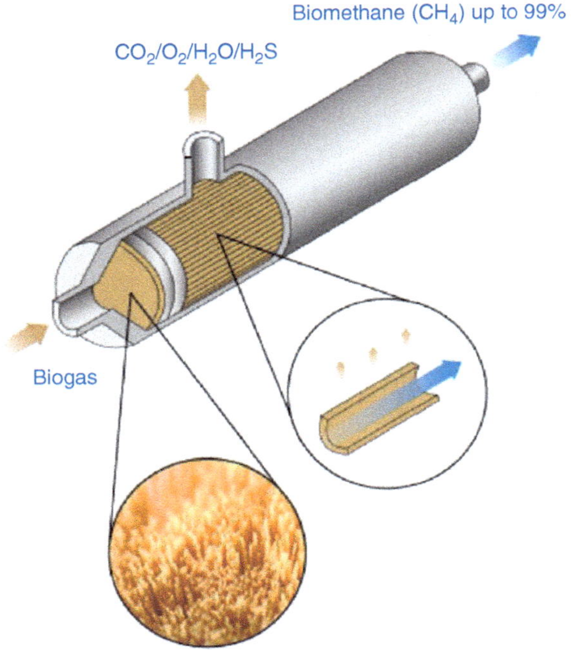

Figure 7.8 Schematic representation and operational principle of a tubular hollow fiber membrane separator.

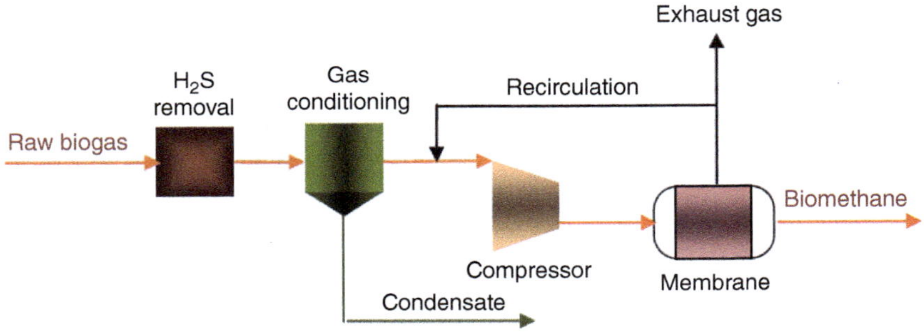

Figure 7.9 Schematic representation of a simple membrane upgrading system.

at different speeds. The driving forces behind the permeation of a gas through a membrane are differences in concentration, pressure, and temperature as well as the difference on the electric charges of the various species. The impurities initially dissolve in the membrane, then diffuse inside it because of the differences in concentration, and are finally separated because of the pressure flow by the pores of the membrane. The retained impurities are then removed from the biogas stream as an exhaust gas. Higher methane purities may be achieved using two or three membrane separators in a cascade setup with or without recirculation [109], as shown in Figure 7.10.

The disadvantages of this technology include the high-pressure losses and the high energy demands of 0.18–0.33 kWh m^{-3} of biogas. About 0.5–2% of the CH_4 is lost

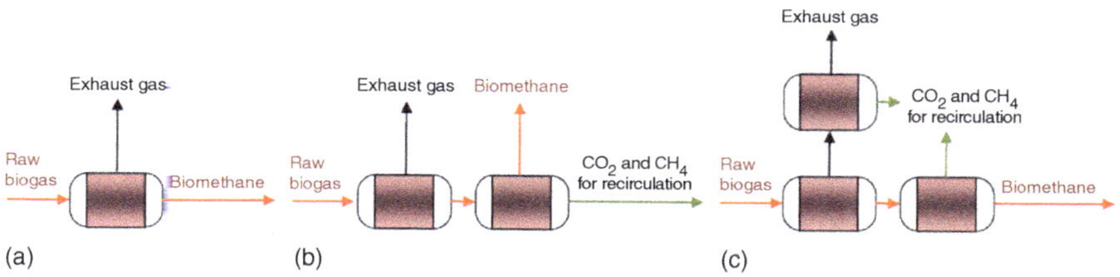

Figure 7.10 Three alternative designs of membrane stages: (a) without biogas recirculation, (b) the exhaust gases of the first stage are removed, and the exhaust flow of the second stage is recirculated back to the compressor to reduce the methane losses, and (c) the biogas outflow from the first stage is polished in the second membrane stage to obtain a product gas with a purity of more than 97% methane. Additionally, the exhaust gases of the first stage are polished in a third membrane stage to minimize the CH_4 concentration in the off-gas and the volume of gas circulated back to the compressor. The exhaust flow of the second stage and the outflow of the third stage are combined and recycled to the compressor.

within the exhaust gas and therefore afterburning is necessary. The most common polymeric membranes for the upgrading of biogas are cellulose acetate and polyimide [106]. However, both suffer from plasticization at high pressures [110], and although polyimide is more resistant to plasticization, it is also more expensive.

7.3.6 Cryogenic Treatment

Thermodynamics dictate that at low temperature and/or high pressure, a gas can condense into liquid or re-sublimate into solid. These phase changing processes may be used for the separation of a gaseous impurity from the biogas stream through liquefaction or solidification. Using a countercurrent refrigeration and/or compression column, this technology is able to provide high-purity biomethane (with more than 99.9 vol% CH_4) at the top and high-purity CO_2 exhaust flow (above than 98 vol% CO_2) at the bottom. This means that the cryogenic treatment has very high purification efficiency and selectivity with very low CH_4 losses. It also facilitates the easy liquefaction of biomethane when this is desired and the CO_2 may be taken in high-purity liquid or solid phase. Despite these advantages, the cryogenic treatment is a new upgrading method and efforts for the effective development of this technology are still underway [8, 16]. Serious disadvantages are the high-energy demand of the cryogenic conditions and the necessity to handle the biogas after a deep pre-cleaning because the solidification of H_2O and H_2S will plug the tubes and the equipment of the plant.

A typical process diagram of cryogenic separation is shown in Figure 7.11. At a first purification stage, the water vapors, H_2S, halogens, siloxanes, and solid dust particles are removed, and then, the biogas is compressed at 10 bar and cooled in three stages, initially at −25 °C and then at −55 and −85 °C, respectively. In the second cooling stage, part of the CO_2 is liquefied, and in the third cooling stage, the remaining CO_2 is removed by solidification [13]. In other designs, cooling takes place in two stages, at the one down to about −50 to −59 °C and at the other down to a temperature

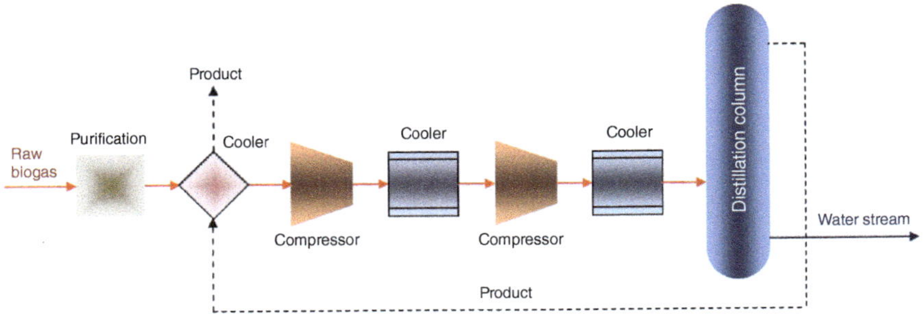

Figure 7.11 Schematic representation of a simple three-stage cryogenic biogas upgrading system.

that causes CO_2 solidification. In these designs, the first cooling stage removes about 30–40% of the CO_2 of biogas and the second stage removes the rest [8].

7.4 Conclusions

Every single sweetening technology has its own advantages and disadvantages and the selection of the most appropriate technology also depends on the size and the economic policy of the biogas plant. Some of them are still at the stage of research or development while others are already mature and find common practice in existing small-, medium-, or large-scale biogas production facilities. In the future, the biogas and biomethane markets are expected to expand, as they offer sustainability, they are profitable, they replace fossil natural gas, and increase energy security. Toward this new era, both the researchers and the biogas producers are searching for ways that will optimize the purification and upgrading processes with higher efficiency and lower cost. Moreover, as the production volumes increase, market forces push for the development of more advanced and large-scale sweetening methods. With these considerations in mind, the present chapter has provided an overview of the most important biogas sweetening technologies and discussed the most promising relevant research advances.

References

1 Guo, M., Sing, W., and Buhain, J. (2015). Bioenergy and biofuels: history, status and perspective. *Renewable Sustainable Energy Rev.* 42: 712–725.

2 Khalid, A., Arshad, M., Anjum, M. et al. (2011). The anaerobic digestion of solid organic waste. *Waste Manage. (Oxford)* 31: 1737–1744.

3 Charisiou, N.D., Douvartzides, S.L., Siakavelas, G.I. et al. (2019). The relationship between reaction temperature and carbon deposition on nickel catalysts based on Al_2O_3, ZrO_2 or SiO_2 supports during the biogas dry reforming reaction. *Catalysts* 9: 676.

4 Xue, S., Song, J., Wang, X. et al. (2020). A systematic comparison of biogas development and related policies between China and Europe and corresponding insights. *Renewable Sustainable Energy Rev.* 117: 109474.

5 Papurello, D., Silvestri, A., and Lanzini, A. (2019). Biogas cleaning: trace compounds removal with model validation. *Sep. Purif. Technol.* 210: 80–92.

6 Charisiou, N.D., Baklavaridis, A., Papadakis, V.G., and Goula, M.A. (2016). Synthesis gas production via the biogas reforming reaction over $Ni/MgO–Al_2O_3$ and $Ni/CaO–Al_2O_3$ catalysts. *Waste Biomass Valorization* 7: 725–736.

7 Rasi, S., Veijanen, A., and Rintala, J. (2007). Trace compounds of biogas from different biogas production plants. *Energy* 32: 1375–1380.

8 Petersson, A. and Weillinger, A. (2009). Biogas upgrading technologies – development and innovations. *IEA Bioenergy Task* 37: 6–9.

9 Abatzoglou, N. and Boivin, S. (2009). A review of biogas purification processes. *Biofuels, Bioprod. Biorefin.* 3: 42–71.

10 Yentekakis, I.V. and Goula, G. (2017). Biogas management: advanced utilization for production of renewable energy and added-value chemicals. *Front. Environ. Sci.* 5: 7.

11 Rodero, M.R., Angeles, R., Marin, D. et al. (2018). Biogas purification and upgrading technologies. In: *Biogas: Fundamentals, Process and Operation*, Biofuels and Biorefinery Technologies, vol. 6, 35–49. Heidelberg Germany: Springer International Publishing AG.

12 Zhou, W.J., Zhou, B., Li, W.Z. et al. (2004). Performance comparison of low-temperature direct alcohol fuel cells with different anode catalysts. *J. Power Sources* 126: 16–22.

13 Khan, I.U., Othman, M.H.D., Hashim, H. et al. (2017). Biogas as a renewable energy fuel – a review of biogas upgrading, utilization and storage. *Energy Convers. Manage.* 150: 277–294.

14 Miltner, M., Makaruk, A., and Harasek, M. (2017). Review on available biogas upgrading technologies and innovations towards advanced solutions. *J. Cleaner Prod.* 161: 1329–1337.

15 Awe, O.W., Zhao, Y., Nzihou, A. et al. (2017). A review of biogas utilization, purification and upgrading technologies. *Waste Biomass Valorization* 8: 267–283.

16 Ryckebosch, E., Drouillon, M., and Verhaaven, H. (2011). Techniques for transformation of biogas to biomethane. *Biomass Bioenergy* 35: 1633–1645.

17 Bauer, F., Persson, T., Hulteberg, C., and Tamm, D. (2013). Biogas upgrading – technology overview, comparison and perspectives for the future. *Biofuels, Bioprod. Biorefin.* 7: 499–511.

18 Muñoz, R., Meier, L., Diaz, I., and Jeison, D. (2015). A review on the state-of-the-art of physical and biological technologies for biogas upgrading. *Rev. Environ. Sci. Biotechnol.* 14: 727–759.

19 Angelidaki, I., Treu, L., Tsapekos, P. et al. (2018). Biogas upgrading and utilization: current status and perspectives. *Biotechnol. Adv.* 36: 452–466.

20 Srirangan, K., Akawi, L., Moo-Young, M., and Chou, C.P. (2012). Towards sustainable production of clean energy carriers from biomass resources. *Appl. Energy* 100: 172–186.

21 Prussi, M., Padella, M., Conton, M. et al. (2019). Review of technologies for biomethane production and assesment of EU transport share in 2030. *J. Cleaner Prod.* 222: 565–572.

22 Lee, D.H. (2017). Evaluation the financial feasibility of biogas upgrading to biomethane, heat, CHP and AwR. *Int. J. Hydrogen Energy* 42: 27718–27731.

23 Collet, P., Flottes, E., Favre, A. et al. (2017). Techno-economic and life cycle assessment of methane production via biogas upgrading and power to gas technology. *Appl. Energy* 192: 282–295.

24 Curto, D. and Martín, M. (2019). Renewable based biogas upgrading. *J. Cleaner Prod.* 224: 50–59.

25 Sarker, S., Lamb, J.J., Hjelme, D.R., and Lien, K.M. (2018). Overview of recent progress towards in-situ biogas upgradation techniques. *Fuel* 226: 686–697.

26 Duran, I., Alvarez-Gutierrez, N., Rubiera, F., and Pevida, C. (2018). Biogas purification by means of adsorption on pine sawdust-based activated carbon: impact of water vapor. *Chem. Eng. J.* 353: 197–207.

27 Noorain, R., Kindaichi, T., Ozaki, N. et al. (2019). Biogas purification performance of new water scrubber packed with sponge carriers. *J. Cleaner Prod.* 214: 103–111.

28 Sloan, E.D., Koh, C.A., and Sum, A. (2010). *Natural Gas Hydrates in Flow Assurance*. Gulf Professional Publishing.

29 Allegue, L.B. and Hinge, J. (2014). *Biogas Upgrading: Evaluation of Methods for H_2S Removal*. Danish Technological Institute.

30 Netusil, M. and Ditl, P. (2011). Comparison of three methods for natural gas dehydration. *J. Nat. Gas Chem.* 20: 471–476.

31 Rachbauer, L., Gabauer, W., Scheidi, S. et al. (2015). Closing the nutrient cycle in two-stage anaerobic digestion of industrial waste streams. *Energy Fuels* 29: 4052–4057.

32 Ciccoli, R., Cigolotti, V., Lo, P.R. et al. (2010). Molten carbonate fuel cells fed with biogas: combating H_2S. *Waste Manage. (Oxford)* 30: 1018–1024.

33 Charisiou, N.D., Siakavelas, G., Papageridis, K.N. et al. (2016). Syngas production via the biogas dry reforming reaction over nickel supported on modified with CeO_2 and/or La_2O_3 alumina catalysts. *J. Nat. Gas Sci. Eng.* 31: 164–183.

34 Zulkefli, N.N., Masdar, M.S., Jahim, J., and Majlan, E.H. (2016). Overview of H_2S removal technologies from biogas production. *Int. J. Appl. Eng. Res.* 11: 10060–10066.

35 Miltner, M., Makaruk, A., Krischan, J., and Harasek, M. (2012). Chemical-oxidative scrubbing for the removal of hydrogen sulphide from raw biogas: potential and economics. *Water Sci. Technol.* 66: 1354–1360.

36 Surra, E., Nogueira, M.C., Bernardo, M. et al. (2019). New adsorbents from maize cob wastes and anaerobic digestate for H_2S removal from biogas. *Waste Manage. (Oxford)* 94: 136–145.

37 Daneshyar, A., Ghaedi, M., Sabzehmeidani, M.M., and Daneshyar, A. (2017). H_2S adsorption onto Cu–Zn–Ni nanoparticles loaded activated carbon and Ni–Co nanoparticles loaded γ-Al_2O_3: optimization and adsorption isotherms. *J. Colloid Interface Sci.* 490: 553–561.

38 de Falco, G., Montagnaro, F., Balsamo, M. et al. (2018). Synergic effect of Zn and Cu oxides dispersed on activated carbon during reactive adsorption of H_2S at room temperature. *Microporous Mesoporous Mater.* 257: 135–146.

39 Sitthikhankaew, R., Chadwick, D., Assabumrungrat, S., and Laosiripojana, N. (2014). Effects of humidity, O_2, and CO_2 on H_2S adsorption onto upgraded and KOH impregnated activated carbons. *Fuel Process. Technol.* 124: 249–257.

40 Barelli L., Bidini, G., de Arespacochaga, N. et al. (2017). Biogas use in high temperature fuel cells: enhancement of KOH–KI activated carbon performance toward H_2S removal. *Int. J. Hydrogen Energy* 42: 10341–10353.

41 Farooq, M., Almustapha, M.N., Imran, M. et al. (2018). In-situ regeneration of activated carbon with electric potential swing desorption (EPSD) for the H_2S removal from biogas. *Bioresour. Technol.* 249: 125–131.

42 Bandosz, T.J. (2002). On the adsorption/oxidation of hydrogen sulfide on activated carbons at ambient temperatures. *J. Colloid Interface Sci.* 246: 1–20.

43 Bagreev, A. and Bandosz, T.J. (2002). A role of sodium hydroxide in the process of hydrogen sulfide adsorption/oxidation on caustic-impregnated activated carbons. *Ind. Eng. Chem. Res.* 41: 672–679.

44 Bagreev, A., Katikaneri, S., Parab, S., and Bandosz, T.J. (2005). Desulfurization of digester gas: prediction of activated carbon bed performance at low concentrations of hydrogen sulfide. *Catal. Today* 99: 329–337.

45 Lau, L.C., Nor, N.M., Mohamed, A.R., and Lee, K.T. (2013). Adsorption of hydrogen sulfide using palm shell activated carbon: an optimization study using statistical analysis. *Int. J. Res. Eng. Technol.* 2: 302–311.

46 Tsai J H, Jeng F.T., Chiang H.L. Removal of H_2S from exhaust gas by use of alkaline activated carbon. *Adsorption* 7 (2001) 357-366.

47 Sisani, E., Cinti, G., Discepoli, G. et al. (2014). Adsorptive removal of H_2S in biogas conditions for high temperature fuel cell systems. *Int. J. Hydrogen Energy* 39: 21753–21766.

48 Feng, W., Kwon, S., Borguet, E., and Vidic, R. (2005). Adsorption of hydrogen sulfide onto activated carbon fibers: effect of pore structure and surface chemistry. *Environ. Sci. Technol.* 39: 9744–9749.

49 Ozerkmekci, M., Salkic, G., and Fellah, M.F. (2015). Use of zeolites for the removal of H_2S: a mini review. *Fuel Process. Technol.* 139: 49–60.

50 Barelli, L., Bidini, G., Micoli, L. et al. (2018). 13X Ex–Cu zeolite performance characterization towards H_2S removal for biogas use in molten carbonate fuel cells. *Energy* 160: 44–53.

51 Sigot, L., Ducom, G., and Germain, P. (2016). Adsorption of hydrogen sulfide (H_2S) on zeolite (Z): retention mechanism. *Chem. Eng. J.* 287: 47–53.

52 Ko, D., Siriwardane, R., and Biegler, L.T. (2002). Optimization of a pressure-swing adsorption process using zeolite 13X for CO_2 sequestration. *Ind. Eng. Chem. Res.* 42: 339–348.

53 de Oliveira, L.H., Meneguin, J.G., Pereira, M.V. et al. (2019). H_2S adsorption on NaY zeolite. *Microporous Mesoporous Mater.* 284: 247–257.

54 Xu, X., Song, C., Andresen, J.M. et al. (2002). Novel polyethylenimine-modified mesoporous molecular sieve of MCM-41 type as high-capacity adsorbent for CO_2 capture. *Energy Fuels* 16: 1463–1469.

55 Ma, X., Wang, X., and Song, C. (2009). Molecular basket sorbents for separation of CO_2 and H_2S from various gas streams. *J. Am. Chem. Soc.* 131: 5777–5783.
56 Chaemchuen, S., Zhou, K., and Verpoort, F. (2016). From biogas to biofuel: materials used for biogas cleaning to biomethane. *ChemBioEng Rev.* 6: 250–265.
57 Mycock, J.C., McKenna, J.D., and Theodore, L. (1995). *Handbook of Air Pollution Control Engineering and Technology*. CRC Press.
58 Chandra, R., Vijay, V.K., and Subbarao, P.M.V. (2012). Vehicular quality biomethane production from biogas by using an automated water scrubbing system. *ISRN Renewable Energy*. Article ID: 904167.
59 Nie, H., Jiang, H., Chong, D. et al. (2013). Comparison of water scrubbing and propylene carbonate absorption for biogas upgrading process. *Energy Fuels* 27: 3239–3245.
60 Tabassum, M.R., Xia, A., and Murphy, J.D. (2018). Biomethane production from various segments of brown seaweed. *Energy Convers. Manage.* 174: 855–862.
61 Tippayawong, N. and Thanompongchart, P. (2010). Biogas quality upgrade by simultaneous removal of CO_2 and H_2S in a packed column reactor. *Energy* 35: 4531–4535.
62 Mamrosh, D.L., McIntush, K.E., and Fisher, K. (2014). Caustic scrubber designs for the H_2S removal from refinery gas streams. *Proceedings of the 2014 AFPM Annual Meeting, AM-14-48* (23–25 March 2014). Orlando, Florida: American Fuel and Petrochemical Manufacturers (AFPM).
63 Mamrosh D.L., Fisher K.S., Beitler C.M. Caustic scrubber designs for refinery fuel gas, sour water stripper gas, and other refinery applications, *Proceedings of the Brimstone Sulfur Symposium* (11–14 September 2012). Vail, Colorado, USA: Brimstone STS Limited.
64 Bontozoglou, V. and Karabelas, A.J. (1991). Numerical calculation of simultaneous absorption of H_2S and CO_2 in aqueous hydroxide solution. *Ind. Eng. Chem. Res.* 30: 2598–2603.
65 Bontozoglou, V. and Karabelas, A.J. (1993). Simultaneous absorption of H_2S and CO_2 in NaOH solutions: experimental and numerical study of the performance of a short-time contactor. *Ind. Eng. Chem. Res.* 32: 165–172.
66 Zarczynski, A., Rosiak, K., Anielak, P., and Wolf, W. (2014). Practical methods of cleaning biogas from hydrogen sulphide. Part 1: application of solid sorbents. *Acta Innovations* 12: 24–34.
67 Alinezhad, E., Haghighi, M., Rahmani, F. et al. (2019). Technical and economic investigation of chemical scrubber and bio-filtration in removal of H_2S and NH_3 from wastewater treatment plant. *J. Environ. Manage.* 241: 32–43.
68 San-Valero, P., Penya-roja, J.M., Alvarez-Hornos, F.J. et al. (2019). Fully aerobic bioscrubber for the desulfurization of H_2S-rich biogas. *Fuel* 241: 884–891.
69 Promnuan, K. and O-Thong, S. (2017). Efficiency evaluation of biofilter for hydrogen sulfide removal from palm oil mill biogas. *Energy Procedia* 138: 564–568.
70 Li, L., Han, Y., Yan, X., and Liu, J. (2013). H_2S removal and bacterial structure along a full-scale biofilter bed packed with polyurethane foam in a landfill site. *Bioresour. Technol.* 147: 52–58.

71 Rabbani, K.A., Charles, W., Kayaalp, A. et al. (2016). Pilot-scale biofilter for the simultaneous removal of hydrogen sulphide and ammonia at a wastewater treatment plant. *Biochem. Eng. J.* 107: 1–10.

72 Chen, Y., Xie, L., Cai, W.F., and Wu, J.W. (2019). Pilot-scale study using biotrickling filter to remove H_2S from sewage liftstation: experiment and CFD simulation. *Biochem. Eng. J.* 144: 177–184.

73 Reddy, C.N., Bae, S., and Min, B. (2019). Biological removal of H_2S gas in a semi-pilot scale biotrickling filter: optimization of various parameters for efficient removal at high loading rates and low pH conditions. *Bioresour. Technol.* 285: 121328.

74 Dumont, E. (2017). Validation of a rapid procedure to determine biofilter performances. *J. Environ. Chem. Eng.* 5: 2668–2680.

75 Lee, E.Y., Cho, K.S., and Ryu, H.W. (2005). Simultaneous removal of H_2S and NH_3 in biofilter inoculated with Acidithiobacillus thiooxidans TAS. *J. Biosci. Bioeng.* 99: 611–615.

76 Georgiadis, A.G., Charisiou, N.D., and Goula, M.A. (2020). Removal of hydrogen sulfide from various industrial gases: a review of the most promising adsorbing materials. *Catalysts* 10: 521.

77 Soreanu, G., Beland, M., Falletta, P. et al. (2011). Approaches concerning siloxane removal from biogas – a review. *Can. Biosyst. Eng.* 53: 8.1–8.18.

78 Maxwell, G.R. (2004). *Synthetic Nitrogen Products: A Practical Guide to the Products and Processes.* Academic/Plenum Publishers.

79 Schweigkofler, M. and Niessner, R. (2001). Removal of siloxanes in biogases. *J. Hazard. Mater.* 83: 183–196.

80 Yang, C., Yang, S., Fan, H. et al. (2019). Tuning the ZnO-activated carbon interaction through nitrogen modification for enhancing the H_2S removal capacity. *J. Colloid Interface Sci.* 555: 548–557.

81 Bazan-Wozniak, A., Nowicki, P., and Pietrzak, R. (2017). The influence of activation procedure on the physicochemical and sorption properties of activated carbons prepared from pistachio nutshells for removal of NO_2/H_2S gases and dyes. *J. Cleaner Prod.* 152: 211–222.

82 Balsamo, M., Cimino, S., de Falco, G. et al. (2016). ZnO–CuO supported on activated carbon for H_2S removal at room temperature. *Chem. Eng. J.* 304: 399–407.

83 Bak, C., Lim, C.J., Lee, J.G. et al. (2019). Removal of sulfur compounds and siloxanes by physical and chemical sorption. *Sep. Purif. Technol.* 209: 542–549.

84 Tran, V.T.L.T., Gelin, P., Ferronato, C. et al. (2019). Adsorption of linear and cyclic siloxanes on activated carbons for biogas purification: sorbents regenerability. *Chem. Eng. J.* 378: 122152.

85 Pandey, P. and Chautan, R.S. (2001). Membranes for gas separation. *Prog. Polym. Sci.* 26: 853–893.

86 Boada, E., Santos-Clotas, E., Bertran, S. et al. (2020). Potential use of Methylibium sp. as a biodegradation tool in organosilicon and volatile compounds removal for biogas upgrading. *Chemosphere* 240: 124908.

87 Santos-Clotas, E., Cabrera-Codony, A., Boada, E. et al. (2019). Efficient removal of siloxanes and volatile organic compounds from sewage biogas by an anoxic biotrickling filter supplemented with activated carbon. *Bioresour. Technol.* 294: 122136.

88 Nielsen, A.M., Christensen, K.V., and Møller, H.B. (2013). Inline NH_3 removal from biogas digesters. *Biomass Bioenergy* 50: 10–18.

89 Kapoor, R., Subbarao, P.M.V., and Vijay, V.K. (2019). Integration of flash vessel in water scrubbing biogas upgrading system for maximum methane recovery. *Bioresource Technol. Rep.* 7: 100251.

90 Budzianowski, W.M., Wylock, C.E., and Marciniak, P.A. (2017). Power requirements of biogas upgrading by water scrubbing and biomethane compression: comparative analysis of various plant configurations. *Energy Convers. Manage.* 141: 2–19.

91 Capra, F., Fettarappa, F., Magli, F. et al. (2018). Biogas upgrading by amine scrubbing: solvent comparison between MDEA and MDEA/MEA blend. *Energy Procedia* 148: 970–977.

92 Magli, F., Capra, F., Gatti, M., and Martelli, E. (2018). Process selection, modelling and optimization of a water scrubbing process for energy-self-sufficient biogas upgrading plants. *Sustainable Energy Technol. Assess.* 27: 63–73.

93 Jamal, A., Meisen, A., and Lim, C.J. (2006). Kinetics of carbon dioxide absorption and desorption in aqueous alkanolamine solutions using a novel hemispherical contactor. I. Experimental apparatus and mathematical modeling. *Chem. Eng. Sci.* 61: 6571–6589.

94 Kismurtono, M. (2011). Upgrade biogas purification in packed column with chemical absorption of CO_2 for energy alternative of small industry (UKM-Tahu). *Int. J. Eng. Technol.* 11: 59–62.

95 Privalova, E., Rasi, S., Maki-Arvela, P. et al. (2013). CO_2 capture from biogas: adsorbent selection. *RSC Adv.* 3: 2979–2994.

96 Abry, R.G. and DuPart, R.S. (1995). Amine plant troubleshooting and optimization. *Hydrocarbon Proc.* 74: 41–50.

97 Grande, C.A. and Rodrigues, A.E. (2007). Biogas to fuel by vacuum pressure swing adsorption. I. Behavior of equilibrium and kinetic-based adsorbents. *Ind. Eng. Chem. Res.* 46: 4595–4605.

98 Cavenati, S., Grande, C.A., and Rodriguez, A.E. (2005). Upgrade of methane from landfill gas by pressure swing adsorption. *Energy Fuels* 19: 2545–2555.

99 Ho, M.T., Allinson, G.W., and Wiley, D.E. (2008). Reducing the cost of CO_2 capture from flue gases using pressure swing adsorption. *Ind. Eng. Chem. Res.* 47: 4883–4890.

100 Hjuler, K. and Aryal, N. (2017). *Review on Biogas Upgrading, FutureGas Project Report*. Danish Gas Technology Centre.

101 Zhou, W.H., Guo, J.P., and Tan, H.Y. (2011). Upgrading of methane from biogas by pressure swing adsorption. *Adv. Mater. Res.* 236–238: 268–271.

102 Mason, J.A., Sumida, K., Herm, Z.R. et al. (2011). Evaluating metal-organic frameworks for post-combustion carbon dioxide capture via temperature swing adsorption. *Energy Environ. Sci.* 4: 3030–3040.

103 Moon, S.H. and Shim, J.W. (2006). A novel process for CO_2/CH_4 gas separation on activated carbon fibers-electric swing adsorption. *J. Colloid Interface Sci.* 298: 523–523.

104 Grande, C.A., Ribeiro, R.P.L., Oliveira, E.L.G., and Rodriguez, A.E. (2009). Electric swing adsorption as emerging CO_2 capture technique. *Energy Procedia* 1: 1219–1225.

105 An, H., Feng, B., and Su, S. (2011). CO_2 capture by electrothermal swing adsorption with activated carbon fibre materials. *Int. J. Greenhouse Gas Control* 5: 16–25.

106 Scholz, M., Melin, T., and Wessling, M. (2013). Transforming biogas into biomethane using membrane technology. *Renewable Sustainable Energy Rev.* 17: 199–212.

107 Harasimowicz, M., Orluk, P., Zakrewska-Tznadel, G., and Chmielewsi, A.G. (2007). Application of polyimide membranes for biogas purification and enrichment. *J. Hazard. Mater.* 144: 698–702.

108 Jiang, L.Y., Chung, T.S., and Kulpathipanya, S. (2006). An investigation to revitalize the separation performance of hollow fibers with a thin mixed matrix composite skin for gas separation. *J. Membr. Sci.* 276: 113–125.

109 Makaruk, A., Miltner, M., and Harasek, M. (2010). Membrane biogas upgrading processes for the production of natural gas substitute. *Sep. Purif. Technol.* 74: 83–92

110 Bos, A., Punt, I.G.M., Wessling, M., and Strathmann, H. (1999). CO_2 - induced plasticization phenomena in glassy polymers. *J. Membr. Sci.* 155: 67–78.

8

CO_2 Conversion to Value-Added Gas-Phase Products: Technology Overview and Catalysts Selection

Qi Zhang[1], Laura Pastor-Pérez[1], Xiangping Zhang[2], Sai Gu[1], and Tomas R Reina[1]

[1] *University of Surrey, Department of Chemical and Process Engineering, Stag Hill, Guildford, GU2 7XH UK*
[2] *Chinese Academy of Sciences, Institute of Process Engineering, North second street, Zhongguancun, Beijing 100190 China*

8.1 Chapter Overview

Nowadays, CO_2 utilization reactions represent a research priority within the scientific community. There are two main reasons for this. Firstly, climate change became one of the most serious threats in the world, and the overmuch greenhouse gas emissions bear much responsibility [1]. As the principal greenhouse gas, the concentration of CO_2 in the atmosphere has consistently increased. Secondly, the demand for electricity storage is increasing because of the increasing shares of wind and solar power (power-to-gas, PtG), and CO_2 utilization as some chemicals, such as formic acid, methanol, and syngas, is a highly effective way to store energy produced by a renewable source. Therefore, research into CO_2 utilization processes has aroused a high degree of attention.

Among these CO_2 utilization processes, reducing CO_2 with H_2 as a co-reactant has been studied extensively. From the perspective of products, three different valuable molecules can be obtained by three different reactions: CH_4 through CO_2 methanation, CO via the reverse water gas shift (RWGS) reaction, and methanol through CO_2-selective hydrogenation. From the energetic point of view, methane is an excellent fuel, with higher volumetric energy content than hydrogen [2]. CO can be added to hydrogen (syngas) to form methanol or liquid hydrocarbons by Fischer–Tropsch synthesis (FTS). Then, the products of FT can be used as a fuel [3], and methanol can be used as one of the alternatives to fossil fuels and help decrease the pollutant emissions [4].

Another commonly used technology for CO_2 utilization is the dry reforming of methane (DRM). The product of this reaction is the valuable syngas (CO and H_2), which can be used as feedstocks for FT to synthesize liquid fuels and valuable chemicals [5]. When an FT application is envisaged, the H_2/CO ratio of syngas is one of the most important factors affecting the yield toward final products. Hence, it is significant to control it. For DRM, CO_2 is used as an oxidant, and the stoichiometric

Engineering Solutions for CO_2 Conversion, First Edition.
Edited by Tomas R. Reina, José A. Odriozola, and Harvey Arellano-Garcia.

ratio of H_2/CO (syngas) produced by this reaction is 1. However, the H_2/CO ratio of syngas is normally required to be 2 in FTS reaction. In order to get more H_2-rich syngas, researchers combined DRM with the steam reforming of methane (SRM), known as the bi-reforming of methane (BRM), reaching stoichiometric syngas with a H_2/CO ratio of 2 [6].

For the above reactions, the stable character of CO_2 hindered a large-scale industrial application for CO_2 utilization. In order to overcome this problem, many efforts have been made to develop highly active, selective, and stable catalysts. In this chapter, an overview of gas-phase CO_2 conversion processes including CO_2 methanation, RWGS reaction, and reforming of methane is presented. We discuss these processes, the catalysts applied in these reactions, their applicability, limitations, and scope for further development. Hence, a better understanding can be given before researchers start to design a new catalyst for chemical CO_2 recycling in value-added gas-phase products.

8.2 CO_2 Methanation

8.2.1 Background

The chemical methanation of CO_2 is considered to be a particularly promising reaction for the production of substitute or synthetic natural gas (SNG) [7]. The reaction is in fact a well-known process discovered in 1902 by Paul Sabatier and Jean–Baptiste Senderens, which has regained attention because of its potential impact as a straightforward CO_2 conversion route [8].

In recent decades, the CO_2 methanation is widely used in environmental protection industry and aerospace field. Among these applications, the combination of CO_2 methanation with water electrolysis is the most common one, called PtG technology. Hashimoto et al. [9] proposed it as a possibility to recycle CO_2 in the 1980–1990s for the sake of environmental protection. In 2009, this idea was revived by Sterner [10], again for the prevention of climate change and the storage of electric energy. Now, according to the objective set by the European Commission, at least 20% of the EU's final energy consumption should be renewable energy sources before 2020 [11]. Therefore, wind and solar energy are considered as necessary sources of energy. However, wind and solar energy are fluctuating and intermittent and must be balanced for electric grid stability purposes. Based on these features, the PtG technology can be an effective way to solve the problem by converting the surplus power to grid-compatible gas such as methane. Overall, the process can be divided into two separate steps: H_2 production by water electrolysis and H_2 conversion with an external CO or CO_2 source to CH_4 via methanation (see Figure 8.1).

In this context, the CO_2 methanation represents a versatile route to convert CO_2 from industrial flue gases into value-added methane, which can be used for heat and power applications. Beyond conventional applications in the chemical and industrial sector, the National Aeronautics Space Administration (NASA) is interested on implementing the Sabatier reaction for future manned space expeditions

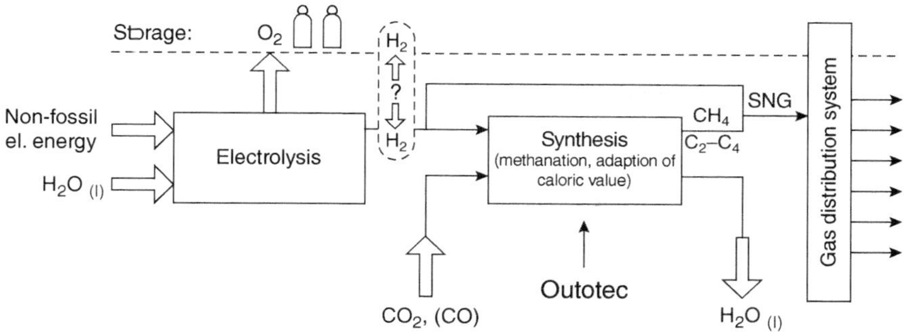

Figure 8.1 Concept for the storage of renewable energy in gas distribution system. Source: Adapted from Poleunis et al. [12].

to Mars. Bringing terrene hydrogen to Mars will make it possible to convert the Martian carbon dioxide atmosphere into methane and water for fuel and astronaut life support systems [13].

8.2.2 Fundamentals

The chemical methanation of carbon dioxide is a gas-phase catalytic process (Eq. (8.1)).

$$CO_2 + 4H_2 \longleftrightarrow CH_4 + 2H_2O \quad \Delta H = -165\ \text{kJ mol}^{-1} \tag{8.1}$$

The reaction is highly exothermic, typically operating between 200 and 450 °C, depending on the catalyst and experimental conditions [2, 14]. Although the CO_2 methanation is thermodynamically favorable, it is still hard to achieve because of the high kinetic barriers of the eight-electron reduction process [2]. A lot of studies have been done with the aim to figure out what is the mechanism of CO_2 methanation. However, there is still no general consensus on the mechanism of the reaction, and it can be classified into two categories. The main difference between these two categories is the intermediate compound involved in the rate-determining step [15]. The first path considers CO as the intermediate and the second one involves the formation of carbonate, formate, or methanol species (not CO) as the main intermediates during the reaction. The mechanism taking place is mainly caused by different experimental operation conditions or model catalysts implemented.

It is generally admitted that CO is the most important intermediate during CO_2 methanation. It can be described as a two-step mechanism. In the first step, carbon dioxide and hydrogen are converted to carbon monoxide and water via the RWGS reaction (Eq. (8.2)).

$$CO_2 + H_2 \longleftrightarrow CO + H_2O \quad \Delta H = 41\ \text{kJ mol}^{-1} \tag{8.2}$$

In the subsequent reaction (Eq. (8.3)), methane is formed from carbon monoxide and hydrogen:

$$CO + 3H_2 \longleftrightarrow CH_4 + H_2O \quad \Delta H = -206\ \text{kJ mol}^{-1} \tag{8.3}$$

This CO mechanism can be normally applied to noble metal-based catalysts. CO_2 is dissociated into CO, and after that, the formed CO would be dissociated into C and O atoms on the metal sites and further hydrogenated into methane by dissociated H_2 still on the metal particles [15, 16].

Another well-accepted alternative mechanism assumes that formate species are the main intermediates. This is in fact supported by kinetics studies [17] or density functional theory (DFT) calculations [18]. However, it can also be evidenced by some experiments. For example, Karelovic and Ruiz showed that CO was never detected in the gas phase in the whole methanation process using the Rh/TiO_2 catalyst [19].

8.2.3 Catalysts

In the past decades, a plethora of works dealing with CO_2 methanation have been reported, aiming at the designing of promising catalysts. Typical methanation catalysts consist of metal nanoparticles (NPs) supported on the carrier materials. In this chapter, we comprehensively discussed the recent progress in these metal-based heterogeneous catalysts and their performances in CO_2 methanation. It has been proven that many metals of group VIIIB in the periodic table are active for the CO_2 methanation [19–21]. Graf et al. ranked the activity for CO_2 methanation of the metals as follows: Ru > Fe > Ni > Co > Rh > Pd > Pt > Ir [22, 23]. Among these catalysts, the most active one (Ru) and the most popular choice (Ni) have been chosen as representative examples for this overview chapter.

8.2.3.1 Ruthenium-Based Catalysts

As can be seen from the rank above, Ru is one of the most highly active phases for CO_2 methanation. Ru outstanding performances in this reaction are not only in terms of CO_2 conversion but also in terms of high CH_4 selectivity and stability.

However, these positive characters are highly dependent on the dispersion of the metallic phase, the type of the support, the morphology of the catalyst, and the addition of promoters that can chemically interact with the metal [24, 25]. Furthermore, the activity of a catalyst can be influenced by the support used. Selecting the right support material is thus an important factor for efficient methanation of CO_2 [22]. Regarding this, Ru catalysts normally have been supported principally on a number of oxide materials, such as Al_2O_3, TiO_2, SiO_2, MgO, CeO_2, and ZrO_2 [26, 27].

Among these support materials, TiO_2 has been demonstrated to be a promising choice. Ravindranathan Thampi et al. reported that Ru NP-loaded TiO_2 prepared by a wet process was effective in lowering the reaction temperature, with the reaction starting from room temperature at atmospheric pressure [28, 29]. However, the reproducibility of these samples under the same catalytic conditions is not easy because of the many steps and parameter to control in the wet process [18]. In order to solve the problem, a method, called a barrel-sputtering route, was developed to boost Ru NPs dispersion on the surface of the TiO_2 support. In addition, this method is also helpful to control the NP size and their deposition density on the support [30]. Besides, the dispersion of Ru on the surface of the catalyst is significantly influenced by the TiO_2 crystal-phase structure. Lin et al. investigated the different performances of CO_2 methanation over Ru/rutile-type TiO_2 (r-TiO_2) and Ru/anatase-type TiO_2 (a-TiO_2) [31]. It turns out that r-TiO_2 appears to be a promising support for the preparation of a highly dispersed Ru catalyst with a narrow size distribution. It should be attributed to the strong interaction between RuO_2 and r-TiO_2 during the calcination, which prohibited the aggregation of RuO_2.

According to previous reports, CeO_2 is also a promising support for Ru [32–34] because of the abundant oxygen vacancies in CeO_2, which can increase the reaction rate by adsorbing and activating the carbon–oxygen bond. Sharma et al. investigated the conversion and selectivity of CO_2 methanation using catalysts consisting of CeO_2 doped with Ni, or Co, or Pd, or Ru. Among these, the Ru-doped ceria was the most active and selective one. 55% of CO_2 was converted and the selectivity of methane reached was 99% at 450 °C, indicating a very promising behavior [33].

Al_2O_3 is also a common support used in this reaction. A number of works show good performance using Ru/Al_2O_3 catalysts. Garbarino et al. compared the CO_2 methanation performance on 3% Ru/Al_2O_3 and 20% Ni/Al_2O_3. The experiment was conducted at 300 °C at 1500 h^{-1} gas hourly space velocity (GHSV) in excess of hydrogen. 96% methane yield has been achieved without any CO coproduction over 3% Ru/Al_2O_3 and only 80% methane yield with some CO coproduction was obtained over 20% Ni/Al_2O_3. The results confirmed that the catalytic performance of 3% Ru/Al_2O_3 catalyst outperformed the reference Ni/Al_2O_3 catalyst [35].

Recently, a highly active Ru/ZrO_2 has been synthesized by using Ru impregnated on UiO-66 (one zirconium terephthalate metal–organic framework [MOF] material). During the process of Ru/UiO-66 activation, the MOF structure collapsed at 330 °C and an amorphous phase appeared. This disordered phase, caused by the loss of the organic linkers, crystallized into the ZrO_2 tetragonal and monoclinic phases. The nanocrystalline ZrO_2 presents an excess of oxygen vacancies that are responsible for the adsorption of CO_2 in the methanation process. Besides, the high level of Ru NP dispersion in the final catalyst should be attributed to the uniform dispersion of Ru in the precursor MOF. After the activation process, Ru migrated to the surface of ZrO_2 and formed evenly dispersed Ru NPs. Therefore, an ideal CO_2 conversion (96–98%) was displayed when the CO_2 methanation operated under high gas flow rates over this catalyst. In addition, the Ru/ZrO_2 catalyst showed remarkable stability and high selectivity to methane (99%) for more than 160 hours of testing [36].

8.2.3.2 Nickel-Based Catalysts

Although Ru catalysts have been extensively studied because of their high performance in the methanation reaction, the high cost hinders the application for these materials [7]. Nickel has an extraordinary activity and high efficiency in CH_4 production. Also, it has a comparatively low price, which is another reason for Ni to become the most investigated material for CO_2 methanation.

Even though unsupported Ni NPs are active for CO_2 methanation [37], supports still play critical roles for this active phase by stabilizing the Ni particles as well as enhancing the adsorption of the key reactants [38]. For this reason, in order to develop highly effective Ni-based catalysts, many supports have been investigated. Among these materials, Al_2O_3, SiO_2, ZrO_2, CeO_2, and some MOF supports are the most commonly used supports for Ni-based catalysts in this reaction. In this section, we will summarize how these supports work when they interact with the Ni NPs.

Common supports for methanation catalysts are metal oxides with a large surface area. Among these supports, γ-Al_2O_3 is one of the most used ones for CO_2 methanation because of its high surface area, developed pore structure, and well-characterized surface acid–base properties [39]. Riani and coworkers compared the catalytic behavior of Ni NPs in the methanation of CO_2 with that of high loading (125% Ni $wt_{Ni}/wt_{support}$) Ni/Al_2O_3. As a result, Ni/Al_2O_3 showed far better activity than Ni NPs. Because both of these two catalysts containing cubic Ni, the CO_2 methanation in large part occurs on the Al_2O_3 support becomes a possible explanation. The support played a significant role in participating in the carbon dioxide adsorption step [37]. Rahmani et al. compared the influence of different Ni contents (10–25 wt%) in Ni/Al_2O_3 catalysts for several aspects. They found that the catalyst with 20 wt% of Ni possessed high activity and stability in CO_2 methanation reaction. For high Ni-loadings (i.e. higher than 20%), further increase of Ni content decreased the Ni dispersion on the surface of the catalyst because of the bigger crystallite size and thus the activity of the sample is negatively affected [40]. Also, different promoters have been used for improving the performance of Ni-supported catalysts. Guilera et al. explored the influence on Ni/Al_2O_3 when different metal oxide promoters (e.g. CeO_2, La_2O_3, Sm_2O_3, Y_2O_3, and ZrO_2) were added into the catalyst. The result shows that the addition of all these metal oxide promoters is beneficial, and Ni–La_2O_3/Al_2O_3 showed the highest catalytic activity. The enhancement of the interaction between CO_2 and catalysts is the reason why all these oxide promoters are beneficial to the reaction, which indicates that the amount of CO_2 adsorbed on the promoted catalysts was clearly higher than that of the unpromoted one. For Ni–La_2O_3/Al_2O_3, the highest catalytic performance was related to the significant increase of nickel reducibility and dispersion, more nickel active sites were available for the reaction [41]. Also, plasma technology can be applied to the preparation of Ni/γ-Al_2O_3. Guo et al. found that the conversion in CO_2 methanation toward plasma-treated Ni/γ-Al_2O_3 reached 84.6% at 250 °C, which was 27.2% higher than that exhibited by an untreated catalyst. This result can be attributed to the enlarged surface area of the sample, the formation of small size crystals, and the high dispersion and enrichment of the Ni element on the catalyst surface, which are promoted by plasma treatment [42].

Alongside with the Al_2O_3 support, silicon supports are also commonly studied materials for the Ni-based catalysts used in CO_2 methanation. There are mainly two categories: SiO_2 and SiC. As the supports, they both can play roles in dispersing the active component, improving the stability of the active-phase NPs by providing strong metal–support interaction. However, considering the difference between these two supports, for SiC, it shows higher thermal conductivity than SiO_2. Because CO_2 methanation is a highly exothermic reaction, this character is beneficial to avoid hot spots in the catalyst bed or a significant radial temperature gradient in the reactor. However, in view of CO_2 adsorption, the adsorption of CO_2 on the Ni–SiO_2 catalysts is stronger than that on the Ni–SiC catalysts [43]. On the whole, Le et al. compared the performance between SiC and SiO_2 as supports in the experiment and it turned out that Ni/SiC exhibited comparable catalytic activity to Ni/SiO_2 in CO_2 methanation. Also, they pointed out that the catalyst preparation method exerts an important effect on catalytic activity. Ni/SiO_2 and Ni/SiC catalysts in this research have been prepared both by wet impregnation (WI) and deposition–precipitation (DP), and H_2-temperature-programmed reduction analysis revealed that the catalysts prepared by DP exhibited stronger interaction between nickel oxide and supports [44]. Zhang et al. found that the reactivity of Ni/SiO_2 catalyst and the dispersion of the active component on Ni/SiO_2 catalyst can be remarkably improved by the treatment with nitrogen plasma and then hydrogen plasma and next combined with calcinations at 350 °C. In the long-term tests of 100 hours, the conversion of CO_2 over the plasma-treated catalyst only decreased by 15.54% or less. Meanwhile, the conversion of CO_2 toward conventional catalysts decreased by 32.05%. This phenomenon can be attributed to the enhanced number of active centers on the surface of catalysts by the plasma treatment. An enhancement of active sites will promote the dissociation of CO_2 and the conversion in the reaction [45]. Mesostructured silica nanoparticle (MSN) is another silicon support of interest because of its distinctive properties such as nanosize, extremely high surface area (>1000 $m^2\,g^{-1}$), large pore volume, and well-defined and tunable pore size (1.5–10 nm). Aziz et al. loaded Ni on MSN for the CO_2 methanation by the sol–gel and impregnation methods. They compared the activity of CO_2 methanation toward different Ni catalysts and the results followed the order Ni/MSN > Ni/MCM-41 > Ni/HY > Ni/SiO_2 > Ni/r-Al_2O_3. This indicates that MSN possesses high potential as a support for Ni catalysts in the methanation reaction [46].

In order to develop highly dispersed Ni-supported catalysts, many efforts have been devoted to finding some supports that can incorporate Ni into the framework of the support itself. Inside the framework, Ni can maintain a high dispersion on the surface without serious aggregation because of the surface anchoring effect. Hence, a highly stable physical stability can be achieved. MCM-41 is proved to be one of the promising supports for this target. Du et al. compared Ni-incorporated MCM-41 catalysts with different amounts of Ni (1% and 3%). As a result, the increased Ni loading in Ni–MCM-41 catalysts enhanced both the CO_2 conversion and the CH_4 selectivity because of the increased Ni active sites. High selectivity (96.0%) and space–time yield (STY, 91.4 $g\,kg^{-1}\,h^{-1}$) were achieved with 3 wt% Ni/MCM-41 at a reaction temperature as low as 300 °C, superior to those of Ni/SiO_2 catalysts, for the following reason: metallic Ni clusters are the active sites; hence, it is necessary to do a reduction before methanation. By reducing the Ni^{2+} to Ni^0 in situ, allowing Ni migration from the framework of the MCM-41 to the pore wall surface, in this way, Ni catalyst with a better dispersion of active sites than the Ni/SiO_2 catalyst likely can be obtained by anchoring Ni to the partially reduced Ni ions on the surface. Besides, the Ni–MCM-41 catalysts retained a highly ordered hexagonal structure after seven hours of stability test without serious aggregation [47].

As it is one of the supports that can effectively promote the activation of CO_2, CeO_2 also drew special attention as an active support for the methanation process. Zhou et al. prepared mesoporous Ni/CeO_2 catalyst by excess impregnation method [48]. CO_2 conversion and CH_4 selectivity reached 91% and 100% at 340 °C, respectively. This extraordinary result should be attributed to the advanced redox properties on the surface of the catalysts that can effectively activate CO_2 molecules to improve CO_2 methanation. L. Atzori et al. synthesized mesostructured NiO–CeO_2 mixed oxides with Ni loadings in the range 5–35 wt% by hard template procedure. Besides, two NiO–Al_2O_3 catalysts (15 and 35 wt% Ni loading) were synthesized for comparison. The testing result is that NiO–CeO_2 exhibited better performance, CO_2 conversions up to 76 and 93 mol% CH_4 selectivity values were achieved toward 35 wt% NiO–CeO_2 under atmospheric pressure, 300 °C, 72 000 $cm^3\,h^{-1}$ $g\,cat^{-1}$. However, for NiO–Al_2O_3 catalysts, CO_2 conversion was lower than 20 mol%. The authors proposed that because CeO_2 increases the reducibility of the NiO species, it also prevents the agglomeration of Ni^0 particles, which are formed during the reduction pretreatment. Under this situation, a high concentration of highly dispersed active sites on the support can be attained accounting for carbon dioxide adsorption and activation [49].

Ni supported on ZrO_2 is one of the most active systems for the CO_2 methanation because of its acidic/basic features and CO_2 adsorption abilities [20, 50]. Many different preparation methods have been used to synthesize Ni/ZrO_2 catalysts and some excellent results such as solution combustion method (SCM) [51], impregnation [20], coprecipitation [52], and hydrogel [53] have been obtained. Zhao et al. synthesized a series of Ni/ZrO_2 catalysts by combustion method with different combustion mediums (urea, glycerol, glycol, ethanol, and *n*-propanol). They pointed out that the type of combustion mediums significantly affected the catalyst structure and the subsequent CO_2 methanation catalytic performance. The highest efficient Ni/ZrO_2 catalyst was developed by using urea as the medium. 60% conversion of CO_2 was observed at 300 °C, 0.1 MPa, and a weight hourly space velocity (WHSV) of 4800 $ml\,g^{-1}\,h^{-1}$. These results can be attributed to higher reducibility and Ni dispersion. The smallest Ni particle size of the Ni/ZrO_2 catalyst was obtained when urea was used as the medium [51].

8.2.3.3 Rhodium and Palladium-Based Catalysts

Rhodium and palladium catalysts are also active in CO_2 methanation reaction. For rhodium, many solid supports have been tested in CO_2 methanation, such as Al_2O_3 [12], SiO_2 [54], MgO [55], or TiO_2. Solymosi et al. showed that among the investigated supports, the most effective one was TiO_2 and the least effective one was SiO_2; this result can be ascribed to the different extents of electronic interaction between Rh and the support, influencing the bonding and the reactivity of the chemisorbed species [56]. Palladium is an interesting choice because it can dissociate molecular hydrogen hence speeding up the reaction [7]. Kim et al. prepared Pd–MgO/SiO_2 catalysts for CO_2 methanation and demonstrated that the different roles of Pd and MgO played with. MgO initiates the reaction by binding a CO_2 molecule, and then Pd core dissociates the H_2 molecule and supplies H atoms to complete the reaction at the interface Pt–MgO [57].

8.3 RWGS Reaction

8.3.1 Background

The RWGS reaction was observed by Carl Bosch and Wil-helm Wild in 1914 when they intended to produce H_2 from steam and carbon monoxide on an iron oxide catalyst (water gas shift [WGS] reaction) [58]. It is an endothermic reaction that is favored at high temperatures. For more efficient results, the products are withdrawn to shift the equilibrium to the RWGS side rather than the forward side in some situations [59]. The RWGS reaction represents a direct route for CO_2 upgrading per se but it is also an important reaction because it is an essential intermediate step or competitive process of many vital CO_2 hydrogenation processes such as the CO_2 methanation of the methanol synthesis from CO_2 [60]. The product of RWGS reaction (CO) can be used in MeOH synthesis and downstream FT for chemicals and fuels [61]. For example, Park et al. explored the CAMERE process of producing methanol from CO_2, which included the intellectual RWGS process [62].

In recent years, many efforts have been made for optimum utilization of this reaction in industrial applications. Among them, the RWGS reaction with chemical looping cycles (RWGS-CL) for the intensified conversion of carbon dioxide is one of the highly effective ones (see from Figure 8.2). It is a combination of water electrolysis with RWGS reaction. For the RWGS part, CO_2 is captured from some excessive CO_2 emissions source or separated from the air. For the water electrolysis part, water is split by using renewable energy such as solar energy or wind energy and H_2 and O_2 are collected as products. Then, H_2 and CO_2 are converted to CO and water by RWGS reaction. Water can be recycled to the water electrolysis part and CO can be combined with additional H_2 for liquid fuel production via FT or methanol synthesis [3].

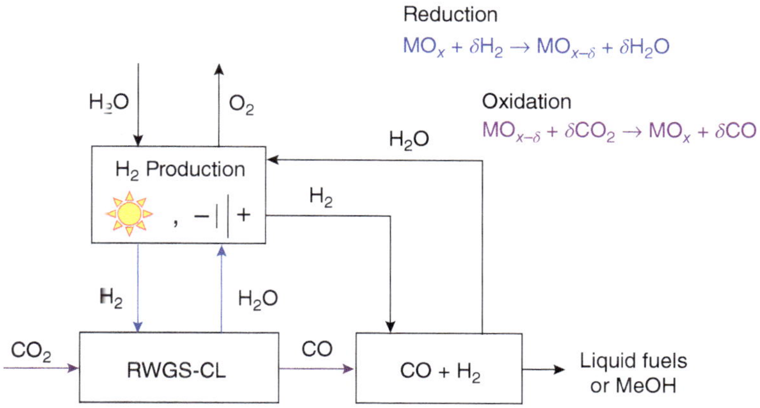

Figure 8.2 Schematic of the intensified reverse water-gas shift-chemical looping (RWGS-CL process. Source: Daza et al. [3].

8.3.2 Fundamentals

The RWGS reaction (Eq. (8.2)) is equilibrium limited and favored at high temperatures because of the endothermic nature of the reaction [63].

Additional side reactions include the Sabatier reaction (Eq. (8.1)) and CO methanation (Eq. (8.3)).

Because RWGS reaction is endothermic, when the reaction is carried out at lower temperatures, the equilibrium will increasingly favor the WGS (reverse of Eq. (8.2)), Sabatier reactions (Eq. (8.1)), and CO methanation (Eq. (8.3)).

According to the reaction stoichiometry, RWGS reaction should be successfully accomplished for H_2:CO_2 ratio of 1.0. However, when it comes to the experiment, the CO_2 conversion is favored at high H_2:CO_2 ratios. The fact that higher hydrogen concentrations favor the process could be related to adsorption capacity and hydrogen coverage on catalyst surface. The preferential adsorption of CO_2 on the surface could result in a CO_2-rich surface for hydrogen-poor reaction mixtures and therefore fewer chances for the reactants to interact. This situation is alleviated when the partial pressure of hydrogen is increased [64].

Regarding the mechanism of the RWGS reaction, there is no consensus. Currently, it is widely accepted that there are two main mechanisms for CO formation from this reaction. One is a redox mechanism, which has been supported by FTIR spectroscopy studies over a Cu/ZnO catalyst. This study suggests that CO_2 dissociates to CO by oxidizing Cu^0 to Cu^{1+}, and H_2 reduces Cu^{1+} to form H_2O [65]. Another commonly accepted mechanism is the formate decomposition mechanism. For this pathway, CO_2 is hydrogenated to formate firstly, followed by cleavage of the C=O bond [61].

From the literature, we can find that for some catalysts and reaction conditions, only one mechanism is taking place in the reaction pathway. For example, Loiland et al. found that the redox mechanism is the only pathway on Fe/Al_2O_3 catalysts. However, it is reported by many authors that one single mechanism does not always exist alone in a reaction, the combination of both mechanisms can be found in some examples. Loiland et al. showed that after adding potassium to Fe/Al_2O_3, K activates the secondary pathway for CO formation and the redox mechanism becomes the predominant one [66]. Chen et al. considered that only one mechanism cannot explain the experimental phenomenon, which was observed in a reaction using Cu/Al_2O_3 catalyst [67]. Clarke and Bell also indicated the coexistence of both mechanisms, redox and formate, for potassium-promoted Cu/SiO_2 catalysts [68].

8.3.3 Catalysts

As noted earlier, the RWGS reaction is favored at high temperatures. Normally, a huge amount of heat is needed to produce CO. Besides, some unwanted by-products can be produced during the reaction (Eqs. (8.1) and (8.3)). Hence, developing highly active, selective, and stable catalysts is important. Based on the proposed mechanisms, a highly effective catalyst should consist of a well-dispersed active metal and metal oxide support, which can participate in the reaction steps [69]. Concerning metal sites. copper and some noble metals (Pt, Pd, and Rh) have been studied extensively. For the support, CeO_2 has been one of the most widely used for this reaction because of its interesting redox properties. Some of the most outstanding studies will be described in this section both from the point of view of the metals and also the most used supports because of their favoring properties for this reaction.

8.3.3.1 Noble Metal-Based Catalysts

Pt-based catalysts are highly popular in this reaction because of their high hydrogenation activity. Their high activity can be attributed to the active site, which appeared in the interface; hence, the interaction between metal site and support is of importance. Kim et al. compared the conversion of CO_2 in RWGS reaction over Pt/TiO_2 and Pt/Al_2O_3 samples. They found that the catalytic activity of the RWGS reaction over these two catalysts was dependent on the type of support. Redox mechanism has been proved to be the mechanism of RWGS reaction toward Pt/TiO_2 and Pt/Al_2O_3. Oxidation and reduction happened on the Pt sites and reducible supports sites on the surface of the catalysts. In this case, TiO_2 is a reducible support and Al_2O_3 is an irreducible support. For $Pt–TiO_2$, new active sites ($Pt–Ov–Ti^{3+}$) located at the metal–support interface were found because of the reducible characteristics of the TiO_2 support. These sites can act as active sites and can also be oxidized or reduced easily by hydrogen atoms and electrons. However, because the Pt/Al_2O_3 catalyst cannot be oxidized or reduced on the Al_2O_3 support, the reaction can proceed through only Pt-carbonyl species. Therefore, $Pt–TiO_2$ catalyst had a higher CO_2 conversion rate than the Pt/Al_2O_3 catalyst. [70]. Porosoff et al. found that modifying $Pt–Mo_2C$ catalyst with Co can further improve the activity, selectivity, and stability of the catalyst because Co species can help maintain Mo_2C in carburized states that are the active states for RWGS reaction during the reaction. Besides, $Co–Mo_2C$ can dissociate CH_4 into H_2 and C, with amorphous $CoMoC_yO_z$ being identified as the critical active phase that dissociates CH_4 [71].

Rh and Pd are also active for RWGS reaction. Researchers have tried a different kind of supports for Rh catalysts, such as SiO_2 [72], Al_2O_3 [73], TiO_2 [74], and Ce_2O_3 [75]. Gogate and Davis tested Rh/TiO_2 at the nominal conditions of 20 atm, 270 °C, and a WHSV of 8000 $cm^3\,g\,cat^{-1}\,h^{-1}$ and got a 7.89% CO_2 conversion at this low temperature, highlighting that most of the products were methane (72.7%) and only 14.5% of products were CO [74]. Pettigrew and Cant synthesized Pd/Al_2O_3 and $Pd/Ce_2O_3/Al_2O_3$ and tested in RWGS reaction. The result shows that the rates of carbon dioxide conversion over ceria-containing catalysts were much faster than Pd/Al_2O_3. It can be explained by the interaction between noble metal and ceria. The role of Pd was to catalyze the reduction of ceria. Then, carbon dioxide can be expected to reoxidize the ceria to form carbon monoxide [75].

8.3.3.2 Copper-Based Catalysts

Despite the active and selective character of noble metals, their elevated cost imposes limitations for large-scale applications. In this sense, copper is another promising metal for developing a highly active and selective catalyst for RWGS reaction. It has comparatively low price than noble metals; moreover, the presence of Cu can make the process start at a low temperature (165 °C). This is because the number of metallic Cu sites on the surface of Cu catalysts can be oxidized to Cu^{1+} oxide at 165 °C; at the same time, CO_2 dissociated to CO [65]. Hence, the Cu catalysts have been studied extensively with many different supports and by incorporation of several promoters into the catalytic system.

Potassium is one of the most frequently used promoters for this reaction. Chen et al. synthesized the $Cu/K_2O/SiO_2$ catalyst offering a much better catalytic activity than Cu/SiO_2. The CO_2 conversion using this catalyst containing 1.9% K is 12.8% at 600 °C, which is more than doubled compared to that exhibited by the unpromoted Cu/SiO_2 (5.8%) at the same reaction condition [76]. This result can be ascribed to the new active sites, which were created at the interface between K and Cu. The main role of K_2O was its participation and good catalytic activity for formate decomposition, besides acting as a promoter for CO_2 adsorption. After that, the group explored the influence of Fe promoter for RWGS reaction in the same Cu/SiO_2 system. It turns out that the addition of a small amount of iron (0.3% Fe to 10% Cu/SiO_2) can improve the catalytic activity and stability of Cu/SiO_2 at high temperatures effectively. The good performance of $Cu–Fe/SiO_2$ catalyst can be assigned to the high copper surface area because of the addition of iron. The formation of small particles of iron species around Cu particles effectively prevents sintering of Cu at high temperatures [77].

The choice of an adequate support is another key factor to tune the performance of copper catalysts. Traditional Cu-oxide catalysts have a tendency to deactivate under operating condition because the copper particles are easy to aggregate at high temperature. Aiming to solve this problem, Zhang et al. developed a highly dispersed copper catalyst over β-Mo_2C as a stable catalyst for RWGS reaction. The conversion of CO_2 on 1% Cu/β-Mo_2C reached at 40% at 600 °C with a ratio of $H_2/CO_2 = 2$ and WHSV = 300 000 ml g^{-1} h^{-1}. This material also showed extraordinary stability in comparison with the commercial 36 wt% $Cu/ZnO/Al_2O_3$ catalyst. This phenomenon can be attributed to the Cu–Mo_2C interaction, which effectively anchors the Cu species, making the catalyst more stable [78].

8.3.3.3 Ceria-Based Support Catalysts

Ce-based support catalyst attracted a lot of attention because of its high oxygen storage capability and high intrinsic activity toward CO_2 adsorption.

Researchers combined some metals with the CeO_2 supports for getting outstanding results. Co is one of the most common metals combined with CeO_2. Wang and Liu prepared a series of Co–CeO_2 catalysts with different Co contents by a colloidal SCM and used for RWGS reaction. Characterizations demonstrated that the spherical pore wall was formed because of the intimate connection between small CeO_2 and Co_3O_4 NPs. Hence, the Co–CeO_2 catalysts had uniform mesoporous structures, high specific surface areas, and pore volumes. Among these synthesized catalysts, the mesoporous 5% Co–CeO_2 catalyst possessed high activity and selectivity in RWGS reaction, reaching 35% of CO_2 conversion and maintains 95.5% of the initial value after 10 hours at 600 °C [79]. In addition to Co, Ni is also an active metal site for CeO_2-based support catalyst in RWGS reaction. Lu and Kawamoto developed a series of NiO/CeO_2 catalysts with different amounts of NiO (1, 3, 4, and 5 wt%) and proved that the CO_2 conversion increased with increasing NiO amount, the 4 wt% NiO/CeO_2 has the highest activity for RWGS reaction. However, the CO selectivity decreased with the increasing NiO amount, 100% CO selectivity can be achieved with less than 3 wt% NiO because a comparatively large amount of NiO is likely to cause aggregation [69].

Samarium-doped ceria (SDC) support is another active support for RWGS. Panaritis et al. investigated the catalytic conversion of CO_2 over Ru–Fe NPs supported on the SDC support. Catalysts with different Ru_xFe_{100-x} compositions ($x = 100$, 80, 45, 20, and 0 at %) have been synthesized for the experiment. It turned out that $Ru_{45}Fe_{55}$/SDC displayed the overall best activity and CO selectivity among all the synthesized samples. CO yield peaked at 47.5% at 800 °C and CO selectivity reached at 100% above 500 °C. The great performance of this catalyst can be attributed to the metal–support interaction. Compared with $Ru_{45}Fe_{55}/CeO_2$, the CO_2 conversion in $Ru_{45}Fe_{55}$/SDC was lower when the temperature is lower than 500 °C but higher at high temperature (>500 °C) [80]. Zhou et al. investigated the effect of Ce/Cu ratio in the CeCu catalyst. They pointed out that the oxygen vacancies and active Cu^0 species were formed in the CeCu catalysts by the H_2 reduction at 400 °C. It can enhance the adsorption performance for the reactant CO_2 and H_2 of the related catalysts and then lead to extraordinary catalytic performance. Among these different Ce/Cu ratio catalysts, the highest CO_2 conversion and outstanding stability have been shown toward $Ce_{1.1}Cu_{1.0}$ [81].

8.3.3.4 Carbide Support Catalysts

In addition to the wide use of the metal oxide supports for this reaction, transition metal carbides (TMCs) seem to be appealing alternative because of their properties similar to precious metals [57]. Besides, they are capable of dispersing metallic particles remarkably [82]. Among these TMC materials, molybdenum carbide became the most popular one because of its low cost and dual functionality for H_2 dissociation and C=O bond scission [61]. Normally, it has a variety of crystal structures, but the most common types are β-MoC_y ($y = 0.5$) with a hexagonal closed packed structure and α-MoC_{1-x} ($x < 0.5$) with a face-centered cubic structure [83, 84]. Both of these two types are active for RWGS reaction. However, β-Mo_2C displays higher conversion of CO_2 because the Mo/C ratio of β-Mo_2C is higher than α-MoC_{1-x}. DFT calculation shows that the lower value of the Mo/C ratio, the less C—O bonds are cleaved (when the ratio is equal to 1, the MoC surface shows the only chemisorption of CO_2 molecule without cleavage of the C—O bonds) [85].

According to the results from Porosoff's group, Mo_2C itself can be an active catalyst for RWGS reaction [71]. However, in order to improve the performance of this catalyst, researchers are still developing metal-supported Mo_2C materials. Xu et al. prepared Cu/Mo_2C, Ni/Mo_2C, and Co/Mo_2C and compared the CO_2 conversion and CO selectivity in these catalysts. The CO selectivity decreases following the sequence: Cu/Mo_2C > β-Mo_2C > Ni/Mo_2C > Co/Mo_2C. On the contrary, the conversion of CO_2 on these four catalysts increased following the sequence: Cu/Mo_2C < β-Mo_2C < Ni/Mo_2C < Co/Mo_2C. With regard to this totally opposite trend, the author considers that there is a clear tendency toward the cleavage of both C—O bonds in the molecule when Ni/Mo_2C and Co/Mo_2C catalysts have been used in this reaction. Hence, the CO_2 conversion and CH_4 selectivity increased but the selectivity of CO decreased. The function of Cu in the carbide catalyst is to prevent the cleavage of both C—O bonds in the reactant molecule. Therefore, the CO selectivity toward Cu/Mo_2C is the highest compared to the rest three [86].

8.4 CO_2 Reforming Reactions

8.4.1 Background

With respect to reducing the content of greenhouse gases in the atmosphere, DRM is a very attractive route because both reactants of the reforming process are greenhouse gases and the product is syngas (CO and H_2), which is a flexible platform to produce liquid fuels and value-added chemicals via FTS [5]. As feedstocks for FT, the H_2/CO ratio of syngas is one of the most important factors that can influence the products of FT. Hence, it is significant to control it. Normally, the H_2/CO ratio of syngas is required to be 2 in FT, lower H_2/CO ratios can improve the selectivity toward long-chain hydrocarbons [6, 87].

The different product ratios of syngas are mainly dependent on the different kinds of methane reforming. There are several types of methane reforming: DRM (Eq. (8.4)), SRM (Eq. (8.5)), partial oxidation of methane (POM, Eq. (8.6)), BRM (Eq. (8.7)), and tri-reforming of methane (TRM, Eq. (8.8)). These reactions differ in the oxidant used, final H_2/CO product ratio, and the kinetics and energetics of the reaction [88].

Historically, DRM was first explored by Fischer and Tropsch using Ni and Co catalysts in 1928. Severe deactivation was observed because of the carbon deposition [89]. After that, many efforts have been made to solve this problem. SRM was introduced at industrial level in the 1930s. However, the reaction was not studied extensively at that time. An industry breakthrough of the steam reforming has been made in 1962 after ICI (imperial chemical industries) succeeded in starting two tubular reformers operating at pressure (15 bar) and using naphtha as a feedstock. This breakthrough significantly decreased the cost and energy consumption of downstream processes. Since then, research on SRM has undergone extensive development [90].

For industrial applications, DRM uses CO_2 as an oxidant and yields syngas with low H_2/CO ratio (maximum 1). In view of environmental protection, it is an effective way to solve the climatic issues because both of the two reactant gases (CO_2 and CH_4) are greenhouse gases. However, DRM can only produce syngas with low H_2/CO ratio and as stated previously, value ~2 is supposed to be a suitable H_2/CO ratio for FT. Meanwhile, SRM can produce a hydrogen-rich syngas with a H_2/CO ratio of about 3. Therefore, the combination of SRM and DRM, known as the BRM (Eq. (8.7)), is of interest for producing H_2-rich syngas using both CO_2 and CH_4 [91]. The combination of these two processes can be seen below (Figure 8.3).

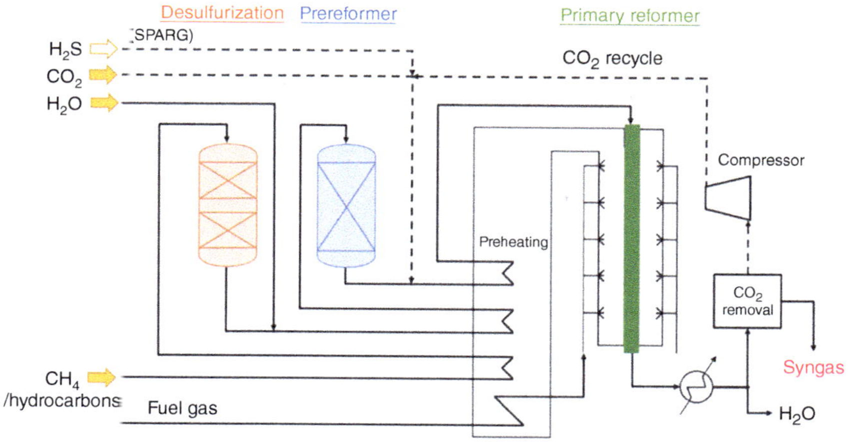

Figure 8.3 Flow sheet of a reforming process with optional configurations (dotted lines). Source: Adapted from Mortensen and Dybkjær [92].

8.4.2 Fundamentals

BRM (Eq. (8.7)) is an approach to produce syngas from CO_2 and methane among these CO_2-consuming reforming processes. Just as previous catalysts, BRM had some side reactions that may cause carbon deposition, such as Boudouard reaction (Eq. (8.10)). In terms of catalyst deactivation by carbonization, a significant advantage has been offered by BRM compared with DRM and SRM. Because the appearance of steam can oxidize the carbon (Eq. (8.9)), the resistance of deactivation is expected to be maximized; hence, the catalyst activity could be maintained over a practical period of time [93].

DRM (Eq. (8.4)) and SRM (Eq. (8.5)) are extremely endothermic reactions. Therefore, a large amount of heat is required for producing syngas with high selectivity [89]. In contrast, POM is an exothermic reaction (Eq. (8.6)).

$$CO_2 + CH_4 \longleftrightarrow 2CO + 2H_2 \quad \Delta H = 247.3\ \text{kJ mol}^{-1} \tag{8.4}$$

$$CH_4 + H_2O \longleftrightarrow CO + 3H_2 \quad \Delta H = 206.8\ \text{kJ mol}^{-1} \tag{8.5}$$

$$CH_4 + 1/2O_2 \longleftrightarrow CO + 2H_2 \quad \Delta H = -35.6\ \text{kJ mol}^{-1} \tag{8.6}$$

Hence, Tomishige et al. suggested combining POM with DRM and SRM, the combination is called TRM (Eq. (8.8)) [94]. For the drawbacks of TRM, the reaction rate of POM is extremely high, and it causes the high temperature near the inlet of the catalyst bed very soon [95]. Besides, the temperature becomes lower in the catalyst bed for methane reforming. This hot spot makes the whole process hard to control and increases the dangers of an explosion. However, the advantages of this tri-reforming combination are also superior; just like BRM, TRM can not only produce synthesis gas ($CO + H_2$) with desired H_2/CO ratios (1.5–2.0) but could also eliminate carbon formation, which is usually a serious problem in the CO_2 reforming of methane [96].

$$3CH_4 + 2CO_2 + H_2O \longleftrightarrow 4CO + 8H_2 \quad \Delta H = 660.9\ \text{kJ mol}^{-1} \tag{8.7}$$

$$20CH_4 + H_2O + 3O_2 + CO_2 \longleftrightarrow 21CO + 41H_2 \quad \Delta H = 12.9\ \text{kJ mol}^{-1} \tag{8.8}$$

$$C + H_2O \longleftrightarrow CO + H_2 \quad \Delta H = 131\ \text{kJ mol}^{-1} \tag{8.9}$$

$$2CO \longleftrightarrow CO_2 + C \quad \Delta H = -171\ \text{kJ mol}^{-1} \tag{8.10}$$

Considered that our project is about CO_2 upgrading, the CO_2-consuming process has been emphasized on discussing, which are DRM, BRM, and TRM.

8.4.3 Catalysts

As discussed previously, DRM (Eq. (8.4)) is an endothermic reaction; hence, it requires the reaction temperature reach at 900 °C to obtain high syngas yield [97]. However, the unstable character of catalysts hinders the accomplishment of the reaction because the metal site of the catalysts is likely to deactivate because of sintering or react with the support. Besides, carbon deposition induced by methane decomposition is also favored. Therefore, thermal stability is a significant character for DRM catalysts that needs to be considered [98]. For the BRM (Eq. (8.7)) and the TRM (Eq. (8.8)), their main advantage over DRM (Eq. (8.4)) is the stoichiometric H_2/CO ratio of 1.5–2. This is a more favorable syngas composition for most downstream operations, such as conversion to liquid fuels via FTS [99]. However, the same as the situation with DRM catalysts, one of the major challenges is to avoid catalyst deactivation via carbon deposition [100]. Many factors influence the activity of catalysts, such as the type of the metal, the support, the interaction between metal and support, surface area, and so on [101]. In order to improve catalysts' performance, scientists and researchers have been making efforts on developing some catalysts that have better stability and activity.

Focusing on DRM catalysts, most of the group VIII elements are identified as active for the dry reforming. Among them, noble metals, in particular, Rh [102] and Ru [103, 104], are highly active and have more resistance to carbon deposition.

8.4.3.1 Noble Metal-Based Catalysts

Rhodium dispersed on silica (Rh/SiO_2) and vanadia-promoted silica ($Rh/VO_x/SiO_2$) have been synthesized and studied by Sigl et al. It has been shown that the promotion of Rh/SiO_2 with vanadium oxide remarkably enhanced both the activity and turnover frequency by 15–20 fold in DRM reaction. This can be ascribed to the formation of a partial VO_x over layer on the Rh surface, which reduces the size of accessible ensembles of Rh atoms required for coke formation and creates new sites at the Rh–VO_x interfacial region that are considered to be active sites for the activation/dissociation of carbon dioxide [105]. Based on this research, Sarusi et al. investigated DRM reaction toward Rh/Al_2O_3 promoted with TiO_2 and V_2O_5. Before the measurement, a pretreatment process was applied. All the catalysts were oxidized at 400 °C in O_2 flow for 30 minutes and reduced at 500 °C in H_2 for 60 minutes. During the pretreating process, Rh reduced to metallic states, V_2O_5 significantly reduced to V^{4+}, and TiO_2 also slightly reduced below 327 °C. For the performances of catalysts, the conversion of CO_2 toward $Rh/V_2O_5–Al_2O_3$ (18.5%) and $Rh/TiO_2–Al_2O_3$ (19.4%) is higher than Rh/Al_2O_3 (13.5%) at a low temperature (500 °C) in DRM reaction. The increases can be attributed to the oxygen vacancies, which formed on the additives during the pretreatment and the reaction [106]. Apart from that, an active structure–reactivity relationship has been proved to be one of other factors that can influence the stability of catalysts. Liu et al. reported the high activity/stability of a ceria-supported Ru nanocluster (0.5 wt% Ru (NC)–CeO_2) catalyst, which was synthesized by the impregnation method. For comparison, a Ru NP catalyst (Ru (NP)–CeO_2) was also prepared. Less than 10% conversion decrease was observed for the Ru (NC)–CeO_2 catalyst at 700 °C and a high space velocity (180 000 ml g cat^{-1} h^{-1}) for 25 hours. In the beginning, Ru (NP)–CeO_2 exhibited similar activity (around 50% CO_2 conversion), but it decreased more than 40% after the period of time studied. The excellent stability of the Ru (NC)–CeO_2 catalyst should be assigned to thermally stable Ru clusters (<1 nm) and improved oxygen mobility produced by metal–support interactions, which deteriorated with an increase of the Ru particle size [107].

8.4.3.2 Ni-Based Catalysts

Noble metals are expensive, and considering their low cost and high activity, Ni-based catalysts have been studied extensively for DRM. However, severe carbon deposition that can lead to catalyst deactivation is still a major concern about Ni-based catalysts. To solve this problem, a second metal has been introduced to obtain a bimetallic Ni-based catalyst because it turns out that the interaction between some second metal and Ni can increase the carbon resistance of the catalyst [108]. Li et al. synthesized a series of Ni/Pt bimetallic catalysts (Ni/Al_2O_3, NiPt (0.18)/Al_2O_3, NiPt (0.66)/Al_2O_3, NiPt (0.71)/Al_2O_3, and Pt/Al_2O_3, where the values ~0.18, 0.66, and 0.71 refer to the molar ratio of the Pt atoms to the surface Ni atoms) with controlled metal surface composition and structure. The relationship between controlled surface segregation of Ni–Pt bimetallic catalysts with its carbon resistance has been investigated in this work. Dry reforming tests showed that the surface modification of the Ni-based catalysts by adding Pt atoms can effectively enhance the catalytic activities and resistance toward carbon formation. After 18 hours test at 700 °C, the CO_2 conversion toward Ni/Al_2O_3 decreased from 62.7% to 48%, whereas the CO_2 conversion toward NiPt (0.71)/Al_2O_3 increased from 90% to 91.2%. This is because that the structure of the bimetallic catalyst is evolved from Pt monolayer island-modified Ni NPs to core–shell bimetallic NPs composed of a Ni-rich core and a Ni/Pt alloy shell upon thermal treatment. The presence of Pt on the Ni surface facilitated the CH oxidation pathway and inhibited the carbon oxidation pathway [109]. Guharoy et al. used Sn as a promoter of Ni and then synthesized bimetallic catalysts with enhanced stability. The CO conversion toward $NiSn_{0.02}Al$ decreased by about 10% after 20 hours of stability test at 700 °C (from 75% to 65%). At the same condition, the CO_2 conversion toward the NiAl catalyst decreased more than 25%. The positive effect of Sn is credited to Sn atoms, occupying C nucleation sites in the vicinity on Ni atoms, slowing coke formation (i.e. increasing the energy barrier for coke nucleation) [110].

Material strategies have also been used to solve the deactivated problem of Ni catalysts. Le Sache et al. stabilized Ni in a pyrochlore–perovskite structure to prevent fast deactivation. Substitution of zirconium by Ni at different loadings in the lanthanum zirconate pyrochlore $La_2Zr_2O_7$ was investigated. For simplicity, the 0, 2, 5, and 10 wt% Ni-loaded catalysts have been referred to as LZ, LNZ2, LNZ5, and LNZ10, respectively. Among these synthesized catalysts, LNZ10 is found to be very active, even at low temperature, with CH_4 and CO_2 conversions of 53% and 69%, respectively, at 600 °C. Moreover, it displayed outstanding stability over a long-term stability test of 360 hours at 700 °C. This is because nickel was stabilized within a thermally stable mixed oxide structure to prevent Ni from sintering and therefore from forming large clusters that favor carbon formation [111]. Kaydouh et al. explored the performance of SBA-15 (hexagonal mesoporous silica material) catalysts containing 5 wt% Ni and 6 wt% Ce in DRM reaction. They pointed out that the order of Ni and Ce impregnation during the preparation is of significant influence. The catalyst prepared by impregnating cerium first and then nickel (Ni–Ce/SBA-15) is more active and stable than the one prepared by impregnating nickel first, cerium second (Ce–Ni/SBA-15), and the one only contains Ni (Ni/SBA-15). 90% conversion of CO_2 was achieved toward Ni–Ce/SBA-15 at 600 °C with a ratio of $H_2/CO_2 = 1$ and GHSV = 264 000 $ml^{-1}\,g^{-1}\,h^{-1}$. This extraordinary performance should be attributed to the smaller particle size obtained on the Ni–Ce/SBA-15 sample and to its higher reducibility. On the contrary, the impregnation of Ce after Ni caused additional blocking of the pores and formation of larger NiO particles, hindering the accessibility of the reactants to the active sites and lowering the activity in DRM reaction [112].

Because Ni-based catalysts have shown excellent activity toward DRM and SRM, they are still the most popular type of catalysts for BRM because of the active and selective character and less expensive price compared to noble metal. Concerning the support, alumina support is a classical choice because it can operate stably with high activity under excess of steam as a consequence of its excellent thermal and mechanical properties. Stroud et al. improved nickel-based catalysts (10 wt%) supported on alumina (Al_2O_3) by the addition of ceria and tin. The most appropriate Sn/Ni ratio has been explored and the alumina base has been promoted with 20 wt% of CeO_2. A value of 0.2 has been proved to be the optimum ratio of Sn/Ni of these Ni-based alumina catalysts. High conversion of methane toward $Sn_{0.02}Ni/Ce–Al$ was exhibited, remaining at a constant level of 70% during the 24 hour BRM test. The excellent performance was assigned to Sn atoms occupying C nucleation sites in the vicinity on Ni atoms, slowing coke formation mechanisms in the process. Meanwhile, the presence of ceria in the catalyst formulation enhances Ni dispersion [100]. Ni–Ce/$MgAl_2O_4$ catalyst was synthesized by the co-impregnation method by Koo et al. The Ce/Ni ratio in the catalysts has been adjusted in a range of 0.0–1.0. Among these different Ce/Ni ratio catalysts, Ni–Ce/$MgAl_2O_4$ (Ce/Ni = 0.25) catalyst shows the highest activity (CH_4 conversion: 81.3%, CO_2 conversion: 65.9% at 700 °C) and coke resistance because of the improvement of metal dispersion, excellent reducibility, and effective surface oxygen transfer [113].

As for the TRM, high CH_4 and CO_2 conversions can be achieved with desired H_2/CO ratios of 1.5–2.0 over supported nickel catalysts at 800–850 °C under atmospheric pressure without the problem of carbon formation on the catalyst. Among these nickel catalysts, their ability to enhance the conversion of CO_2 follows the order of Ni/MgO > Ni/MgO/CeZrO > Ni/CeO_2 ≈ Ni/ZrO_2 ≈ Ni/Al_2O_3 > Ni/CeZrO. Over 97%, CH_4 conversion and about 80% CO_2 conversion can be obtained in TRM toward Ni/MgO catalysts with the molar ratio of CH_4:CO_2:H_2O:O_2 = 1 : 0.48 : 0.54 : 0.1. The higher CO_2 conversion over Ni/MgO in tri-reforming can be attributed to the stronger interaction of CO_2 with MgO and a more available interface between Ni and MgO resulting from the formation of NiO/MgO solid solution [98]. However, for Ni–CeO_2 catalyst, the activity can be enhanced by an appropriate amount of La doping (10 at%), the CH_4 and CO_2 conversion increased from 93% and 83% to 96% and 86.5%, respectively. The activity improvement can be assigned to the synergic effect of nickel–lanthana–surface oxygen vacancies of ceria via interfacial active sites that contribute to increase the nickel dispersion and create a basic site distribution on the surface able to interact with CO_2 [114]. Majewski and Wood synthesized a nickel–silica core@shell catalyst and explored the optimal conditions of TRM toward this catalyst. They controlled the syngas H_2:CO molar ratios by changing the feedstock gas composition. Increasing the amount of oxygen in the feedstock improved methane conversion to 90% and reduced coke deposition. Decreasing steam partial pressure to zero reduced the H_2:CO molar ratio to value ~2 in produced syngas and increased CO_2 conversion to over 90% but decreased the methane conversion. From all the tested reaction conditions, the optimal for tri-reforming over the 11%Ni@SiO_2 catalyst was a feed composition with a molar ratio of CH_4:CO_2:H_2O:O_2:He = 1 : 0.5 : 0.5 : 0.1 : 0.4 at 750 °C. Over 70% CH_4 conversion, about 60% CO_2 conversion and the H_2:CO molar ratio of 2.5 were achieved [115].

8.4.3.3 Catalytic Supports

From the point of view of the support, a lot of studies have been carried out for exploring the mechanism of catalyst deactivation in order to conquer the carbon deposition problem [116, 117]. It is shown that there are many characters of supports that can influence the activity and stability of catalysts, such as acid–base property, morphology, promoters, and so on. Various supports have been investigated for the above metal sites, including SiO_2, La_2O_3, ZrO_2, TiO_2, CeO_2, Al_2O_3, and MgO. According to the acid–base properties, these supports have been divided into three categories: inert material (e.g. SiO_2), acid support (e.g. Al_2O_3), and base support (e.g. La_2O_3 and CeO_2). For catalyst supported on inert materials, both reactants (CO_2 and CH_4) are activated by the metal sites alone. For the acid- or base-supported catalysts, CH_4 is activated on the metal site and CO_2 is activated on supports [118, 119]. Based on this mechanism, on acidic supports, CO_2 is activated by formates and surface hydroxyls, and on basic supports, it is activated by forming oxy-carbonates. When the rate of methane decomposition and carbonate reduction are in balance, the catalytic activity remains stable. However, on inert supports, once carbon formation occurs by dehydrogenation of methane, subsequent activation of CO_2, and reaction with carbon is limited, leading to deactivation. Hence, inert supports such as SiO_2 have a comparatively lower stable character compared to acid and basic supports.

Morphology is also a crucial factor that can influence the performance of catalysts. He et al. compared different catalytic activities of 3 wt% Ru/CeO_2 in DRM toward three different morphologies of supports: nanorods (CeO_2-NRs), nanocubes (CeO_2-NCs), and nanopolyhedra (CeO_2-NPs) (the morphologies of supports and catalysts can be seen in Figure 8.4). The performance was tested at 500 °C with the GHSV of 24 000 $ml\,h^{-1}\,g^{-1}$ at CO_2:CH_4 = 1 ratio. The conversion trend of CO_2 is followed by Ru/CeO_2-NRs (26.5%) ≈ Ru/CeO_2-NCs (26.0%) > Ru/CeO_2-NPs (19.3%) after a 240 minutes reaction process. According to the results of characterization, only the (100) crystal facet exists in the CeO_2-NCs, and the (110), (111) crystal facets are the mainly exposed planes for CeO_2-NRs and CeO_2-NPs, respectively. The good performance of Ru/CeO_2-NRs and Ru/CeO_2-NCs can be attributed to the strong interaction between Ru species and supports exposed (110) and (100) facets, and the (110) and (100) facets on the CeO_2 with high oxygen vacancy concentration benefit the activation of CO_2 during the reaction [120]. The same thing happened in Ni/CeO_2-NR and Ni/CeO_2-NP. The Ni/CeO_2-NR catalysts also displayed better catalytic activity and higher coke resistance compared with the Ni/CeO_2-NP. The better performance of Ni/CeO_2-NR should be assigned to the predominant (110) and (100) planes on the CeO_2-NR. In comparison with the (111) planes on the CeO_2-NP, the

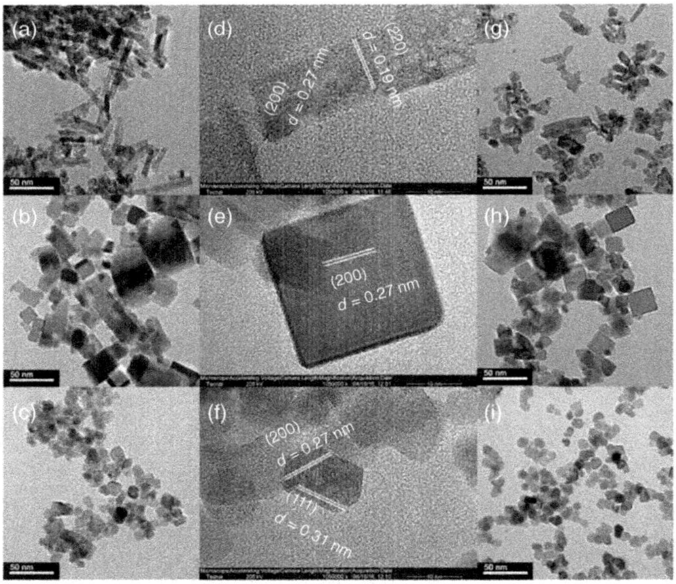

Figure 8.4 TEM images of (a) CeO_2-NRs, (b) CeO_2-NCs, and (c) CeO_2-NPs; high-resolution transmission electron microscopy (HRTEM) images of (d) CeO_2-NRs, (e) CeO_2-NCs, and (f) CeO_2-NPs; and TEM images of (g) Ru/CeO_2-NRs, (h) Ru/CeO_2-NCs, and (i) Ru/CeO_2-NPs. Source: He et al. [120].

(110) and (100) planes show great superiority for the anchoring of Ni NPs, which can be helpful to prevent sintering of Ni particles [48].

Additionally, the addition of suitable promoters for supports is another way to improve the performance of catalysts. Aghamohammadi et al. synthesized $Ni/Al_2O_3-CeO_2$ (10 wt% Ni and 10 wt% CeO_2 in support) and reported that CeO_2 addition as a support modifier to the Ni/Al_2O_3 catalysts improved catalytic activity and stability. The conversion of CO_2 on $Ni/Al_2O_3–CeO_2$ reached at 55% at 550 °C with a ratio of $H_2/CO_2 = 1$ and GHSV = 24 000 $ml^{-1}\,g^{-1}\,h^{-1}$. This is because CeO_2 as a good textual promoter, leading to a uniform particle size distribution [121].

8.5 Conclusions and Final Remarks

Gas-phase CO_2 conversion is a niche area of heterogeneous catalysis, which has been broadly explored and still retain the attention of the scientific community. A lot of effort has been made on CO_2 upgrading from both process and catalysis side. Plenty of highly active, selective, and stable catalysts have been developed. Noble metal catalysts are active for both methanation, RWGS reaction, and reforming of methane but their low availability and high cost limit their utilization at large scale. Hence, transition metal options are more appealing, and among them, nickel-based catalysts are the most investigated catalysts for these CO_2-valorization reactions. Carbon deposition and metallic sintering represent the major culprits of Ni-based catalysts in these processes. Nevertheless, multiple strategies have been developed to mitigate Ni deactivation. This include, for example, the use of promoters, bimetallic formulations, core–shell/yolk–shell architectures, or stabilization of Ni on inorganic structures, resulting in the design of robust catalysts for gas-phase CO_2 conversion.

Although the catalysis community remains ambitious targeting C_2 and C2+ products from CO_2, presently, there no commercial direct routes to produce C2+ products from CO_2. Hence, gas-phase CO_2 recycling via RWGS, CO_2 methanation, and reforming reactions represent an immediate solution to deal with flue gases in heavy carbon industries. As an additional advantage, most of the heavy CO_2 emitters (i.e. refineries, steel-making factories, or cement manufacturers) have the infrastructure in place to implement CO_2 gas-phase upgrading technologies, making this approach a realistic short-term solution to mitigate localized CO_2 emissions.

References

1 Dell'Amico, D.B., Calderazzo, F., Labella, L. et al. (2003). Converting carbon dioxide into carbamato derivatives. *Chem. Rev.* 103 (10): 3857–3898.

2 Su, X., Xu, J., Liang, B. et al. (2016). Catalytic carbon dioxide hydrogenation to methane: a review of recent studies. *J. Energy Chem.* 25 (4): 553–565.

3 Daza, Y.A., Kent, R.A., Yung, M.M., and Kuhn, J.N. (2014). Carbon dioxide conversion by reverse water–gas shift chemical looping on perovskite-type oxides. *Ind. Eng. Chem. Res.* 53 (14): 5828–5837.

4 Verhelst, S., Turner, J.W., Sileghem, L., and Vancoillie, J. (2019). Methanol as a fuel for internal combustion engines. *Prog. Energy Combust. Sci.* 70: 43–88.
5 Jahangiri, H., Bennett, J., Mahjoubi, P. et al. (2014). A review of advanced catalyst development for Fischer–Tropsch synthesis of hydrocarbons from biomass derived syn-gas. *Catal. Sci. Technol.* 4: 2210–2229.
6 Arora, S. and Prasad, R. (2016). An overview on dry reforming of methane: strategies to reduce carbonaceous deactivation of catalysts. *RSC Adv.* 6 (110): 108668–108688.
7 Frontera, P., Macario, A., Ferraro, M., and Antonucci, P. (2017). Supported catalysts for CO_2 methanation: a review. *Catalysts* 7 (12): 59.
8 Rönsch, S., Schneider, J., Matthischke, S. et al. (2016). Review on methanation – from fundamentals to current projects. *Fuel* 166: 276–296.
9 Hashimoto, K., Yamasaki, M., Fujimura, K. et al. (1999). Global CO_2 recycling – novel materials and prospect for prevention of global warming and abundant energy supply. *Mater. Sci. Eng., A* 267 (2): 200–206.
10 Sterner, M. (2009). Bioenergy and renewable power methane in integrated 100% renewable energy systems – limiting global warming by transforming energy systems. PhD thesis, University of Kassel.
11 Zervos, A., Lins, C., Tesniere, L., and Smith, E. (2011). Mapping renewable energy pathways towards 2020. *Eur. Renewable Energy Counc.* 28: 1–103.
12 Poleunis, C., Ruiz, P., Bertrand, P. et al. (2012). CO_2 methanation on Rh/γ-Al_2O_3 catalyst at low temperature: "in situ" supply of hydrogen by Ni/activated carbon catalyst. *Appl. Catal., B* 125: 41–50.
13 VanderWiel, D.P., Zilka-Marco, J.L., Wang, Y., et al. (2000). AIChe 2000, Spring National Meeting, Atlanta, GA, 5–9 March.
14 Schaaf, T., Grünig, J., Schuster, M.R. et al. (2014). Methanation of CO_2 storage of renewable energy in a gas distribution system. *Energy Sustain. Soc.* 4 (1): 1–14.
15 Aldana, P.A.U., Ocampo, F., Kobl, K. et al. (2013). Catalytic CO_2 valorization into CH_4 on Ni-based ceria-zirconia. Reaction mechanism by operando IR spectroscopy. *Catal. Today* 215: 201–207.
16 Muroyama, H., Tsuda, Y., Asakoshi, T. et al. (2016). Carbon dioxide methanation over Ni catalysts supported on various metal oxides. *J. Catal.* 343: 178–184.
17 Ibraeva, Z.A., Nekrasov, N.V., Gudkov, B.S. et al. (1991). Kinetics of methanation of carbon dioxide. *Sciences-New York* 26 (5): 620–624.
18 Akamaru, S., Shimazaki, T., Kubo, M., and Abe, T. (2014). Density functional theory analysis of methanation reaction of CO_2 on Ru nanoparticle supported on TiO_2 (101). *Appl. Catal., A* 470: 405–411.
19 Karelovic, A. and Ruiz, P. (2013). Mechanistic study of low temperature CO_2 methanation over Rh/TiO_2 catalysts. *J. Catal.* 301: 141–153.
20 Da Silva, D.C.D., Letichevsky, S., Borges, L.E.P., and Appel, L.G. (2012). The Ni/ZrO_2 catalyst and the methanation of CO and CO_2. *Int. J. Hydrogen Energy* 37 (11): 8923–8928.
21 Lunde, P.J. and Kester, F.L. (1973). Rates of methane formation from carbon dioxide and hydrogen over a ruthenium catalyst. *J. Catal.* 30 (3): 423–429.

22 Graf, F., Götz, M., Henel, M. et al. (2014). Technoöko-nomische Studie von Power-to-Gas-Konzepten. Final Report, Deutscher Verein des Gas- und Wasserfaches (DVGW), Bonn.
23 Ghaib, K., Nitz, K., and Ben-Fares, F.Z. (2016). Chemical methanation of CO_2: a review. *ChemBioEng Rev.* 3 (6): 266–275.
24 Garbarino, G., Bellotti, D., Finocchio, E. et al. (2016). Methanation of carbon dioxide on Ru/Al_2O_3: catalytic activity and infrared study. *Catal. Today* 277: 21–28.
25 Toemen, S., Abu Bakar, W.A.W., and Ali, R. (2016). Effect of ceria and strontia over $Ru/Mn/Al_2O_3$ catalyst: catalytic methanation, physicochemical and mechanistic studies. *J. CO_2 Util.* 13: 38–49.
26 Karpinski, Z., Kowalczyk, Z., Miśkiewicz, E. et al. (2008). Supported ruthenium catalysts for selective methanation of carbon oxides at very low CO_x/H_2 ratios. *Appl. Catal., A* 342 (1–2): 35–39.
27 Li, T., Wang, S., Gao, D., and Wang, S. (2015). Effect of support calcination temperature on the catalytic properties of $Ru/Ce_{0.8}Zr_{0.2}O_2$ for methanation of carbon dioxide. *J. Fuel Chem. Technol.* 42 (12): 1440–1446.
28 Ravindranathan Thampi, K., Kiwi, J., and Grätzel, M. (1987). Methanation and photo-methanation of carbon dioxide at room temperature. *Nature* 327: 506–508.
29 Liu, Y. and Liu, D. (1999). Study of bimetallic Cu–Ni/γ-Al_2O_3 catalysts for carbon dioxide hydrogenation. *Int. J. Hydrogen Energy* 24 (4): 351–354.
30 Abe, T., Tanizawa, M., Watanabe, K., and Taguchi, A. (2009). CO_2 methanation property of Ru nanoparticle-loaded TiO_2 prepared by a polygonal barrel-sputtering method. *Energy Environ. Sci.* 2 (3): 315–321.
31 Lin, Q., Liu, X.Y., Jiang, Y. et al. (2014). Crystal phase effects on the structure and performance of ruthenium nanoparticles for CO_2 hydrogenation. *Catal. Sci. Technol.* 4 (7): 2058–2063.
32 Wang, F., Li, C., Zhang, X. et al. (2015). Catalytic behavior of supported Ru nanoparticles on the {1 0 0},{1 1 0}, and {1 1 1} facet of CeO_2. *Catalyst* 329: 177–186.
33 Sharma, S., Hu, Z., Zhang, P. et al. (2011). CO_2 methanation on Ru-doped ceria. *Catalyst* 278: 297–309.
34 Wang, F., He, S., Chen, H. et al. (2016). Active site dependent reaction mechanism over Ru/CeO_2 catalyst toward CO_2 methanation. *J. Am. Chem. Soc.* 138 (19) 6298–6305.
35 Garbarino, G., Riani, P., Busca, G. et al. (2015). Methanation of carbon dioxide on Ru/Al_2O_3 and Ni/Al_2O_3 catalysts at atmospheric pressure: catalysts activation, behaviour and stability. *Int. J. Hydrogen Energy* 40 (30): 9171–9182.
36 Lippi, R., Howard, S.C., Barron, H. et al. (2017). Highly active catalyst for CO_2 methanation derived from a metal organic framework template. *J. Mater. Chem. A* 5 (25): 12990–12997.
37 Canepa, F., Riani, P., Busca, G. et al. (2013). Unsupported versus alumina-supported Ni nanoparticles as catalysts for steam/ethanol conversion and CO_2 methanation. *J. Mol. Catal. A: Chem.* 383–384: 10–16.

38 Le, T.A., Kim, M.S., Lee, S.H. et al. (2017). CO and CO_2 methanation over supported Ni catalysts. *Catal. Today* 293–294: 89–96.

39 Gao, J., Liu, Q., Gu, F. et al. (2015). Recent advances in methanation catalysts for the production of synthetic natural gas. *RSC Adv.* 5 (29): 22759–22776.

40 Rahmani, S., Rezaei, M., and Meshkani, F. (2014). Preparation of highly active nickel catalysts supported on mesoporous nanocrystalline γ-Al_2O_3 for CO_2 methanation. *J. Ind. Eng. Chem.* 20 (4): 1346–1352.

41 Guilera, J., Del Valle, J., Alarcón, A. et al. (2019). Metal-oxide promoted Ni/Al_2O_3 as CO_2 methanation micro-size catalysts. *J. CO_2 Util.* 30: 11–17.

42 Guo, F., Chu, W., Xu, H.Y., and Zhang, T. (2007). Glow discharge plasma-enhanced preparation of nickel-based catalyst for CO_2 methanation. *Chin. J. Catal.* 28 (5): 429–434.

43 Duong-Viet, C., Ba, H., El-Berrichi, Z. et al. (2016). Silicon carbide foam as a porous support platform for catalytic applications. *New J. Chem.* 40 (5): 4285–4299.

44 Le, T.A., Kang, J.K., and Park, E.D. (2018). CO and CO_2 methanation over Ni/SiC and Ni/SiO_2 catalysts. *Top. Catal.* 61 (15–17): 1537–1544.

45 Zhang, X., Sun, W., and Chu, W. (2013). Effect of glow discharge plasma treatment on the performance of Ni/SiO_2 catalyst in CO_2 methanation. *J. Fuel Chem. Technol.* 41 (1): 96–101.

46 Aziz, M.A.A., Taufiq-Yap, Y.H., Sazegar, M.R. et al. (2013). Highly active Ni-promoted mesostructured silica nanoparticles for CO_2 methanation. *Appl. Catal., B* 147: 359–368.

47 Du, G., Lim, S., Yang, Y. et al. (2007). Methanation of carbon dioxide on Ni-incorporated MCM-41 catalysts: the influence of catalyst pretreatment and study of steady-state reaction. *J. Catal.* 249 (2): 370–379.

48 Zhou, G., Liu, H., Cui, K. et al. (2016). Role of surface Ni and Ce species of Ni/CeO_2 catalyst in CO_2 methanation. *Appl. Surf. Sci.* 383: 248–252.

49 Atzori, L., Cutrufello, M.G., Meloni, D. et al. (2017). CO_2 methanation on hard-templated NiO–CeO_2 mixed oxides. *Int. J. Hydrogen Energy* 42 (32): 20689–20702.

50 Yamasaki, M., Habazaki, H., Asami, K. et al. (2006). Effect of tetragonal ZrO_2 on the catalytic activity of Ni/ZrO_2 catalyst prepared from amorphous Ni–Zr alloys. *Catal. Commun.* 7: 24–28.

51 Zhao, K., Wang, W., and Li, Z. (2016). Highly efficient Ni/ZrO_2 catalysts prepared via combustion method for CO_2 methanation. *J. CO_2 Util.* 16: 236–244.

52 Huang, Y.H., Wang, J.J., Liu, Z.M. et al. (2013). Highly efficient Ni-ZrO_2 catalyst doped with Yb_2O_3 for co-methanation of CO and CO_2. *Appl. Catal., A* 466: 300–306.

53 Liu, Q., Ren, J., Qin, Z. et al. (2013). Catalytic performance of the Ni/ZrO_2 catalysts prepared with the hydrogel method in methanation of CO_2. *Chem. J. Chin. Univ.* 34 (9): 2171–2177.

54 Martin, N.M., Hemmingsson, F., Wang, X. et al. (2018). Structure–function relationship during CO_2 methanation over Rh/Al_2O_3 and Rh/SiO_2 catalysts under atmospheric pressure conditions. *Catal. Sci. Technol.* 8 (10): 2686–2696.

55 Bacariza, M.C., Graça, I., Bebiano, S.S. et al. (2017). Magnesium as promoter of CO_2 methanation on Ni-based USY zeolites. *Energy Fuels* 31 (9): 9776–9789.

56 Solymosi, F., Erdöhelyi, A., and Bansagi, T. (1981). Methanation of CO_2 on supported rhodium catalyst. *J. Catal.* 68: 371–382.

57 Kim, H.Y., Lee, H.M., and Park, J.N. (2010). Bifunctional mechanism of CO_2 methanation on Pd-MgO/SiO_2 catalyst: independent roles of MgO and Pd on CO_2 methanation. *J. Phys. Chem. C* 114 (15): 7128–7131.

58 Bosch, C. and Wild, W. (1914). Producing hydrogen. US patent US1115776A, filed 30 June 1913 and issued 3 November 1914.

59 Centi, G. and Perathoner, S. (2009). Opportunities and prospects in the chemical recycling of carbon dioxide to fuels. *Catal. Today* 148 (3–4): 191–205.

60 Chen, X., Su, X., Liang, B. et al. (2016). Identification of relevant active sites and a mechanism study for reverse water gas shift reaction over Pt/CeO_2 catalysts. *J. Energy Chem.* 25 (6): 1051–1057.

61 Porosoff, M.D., Yan, B., and Chen, J.G. (2016). Catalytic reduction of CO_2 by H_2 for synthesis of CO, methanol and hydrocarbons: challenges and opportunities. *Energy Environ. Sci.* 9 (1): 62–73.

62 Park, S.W., Joo, O.S., Jung, K.D. et al. (2001). Development of ZnO/Al_2O_3 catalyst for reverse-water-gas-shift reaction of CAMERE (carbon dioxide hydrogenation to form methanol via a reverse-water-gas-shift reaction) process. *Appl. Catal., A* 211 (1): 81–90.

63 Daza, Y.A. and Kuhn, J.N. (2016). CO_2 conversion by reverse water gas shift catalysis: comparison of catalysts, mechanisms and their consequences for CO_2 conversion to liquid fuels. *RSC Adv.* 6 (55): 49675–49691.

64 Zhang, Q., Pastor-pérez, L., Jin, W. et al. (2019). Understanding the promoter effect of Cu and Cs over highly effective β-Mo_2C catalysts for the reverse water-gas shift reaction. *Appl. Catal., B* 244 (2018): 889–898.

65 Fujita, S.I., Usui, M., and Takezawa, N. (1992). Mechanism of the reverse water gas shift reaction over Cu/ZnO catalyst. *J. Catal.* 134 (1): 220–225.

66 Loiland, J.A., Wulfers, M.J., Marinkovic, N.S., and Lobo, R.F. (2016). Fe/γ-Al_2O_3 and Fe–K/γ-Al_2O_3 as reverse water–gas shift catalysts. *Catal. Sci. Technol.* 6 (14): 5267–5279.

67 Chen, C.S., Cheng, W.-H., and Lin, S.-S. (2000). Mechanism of CO formation in reverse water–gas shift reaction over Cu/Al_2O_3 catalyst. *Catal. Lett.* 68: 45–48.

68 Clarke, D.B. and Bell, A.T. (1995). An infrared study of methanol synthesis from CO_2 on clean and potassium-promoted Cu/SiO_2. *J. Catal.* 154 (2): 314–328.

69 Lu, B. and Kawamoto, K. (2014). Preparation of mesoporous CeO_2 and monodispersed NiO particles in CeO_2, and enhanced selectivity of NiO/CeO_2 for reverse water gas shift reaction. *Mater. Res. Bull.* 53: 70–78.

70 Kim, S.S., Lee, H.H., and Hong, S.C. (2012). A study on the effect of support's reducibility on the reverse water-gas shift reaction over Pt catalysts. *Appl. Catal., A* 423–424: 100–107.

71 Porosoff, M.D., Yang, X., Boscoboinik, J.A., and Chen, J.G. (2014). Molybdenum carbide as alternative catalysts to precious metals for highly selective reduction of CO_2 to CO. *Angew. Chem. Int. Ed.* 53 (26): 6705–6709.

72 Kusama, H., Bando, K.K., Okabe, K., and Arakawa, H. (2001). CO_2 hydrogenation reactivity and structure of Rh/SiO_2 catalysts prepared from acetate, chloride and nitrate precursors. *Appl. Catal., A* 205 (1–2): 285–294.

73 Karakaya, C., Otterstätter, R., Maier, L., and Deutschmann, O. (2014). Kinetics of the water-gas shift reaction over Rh/Al_2O_3 catalysts. *Appl. Catal., A* 470: 31–44.

74 Gogate, M.R. and Davis, R.J. (2010). Comparative study of CO and CO_2 hydrogenation over supported Rh–Fe catalysts. *Catal. Commun.* 11 (10): 901–906.

75 Pettigrew, D.J. and Cant, N.W. (1994). The effects of rare earth oxides. *Catal. Lett.* 28 (2–4): 313–319.

76 Chen, C.S., Cheng, W.H., and Lin, S.S. (2002). Study of reverse water gas shift reaction by TPD, TPR and CO_2 hydrogenation over potassium-promoted Cu/SiO_2 catalyst. *Appl. Catal., A* 238 (1): 55–67.

77 Chen, C.S., Cheng, W.H., and Lin, S.S. (2004). Study of iron-promoted Cu/SiO_2 catalyst on high temperature reverse water gas shift reaction. *Appl. Catal., A* 257 (1): 97–106.

78 Zhang, X., Zhu, X., Lin, L. et al. (2017). Highly dispersed copper over β-Mo_2C as an efficient and stable catalyst for the reverse water gas shift (RWGS) reaction. *ACS Catal.* 7 (1): 912–918.

79 Wang, L. and Liu, H. (2018). Mesoporous Co-CeO_2 catalyst prepared by colloidal solution combustion method for reverse water-gas shift reaction. *Catal. Today* 316 (2017): 155–161.

80 Panaritis, C., Edake, M., Couillard, M. et al. (2018). Insight towards the role of ceria-based supports for reverse water gas shift reaction over RuFe nanoparticles. *J. CO_2 Util.* 26: 350–358.

81 Zhou, G., Dai, B., Xie, H. et al. (2017). CeCu composite catalyst for CO synthesis by reverse water–gas shift reaction: effect of Ce/Cu mole ratio. *J. CO_2 Util.* 21: 292–301.

82 Li, J., Liu, L., Liu, Y. et al. (2014). Direct conversion of cellulose using carbon monoxide and water on a Pt-Mo_2C/C catalyst. *Energy Environ. Sci.* 7 (1): 393–398.

83 Sathish, C.I., Guo, Y., Wang, X. et al. (2012). Superconducting and structural properties of δ-$MoC_{0.681}$ cubic molybdenum carbide phase. *J. Solid State Chem.* 196: 579–585.

84 Hugosson, H.W., Korzhavyi, P., Jansson, U. et al. (2001). Phase stabilities and structural relaxations in substoichiometric TiC_{1-x}. *Phys. Rev. B-Condens. Matter Mater. Phys.* 63: 1–8.

85 Xu, W., Ramirez, P.J., Stacchiola, D., and Rodriguez, J.A. (2014). Synthesis of α-MoC_{1-x} and β-MoC_y catalysts for CO_2 hydrogenation by thermal carburization of Mo-oxide in hydrocarbon and hydrogen mixtures. *Catal. Lett.* 144 (8): 1418–1424.

86 Xu, W., Ramírez, P.J., Stacchiola, D. et al. (2015). The carburization of transition metal molybdates (M_xMoO_4, M = Cu, Ni or Co) and the generation of highly active metal/carbide catalysts for CO_2 hydrogenation. *Catal. Lett.* 145 (7): 1365–1373.

87 Pakhare, D. and Spivey, J. (2014). A review of dry (CO_2) reforming of methane over noble metal catalysts. *Chem. Soc. Rev.* 43 (22): 7813–7837.

88 Gadalla, A.M. and Sommer, M.E. (1989). Carbon dioxide reforming of methane on nickel catalysts. *Chem. Eng. Sci.* 44 (12): 2825–2829.

89 Fischer, F. and Tropsch, H. (1928). *Brennst-Chem* 9: 39.

90 Reitmeier, R.E., Atwood, K., Bennett, H., and Baugh, H. (2005). Production of synthetic gas – reaction of light hydrocarbons with steam and carbon dioxide. *Ind. Eng. Chem.* 40 (4): 620–626.

91 Olah, G.A., Goeppert, A., Czaun, M., and Prakash, G.K.S. (2013). Bi-reforming of methane from any source with steam and carbon dioxide exclusively to metgas (CO-$2H_2$) for methanol and hydrocarbon synthesis. *J. Am. Chem. Soc.* 135 (2): 648–650.

92 Mortensen, P.M. and Dybkjær, I. (2015). Industrial scale experience on steam reforming of CO_2 rich gas. *Appl. Catal., A* 495: 141–151.

93 Kumar, N., Shojaee, M., and Spivey, J.J. (2015). Catalytic bi-reforming of methane: from greenhouse gases to syngas. *Curr. Opin. Chem. Eng.* 9: 8–15.

94 Tomishige, K., Nurunnabi, M., Maruyama, K., and Kunimori, K. (2004). Effect of oxygen addition to steam and dry reforming of methane on bed temperature profile over Pt and Ni catalysts. *Fuel Process. Technol.* 85 (8–10): 1103–1120.

95 Dissanayake, D., Rosynek, M.P., and Lunsford, J.H. (1993). Are the equilibrium concentrations of CO and H_2 exceeded during the oxidation of CH_4 over a Ni/Yb_2O_3 catalyst. *J. Phys. Chem.* 97: 3644–3646.

96 Song C. and Pan, W. (2004). Tri-reforming of methane: a novel concept for catalytic production of industrially useful synthesis gas with desired H_2/CO ratios. *Catal. Today* 98 (2004): 463–484.

97 Bradford, M.C.J. and Vannice, M.A. (1996). Catalytic reforming of methane with carbon dioxide over nickel catalysts II. Reaction kinetics. *Appl. Catal., A* 142 (1): 97–122.

98 García-Diéguez, M., Pieta, I.S., Herrera, M.C. et al. (2010). Transient study of the dry reforming of methane over Pt supported on different γ-Al_2O_3. *Catal. Today* 149 (3–4): 380–387.

99 Son, I.H., Lee, S.J., Soon, A. et al. (2013). Steam treatment on Ni/γ-Al_2O_3 for enhanced carbon resistance in combined steam and carbon dioxide reforming of methane. *Appl. Catal., B* 134–135: 103–109.

100 Stroud, T., Smith, T.J., Le Saché, E. et al. (2018). Chemical CO_2 recycling via dry and bi reforming of methane using Ni–Sn/Al_2O_3 and Ni–Sn/CeO_2–Al_2O_3 catalysts. *Appl. Catal., B* 224 (2017): 125–135.

101 Ballarini, A.D., De Miguel, S.R., Jablonski, E.L. et al. (2005). Reforming of CH_4 with CO_2 on Pt-supported catalysts: effect of the support on the catalytic behaviour. *Catal. Today* 107–108: 481–486.

102 Yentekakis, I.V., Goula, G., Hatzisymeon, M. et al. (2019). Effect of support oxygen storage capacity on the catalytic performance of Rh nanoparticles for CO_2 reforming of methane. *Appl. Catal., B* 243 (2018): 490–501.

103 Carrara, C., Múnera, J., Lombardo, E.A., and Cornaglia, L.M. (2008). Kinetic and stability studies of Ru/La_2O_3 used in the dry reforming of methane. *Top. Catal.* 51 (1–4): 98–106.

104 Ferreira-Aparicio, P., Marquez-Alvarez, C., Rodriguez-Ramos, A. et al. (1999). A transient kinetic study of the carbon dioxide reforming of methane over supported Ru catalysts. *J. Catal.* 184 (1): 202–212.

105 Sigl, M., Bradford, M.C.J., Knözinger, H., and Vannice, M.A. (1999). CO_2 reforming of methane over vanadia-promoted Rh/SiO_2 catalysts. *Top. Catal.* 8: 211–222.

106 Sarusi, I., Fodor, K., Baán, K. et al. (2011). CO_2 reforming of CH_4 on doped Rh/Al_2O_3 catalysts. *Catal. Today* 171 (1): 132–139.

107 Liu, Z., Zhang, F., Rui, N. et al. (2019). Highly active ceria-supported Ru catalyst for the dry reforming of methane: in situ identification of $Ru^{\delta+}$–Ce^{3+} interactions for enhanced conversion. *ACS Catal.* 9 (4): 3349–3359.

108 Bian, Z., Das, S., Wai, M.H. et al. (2017). A review on bimetallic nickel-based catalysts for CO_2 reforming of methane. *ChemPhysChem* 18 (22): 3117–3134.

109 Li, L., Zhou, L., Ould-Chikh, S. et al. (2015). Controlled surface segregation leads to efficient coke-resistant nickel/platinum bimetallic catalysts for the dry reforming of methane. *ChemCatChem* 7 (5): 819–829.

110 Guharoy, U., Le Saché, E., Cai, Q. et al. (2018). Understanding the role of Ni–Sn interaction to design highly effective CO_2 conversion catalysts for dry reforming of methane. *J. CO_2 Util.* 27: 1–10.

111 Le Saché, E., Pastor-Pérez, L., Watson, D. et al. (2018). Ni stabilised on inorganic complex structures: superior catalysts for chemical CO_2 recycling via dry reforming of methane. *Appl. Catal., B* 236 (May): 458–465.

112 Kaydouh, M.N., El Hassan, N., Davidson, A. et al. (2015). Effect of the order of Ni and Ce addition in SBA-15 on the activity in dry reforming of methane. *C.R. Chim.* 18 (3): 293–301.

113 Koo, K.Y., Lee, S.H., Jung, U.H. et al. (2014). Syngas production via combined steam and carbon dioxide reforming of methane over Ni–Ce/$MgAl_2O_4$ catalysts with enhanced coke resistance. *Fuel Process. Technol.* 119: 151–157.

114 Pino, L., Vita, A., Cipitì, F. et al. (2011). Hydrogen production by methane tri-reforming process over Ni-ceria catalysts: effect of La-doping. *Appl. Catal., B* 104 (1–2): 64–73.

115 Majewski, A.J. and Wood, J. (2014). Tri-reforming of methane over Ni@SiO_2 catalyst. *Int. J. Hydrogen Energy* 39 (24): 12578–12585.

116 Hou, Z., Yokota, O., Tanaka, T., and Yashima, T. (2003). Characterization of Ca-promoted Ni/α-Al_2O_3 catalyst for CH_4 reforming with CO_2. *Appl. Catal., A* 253 (2): 381–387.

117 Gallego, G.S., Mondragón, F., Barrault, J. et al. (2006). CO_2 reforming of CH_4 over La–Ni based perovskite precursors. *Appl. Catal., A* 311 (1–2): 164–171.

118 Bitter, J.H., Seshan, K., and Lercher, J.A. (1998). Mono and bifunctional pathways of CO_2/CH_4 reforming over Pt and Rh based catalysts. *J. Catal.* 176 (1): 93–101.

119 Ferreira-Aparicio, P., Rodriguez-Ramosa, I., Andersonb, J.A., and Guerrero-Ruiz, A. (2000). Mechanistic aspects of the dry reforming of methane over ruthenium catalysts. *Appl. Catal., A* 202: 183–196.

120 He, L., Ren, Y., Fu, Y. et al. (2019). Morphology-dependent catalytic activity of Ru/CeO_2 in dry reforming of methane. *Molecules* 24 (3): 13–15.

121 Aghamohammadi, S., Haghighi, M., Maleki, M., and Rahemi, N. (2017). Sequential impregnation vs. sol–gel synthesized Ni/Al_2O_3–CeO_2 nanocatalyst for dry reforming of methane: effect of synthesis method and support promotion. *Molecular Catalysis* 431: 39–48.

9

CO_2 Utilization Enabled by Microchannel Reactors

Luis F. Bobadilla[1], Lola Azancot[1], and José A. Odriozola[1,2]

[1] *Instituto de Ciencia de Materiales de Sevilla, Centro Mixto CSIC – Universidad de Sevilla, Departamento de Química Inorgánica, Av. Américo Vespucio 49, 41092, Seville, Spain*
[2] *University of Surrey, Department of Chemical & Process Engineering, Guilford, GU27HX, UK*

9.1 Introduction

The vast majority of processes in the chemical industry are currently carried out in conventional reactors, i.e. tubular fixed bed, circulating fluidized, or slurry bubble column reactors. However, structured catalysts, mainly monolithic ones, have emerged as the most common small reactors because of their environmental applications, i.e. cleaning of automotive exhaust gases or of flue gases. Moreover, in recent years, alternative reactors based on microchannel technology have appeared in a full panoply of chemical processes. The microchannel reactor (MR) term is typically used to define three-dimensional (3-D) structured devices formed by channels with inner dimensions in the micrometric range from ten to few hundred micrometers [1]. Structured reactors are characterized by high porosities, typically almost twice that of packed beds, and pressure drops two to four times lower than in conventional fixed beds. Size reduction results in a decrease in linear dimensions and the increase of the surface-to-volume ratio, volume reduction, and production flexibility. Therefore, faster transfer of results from research to production, faster start-up of production at lower costs, easier scaling up of production, smaller plant size, lower materials, transport and energy costs, and higher flexibility to market demands are benefits of the MR technology. However, although the advantages of this technology are well known and emphasized by industrial users, just a reduced number of processes have been implemented at the production scale.

Moreover, this approach plays a key role in process intensification [2]. Despite the multiple aspects of process intensification (PI) since the first works more than 30 years ago [3], miniaturization remains its fundamental basis, with microreactor being the most typical example. However, "improvements in chemical manufacturing and processing, substantially decreasing equipment volume, energy consumption, or waste formation, and ultimately leading to cheaper, safer, sustainable technologies broadens" the PI concept [4].

Engineering Solutions for CO_2 Conversion, First Edition.
Edited by Tomas R. Reina, José A. Odriozola, and Harvey Arellano-Garcia.

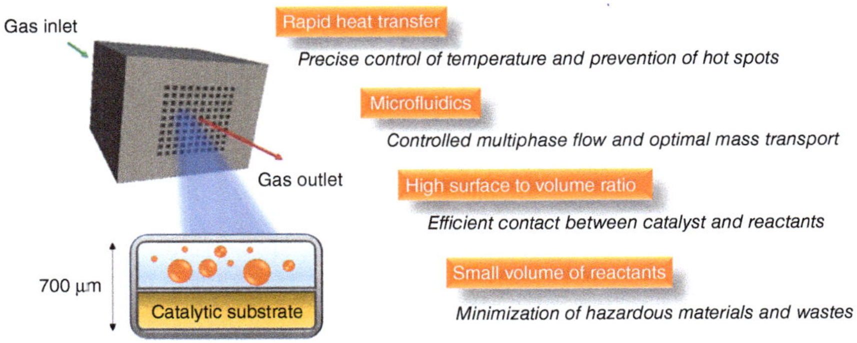

Figure 9.1 Some benefits of process intensification. Source: Reproduced with permission from Christina Wulf, Jochen Linben, Petra Zapp, Review of power to gas projects in europe, Elsevier.

The reduction of scale in MRs leads to certain unique properties that are schematized in Figure 9.1. Microreactors, as multitubular reactors, increase heat transfer by increasing the heat transfer surface to the reactor volume ratio. On the other hand, metallic honeycomb monoliths and open-cell foams, which are monolithic materials with thermally connected structures, offer similar advantages to those of microreactors with high radial heat transfer rates because of the enhanced heat conduction within the thermally connected solid matrix. However, while heat and mass transfer rates are similar in honeycomb monoliths and open-cell foams, the pressure drop is greater [5]. The small channel inner diameter results in high heat transfer performances, enabling rapid heating/cooling allowing temperature control of the structured device that become almost isothermal. The existence of hot/cold spots is therefore avoided and undesired reactions suppressed improving reaction selectivity.

Therefore, MRs are suitable to perform strong exothermic or endothermic chemical reactions [6]. Moreover, the small dimensions of the reactor channels reduces considerably the diffusion time of the reactants, resulting in short radial diffusion times, which allows obtaining selectively intermediate compounds in a reaction sequence because of the narrow residence time distribution allowed by the short radial diffusion time. Typically, this avoids mass transfer limitations and minimizes the influence of mass transport in the reaction rate [7].

The conditions of hydrodynamic flow in the microchannels usually favor laminar flow of high uniformity, which ensure an optimal mixing in very short time [8]. Moreover, the heat transfer from the reaction medium is inversely proportional to the channel diameter; therefore, the heat transfer coefficient is one order higher than in conventional reactors. MRs are highly promising for process intensification of solid-catalyzed gas–liquid and liquid–liquid reactions because of the high order between phases in multiphase flow exhibited by microfluidic devices [9].

Miniaturizing the reactor down to submillimetric dimensions leads to higher surface-to-volume ratios than in conventional reactors, which notably improves coated catalyst and reactants contact, allowing to intensify chemical reactions that occur 10–1000 times faster than in conventional reactors [10]. Furthermore, MRs generate unique opportunities to greatly improve the existing chemical routes in terms of safety and waste minimization. The small volume of reactants reduces the amounts of toxic substances or explosive chemicals and minimizes the associated risks during transportation and storage [11].

The MRs are made out of a wide variety of substrate materials such as silicon, aluminum, copper, stainless steels, polymers, perfluoroalkoxy (PFA) ceramics, and glasses [12]. The optimum material choice is determined by the chemical compatibility of the microchannel surface and the catalyst to be coated as well as by materials properties under operating conditions. Among these properties, mechanical (creep behavior and brittle phase formation), thermal, and chemical resistance are particularly relevant [13]. Depending on the selected material substrate and the desired complexity of the MR, the manufacturing method may be selected among an endless catalog: micromachining, injection molding, soft lithography, photolithography, laser ablation, nanocasting, hot embossing, etching, welding, or electroforming, [12, 14] and more recently, additive manufacturing methods [15] are among the reported choices. Finally, coating of the microchannel walls with the selected catalysts must be considered. Meille, in a comprehensive review [16], provided a detailed description methods usually employed to coat structured substrates, such as physical vapor deposition (PVD), dip coating, flame or plasma spraying, sol–gel, chemical vapor deposition (CVD), direct synthesis, impregnation, or electrochemical deposition. However, wash coating is the most common method allowing to control catalyst thickness [13, 17].

This chapter addresses CO_2 capture and its transformation into value-added chemicals using MRs, an approach leading to comprehensive valorization of CO_2 resulting in the next generation of zero emissions industries because MR technology allows distributed production technologies favoring the Millennium Development Goals. Future industrial processes must be designed bearing in mind the optimum carbon uptake. Therefore, catalytic processes based on CO_2 conversion are essential to hit emission targets and fundamental to enable the transition toward a low-carbon society. This chapter starts with a brief overview of the effects of heat and mass transfer in MRs because chemical reactions involved in CO_2 utilization require optimal temperature and mass transfer control. Then, the utilization of MRs in different processes involving CO_2 sequestration and utilization exhibited using selected examples of promising potential application such as carbon capture, utilization, and storage (CCU/CCUS) processes helped by MRs is further described, and, finally, future trends and application prospects from the industrial point of view are highlighted.

9.2 Transport Phenomena and Heat Exchange in Microchannel Reactors

As stated in Section 9.1, the use of MRs in industrial chemical processes may not only improve selectivity (higher quality of products) but also reduce energy consumption and wastes. From an industrial viewpoint, the main advantages, which make the

MRs competitive, are the high efficiency for conducting fast reactions because of the inhibition of the mass transport limitations and the high ability for heat exchange in the chemical reaction.

To avoid mass transfer influence, the characteristic mass transfer time, t_m, should be at least one order of magnitude smaller than the characteristic reaction time. The high MR surface-to-volume ratio, A/V_R, results in low characteristic mass transfer times, up to two orders of magnitude lower than in conventional reactors (Eq. (9.1)):

$$t_m = \frac{1}{k_{fl}} \cdot \frac{V_R}{A} \tag{9.1}$$

Temperature control and the pattern flow regulation are crucial to control the selectivity of chemical processes and to maintain safety [18]. Figure 9.2 compares heat and mass transfer characteristics of commonly used reactors. MRs lie on the upper right corner of the figure stressing their superior performances in relation to conventional reactors. This section concisely describes the basic operation principles of mass transport under microfluidic conditions and the associated heat transfer phenomena taking place in MRs.

9.2.1 Microfluidics and Mixing Flow

Microfluidics involves the study of fluid flows along channels with micron-sized geometrical dimensions [20]. Achieving optimized designs of microfluidic devices is mandatory to describe flow characteristics–pressure drop relationships in the microchannels. Reynolds number (Re) defines flow patterns in the channels; this dimensionless number is given by the following equation:

$$\text{Re} = \frac{u\rho d_h}{\mu} \tag{9.2}$$

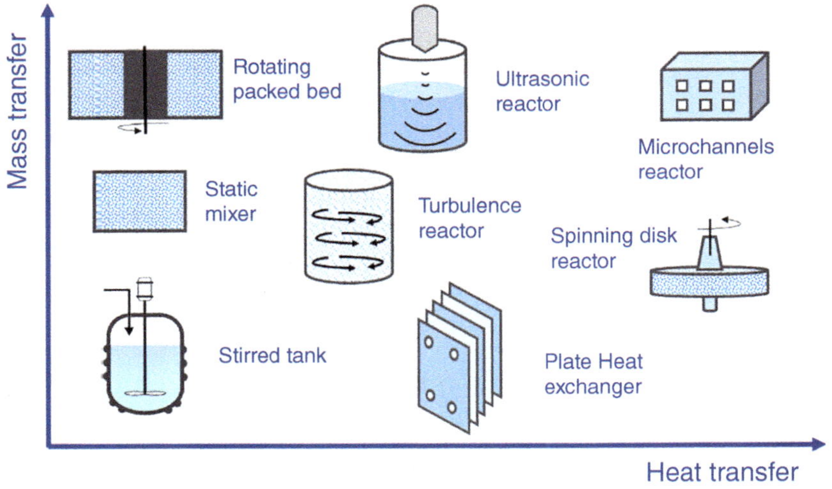

Figure 9.2 Heat and mass transfer performance of various types of reactors. Source: Adapted from Pask et al. [19].

where u is the linear velocity, ρ is the density, d_h is the hydraulic or equivalent diameter of the channel, and μ is the viscosity of the fluid. Physically, the Reynolds number expresses the ratio of inertia forces that are resistant to changes produced by random vortices or other flow instabilities [21]. The flow will be turbulent when the Reynolds number is above 4000, and the operation will be in the laminar flow regime if it is below ~2200 [22]. Depending on the linear flow velocity and the channel dimensions, typical Reynolds numbers found in microchannels reactors are in the 1–1000 range prevailing the laminar flow regime [6]. Because of the low Reynolds number, the mixing of reactants in MRs is controlled by molecular diffusions. Such the small diffusion paths as the very short mixing time conduct to an efficient mixing and the mass transport phenomena are greatly improved [23].

Many results referring to single-phase fluid flows in microchannel have been reported and applied in microfluidic engineering [24]. Almost one decade ago, Kumar et al. [24a] reported a comprehensive review that includes the fundamental and basic understanding of hydrodynamics of single-phase fluid flows and mixing process in microchannels. In comparison with single-phase flows, multiphase flows are generated when two partially miscible or immiscible fluids are mixed. Liquid–liquid and gas–liquid are the typical mixtures involved in many chemical processes such as gas–liquid catalytic hydrogenation reactions or synthesis of biodiesel using biphasic transesterification [9a], as well as in other applications such as separation, purification, and absorption [25]. MRs afford an innovative solution to control the multiphase contact and obtaining a high interfacial area with the desirable mass transfer [9a, 26]. As shown in Figure 9.3, the hydrodynamic behavior of biphasic flows in microreactors, for example, gas–liquid, can be mainly divided into two regimes: (i) *slug flow*, also known as the *Taylor flow*, characterized by alternated elongated bubbles and liquid slugs passing through the coated microchannels, and (ii) *annular flow*, described as a continuous flow of gas in the core surrounded by a continuous liquid film flow around the catalyst coating. The gas–liquid slug flow dominates at low gas–liquid flow ratios, whereas annular flow is typical at very high gas–liquid flow ratios, and a transitional slug–annular flow occurs at intermediate gas–liquid flow ratios [9a, 27]. Additionally, as both liquid and gas phases reached high flow rates, a flow regime known as "*churn flow*" appears, in which the two phases are highly inter-dispersed with rapidly changing phase distribution patterns [28]. Eventually, another type of flow regime designated as "*bubbly flow*" is also possible, in which many small gas bubbles are uniformly dispersed in the liquid phase [7a]. This pattern regime flow is typically observed at a relatively high linear velocity of liquid and low gas velocity. Figure 9.3b includes a representation of the different types of flow regimes observed on a biphasic flow of water and CO_2 when the liquid/gas flow ratios are varied [27]. As can be observed, flow regime patterns such as bubbly flow, slug flow, slug–annular flow, annular flow, and churn flow were identified. In this study, bubbly flow was evidenced at relatively high liquid flow and the lowest gas flow, which was characterized by the dispersion of spherical bubbles in a continuous liquid phase. The small bubbles tended to coalesce because of the velocity differences between gas and liquids. When the gas flow increased, the size of the bubble approached the microchannel width achieving an annular flow.

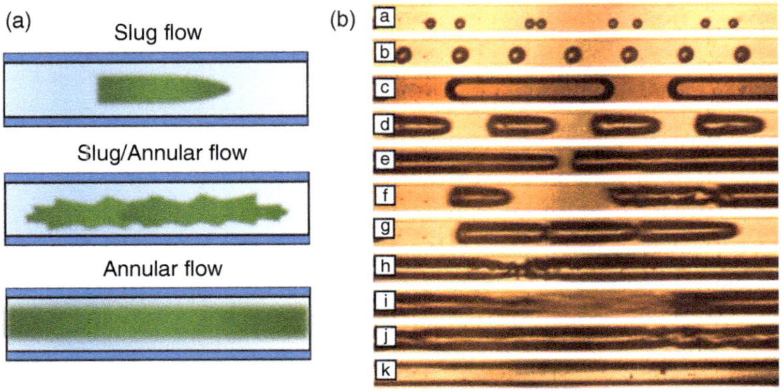

Figure 9.3 (a) Schematic representation of characteristic regimes observed in biphasic flow microreactors. (b) Representative images captured during CO_2–water flow in the microchannel: a – bubbly flow (u_{CO_2} = 0.16 m s^{-1} and u_{H_2O} = 1.0 m s^{-1}), b – bubbly flow (u_{CO_2} = 0.29 m s^{-1} and u_{H_2O} = 1.0 m s^{-1}), c – Taylor flow (u_{CO_2} = 0.16 m s^{-1} and u_{H_2O} = 0.04 m s^{-1}), d – Taylor flow (u_{CO_2} = 1.28 m s^{-1} and u_{H_2O} = 1.0 m s^{-1}), e – unstable slug flow (u_{CO_2} = 1.74 m s^{-1} and u_{H_2O} = 0.51 m s^{-1}), f – unstable slug flow (u_{CO_2} = 2.14 m s^{-1} and u_{H_2O} = 0.15 m s^{-1}), g – train slug flow (u_{CO_2} = 2.07 m s^{-1} and u_{H_2O} = 1.0 m s^{-1}), h – slug/annular flow (u_{CO_2} = 7.51 m s^{-1} and u_{H_2O} = 0.2 m s^{-1}), i – churn flow (u_{CO_2} = 12.7 m s^{-1} and u_{H_2O} = 1.0 m s^{-1}), j – churn flow (u_{CO_2} = 31 m s^{-1} and u_{H_2O} = 0.51 m s^{-1}), and k – annular flow (u_{CO_2} = 21.5 m s^{-1} and u_{H_2O} = 0.02 m s^{-1}). Source: Adapted from Yue et al. [27].

The analysis of fluid dynamics requires numerical solutions and algorithms to solve the problems involved in these processes. Computational fluid dynamics (CFDs) is a useful tool to model the hydrodynamic behavior of fluids [29]. CFD modeling has been applied by many researchers to investigate the pattern flow in microreactors. For instance, Ho et al. [30] performed a 3-D and two-dimensional (2-D) CFD simulations, in which the hydrodynamic characteristics of a biphasic flow in a falling film microreactor were analyzed. 3-D simulation provides comprehensive information on the flow profiles, while 2-D simulation is employed for parametric studies of fluid properties and flow conditions. CFD simulations are essential to gain hydrodynamic details and phenomena transport mechanism, which are hardly experimentally accessible.

9.2.2 Heat Exchange and Temperature Control

The heat management in both endothermic and exothermic chemical processes is crucial to utilize the released heat effectively. Temperature gradient generation and hot spot formation in fixed bed reactors are difficult to prevent because of heat exchange limitations. Microreaction technology has been successfully applied to solve many of the problems associated with heat transfer issues [31]. Main MR characteristics are high heat transfer coefficients and large surface area-to-volume ratio achieved on decreasing the hydraulic diameter of channels. In MRs, the heat transfer coefficient can reach values of 20 kW m^{-2} K^{-1}, while in a conventional reactor, it barely exceeds 2 kW m^{-2} K^{-1} [32]. Thus, MRs' heat transfer efficiency is greatly improved and virtually ensures isothermal reactor operation [31a]. The heat transfer capability of a microreactor depends basically on the selection of the adequate material for achieving a high thermal conductivity. In addition, the wall thickness is especially important to remove the heat of the reaction generated in highly exothermic reactions, preventing thermal runaway and gaining operational safety.

Nusselt number (*Nu*), a dimensionless number that represents the ratio between convective and conductive heat transfer (Eq. (9.3)), describes heat transfer performance of reactors,

$$\mathrm{Nu} = \frac{\alpha d_h}{\lambda} \tag{9.3}$$

where α is the heat transfer coefficient, λ is the fluid thermal conductivity, and d_h is the hydraulic diameter of the channel [22, 33]. For a macrofluidics system in a laminar flow regime (Re < 2300), the Nusselt number is typically constant, while in the transitional (2300 < Re < 10 000) and turbulent (Re > 10 000) flow regimes, Nu can be described as a function of Re. Depending on the characteristics of the transitional and turbulent flow regimes, several empirical correlations between Nu and Re have been proposed in the literature as, for example, the Sieder–Tate equation, Dittus–Boelter equation, or the Gnielinski correlation [34].

As mentioned above, the flow profile typically observed in an MR is laminar, and consequently, the Nusselt number should be constant and independent of the flow rate under these conditions [11a, 35]. However, the available experimental data on laminar flow heat transfer show a certain inconsistency with the values predicted by the molecular diffusion theory, and an increase of Nusselt number with increasing Reynolds number is generally observed, as can be seen in Figure 9.4a. Because of the intrinsic microfluidics properties of the microchannels, eddies can be randomly formed and secondary flow patterns can become turbulent at Reynolds numbers below the typical values for laminar to turbulent flow transitional condition [36]. The formation of these local chaotic flow patterns is one of the causes that explain the dependence of Nusselt number with the Reynolds number observed in MRs rather than Nu remaining constant as would be expected. Mei et al. [37] analyzed the dependence of Nu with Re measured in laminar flow regime for three Al- and Cu-based MRs with different hydraulic diameters and adjusted the observed tendency using the empirical correlations valid for transitional and turbulent flow regime in macroscopic systems. Additionally, it should be considered that the microchannels have relatively short lengths and the influence of the entrance region cannot be neglected. Figure 9.4b shows that in a microchannel with a hydraulic diameter channel of 10 μm, the Nusselt number is higher at the inlet and gradually decreases until reaching a constant value. Chan et al. [38] analyzed this effect by CFD modeling considering as boundary condition a constant wall temperature of 350 K and developing the temperature profile from channel inlet until reaching the outlet. They observed an evident temperature variation at the inlet, but as the flow passes through the channel, almost isothermal profile is reached as bulk fluid temperature

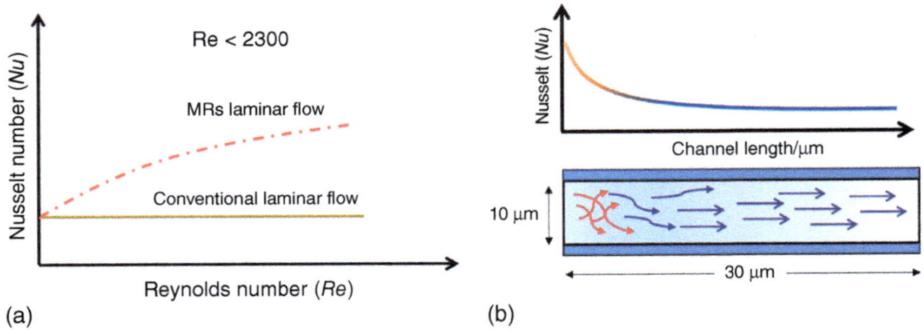

Figure 9.4 Heat transfer characteristic data: (a) Illustrative representation of Nusselt number versus Reynolds number for a conventional system and MRs under laminar flow regime and (b) pictorial depiction of the variation of Nusselt number with the channel length in MRs.

approaches the wall surface temperature. Therefore, it is remarkable that the importance of the entrance effect where heat transfer takes place along the microchannel generating a temperature gradient despite laminar flow is fully developed and demonstrates that the heat transfer performance is greatly improved in MRs.

9.3 Application of Microreactors in CO_2 Capture, Storage, and Utilization Processes

Recent advances in CCUS using MRs are summarized in this section. It should be emphasized that knowledge and development of CO_2 capture/utilization processes in MRs is an active research area starting to be expanded with a still limited number of publications. Herein, some of the most relevant examples of microchannel technology application in CO_2 capture/utilization published so far are described.

9.3.1 CO_2 Capture and Storage (CCS)

To mitigate CO_2 emissions from heavy carbon industries, different carbon dioxide capture and storage (CCS) approaches have been taken into account at laboratory scale and even industrially explored. The CO_2 capture methods most widely used include chemical absorption [39], adsorption [40], membrane [41], or molecular sieve [42] separation, mineral carbonation [43], chemical looping combustion [44], and cryogenic separation [45]. One of the industrially preferred pathway for CCS is the regenerative chemical absorption of CO_2 in aqueous solutions of alkanolamines. Owing to its excellent properties, monoethanolamine (MEA) is the most widely used absorbent [46].

Absorption by amines is typically used for CO_2 removal at large scale in many heavy carbon industries (coal gasification, natural gas processing, combustion fuels, or oil refining); however, this capture process is still highly costly and ineffective because of the large volume of exhaust gas involved and problems associated with mass transport limitation existing in conventional gas–liquid contactors. Other drawbacks of amine-based absorption methods are corrosiveness, solvent volatility/degradation, toxicity of amines, large consumption of energy for regeneration, or stoichiometry-limited CO_2 absorption capacity, 1 : 2 CO_2:amine molar ratio [47].

MRs may become an interesting, less costly, and more efficient alternative for CO_2 absorption by amines. This has driven studies with a variety of amines and operation variables. Multiphase MRs provide important advantages: large and well-defined gas–liquid interfacial areas, fast mixing, and reduced gas–liquid mass–transfer limitations. Numerous studies have been conducted to confirm the enhancement of gas/liquid mass transfer intensifying the absorption process in MRs [47, 48]. As a general output all of them found and enhancement of the absorption capacity, generally higher than 90%; on using MRs [47, 48c, f, 49]. The absorption temperature has a limited influence on the absorption capacity, but on increasing pressure, an enhancement of CO_2 capture is observed [48c, 49a, 50]. Ganapathy et al. [48a] reported absorption efficiencies for CO_2 close to 100% under certain operating conditions for CO_2–N_2 flows in aqueous diethanolamine. Mass transfer enhancement in MR is responsible for the observed high efficiency. CO_2 reaction with dialkylamines results in ionic liquid formation; upon heating, CO_2 is released and the high polar ionic liquid switches back to a low polarity liquid (1 and 2). This approach is undertaken by Li et al. [48b] to study the kinetics of fast gas–liquid reactions in MRs, Figure 9.5

$$2R_1NHR_2 + CO_2 \rightarrow (R_1R_2NH_2)^+(R_1R_2NCO_2)^- \tag{9.4}$$

$$(R_1R_2NH_2)^+(R_1R_2NCO_2)^- \xrightarrow{60^\circ C} 2R_1NHR_2 + CO_2 \tag{9.5}$$

After mixing CO_2 gas and the dialkylamine solution in acetonitrile at the entrance Y-junction of the MR, the gas flow breaks and generates CO_2 plugs periodically separated by liquid. The carbamate formation reaction starts at the Y-junction and continues until the heated zone of the MR that switches back the reaction to the formation of gas-phase CO_2. As can be observed in Figure 9.5, the volume of plugs continuously reduces while traveling along the non-heated region because of the absorption of CO_2 by alkylamines. This behavior is reversed in the heated zone as the carbamate salt decomposes.

In an interesting study, Li et al. [51] demonstrated the importance of the flow regime in the CO_2 absorption capacity. For relatively low gas flows, the absorption capacity increases on increasing gas flow until a critical value. At his critical flow, the Taylor flow transfers to the annular flow regime, and from this value, the effect of the gas flow on the absorption capacity is negligible. These results seem to be contradictory with other reported data that found a decrease of the absorption capacity on increasing the gas flow rate [49b], but attention has to be paid to the modification of the flow regime on changing the variables of operation. An interesting result comes from the use of 3-D-printed fractal MRs. By decreasing flow velocity since channel bifurcations the residence time increases and the slug flow is maintained that increases contact area and time intensifying the CO_2-MEA interaction [52]. The complexity of mass transfer problem has been recently addressed in an excellent

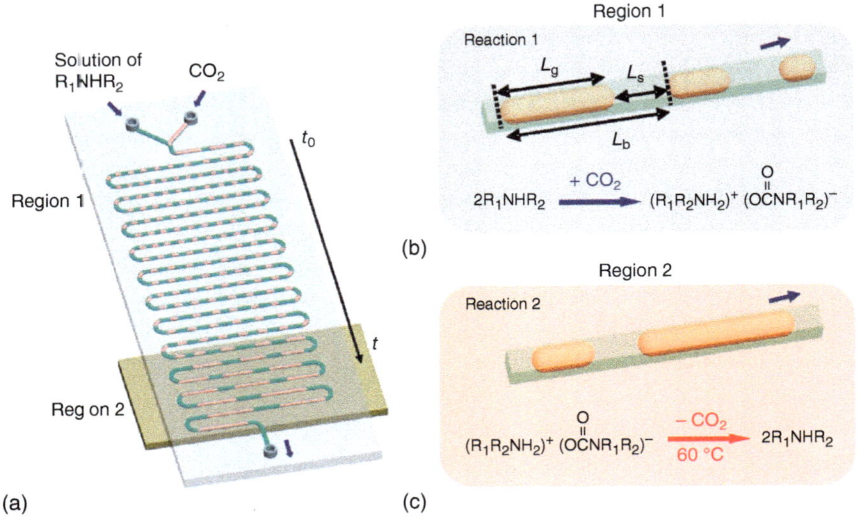

Figure 9.5 (a) Schematic representation of the microchannel reactor including the process that occurs in regions 1 and 2, respectively. (b) Reaction of CO_2 with a secondary amine to yield a carbamate salt (Reaction 1). (c) Release of CO_2 mediated by the increase in temperature (Reaction 2). Source: Li et al. [48b].

review paper by Sattari-Najafabadi et al. [7a]. These studies evidence the ability of MR to conduct efficiently rapid gas–liquid absorption processes.

9.3.2 CO_2 as a Feedstock for Producing Valuable Commodity Chemicals

The crucial step for the recycling of carbon resource is the chemical transformation of CO_2 into valuable products such as polymers and fuels. Overall, the CO_2 transformation reactions can be classified into three categories: (i) CO_2 reduction to chemical building blocks such as CO, methane, methanol, formaldehyde, formic acid, and among others [53]; (ii) production of synthetic fuels [54]; and (iii) synthesis of cyclic carbonates, inorganic carbamates, and polymers [55]. Figure 9.6 schematizes the concept of capture/sequestration of CO_2 and subsequent transformation into valuable commodity using microreactors. A number of examples are given below to illustrate the importance of MRs in these processes.

9.3.2.1 Methanation of Carbon Dioxide (Sabatier Reaction)

CO_2 methanation, a H_2 reduction of CO_2, is a reversible and exothermic reaction:

$$CO_2 + 4H_2 \leftrightarrow CH_4 + 2H_2O \qquad (\Delta H^\circ = -167\ \text{kJ mol}^{-1}) \qquad (9.6)$$

This reaction was first described by the French chemist Paul Sabatier in 1902 using finely divided nickel as a catalyst [56]. In the first half of the twentieth century, the main industrial application of methanation reaction was the removal of CO and CO_2 for producing H_2-rich feed gases in ammonia plants. In the past decade, global

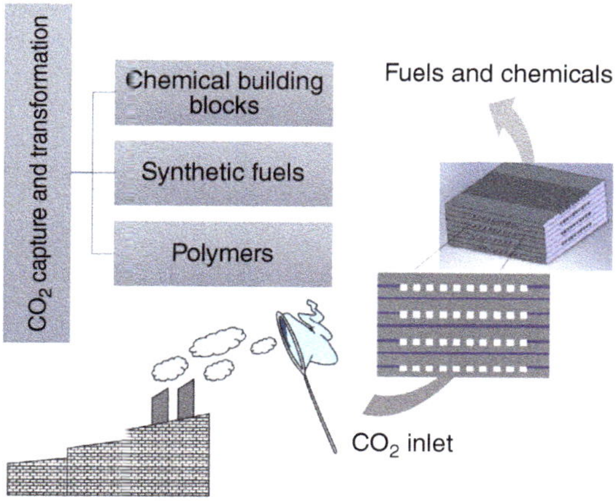

Figure 9.6 An overall schematic representation of the utilization of microchannel reactors for the transformation of CO_2 into feedstock for the chemical/process industry.

warming concerns brings a new scenario for CO_2 methanation, reduction of greenhouse gases become mandatory, and global carbon recycling and the development of cost efficient technologies for renewable hydrogen production boost research works on this reaction. Moreover, the renaissance of the Sabatier reaction in the twenty-first century is also associated with space agencies (NASA or ESA) interest for space-based applications to achieve long-term space exploration missions [57].

The Sabatier process has been commercially implemented using conventional fixed bed reactors. However, the methanation reaction is highly exothermic and requires an efficient heat removal to prevent hot spot formation and catalyst thermal deactivation. In this context, the utilization of MRs offers an optimal heat management and enhanced catalytic performance, a recent review highlight this aspects [58]. Brooks et al. [59] studied the Sabatier reaction using an MR as shown in Figure 9.7, and they demonstrated that a precise thermal control is enabled by an oil counterflow to maintain the desired microchannel wall temperatures (Figure 9.7c). In addition, they established and validated a unidimensional plug flow model to assist the interpretation of the measured conversion rates and selectivity in MR.

Engelbretch et al. [60] also demonstrated that Sabatier reaction is successfully performed in MRs achieving an optimal operating point, which maximizes the methane productivity at 5 bar and 400 °C with a space velocity of 97.8 $l\,{g_{cat}}^{-1}\,h^{-1}$. Furthermore, these authors proposed a CFD model to describe the reaction-coupled transport phenomena and identify the operational limits in the microreactor. Overall, this work gives a fundamental idea to illustrate the further development of MRs for CO_2 methanation in power-to-gas applications.

As mentioned above, the Sabatier process can also be used in long-term space explorations as, for example, trips to Mars. The rocket propellants typically used are HYDRALOX (hydrogen and liquid oxygen) and RP-1 (kerosene and liquid oxygen). Both hydrogen and kerosene contained in the propellants are being phased out by liquid methane for several reasons. Compared with hydrogen, methane is more

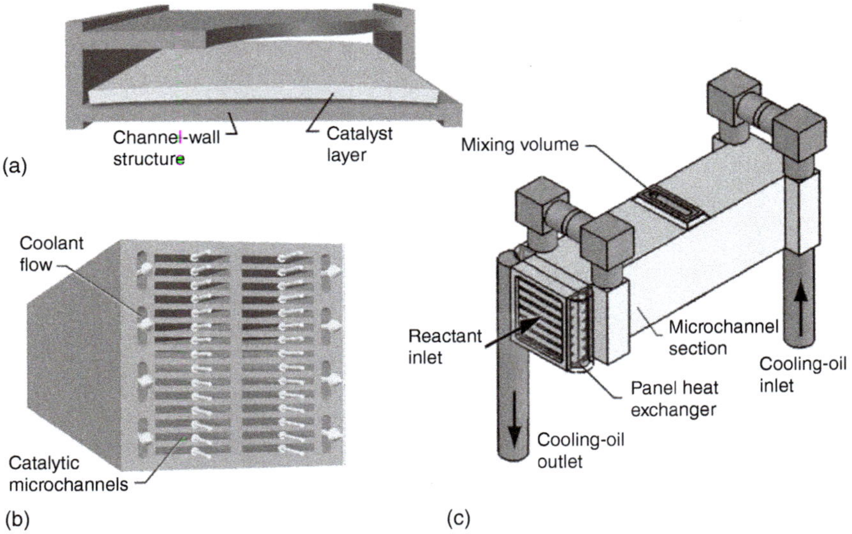

Figure 9.7 (a) Representation of a single microchannel with a deposited catalytic layer, (b) illustration of the transversal section of MR, and (c) engineering scheme of the MR assembly. Source: Reproduced from Brooks et al. [59].

stable in space, and it does not cause metal embrittlement while that in comparison with kerosene methane is lighter, burns at higher temperature, and does not cause coking [57a]. In this regard, Holladay et al. [57b] explored a novel approach using microreactors technology in which the reverse water gas shift reaction (RWGS) and the Sabatier reaction are integrated to manufacture CO_2 from the atmosphere on Mars and produce methane as a propellant. All the process relies on the recycling of CO_2 and H_2O produced in the space shuttle. The process is schematically depicted in Figure 9.8 and can be summarized as follows: produced water vapor is condensed

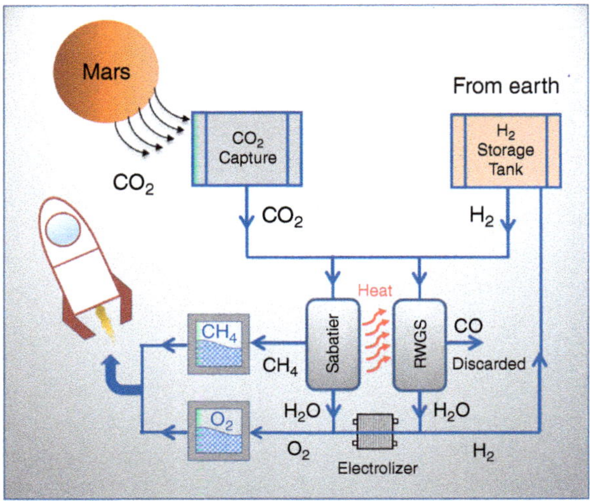

Figure 9.8 Schematic representation of in situ CH_4/O_2 production as a propellant fuel for Mars missions.

and decomposed by electrolysis into oxygen and hydrogen, CO_2 is captured, further mixed with hydrogen, and submitted to the Sabatier reactor to produce methane and water. Propulsion is efficiently achieved using an oxygen-to-methane mass ratio of 3.8. The RWGS reactor helps in fixing the desired O_2-to-CH_4 ratio, CO and H_2O are the products of this reactor, the former is discarded and the latter is electrolyzed to hydrogen and oxygen. Hydrogen is recycled and oxygen is used to adjust the O_2-to-CH_4 mass ratio. Using microchannel architecture, both Sabatier and RWGS reactions are thermally coupled minimizing heat requirements because the heat consumed in the RWGS is produced in the Sabatier reaction. Moreover, to reduce the total energy required, the utilization of integrated MRs is accomplished by maximizing the catalyst activity, increasing CO_2 conversion.

9.3.2.2 CO_2-to-Methanol and Dimethyl Ether (DME) Transformation

Within the CCUS processes, the CO_2 to methanol transformation is considered as one of the most effective for CO_2 utilization [61]. Methanol (CH_3OH) is an important chemical feedstock used for many chemical industrial processes as well as an excellent energy storage material [62]. The synthesis of methanol from captured CO_2 has attracted huge interest, and countries such as Iceland or Japan have already plants for methanol production from captured CO_2 and renewable hydrogen [63]. In 2011, Carbon Recycling International (CRI) started the operation of the first commercial demonstration plant in Iceland achieving a production of 5 Mt_{MeOH} per year. In 2008, Mitsui Chemicals Inc. constructed a pilot plant in Osaka to produce methanol with a capacity of around 100 t_{MeOH} per year using CO_2 emitted from factories and hydrogen obtained from water photolysis [64].

Methanol synthesis from CO_2 hydrogenation is an exothermic reaction equilibrium limited usually governed by the following three main reactions [53c]:

$$CO_2 + 3H_2 \leftrightarrow CH_3OH + H_2O \qquad (\Delta H^\circ = -49\ \text{kJ mol}^{-1}) \tag{9.7}$$

$$CO_2 + 3H_2 \leftrightarrow CO + H_2O \qquad (\Delta H^\circ = 41.2\ \text{kJ mol}^{-1}) \tag{9.8}$$

$$CO + 2H_2 \leftrightarrow CH_3OH \qquad (\Delta H^\circ = -90.7\ \text{kJ mol}^{-1}) \tag{9.9}$$

Thermal management is critical to prevent the accumulation of reaction heat on the catalytic particle surface, which promotes side reactions and decreases methanol selectivity [65]. The process intensification potential of microreactors technology is particularly interesting to solve the problems associated with the limitation of heat transfer and avoid the formation of hot spots and temperature runaway. Bakhtiary-Davijany et al. [66] compared the performance of a multi-slit packed bed microstructured reactor with a conventional fixed bed reactor in the synthesis of methanol. The main difference found in the catalytic performance was ascribed to superior heat exchange properties of the microreactor. The performances of this microstructured reactor were evaluated using a mathematical 3-D pseudo-homogeneous approach that evidenced the superior heat and mass transport characteristics of MRs apart from validating the concept of isothermal operation [67]. The excellent heat exchange properties of MRs have also highlighted by other authors that ascribe the superior performances of these reactors to this property, i.e.

Phan et al. [68] compared a stacked foil microreactor with a conventional laboratory fixed bed one using the same Cu-based catalyst, and they claim that the observed differences are related to heat exchange properties of the catalytic device.

Wall-coated MRs are mostly used for this application, with the main constrain of this approach being the easiness of crushing the catalyst pellets. For solving this and to intensify the methanol synthesis process, Liang et al. [69] efficiently integrated a layer structured Cu-Zn/Al monolithic foam catalyst in a microreactor. As a result, high methanol yield (7.81 $g_{MeOH}\, g_{Cu}^{-1}\, h^{-1}$ at 250 °C and 3 Mpa) was obtained because of the enhanced heat/mass transfer properties.

Dimethyl ether (DME) is another important intermediate in the chemical industry for a wide variety of feedstocks and can be obtained directly by dehydration of methanol over an acidic catalyst. The combination of methanol synthesis and methanol dehydration in a single reactor increases the process intensification potential by shifting methanol synthesis equilibrium limitation [70]. Bansode and Urakawa [71] provided an efficient and highly productive processing single-step strategy for CO_2 conversion into methanol and subsequently in DME. They used a microreactor packed with a mixed bed consisting of a commercial $Cu/ZnO/Al_2O_3$ and H-ZSM-5 catalysts and achieved one-step conversion of CO_2 into DME (>65.8%) with remarkable selectivity (89%). The overall direct synthesis of DME is more exothermic than methanol synthesis alone, and thus, the thermal control is essential. The group of Prof. Venvick has reported several studies on the direct synthesis of DME by combining micro-packed reactors with heat exchangers [70, 72]. In one of the most interesting works reported by themselves [72a], they investigated the performance of three integrated micro-packed bed reactor heat exchangers for direct DME synthesis over physical mixtures of $CuO–ZnO–Al_2O_3$ and γ-Al_2O_3 catalysts and found a minimal pressure drop and isothermal environment free from transport limitations for the direct DME synthesis, in the kinetic regime as well as at equilibrium conversion.

9.3.2.3 CO_2-to-Higher Hydrocarbons and Fuels

The CO_2 conversion into C_{2+} chemicals is an emerging application to produce a wide variety of products including important alcohols, acetic acid, DME, olefins, and gasoline [54a]. Figure 9.9 includes an overview of CO_2 transformation routes

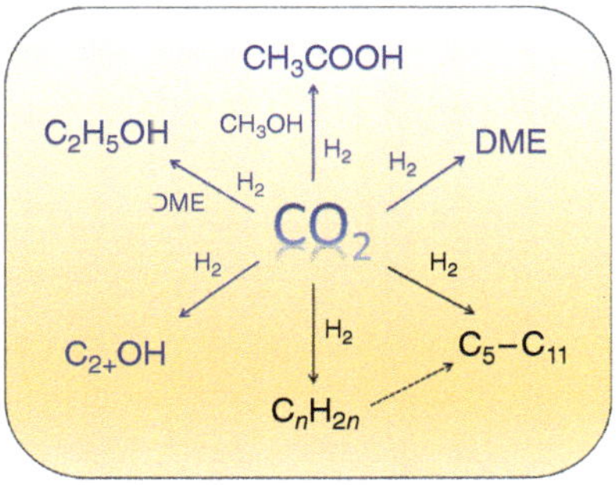

Figure 9.9 Overview of the transformation routes of CO_2 into C_{2+} chemicals. Source: Adapted from Li et al.. [54a].

to C_{2+} oxygenated chemicals and fuels. It has been proposed that CO_2 can be transformed into hydrocarbons or oxygenates mainly toward two routes [73]. The first one is labeled modified Fischer–Tropsch synthesis (FTS) route, which consists in first reducing CO_2 to CO by RWGS reaction and subsequently CO hydrogenation to hydrocarbons via the FTS. In the second route, methanol produced by direct hydrogenation of CO_2 is converted into other longer-chain hydrocarbons.

Nevertheless, because CO_2 is a thermodynamically very stable molecule ($\Delta G°_f = -396\,\text{kJ mol}^{-1}$) and kinetically inert, the C—C bond formation via CO_2 reduction still remains as one of the greatest challenges for CO_2 valorization nowadays [74]. The transformation of CO_2 into more complex C_{2+} chemicals requires high energy input. The search for more efficient catalysts to reduce the formation of C—C bonds high energy kinetic barriers is a hot topic for many researchers [75].

Additionally, the utilization of novel reactor configurations, such as MRs, combined with more efficient optimized catalysts offers a practical approach to address this issue The number of works published using MRs for CO_2 reduction into C_{2+} chemicals is still very limited. Recently, Sun et al. [76] employed a combination of artificial neuron networks (AANs) with response surface methodology (RSM) in a MR using a K-promoted iron-based catalyst to analyze the effect of operating conditions, i.e. gas ratio, temperature, pressure, and space velocity on product distribution for selective CO_2 hydrogenation. Promising selectivities for C_2–C_4 alkenes (14.98%), C_1 (6.99%), C_{5+} (10.88%), and CO (12.91%) working at 300 °C and 3.3 bar with a H_2/CO_2 molar ratio of 1.2 and a gas hourly space velocity (GHSV) of 2303 ml h^{-1}.

9.3.2.4 Production of Cyclic Organic Carbonates

The transformation of CO_2 into five-membered cyclic carbonates is a very promising route for CO_2 fixation. These compounds are widely used as raw materials for the production of polycarbonates and polyurethanes; have important applications as chemical intermediates, polar aprotic solvents, and electrolytes in lithium-ion secondary batteries; and are important intermediates for pharmaceutical industry [77]. The synthesis of cyclic carbonates occurs via coupling of CO_2 and epoxides as shown in the reaction Scheme 9.1 that follows.

To achieve the optimal yield of cyclic carbonates, this reaction must be conducted with the adequate catalyst and requires temperatures above 180 °C and high

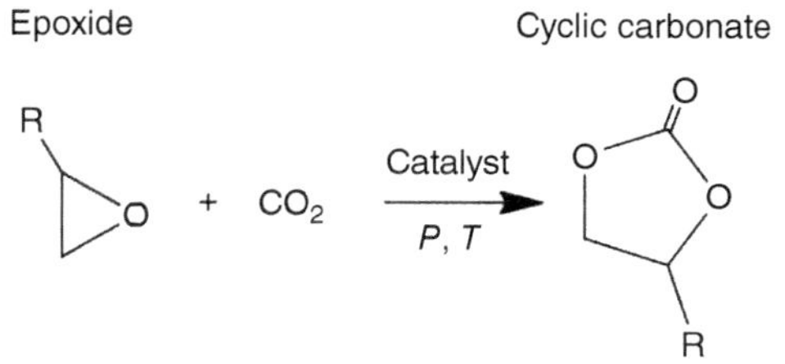

Scheme 9.1 Reaction of CO_2 cycloaddition with epoxides.

pressures (50–80 bar) [78]. In recent years, numerous catalysts have been developed and employed to promote the reaction of CO_2 cycloaddition with epoxides including various homogeneous catalytic systems, such as metal complexes, organocatalysts, quaternary ammonium, and phosphonium salts [79], or solid catalysts, such as metal oxides [80], immobilized complexes or ionic liquids [81], and zeolites [82]. More recently, the utilization of porous metal–organic framework (MOF) materials has also emerged as an interesting catalyst for CO_2 coupling reactions at mild conditions mainly because of its high capacity to capture and activate CO_2 [83].

The production of cyclic carbonates from epoxides and carbon dioxide was patented in 1943 by a German company and has been commercially applied since 1950s [84]. However, large-scale production is still performed using other traditional methods such as the phosgene route [63]. Currently, the annual production of dimethyl carbonate (DMC) is of 0.1 Mtons per year. A market study of TEXACO has evaluated that a potential of 30 Mtons of DMC per year are required to satisfy the potential demand of DMC as a fuel additive [85]. In this regard, it is expected to achieve this production level using a competitive direct industrial CO_2 coupling synthesis process to replace the current phosgene route. Many companies are making important investments and considerable research efforts to upgrade a CO_2-based cost-competitive process. The intensification of processes in terms of yield, energetic efficiency, mass/heat transfer rates, reaction control, and safety that offers the microreaction technology is regarded as one of the most attractive strategies for chemical synthesis of cyclic carbonates using CO_2-based competitive process. In 2013, Zhao et al. [86] reported for the first time a study on the utilization of a microreactor to carry out the cycloaddition of propylene oxide (PO) with CO_2 to give propylene carbonate (PC) using a hydroxyl-functionalized quaternary ammonium salt as a catalyst. They analyzed the effects of different operating conditions (reaction temperature, pressure, residence time, molar ratio of CO_2/PO, and catalyst loading) and demonstrated that the mass transfer capability and the reaction rate in terms of space–time yield (STY) were notably improved in the microreactor compared to a conventional stirred reactor. More recently, Li et al. also explored the application of microreactor technology in the coupling reaction of CO_2 with epoxides using a binary aluminum *salen* complex/quaternary ammonium salt as a catalytic system. To analyze the applicability of the microreactor, a series of monosubstituted terminal epoxides and ethylene oxide were tested, and all the terminal cyclic carbonates were obtained with excellent yields superior to 90% and high selectivity of 99% under 150 °C and 2.0 MPa within the residence time of less than 100 seconds because of the intensification of "electrophile–nucleophile" synergistic effect for epoxides ring-opening.

These examples prove that the microreactor technology is a powerful tool for the suited industrial preparation of cyclic carbonates. To some extent, it validates the concept of "Novel Process Windows" proposed by Hessel, which indicates that the microprocess technology accelerates the kinetics of reactions reducing dramatically the reaction time and increasing the productivity of the target products [86, 87].

9.4 Concluding Remarks and Future Perspectives

Microreactor technology offers an attractive opportunity for the chemical and energy sector. The introduction of MRs with enhanced mass and heat transfer properties opens a new scenario to enhance the chemical processes in the industry in terms of productivity and safety. In this chapter, the most important aspects of the application of microreaction technology in the processes of capture and utilization of CO_2 are critically reviewed. Varied examples on the application of microreactors in CO_2 capture processes and the subsequent transformation into value-added products (methane, methanol, DME, C_{2+} hydrocarbons, and cyclic organic carbonates) are discussed, emphasizing the performances of microreactors in comparison to conventional reactors. However, it is important to note that in spite of the current development and the recent advances, MRs are not inherently free from mass/heat transfer limitations and the assumption of ideal conditions must be verified to perform intrinsic kinetic studies. In addition, the immobilization of the catalyst in the flow channels is another challenge to be overcome. Research in all aspects from catalysis to unit operation design and process configuration is still required to implement this technology.

References

1 Ehrfeld, V.H.W. and Lowe, H. (2000). *Microreactors, New Technology for Modern Chemistry*, vol. D-69469. Weinheim, Germany: Wiley-VCH GmbH.

2 Sitter, S., Chen, Q., and Grossmann, I.E. (2019). *Curr. Opin. Chem. Eng.* 25.

3 Ramshaw, C. (1995). "The Incentive for Process Intensification," Proceedings, 1st Intl. Conf. Proc. Intensif. for Chem. Ind., 18, BHR Group, London, p. 1.

4 Stankiewicz, A. and Moulijn, J.A. (2000). *Chem. Eng. Prog.* 96: 22–33.

5 Montebelli, A., Visconti, C.G., Groppi, G. et al. (2014). *Catal. Sci. Technol.* 4: 2846–2870.

6 Jähnisch, K., Hessel, V., Löwe, H., and Baerns, M. (2004). *Angew. Chem. Int. Ed.* 43: 406–446.

7 (a) Sattari-Najafabadi, M., Nasr Esfahany, M., Wu, Z., and Sunden, B. (2018). *Chem. Eng. Process.* 127: 213–237. (b) Walter, S., Malmberg, S., Schmidt, B., and Liauw, M.A. (2005). *Catal. Today* 110: 15–25.

8 Commenge, J.M., Falk, L., Corriou, J.P., and Matlosz, M. (2002). *AIChE J.* 48: 345–358.

9 (a) Yue, J. (2018). *Catal. Today* 308: 3–19. (b) Woitalka, A., Kuhn, S., and Jensen, K.F. (2014). *Chem. Eng. Sci.* 116: 1–8.

10 Atkinson, D. (2010). *Biofuels, Bioprod. Biorefin.* 4: 12–16.

11 (a) Kockmann, N. and Roberge, D.M. (2011). *Chem. Eng. Process.* 50: 1017–1026. (b) Basavaraju, K.C., Sharma, S., Maurya, R.A., and Kim, D.-P. (2013). *Angew. Chem. Int. Ed.* 52: 6735–6738.

12 (a) Watts, P. and Wiles, C. (2007). *Chem. Commun.*: 443–467. (b) Suryawanshi, P.L., Gumfekar, S.P., Bhanvase, B.A. et al. (2018). *Chem. Eng. Sci.* 189: 431–448.

13 Sanz, O., Echave, F.J., Romero-Sarria, F. et al. (2013). *Renewable Hydrogen Technologies* (Chapter 9) (eds. L.M. Gandía, G. Arzamendi and P.M. Diéguez), 201–224. Amsterdam: Elsevier.

14 (a) Das, S. and Srivastava, V.C. (2016). *Photochem. Photobiol. Sci.* 15: 714–730. (b) Sun, J., Yun, C., Cui, B. et al. (2018). *Polymers* 10: 1209. (c) Cristaldi, D.A., Yanar, F., Mosayyebi, A. et al. (2018). *New Biotechnol.* 47: 1–7.

15 Parra-Cabrera, C., Achille, C., Kuhn, S., and Ameloot, R. (2018). *Chem. Soc. Rev.* 47: 209–230.

16 Meille, V. (2006). *Appl. Catal., A* 315: 1–17.

17 Almeida, L.C., Echave, F.J., Sanz, O. et al. (2010). *Studies in Surface Science and Catalysis*, vol. 175 (eds. E.M. Gaigneaux, M. Devillers, S. Hermans, et al.), 25–33. Elsevier.

18 Yoshida, J., Nagaki, A., Iwasaki, T., and Suga, S. (2005). *Chem. Eng. Technol.* 28: 259–266.

19 Pask, S.D., Cai, Z., Mack, H. et al. (2013). *Macromol. React. Eng.* 7: 98–106.

20 Naher, S., Orpen, D., Brabazon, D. et al. (2011). *Simul. Modell. Pract. Theory* 19: 1088–1095.

21 Kositanont, C., Putivisutisak, S., Tagawa, T. et al. (2014). *Chem. Eng. J.* 253: 332–340.

22 Tanimu, A., Jaenicke, S., and Alhooshani, K. (2017). *Chem. Eng. J.* 327: 792–821.

23 Ober, T.J., Foresti, D., and Lewis, J.A. (2015). *Proc. Natl. Acad. Sci. U.S.A* 112: 12293–12298.

24 (a) Kumar, V., Paraschivoiu, M., and Nigam, K.D.P. (2011). *Chem. Eng. Sci.* 66: 1329–1373. (b) Rebrov, E.V., Schouten, J.C., and de Croon, M.H.J.M. (2011). *Chem. Eng. Sci.* 66: 1374–1393.

25 (a) Yin, Y., Fu, T., Zhu, C. et al. (2019). *Sep. Purif. Technol.* 210: 541–552. (b) Guo, R., Zhu, C., Yin, Y. et al. (2019). *J. Ind. Eng. Chem.* 75: 194–201.

26 Zhao, C.-X. and Middelberg, A.P.J. (2011). *Chem. Eng. Sci.* 66: 1394–1411.

27 Yue, J., Luo, L., Gonthier, Y. et al. (2008). *Chem. Eng. Sci.* 63: 4189–4202.

28 Yang, L., Shi, Y., Abolhasani, M., and Jensen, K.F. (2015). *Lab Chip* 15: 3232–3241.

29 Glatzel, T., Litterst, C., Cupelli, C. et al. (2008). *Comput. Fluids* 37: 218–235.

30 Ho, C.-D., Chang, H., Chen, H.-J. et al. (2011). *Int. J. Heat Mass Transfer* 54: 3740–3748.

31 (a) Arzamendi, G., Diéguez, P.M., Montes, M. et al. (2010). *Chem. Eng. J.* 160: 915–922. (b) Cruz, S., Sanz, O., Poyato, R. et al. (2011). *Chem. Eng. J.* 167: 634–642.

32 Sun, B., Jiang, J., Shi, N., and Xu, W. (2016). *Proc. Safety Prog.* 35: 365–373.

33 Xia, G., Ma, D., Zhai, Y. et al. (2015). *Energy Convers. Manage.* 105: 848–857.

34 Kays, W.M. and Crawford, E.E. (2005). *Convective Heat and Mass Transfer*, 4e. Boston, MA: McGraw-Hill.

35 Kandlikar, S.G. (2014). Chapter 3 - Single-phase liquid flow in minichannels and microchannels. In: *Heat Transfer and Fluid Flow in Minichannels and Microchannels*, 2e (eds. S.G. Kandlikar, S. Garimella, D. Li, et al.), 103–174. Oxford: Butterworth-Heinemann.

36 (a) Schönfeld, F. and Hardt, S. (2004). *AlChE J.* 50: 771–778. (b) Mielke, E., Plouffe, P., Koushik, N. et al. (2017). *React. Chem. Eng.* 2: 763–775.

37 Mei, F., Parida, P., Jiang, J. et al. (2008). *J. Microelectromech. Syst.* 17: 869–881.

38 Chan, S.M., Chong, K.H., and Wong, B.T. (2018). *J. Mech. Eng. Sci.* 12: 3724–3737.

39 (a) Rochelle, G.T. (2009). *Science* 325: 1652–1654. (b) Svendsen, H.F., Hessen, E.T., and Mejdell, T. (2011). *Chem. Eng. J.* 171: 718–724. (c) Raynal, L., Bouillon, P.-A., Gomez, A., and Broutin, P. (2011). *Chem. Eng. J.* 171: 742–752.

40 (a) Cheung, O., Bacsik, Z., Liu, Q. et al. (2013). *Appl. Energy* 112: 1326–1336. (b) González, A.S., Plaza, M.G., Rubiera, F., and Pevida, C. (2013). *Chem. Eng. J.* 230: 456–465.

41 (a) Kovvali, A.S. and Sirkar, K.K. (2002). *Ind. Eng. Chem. Res.* 41: 2287–2295. (b) Duan, S., Taniguchi, I., Kai, T., and Kazama, S. (2012). *J. Membr. Sci.* 423–424: 107–112.

42 Rezaei, M., Ismail, A.F., Bakeri, G. et al. (2015). *Chem. Eng. J.* 260: 875–885.

43 (a) Gutknecht, V., Snæbjörnsdóttir, S.Ó., Sigfússon, B. et al. (2018). *Energy Procedia* 146: 129–134. (b) Naraharisetti, P.K., Yeo, T.Y., and Bu, J. (2019). *Renewable Sustainable Energy Rev.* 99: 220–233.

44 Güleç, F., Meredith, W., Sun, C.-G., and Snape, C.E. (2019). *Fuel* 244: 140–150.

45 (a) Tunnier, M.J., van Sint Annaland, M., Kramer, G.J., and Kuipers, J.A.M. (2010). *Chem. Eng. Sci.* 65: 114–119. (b) Song, C., Liu, Q., Deng, S. et al. (2019). *Renewable Sustainable Energy Rev.* 101: 265–278.

46 Mandal, B.P. and Bandyopadhyay, S.S. (2006). *Chem. Eng. Sci.* 61: 5440–5447.

47 Gao, N.-N., Wang, J.-X., Shao, L., and Chen, J.-F. (2011). *Ind. Eng. Chem. Res.* 50: 6369–6374.

48 (a) Ganapathy, H., Shooshtari, A., Dessiatoun, S. et al. (2014). *Appl. Energy* 119: 43–56. (b) Li, W., Liu, K., Simms, R. et al. (2012). *J. Am. Chem. Soc.* 134: 3127–3132. (c) Yao, C., Zhu, K., Liu, Y. et al. (2017). *Chem. Eng. J.* 319: 179–190. (d) Pan, M.-Y., Qian, Z., Shao, L. et al. (2014). *Sep. Purif. Technol.* 125: 52–58. (e) Liu, H., Yao, C., Zhao, Y., and Chen, G. (2017). *Chem. Eng. J.* 307: 776–784. (f) Ganapathy, H., Shooshtari, A., Dessiatoun, S. et al. (2015). *Chem. Eng. J.* 266: 258–270.

49 (a) Ye, C., Dang, M., Yao, C. et al. (2013). *Chem. Eng. J.* 225: 120–127. (b) Aghel, B., Heidaryan, E., Sahraie, S., and Nazari, M. (2018). *J. CO2 Util.* 28: 264–273.

50 Ye, C., Chen, G., and Yuan, Q. (2012). *Chin. J. Chem. Eng.* 20: 111–119.

51 Li, C., Zhu, C., Ma, Y. et al. (2014). *Int. J. Heat Mass Transfer* 78: 1055–1059.

52 Zhang, S., Lu, Y., Gu, Y. et al. (2018). *Chem. Eng. Process.* 132: 42–47.

53 (a) Hu, B., Guild, C., and Suib, S.L. (2013). *J. CO2 Util.* 1: 18–27. (b) Perathoner, S. and Centi, G. (2014). *ChemSusChem* 7: 1274–1282. (c) Rafiee, A., Rajab Khalilpour, K., Milani, D., and Panahi, M. (2018). *J. Environ. Chem. Eng.* 6: 5771–5794.

54 (a) Li, J., Wang, L., Cao, Y. et al. (2018). *Chin. J. Chem. Eng.* 26: 2266–2279. (b) Wu, J. and Zhou, X.-D. (2016). *Chin. J. Catal.* 37: 999–1015.

55 (a) Hu, J., Ma, J., Zhu, Q. et al. (2016). *Green Chem.* 18: 382–385. (b) Tabanelli, T., Bonincontro, D., Albonetti, S., and Cavani, F. (2019). Chapter 7 - Conversion

of CO_2 to valuable chemicals: organic carbonate as green candidates for the replacement of noxious reactants. In: *Studies in Surface Science and Catalysis*, vol. 178 (eds. S. Albonetti, S. Perathoner and E.A. Quadrelli), 125–144. Elsevier.
56 Sencerens, J.B. and Sabatier, P. (1902). *Comptes Rendus Acad. Sci.* 82: 514–516.
57 (a) Vogt, C., Monai, M., Kramer, G.J., and Weckhuysen, B.M. (2019). *Nat. Catal.* 2: 188–197. (b) Holladay, J.D., Brooks, K.P., Wegeng, R. et al. (2007). *Catal. Today* 120: 35–44.
58 Navarro, J.C., Centeno, M.A., Laguna, O.H., and Odriozola, J.A. (2018). *Catalysts* 8: 578.
59 Brooks, K.P., Hu, J., Zhu, H., and Kee, R.J. (2007). *Chem. Eng. Sci.* 62: 1161–1170.
60 Engelbrecht, N., Chiuta, S., Everson, R.C. et al. (2017). *Chem. Eng. J.* 313: 847–857.
61 Lange, J.-P. (2001). *Catal. Today* 64: 3–8.
62 Räuchle, K., Plass, L., Wernicke, H.-J., and Bertau, M. (2016). *Energy Technol.* 4: 193–200.
63 Quadrelli, E.A., Centi, G., Duplan, J.-L., and Perathoner, S. (2011). *ChemSusChem* 4: 1194–1215.
64 Pérez-Fortes, M., Schöneberger, J.C., Boulamanti, A., and Tzimas, E. (2016). *Appl. Energy* 161: 718–732.
65 Rodriguez, J.A., Liu, P., Stacchiola, D.J. et al. (2015). *ACS Catal.* 5: 6696–6706.
66 Bakhtiary-Davijany, H., Hayer, F., Kim Phan, X. et al. (2011). *Chem. Eng. Sci.* 66: 6350–6357.
67 Bakhtiary-Davijany, H., Hayer, F., Kim Phan, X. et al. (2020). *Catal. Today* 343: 226–233.
68 Phan, X.K., Bakhtiary, H.D., Myrstad, R. et al. (2010). *Ind. Eng. Chem. Res.* 49: 10934–10941.
69 Liang, Z., Gao, P., Tang, Z. et al. (2017). *J. CO2 Util.* 21: 191–199.
70 Venvik, H.J. and Yang, J. (2017). *Catal. Today* 285: 135–146.
71 Bansode, A. and Urakawa, A. (2014). *J. Catal.* 309: 66–70.
72 (a) Hayer, F., Bakhtiary-Davijany, H., Myrstad, R. et al. (2013). *Chem. Eng. Proc.* 70: 77–85. (b) Hayer, F., Bakhtiary-Davijany, H., Myrstad, R. et al. (2011). *Chem. Eng. J.* 167: 610–615.
73 Li, W., Wang, H., Jiang, X. et al. (2018). *RSC Adv.* 8: 7651–7669.
74 Aresta, M., Dibenedetto, A., and Angelini, A. (2014). *Chem. Rev.* 114: 1709–1742.
75 Ma, J., Sun, N., Zhang, X. et al. (2009). *Catal. Today* 148: 221–231.
76 Sun, Y., Yang, G., Wen, C. et al. (2018). *J. CO2 Util.* 24: 10–21.
77 (a) Sakakura, T., Choi, J.-C., and Yasuda, H. (2007). *Chem. Rev.* 107: 2365–2387. (b) North, M., Pasquale, R., and Young, C. (2010). *Green Chem.* 12: 1514–1539.
78 Srivastava, R., Bennur, T.H., and Srinivas, D. (2005). *J. Mol. Catal. A Chem.* 226: 199–205.
79 (a) Martín, C., Fiorani, G., and Kleij, A.W. (2015). *ACS Catal.* 5: 1353–1370. (b) Cokoja, M., Wilhelm, M.E., Anthofer, M.H. et al. (2015). *ChemSusChem* 8: 2436–2454. (c) Caló, V., Nacci, A., Monopoli, A., and Fanizzi, A. (2002). *Org. Lett.*

4: 2561–2563. (d) Kruper, W.J. and Dellar, D.D. (1995). *J. Organomet. Chem.* 60: 725–727.

80 (a) Li, Q., Zhang, W., Zhao, N. et al. (2006). *Catal. Today* 115: 111–116. (b) Yamaguchi, K., Ebitani, K., Yoshida, T. et al. (1999). *J. Am. Chem. Soc.* 121: 4526–4527.

81 Xian-Dong, L., Xiao-Fang, L., and Liang-Nian, H. (2015). *Curr. Org. Chem.* 19: 681–694.

82 Srivastava, R., Srinivas, D., and Ratnasamy, P. (2003). *Catal. Lett.* 89: 81–85.

83 (a) Song, L., Zhang, X., Chen, C. et al. (2017). *Microporous Mesoporous Mater.* 241: 36–42. (b) Maina, J.W., Pozo-Gonzalo, C., Kong, L. et al. (2017). *Mater. Horiz.* 4: 345–361.

84 (a) Peppel, W.J. (1958). *Ind. Eng. Chem.* 50: 767–770. (b) Metcalfe, I.S., North, M., and Villuendas, P. (2013). *J. CO2 Util.* 2: 24–28.

85 (a) Zevenhoven, R., Eloneva, S., and Teir, S. (2006). *Catal. Today* 115: 73–79. (b) Aresta, M. and Dibenedetto, A. (2004). *Catal. Today* 98: 455–462.

86 Zhao, Y., Yao, C., Chen, G., and Yuan, Q. (2013). *Green Chem.* 15: 446–452.

87 Hessel, V. (2009). *Chem. Eng. Technol.* 32: 1655–1681.

10

Analysis of High-Pressure Conditions in CO_2 Hydrogenation Processes

Andrea Álvarez Moreno[1], Esmeralda Portillo[3], and Oscar Hernando Laguna[2]

[1] *Universidad de Boyacá, Departamento de Química y Bioquímica, Carrera 2a Este No. 64-169, Tunja, 150003, Colombia*
[2] *Universidad de Jaén, Departamento de Ingeniería Química, Ambiental y de los Materiales, Escuela Politécnica Superior de Linares, Av de la Universidad s/n, 23700 Linares (Jaén), Spain*
[3] *Universidad de Sevilla, Escuela Técnica Superior de Ingeniería, Departamento de Ingeniería Química y Ambiental, C/Camino de los Descubrimientos s/n, 41092 Seville, Spain*

10.1 Introduction

CO_2 can be seen as the key molecule to proceed from a fossil fuel-based economy toward a more sustainable chemical industry. If analyzed, the majority of bulk chemicals (hydrocarbons, olefins, and alcohols) are organic compounds that contain essentially carbon and hydrogen. Currently, those elements come mainly from fossil feedstocks such as oil and natural gas [1]. However, taking into account that almost 60% of the global CO_2 emissions are related to energy generation and manufacturing processes, a highly advantageous strategy could be the valorization of the carbon dioxide waste stream in order to use it as a C-1 building block [2].

However, even if it is a very appealing strategy, CO_2 is a very stable molecule, and its activation can become quite challenging. Usually, for the large-scale production of useful chemicals and fuels, a highly energy reactant is needed along with continuous processes using heterogeneous catalysis [1, 3]. Hydrogen is usually the preferred energy reactant; nevertheless, in order to achieve an advisable strategy from an energy, environmental, and economic standpoint, H_2 needs to be obtained by renewable sources such as water electrolysis by using solar, geothermal, hydroelectric, or wind power [1, 4–6].

Catalytic heterogeneous conversion of CO_2 by renewable H_2 can generate different products such as formic acid (HCOOH) [7, 8], carbon monoxide (CO) [9, 10], methanol (CH_3OH) [11, 12], and methane (CH_4) [13, 14]. Each of these molecules can be converted to further bulk chemicals. For example, a mixture of carbon monoxide and hydrogen (syngas) can be converted to olefins via the Fischer–Tropsch reaction [2, 5]. Formic acid, apart from being a valuable chemical commonly used as a preservative and an antibacterial agent, is an established hydrogen storage component via its decomposition to CO_2 and H_2 [15–17]. Methanol, besides being one of

Engineering Solutions for CO_2 Conversion, First Edition.
Edited by Tomas R. Reina, José A. Odriozola, and Harvey Arellano-Garcia.

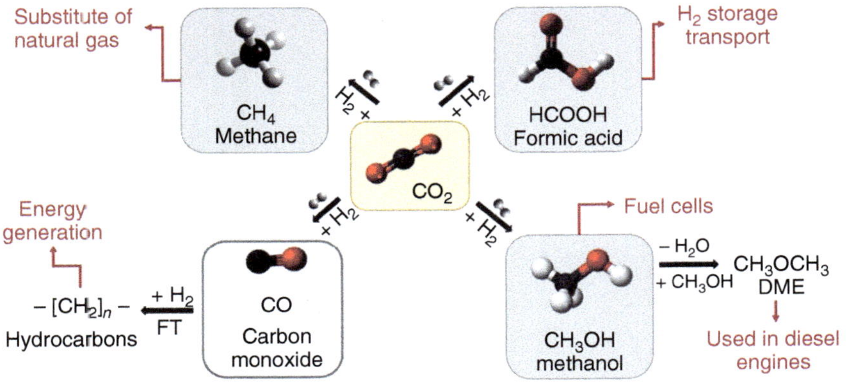

Figure 10.1 Bulk chemicals based in CO_2 hydrogenation reactions. Products with gray background indicate the ones whose yields are improved by the use of high pressure.

the top five commodity chemicals shipped around the world, is an intermediate for the production of formaldehyde, acetic acid, and dimethyl ether [11, 18, 19]. Last but not least, methane is a molecule that can be used as a substitute of natural gas taking advantage of the already well-known infrastructure [2] (Figure 10.1).

Currently, at laboratory scale, in order to enhance the product yield and selectivity, new catalysts are designed and new reaction conditions are optimized. One of those conditions is the use of high pressure. From a chemical point of view, high-pressure conditions promote the selectivity and rate reaction owed to the better thermodynamic conditions related to chemical equilibria. From a physicochemical point of view, the use of high pressure enhances the solubility and phase conditions of the components involved, especially in supercritical state [20].

However, between the high-impact CO_2 hydrogenation reactions shown in Figure 10.1 (also given in Table 10.1), thermodynamic calculations suggest that the use of high pressure does not enhance all chemical reactions. In the case of CO production via reverse water gas shift reaction, thermodynamic calculations showed that even an increase of pressure from 10 to 100 bars does not affect the CO_2 conversion. The critical factor of this reaction is basically temperature [21]. In the generation of formic acid, an enhance in the CO_2 conversion is reported when raising the pressure form 1 bar toward 300 bars [21]. However, the calculated yields toward formic acid are extremely low (less than 0.01%), mainly owed to the kinetic constrains. Regarding methanol, several authors have pointed out how low temperatures and high pressures can significantly boost the methanol production [11, 19, 21–23]. A higher pressure leads to higher CO_2 conversions at the same reaction temperature. It is shown that under atmospheric conditions, CO_2 can be hardly converted toward methanol (conversion <1%); however, a CO_2 conversion around 100% can be achieved under 50 bars at 100 °C and can be maintained at up to 250 °C under both pressures of 200 and 300 bar. Lastly, in the case of methanation reactions, thermodynamic calculations show that low temperatures and high pressures are favorable for CH_4 production because it is a mole reducing and an exothermic reaction [15, 24].

Table 10.1 Gibbs free energy, enthalpy change, and standard equilibrium constants for the hydrogenation of CO_2 to carbon monoxide, methane, methanol, and formic acid.

Bulk chemical	CO_2 hydrogenation reaction	ΔG_r° (298 K) (kJ mol^{-1}) [21]	ΔH_r° (298 K) (kJ mol^{-1}) [21]	K (298 K) [21]	Equilibrium constant expression
CO	$CO_2 + H_2 \leftrightharpoons CO + H_2O$	28.6	41.2	9.67×10^{-6}	$K = \frac{P_{CO} \times P_{H_2O}}{P_{CO_2} \times P_{H_2}}$
CH_4	$CO_2 + 4H_2 \leftrightharpoons CH_4 + 2H_2O$	−113.5	−165.0	7.79×10^{19}	$K = \frac{P_{CH_4} \times P^2_{H_2O}}{P_{CO_2} \times P^4_{H_2}}$
CH_3OH	$CO_2 + 3H_2 \leftrightharpoons CH_3OH + H_2O$	3.5	−49.3	2.45×10^{-1}	$K = \frac{P_{CH_3OH} \times P_{H_2O}}{P_{CO_2} \times P^3_{H_2}}$
HCOOH	$CO_2 + H_2 \leftrightharpoons HCOOH$	43.5	14.9	7.79×10^{-8}	$K = \frac{P_{HCOOH}}{P_{CO_2} \times P_{H_2}}$

This chapter presents the thermodynamic aspects of the high-pressure use in hydrogenation reactions, together with an overview of the current state of this technology in industry and a techno-economic profile of the CO_2 hydrogenation toward methanol from a CCU approach.

10.2 Thermodynamic Aspects

Thermodynamic calculations of chemical reactions are helpful in understanding and predicting catalytic processes [21]. They can provide preliminary information of the thermodynamic stability of a desired chemical specie, the yields, and selectivity of a desired product and the impact of reaction parameters such as temperature, pressure, and reactant ratio.

The vast majority of the thermodynamic calculations related to CO_2 hydrogenation reactions look for the optimization of the reaction parameters such as temperature, pressure, or reactant ratio, in order to maximize the product yield [21, 23–28]. This optimization is performed through equilibrium calculations.

The equilibrium state plays a central role in thermodynamics. This state is characterized by the absence of temperature, pressure, or concentration changes, all flows of heat, mass, or work between the system and its surroundings are zero, as well is zero, the net rate of all chemical reactions [29]. At first, it might appear that the characteristics of the equilibrium state are so restrictive that is rarely achieved. Nevertheless, the opposite is true. The system will always tend to the equilibrium [29].

However, it is worth mentioning that this state does not only apply for closed systems. Systems with open mass, temperature, or pressure flows may also evolve to an equilibrium state. This last case applies to chemical reactions. Chemical reactions tend to move toward a dynamic equilibrium in which both reactants and products are present but have no further tendency to undergo net change.

The equilibrium conditions for chemical reaction equilibria can be derived by using the Gibbs energy (G) of the reaction. According to the second law of thermodynamics, the total Gibbs energy of a closed system at constant temperature and pressure is minimum at equilibrium, but the real question here is: How does pressure affect the equilibrium? Sections 10.2.1 and 10.2.2 are expected to answer this question.

10.2.1 Le Chatelier Principle as a Simple Way to Understand the Effect of Pressure in Chemical Reactions

Let us assume the simplest situation with the hypothetical equilibrium of two gases A and B:

$$A_{(g)} \leftrightarrow B_{(g)} \tag{10.1}$$

The reaction Gibbs energy (ΔG_r) of this system can be defined as the difference between the chemical potentials (μ) of the reactants and products at the composition of the reaction mixture:

$$\Delta G_r = \mu_B - \mu_A \tag{10.2}$$

As thermodynamic implies, the spontaneity of a reaction can be expressed in terms of the reaction Gibbs energy: if $\Delta G_r < 0$, the forward reaction is spontaneous, if $\Delta G_r > 0$, the reverse reaction is spontaneous, and if $\Delta G_r = 0$, the reaction is at equilibrium [30]. This means that, in our hypothetical reaction between A and B, the reaction toward B is spontaneous when $\mu_A > \mu_B$, the reverse reaction is spontaneous when $\mu_A < \mu_B$, and the reaction is at equilibrium when $\mu_A = \mu_B$. Implying that if we find the composition of the reaction mixture that ensures $\mu_A = \mu_B$, then we can identify the composition of the reaction mixture at equilibrium.

If A and B are perfect gases, the chemical potential for each one of them can be expressed as

$$\mu_A = G_A^\circ + RT \ln P_A \qquad \text{where}: \quad P_A = \frac{P_A}{P^\circ} \tag{10.3}$$

$$\mu_B = G_B^\circ + RT \ln P_B \qquad \text{where}: \quad P_B = \frac{P_B}{P^\circ} \tag{10.4}$$

Implying that ΔG_r can be expressed:

$$\Delta G_r = \mu_B - \mu_A = (G_B^\circ + RT \ln P_B) - (G_A^\circ + RT \ln P_A)$$

$$= (G_B^\circ - G_A^\circ) + RT \ln \frac{P_B}{P_A}$$

$$\Delta G_r = \Delta G_r^\circ + RT \ln \frac{P_B}{P_A} \tag{10.5}$$

The ratio of P_B and P_A (partial pressures of A and B at the equilibrium) denotes the reaction quotient (Q), and this ratio becomes the equilibrium constant (K) when the condition of equilibrium is achieved ($\Delta G_r = 0$).

$$0 = \Delta G_r^\circ + RT \ln K$$

$$-\Delta G_r^\circ = \text{RT} \ln K \qquad \text{where}: \quad K = \frac{P_\text{B}}{P_\text{A}} \tag{10.6}$$

Equation (10.6) denotes one of the most important relations in thermodynamics, the link between the thermodynamic data of the reaction (Gibbs energy at standard conditions: ΔG_r°) and the equilibrium constant K. This relation implies that the equilibrium constant depends on the value of ΔG_r°, which is defined at a single standard pressure. The value of ΔG_r° and therefore the value of K are independent of the pressure at which the equilibrium is actually established. Nevertheless, the conclusion that K is independent of pressure does not necessarily means that the *equilibrium composition* is independent of the pressure.

Let us now assume the hypothetical equilibrium of two gases A and B that are being compressed in the reaction:

$$\text{A}_{(g)} \leftrightarrow 2\text{B}_{(g)} \tag{10.7}$$

The equilibrium constant, which also depends on the stoichiometric coefficients, is represented by the ratio of the partial pressures of B and A in the equilibrium:

$$K = \frac{(P_\text{B})^2}{(P_\text{A})} \tag{10.8}$$

If K is the equilibrium constant, the right hand of the equation will only remain constant if an increase in P_A cancels with an increase in the square of P_B. This ratio of increase between P_A and P_B will only occur if the equilibrium composition shifts in favor of A at the expense of B. Then, the number of A molecules will increase as the volume of the container is decreased, implying that its partial pressure will rise more rapidly. The increase in the number of A molecules and the corresponding decrease in the number of B molecules in the proposed reaction is a special case of a principle proposed by the French chemist Henri Le Chatelier, which states that "A system at equilibrium, when subjected to a disturbance, responds in a way that tends to minimize the effect of the disturbance." The principle implies that, if a system at equilibrium is compressed, then the reaction will adjust so as to minimize the increase in pressure. This can be done by reducing the number of particles in the gas phase, which implies A ← 2B.

Following this argument, in the case of the four hydrogenation reactions previously discussed (Figure 10.1), high-pressure conditions will increase the yield toward CH_4, CH_3OH, and HCOOH, which are the mole reducing reactions (Table 10.1), and agree to the previously discussed thermodynamic studies.

However, even if the Le Chatelier principle can give a general tendency of the expected products, in a real industrial scenario, when the composition of an equilibrium mixture is determined by a number of simultaneous reactions, calculations based solely on equilibrium constants become complex and tedious. A more suitable way for general computer solution is based on minimization of the total Gibbs energy [31].

10.2.2 Equilibrium Composition Calculations of High-Pressure Gas Reactions Based on the Computerized Gibbs Energy Minimization

The underlying fundamental principle is that at equilibrium at constant temperature and pressure, the Gibbs energy should be minimum. Therefore, what is done, is to develop a general expression for the Gibbs energy in terms of the number of moles of all the species present (reactants and products) in all phases. The calculations then vary the number of moles of each specie in each phase subject to stoichiometric constrains and find a solution that minimizes the total Gibbs energy of the system giving also the equilibrium composition achieved. In this way, the problem is transformed from solving a specific nonlinear algebraic set of equations to the problem of minimization of a function regardless of how many reactions or phases are involved. In this manner, a general minimization algorithm can be used to solve all chemical reaction equilibrium problems [32].

The first step into developing the function that is going to minimize is to formulate the constraining material–balance equations. This balance is based on conservation of the total number of atoms of each element in a system with "w" elements.

$$\sum_i n_i\, a_{ik} = A_k \quad \text{or} \quad \sum_i n_i\, a_{ik} - A_k = 0 \qquad (k = 1,\ 2, \ldots w) \tag{10.9}$$

Assuming each specie as "$_i$", n_i identifies the number of moles of the species i. Subscript "k" identifies a particular atom, implying that a_{ik} is the number of atoms of the k_{th} element present in each molecule of chemical specie i and A_k identify the total number of atomic masses of the k_{th} element in the feed.

In order to incorporate the constrains related to the conservation of the total amount of individual chemical elements into the body of the problem, a Lagrange (λ_k) multiplier is used:

$$\lambda_k \left(\sum_i n_i\, a_{ik} - A_k \right) = 0 \qquad (k = 1,\ 2, \ldots\ w) \tag{10.10}$$

Summed over k, these equations give:

$$\sum_k \lambda_k \left(\sum_i n_i\, a_{ik} - A_k \right) = 0 \qquad (k = 1,\ 2, \ldots\ w) \tag{10.11}$$

The constrained function (F) to be minimized should be formed by adding the total Gibbs energy (G_T):

$$F = G_T + \sum_k \lambda_k \left(\sum_i n_i\, a_{ik} - A_k \right) = 0 \tag{10.12}$$

Function F is identical with G_T because the summation term is zero.

The minimum value of both, F and G_T, is found when the partial derivatives of F with respect to n_i are set equal to zero. However, these partial derivatives of F and G_T with respect to n_i are different because function F incorporates the constrains of the material balances.

$$\left(\frac{\partial F}{\partial n} \right)_{T,P,n_j} = \left(\frac{\partial G_T}{\partial n} \right)_{T,P,n_j} + \sum_k \lambda_k\, a_{ik} = 0 \tag{10.13}$$

The first term in the right represents the partial molar Gibbs energy and implies the dependence of the Gibbs energy (G_T) on the composition of the mixture (n_i). This relation is also known as the **Chemical Potential** (μ) and was previously mentioned in Eqs. (10.3) and (10.4).

The equation to be optimized now becomes:

$$\left(\frac{\partial F}{\partial n}\right)_{T,P,n_j} = \mu + \sum_k \lambda_k \ a_{ik} = 0 \tag{10.14}$$

The chemical potential has a broader meaning because it measures the potential that a substance has for undergoing physical and chemical change in a system. Following the definition stated in Eqs. (10.3) and (10.4), chemical potential can be expressed as:

$$\mu_i = G_i^\circ + \text{RT} \ln \frac{P_i}{P^\circ} \tag{10.15}$$

This expression is only valid for gas-phase reactions with pure ideal gases at standard states and pressures (P°). However, to describe the processes that take advantage of high-pressure conditions, it should be considered that the basic equations are essentially the same used in the case of low pressure, but the difference is given by the peculiar behavior of fluids when the pressure is relatively high and close to the critical one. In this case, a high-density region is achieved where fluids exhibit properties similar to those of liquids [20]. Gases are far away from their ideal state; then, fugacities (f_i) should be used instead of pressures (P_i) because fugacity takes into account the effect of gas imperfections in a manner that resulted in the least upset of the form of equations.

Fugacity is related to the pressure by: $f_i = y_i \ \phi_i \ P$

where y_i is the mole fraction of specie i, ϕ_i is the fugacity coefficient of the specie i, and P is the pressure of the system.

Taking into account the previous definitions, the chemical potential can be expressed as:

$$\mu_i = G_i^\circ + \text{RT} \ln \frac{y_i \ \phi_i \ P}{P^\circ} \tag{10.16}$$

where G_i° is the standard Gibbs energy formation of species i, R is the ideal gas constant, T is the absolute temperature, and P° stands for the standard pressure.

The combination with Eq. (10.14) gives the final function that is going to be minimized:

$$\left(\frac{\partial F}{\partial n}\right)_{T,P,n_j} = G_i^\circ + \text{RT} \ln \frac{y_i \ \phi_i \ P}{P^\circ} + \sum_k \lambda_k \ a_{ik} = 0 \qquad (i = 1, \ 2, \dots \ N) \tag{10.17}$$

If species i is an element, G° is zero. There are N equilibrium equations, one for each chemical specie, and there are w material balance equations (Eq. (10.9)), one for each element. Consequently, there are $N + w$ equations. Regarding the unknowns, the first one are the moles of specie "i" (n_i), which are related to the mole fraction (y_i) expressed in Eq. (10.16) as $y_i = n_i / \Sigma_i \ n_i$ (N unknowns). The second one is the Lagrange multiplier λ_k (w unknowns). Therefore, there is a total of $N + w$ unknowns and $N - w$ equations, which imply that it is sufficient for the determination of all unknowns.

Eq. (10.17) is derived on the presumption that the fugacity coefficients (ϕ_i) are known. If the phase is an ideal gas, then each ϕ_i is unity; however, because gases at high pressure are far away from their ideal state, fugacity coefficients should be calculated using and equation of state with preliminary values of y_i.

Critical reviews of existing equations of state to model equilibrium compositions at high pressures have been made in recent years. When the critical point is approached, predictions and even correlations of critical curves and solubilities are extremely difficult because of the nonclassical behavior in this region [20]. The most widely used equations of state used to predict the fugacity coefficients are the van der Waals family of cubic equations of state.

The van der Waals equation is a model equation of state for real gases expressed in terms of two parameters, one corresponding to molecular attractions and the other corresponding to molecular repulsions. This equation captures the general features of the behavior of real gases, including their critical behavior.

In a general way, the van der Waals family of cubic equations of state has this form:

$$P = \frac{n\text{RT}}{V - nb} - \frac{an^2}{V^2} \tag{10.18}$$

The first term includes the van der Waals coefficient "b," which represents the strength of repulsive forces, while the second term with the coefficient "a" represents the strength of the attractive forces. Several modifications have been done to this basic equation. The modification by Redlich and Kwong who introduced a different temperature dependence and a slightly different volume dependency in the attractive term is very important because it opened the way to a better description of the temperature-dependent properties. However, it was Soave's modification of the temperature dependence of the "a" parameter, which resulted in accurate vapor pressure predictions (specially at above 1 bar), which lead to cubic equations of state of becoming important tool for the prediction of vapor–liquid equilibria at moderate high pressures [30, 31].

Being able to find the equilibrium composition of a system where multiple reactions are taking place is a very important tool in any chemical process. However, in order to perform these calculations, the basic requirement is the choice of the appropriate thermodynamic model and the knowledge of the parameters used in it. The choice of a set of species is equivalent to the choice of independent reactions among the species; similarly, the selection of the appropriate thermodynamic model can result in a more accurate equilibrium composition under high-pressure conditions.

10.3 Overview of Some Industrial Approaches Focused on the Production of Valuable Compounds form CO_2 Using a Carbon Capture and Utilization (CCU) Approach

Within the different valuable products obtained from CO_2 through hydrogenation cited above (Figure 10.1), special attention has been focused on methanol and methane because currently, the infrastructure for their production presents promising advances. However, considerable efforts are required yet for making these processes profitable, energy efficient, and environmentally friendly [33]. Most importantly, lots of research is needed in order to reach the scenario where all these bulk chemical production processes would have a substantial reduction of the greenhouse gas (GHG) emissions in order to avoid the global warming effects.

In this sense, aspects such as the reactor design, the suitability of new catalytic systems, or the control of the process variables such as the temperature or the pressure have to be reformulated aiming to enhance the yield of any CO_2 hydrogenation product. In case of high-pressure conditions, despite the clear thermodynamic advantages that it presents, it has not been applied in industry owed mainly to budget reasons.

It is clear that the machinery for launching carbon capture and utilization (CCU) technologies requires further developments. However, some examples of initiatives in which CO_2 is used as a raw material to produce fuels or value-added products are already in operation with promising results both at the environmental and economic level. Some of them will be presented below.

10.3.1 Industrial Production of Methanol

Within the scenario of an environmentally friendly production of methanol, the Methanol Institute[1] has established the category of "renewable methanol" to "*...an ultra-low carbon chemical produced from sustainable biomass, often called bio-methanol, or from carbon dioxide and hydrogen produced from renewable electricity*" [34]. It is estimated that renewable methanol can reduce carbon emissions by 65–95% depending on the feedstock and the applied procedure, which represents a superior reduction potential respect to those of other fuels such as gasoline, diesel, coal, and methane [34, 35]. This is based on smart approaches such as that of many companies that are considering to combine clean hydrogen (obtained by electrolysis using excess renewable energy) with excess CO_2 from industrial (such as steel or cement factories) to produce renewable methanol (Figure 10.2). Nevertheless, it has to be noticed that the inclusion of pressurization units is not remarked until now in the features of the developed technologies for producing renewable methanol, implying that this subject could be explored in the near future.

The main initiatives that have commercial developments at the present time were summarized in the most recent report of the methanol institute and are presented in Table 10.2. It is important to highlight that three different categories of methanol have been differentiated depending on the raw material or the final CO_2 footprint. Initially, the *biomethanol* is that obtained from waste biomass. Secondly, the *renewable methanol* is that where the use of waste CO_2 is included in the procedure and finally the *low carbon methanol* refers to technologies that achieve a reduction in the total GHG emissions compared with traditional procedures.

1 The Methanol Institute, founded in 1989, acts as global trade association for the methanol-based industry representing the world's leading producers, distributors and technology companies (https://www.methanol.org/).

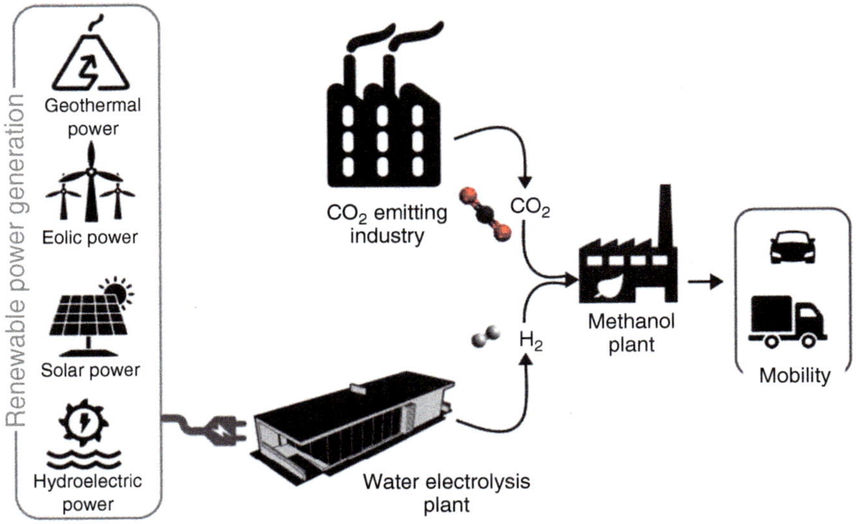

Figure 10.2 Renewable methanol production using renewable hydrogen and waste CO_2. Source: Modified from Hobson [34].

A relevant example of sustainable production of renewable methanol is the "*World's largest CO_2 methanol plant*" as they define themselves located in Iceland [36]. This is the George Olah Renewable Methanol Plant in Svartsengi, Near Grindavik.[2] Currently, they highlight the milestone in carbon CCU achieved in 2015 when the plant capacity was scaled from 1300 to 4000 metric tons per year, which results in a CO_2 recycling capacity of 5500 tons [36]. In their technology, the use of hydrogen obtained by electrolysis using renewable energy from a geothermal power plant is a key point [34, 36]. This demonstrates how a carbon CCU approach can be adapted to the environment using the strategic advantages of some regions such as the geothermal energy in Iceland.

Other case study highlighted by the Methanol Institute is that of Enerkem [37]. This Canadian company based on Montreal promotes the idea of "*manufacturing biofuels and renewable chemical products from non-recyclable waste*" since 2000 [34, 37]. Through their patented technology of four main steps, they are able to chemically extract and reuse carbon from nonrecyclable waste. Firstly, the feedstock is prepared by sorting, shredding, and drying. Then, this is gasified for the generation of syngas. The third step implies the cleaning of the obtained syngas and the fourth stage is the catalytic conversion into valuable chemicals including methanol [34, 37].

The most recent advance in the production of methanol (not included in the Methanol Institute's report [34] but included in Table 10.2) was announced by the company BASF in the 2019 [38]. They have achieved a process for obtaining what they called "climate-friendly" methanol. In their new process designed with the help of BASF's OASE® gas treatment technology, no CO_2 emissions are produced. The main reason for the zero CO_2 emissions is that, although methanol

2 The plant was named after Nobel Laureate George Olah, a pioneer in the concept of the methanol-based economy.

Table 10.2 Companies and institutions involved in the production of methanol in such a way that manages to reduce the emission of CO_2 into the atmosphere.

Methanol category	Commercial		Feasibility and R&D	
Biomethanol	• BASF (GER) • BioMCN (NL) • Enerkem (CAN)	• New fuel (DEN) • Nordic green (DEN)	• Biogo (GER) • Enerkem (NL) • LowLands Methanol Heveskes Energy (NL)	• NREL (USA) • Origin Materials (USA) • Södra (SE)
Renewable methanol	• CRI (IC) • Innogy (GER) • BASF (GER)		• Advanced Chemical Technologies (CAN) • Asahi Kasei (JPN) • Blue Fuel Energy (CAN) • bse Engineering (GER) • Catalytic Innovations (USA) • CRI (CN/GER) • Gensoric (GER) • Infraserv (GER) • Liquid wind (SE)	• MefCO2 (GER) • Neo-H2 (USA) • Port of Antwerp (BE) • Quantiam Technologies (CAN) • STEAG (GER) • Swiss Liquid Future (CH) • Thyssenkrupp (GER) • USC (USA) • ZASt (GER)
Low-carbon methanol	• SABIC (KSA) • QAFAC (QAT)	• Methanex (CAN) • GPIC (BAH)	• Carbon2Chem (GER) • FRESME (SE) • GasTechno (USA) • Haldor Topsoe (DEN)	• Maverick Synfuels (USA) • NCF (CN) • OPTIMeoH (GER)

Source: Modified from Hobson [34].

is synthesized using syngas previously generated by partial oxidation of natural gas, methanol is then distilled and the waste gases (CH_4, CO_2, CO, and H_2) are combusted in an Oxyfuel process with pure oxygen. Consequently, a small portion of flue gas (mainly CO_2) is generated, which is scrubbed and reinjected at the beginning of the process. This implies the use of additional hydrogen, which BASF aims to produce without any CO_2 emissions [38].

10.3.2 Production of Methane

The CO_2 methanation process is part the power-to-gas (PtG) technology, and this has emerged as an alternative to store surplus production in generating plants from solar or wind energy. Following a similar strategy as the one presented in Figure 10.2, the PtG technology connects the renewable power grids with the gas grid converting the surplus energy into clean hydrogen through water electrolysis that is used subsequently for converting CO_2 (preferably residual or captured from the atmosphere) into methane [33, 39–41]. For this purpose, the most common catalytic methanation concept, adiabatic fixed-bed methanation, is applied [33].

The relevance of the PtG technology in the current global scenario has motivated initiatives such as the creation of the European Power to Gas Platform, which is a *"joint body, based on an integrated network of stakeholders, which aims to explore the viability of power to gas in European countries"* [42]. One of the useful tasks developed by this platform is, for example, the monitoring of projects dedicated to PtG technology nowadays. Therefore, a list of the projects followed by the platform devoted to the production of CH_4 or both CH_4 and H_2, is presented in Table 10.3, where Germany's leadership is evident.

Within the different PtG demonstration plants and projects worldwide reviewed by Bailera et al. [43] and Götz et al. [33], some require special attention. In first place, the pioneer PtG pilot plant built in Japan in the Tohoku Institute of Technology in 2003. After the prototype plant developed in 1996, the expansion continued until the creation of an industrial pilot plant, able to produce 4 $Nm^3\ h^{-1}$ of H_2 and 1 $Nm^3\ h^{-1}$ of CH_4. The concept of plant was based on two reactors in series with an intermediate removal of water and an electrolyzer fed by a photovoltaic cell [43]. Secondly, the Audi e-gas plant (Wertle – Germany), which is the biggest PtG plant world. Within this facility, H_2 is generated from three alkaline electrolyzers with a total electrical power of 6 MW, while a nearby plant supplies the CO_2.

It has to be remarked that, as in the case of renewable methanol production at industrial scale cited above, the pressurization is not considered yet in the industry of PtG as a remarkable feature. Until now, the positive effect of pressure on selectivity toward the CH_4 production avoiding the undesired CO formation during the CO_2 methanation has been only demonstrated at laboratory scales [24, 44–46]. Therefore, for a better use of the advantages of performing high-pressure CO_2 methanation at higher production levels, further developments are required during the scaling up of the process in order to maximize the benefits of the plants that are established today.

10.4 Techno-Economic Considerations for the Methanol Production from a CCU Approach with the Use of High Pressure

The core of the carbon CCU technologies is the concept of using CO_2 as a raw material for obtaining valuable products. As it was previously established, this strategy results highly attractive from the environmental point of view because, from one side, it could end the constant dependence on fossil fuels and from other side, it could help to reduce GHG emissions. Despite this, some concerns arise about the energy investment and operation costs required for any CCU approach. In this manner, costs can be excessive, or even worse; the process can cause a CO_2 footprint greater than the problem that is tried to be solved. Therefore, in order to have an idea of the environmental impact and the potential profit that it could generate, it

Table 10.3 PtG (demonstration) projects in Europe reported by the European Power to Gas Platform.

Output product	Project	Country	Involved companies and/or institutions
CH_4	PtG Hungary Ltd.	Hungary	Electrochaea, MVM
	Solothurn (CH) – Regio Energie Solothurn	Switzerland	STORE&GO Project
	Stuttgart (D) – ZSW II	Germany	ZSW, ETOGAS GmbH, Fraunhofer IWES
	Schwandorf (D) – Eucolino: Schmack & Viessmann		MicrobEnergy GmbH, Schmack Biogas GmbH
	Straubing (D)		MicroPyros GmbH
	Alzey (G) – Exytron Null-Emission-Wohnanlage		EXYTRON GmbH, Leibniz-Institus für Katalyse (LIKAT)
	Niederaussem (D)		RWE Power AG
	Falkenhagen (D) – DVGW		DVGW-Forschungsstelle am EBI, Engler-Bunte-Institut am KIT
	Rozenburg (NL)	Netherlands	Stedin, TKI gas, Gemeente Rotterdam, Ressort Wonen, RvO Nederland
	Copenhagen DK Haldor Topsøe	Denmark	Haldor Topsøe
	Avedøre (DK) – BioCatProject		Hydrogenics, Audi, NEAS Energy, HMN Gashandal, BIOFOS, INSERO A/S
	Foulum (DK) – Electrochaea		Electrochaea, Nidus Partners LP, E.On, Erdgas Zürich, ewz, NEAS Energy, Aarhus University, EUDP, Energinet.dk
CH_4 and H_2	Fos-sur-Mer (F) – Jupiter 1000	France	GRTgaz, ATMOSTAT, CEA, CNR, Leroux & Lotz Technologies, Le Port de Marseille Fos, McPhy Energy, TIGF
	Dunkerque (F) – GRHYD project		ENGIE, GrDF, NVERT, AREVA Hydrogène et Stockage de l'énergie, CEA, McPhy Energy, INERIS, CETIAT and CETH2

Table 10.3 (Continued)

Output product	Project	Country	Involved companies and/or institutions
	Stuttgart (D) – ZSW I	Germany	Not specified
	Allendorf, Eder (D) – BioPower2Gas		IdE Institut dezentrale Energietechnologien GmbH, MicrobEnergy GmbH, EnergieNetz Mitte GmbH, EAM EnergiePlus GmbH, CUBE Engineering GmbH
	Emden (D) – I		Stadtwerke Emden GmbH
	Kirchheimbolanden (D) – RegEnKibo		e-rp GmbH, KIT, DVGW-Forschungstelle am Engler-Bunte-Institut des KIT, Transferstelle für Rationelle und Regenerative Energienutzung Bingen (TSB), Viessman Gruppe, Stadt Kirchheimbolanden
	Vienna (AT) – Underground Sun Storage	Austria	Rohöl-Aufsuchungs Aktiengesellschaft (RAG), University of Leoben, University of Natural Resources and Life Siences Vienna, Energy Institute at Johannes Kepler University Linz, Verbund, Axiom Angewandte Prozesstechnik, Nafta, Etogas, German Technical and Scientific Association for Gas and Water or DVGW, Hychico
	Delfzijl (NL)	Netherlands	Torrgas, Siemens, Stedin, Gasunie, A.Hak, Hanze-hogeschool/EnTranCe and stichting Energy Valley

Source: Gas [42].

is important to be aware of the major number of variables in any new process that arises.

This section develops a techno-economic analysis concerning a methanol production plant from a CCU approach with the use of high pressure. As it was mentioned before, methanol is one of the most important bulk chemicals worldwide. It is a chemical feedstock for various products (plastics, synthetic fibers, solvents, paints, etc.), it can be used as a hydrogen carrier for fuel cell applications, or can even be

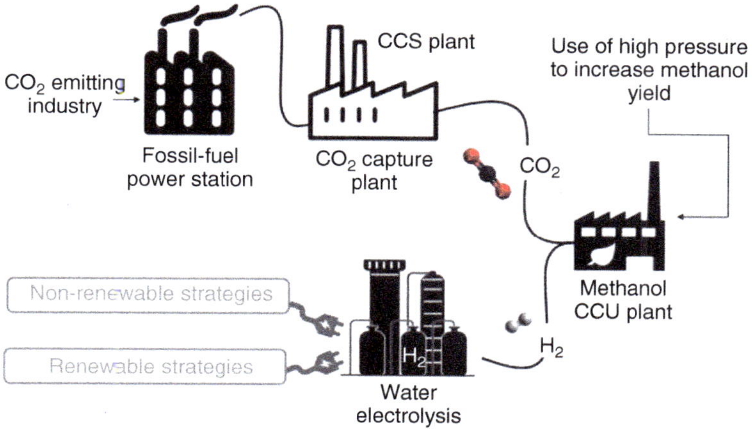

Figure 10.3 Proposed CCU methanol plant for the techno-economic evaluation.

used as a transportation fuel itself [47]; therefore, the evaluation of the feasibility of a CCU methanol plant is highly relevant.

A simplified scheme of the considered process is represented in Figure 10.3. In this scenario, carbon dioxide would be emitted by a fossil fuel power-driven plant. A CO_2 capture plant (CCS plant) is in charge of separating the CO_2 from the flue. Typically, the chemical absorption with a solution with monoethanolamine (MEA) or a sterically hindered amine (KS-1) is the most common process to retire the CO_2 under post-combustion capture. Hydrogen is supplied from a water electrolysis plant using alkaline technology (AEL). Energy for the electrolysis plant can come either from a nonrenewable or renewable source. Finally, both products will go into the CCU methanol plant where methanol production is expected to be boosted by the use of high pressure.

In order to evaluate the feasibility of the hydrogenation of captured CO_2 to methanol from a techno-economic point of view, the analysis is going to be based on several techno-economic studies related to the process [47–55]. The consulted information has been homogenized in order to compare feedstock, CO_2 and H_2 origin sources, energy sources, operating conditions, and aspects related to selling prizes and taxes.

For the analysis, three key indicators are considered: thermodynamic metrics, economic metrics, and environmental metrics (Figure 10.4). These metrics are chosen because they allow an assessment of the potential of the process from thermodynamic, economic, and environmental point of view.

Regarding the thermodynamic metrics, it is vital to evaluate the energy sources that are required for the production process, mainly, because these parameters affect both the economic and environmental metrics. In this sense, the thermal energy item is quantified by the MWh used for the cooling and heating systems to adjust the temperature in each section of the plant (carbon capture, hydrogen synthesis, or methanol production). Likewise, in order to dissipate the generated heat, the facility requires consumption of water, fuel, and steam that will affect the unitary variable

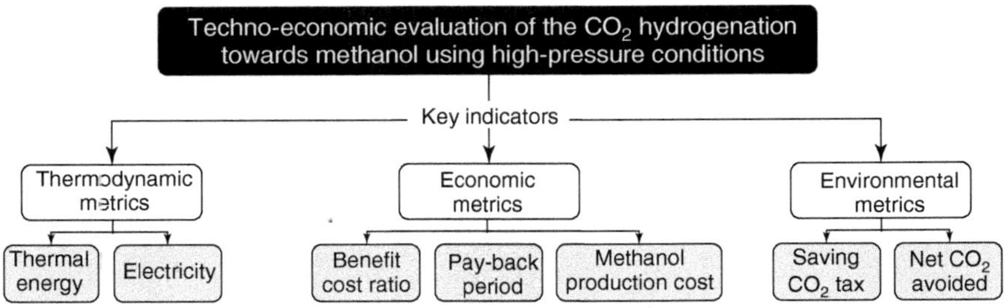

Figure 10.4 Strategy for the techno-economic analysis.

cost. Also, the electricity consumed in the process is another important parameter that should not be neglected. Regarding this, the evaluation was done taking into account whether the electricity source is derived from a power plant based on fossil fuel combustion or whether it was provided from renewable sources. In this sense, this aspect involves indirect CO_2 emissions that affect both the economic and environmental metrics.

Results of the thermodynamic metrics are expressed in Table 10.4 by the mass and energy balance for 1 ton yr^{-1} of methanol from hydrogenation of captured CO_2, considering an operation of 8000 h yr^{-1} and different compression conditions.

Table 10.4 shows that the required CO_2 varies between 1.38 and 1.68 ton CO_2/ton methanol, while hydrogen oscillates around 0.20 ton H_2/ton methanol, implying a CO_2/H_2 ratio higher than seven in most cases. Also, an output of oxygen is evidenced as a result of the alkaline water electrolysis (AEL) process. Concerning the energy balance, the energy sources (thermal energy, TE and electricity energy, EE) from each area of the process are specified, being the hydrogen production system the one that demands the most. In this sense, most cases estimate the electrolysis process consumes between a 96% and 98% of the total required MWh. According to Savaete and coworker [47], the electric energy requirement is lower when a renewable energy system is integrated to the electrolysis process, entailing an energy saving of 21%. Likewise, it is worth mentioning that the higher the pressure conditions (between 76 and 80 bar), the higher the consumption of MWh in the system.

Concerning the economic metrics, the main target is to verify if the methanol plant provides profit after keeping into account variables such as the methanol production cost, the annual net income, prices of feedstock or utilities, the years required to recover the cost of the initial investment (payback period, BBP), and the benefit/cost ratio (Table 10.5). In this sense, each researcher has considered several cases according to the price of electricity in different regions, type energy sources, and dates. In most cases, the selling of methanol and oxygen have been considered as Revenues, fixing their market price in 400 and 150 €/ton, respectively, and the penalty of emitted CO_2 has fixed between 4 and 6 €/ton CO_2 emitted to atmosphere, based on the study proposed by Smejkal and et al. [52]. This is a very conservative approach because, according to the EU Reference Scenario 2016, the expectations for the CO_2 allowance price will range from 85 to 200 €/CO_2-eq by 2050 [53].

Table 10.4 Mass and energy balance for the methanol CCU plant according to bibliographic studies.

Mass balance (ton/ton methanol)																
References	Bellotin [48]		Savaete[b] [47]		Savaete[b] [47]		Nyari [50]		Ericcson [49]		Perez-Fortes [51]		Simoes [54]		Xu [55]	
Compression[a]	30 bar		30 bar		30 bar		50 bar		76 bar		76 bar		78 bar		80	
Methanol production ($ton\ yr^{-1}$)	50 000		59 333		58 333		250 000		537 000		450 000		474 400		437 840	
Section	I	O	I	O	I	O	I	O	I	O	I	O	I	O	I	O
CO_2	1.44	—	1.39	—	1.38	0.26	1.43	—	1.68	—	1.46	0.91	1.48	—	1.40	—
H_2	0.197	—	N/S	N/S	N/S	N/S	0.196	—	0.23	—	N/S	—	0.20	—	0.19	—
Water	N/S		N/S	1.72	0.57	1.94	6	—	N/S		1.99	0.79	N/S		N/S	
MEA	N/S		N/S	0.003	0.003	0.003	0.003	0.001	N/S		N/S		N/S		N/S	
O_2	—	1.57	—	1.51	—	1.50	—	1.56	—	1.83	—	1.62	—	1.62	—	N/S
Energy balance ($MWh\ ton^{-1}$ MeOH)																
References	Bellotin [48]		Savaete[b] [47]		Savaete[b] [47]		Nyari [50]		Ericcson[49]		Perez-Fortes [51]		Simoes [54]		Xu [55]	
Section	TE	EE	TE	EE	TE	EE	TE	EE	TE	EE	TE	EE	TE	EE	TE	EE
CCS plant	1.21	0.07	N/S	N/S	N/S	N/S	N/S	0.20	N/S	0.42	N/S	N/S	1.3	0.07	N/S	N/S
CO_2 compression	—	0.10						0.00017		N/S		N/S	—	0.04		
H_2 compression	—	0.13						N/S		N/S		N/S	—	0.34		
Electrolysis AEL	—	10.33						10.03		11.39		11.78	—	10.93		
Total	1.21	10.63	N/S	11.4	N/S	9.02	N/S	10.23	N/S	11.81	N/S	11.95	1.3	11.36	N/S	13.74

N/S: not specified
I: Input
O: Output
TE: Thermal energy
EE: Electricity energy
MeOH: Methanol
a) Operating pressure fixed during methanol production.
b) Study with renewable and nonrenewable electricity for the process.

Table 10.5 Economic evaluation for the methanol CCU plant according to bibliographic studies.

	Economic evaluation											
References	**Bellotin [48]**			**Savaete [47]**		**Nyary [50]**			**Ericcson [49]; Perez-Fortes [51]**			**Simoes [53]**
Process pressure	30 bar			30 bar		50 bar			76 bar			78 bar
Cases[a)]	1	2	3	1	2	1	2	3	1	2	3	1
Prices electricity (€ MWh^{-1})	10	33.31	53.95	0.02	10	20	30	40	43	50	90	39.42
Selling price (€ ton^{-1} Methanol)	400			400		400			400		350	400
Selling price (€ ton^{-1} O_2)	150			150		100			150		150	150
Tax CO_2 (€ ton^{-1} $CO_{2\ emitted}$) [52]	4–6			4–6		4–6			4–6		4–6	4–6
Revenues (M€ yr^{-1})												
Methanol	21.2	21.2	21.2	23.7	23.3	100	100	100	360	360	229.9	189.8
Oxygen	12.4	12.4	12.4	13.4	13.2	39.0	39.0	39.0	147.3	147.3	388.3	115.2
TIC: Total investment cost (M€)	58.7	85.7	85.7	N/S	N/S	174.4	174.4	174.4	1022	1022	565	N/S
Variable annual cost (M€ yr^{-1})[b)]	5.2	18.6	30.2	0.005	0.007	68	95	121	249	299	494	175
Fixed annual cost (M€ yr^{-1})	4.29	4.29	4.29	N/S	N/S	5.00	5.00	5.00	39.30	39.30	31.00	N/S
Methanol production cost (€ ton^{-1} methanol)	186	433	650	N/S	N/S	524	632	736	690	782	2132	N/S
Annual net income	28.0	14.9	3.4	36.5	37.1	71.1	44.0	18.0	257.9	208.3	124.2	130.2
PBP: payback period (yr)[c)]	3	6	25	N/S	N/S	2	4	10	5	4	25	N/S
TBP: Total unitary benefit of selling (M€ yr^{-1})	34	34	34	37	37	139	139	139	507	507	618	305
BCR: benefit–cost ratio[d)]	0.18	0.08	0.05	N/S	N/S	0.27	0.22	0.19	0.74	0.65	0.29	N/S

N/S: not specified

a) Each research economically analyzes different cases where the difference is related to the prices of electricity in different regions or dates.
b) Depends on the cost of electricity in each case.
c) PBP is the amount of time to recover the total investment (TIC), considering the annual net income (ANI), which is calculated as the difference between the total annual revenues and total annual cost [48] (PBP=TIC/ANI).
d) BCR is defined as the ratio between the total unitary benefit of selling products (TBP) and the total unitary cost to make the product (TCP) [51] (BCR = TBP/TCP).

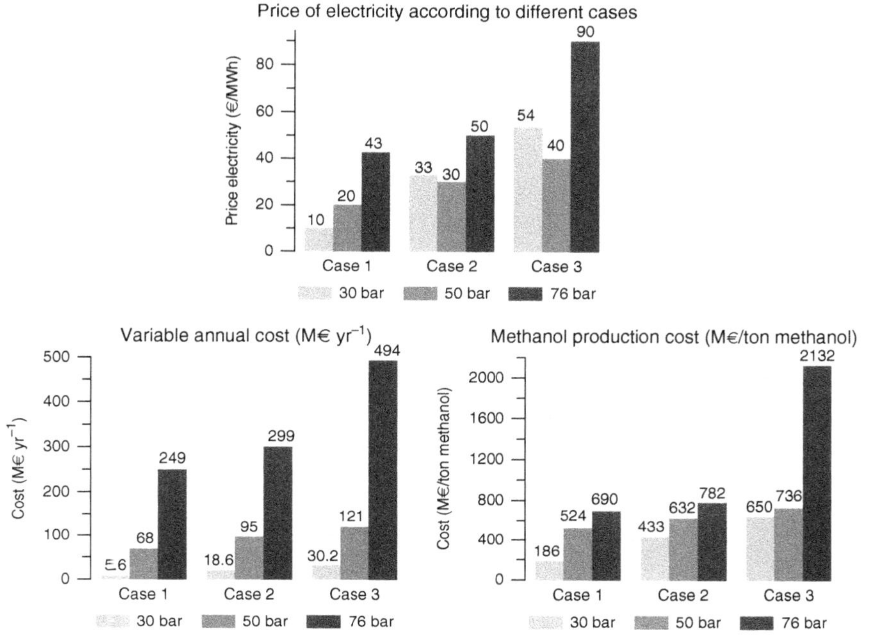

Figure 10.5 Evaluation of variable annual cost (M€ yr^{-1}) and methanol production cost (M€/ton methanol) according to the difference of electricity prices and pressure conditions during methanol synthesis.

Considering the bibliographic sources, the total investment cost (TIC) depends on the total methanol production and operating conditions. For example, **Ericsson** [49] estimated a TIC equal to 2022 M€ for a production of 537 000 ton yr^{-1} of methanol with a pressure of compression of 76 bar, whereas Bellotti et al. [48] calculated 58.7 M€ for a production of 50 000 ton yr^{-1} with a compression of 30 bar. These results reveal that the higher level of production and condition of pressure are, the higher investment cost is required.

As far as the operation cost is concerned, the electricity cost is the most responsible in the variable annual cost, contributing in a 98% in most cases. Considering all these parameters, the methanol production cost has been determined (Figure 10.5). Results show that the variable annual cost and methanol production cost are directly influenced by the electricity price and pressure conditions in each scenario analyzed. In this sense, the more expensive the electricity price is, the higher both variable annual and methanol production costs are.

Payback period and benefit–cost ratio are also influenced by prices of electricity and operating condition (Figure 10.6). For example, Bellotti et al. [48] calculated a loss equal to 88% of annual net income when the price of electricity is 53.9 € MWh^{-1} instead of 10 € MWh^{-1}, requiring 22 years additional to pay back. Therefore, the benefit–cost ratio decreases from 0.18 to 0.05 M€ yr^{-1}.

Regarding the environmental indicators, the amount of CO_2 in the process is the key. The results of the environmental metrics are shown in Table 10.6. *CO_2 outlet*

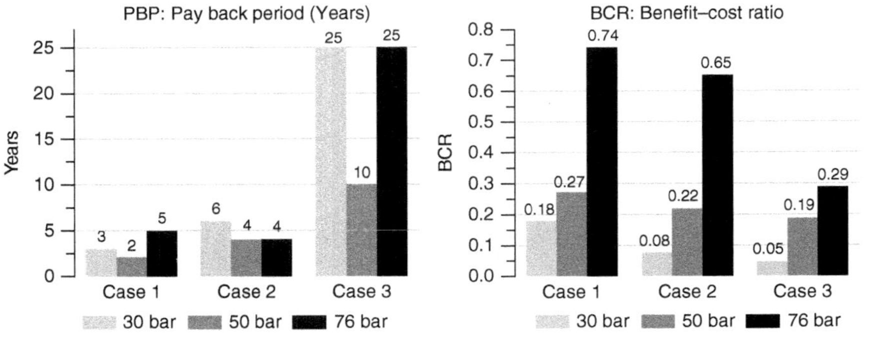

Figure 10.6 Evaluation of payback period (PBP) and benefit–cost ratio (BCR) according to the difference of electricity prices and pressure conditions during methanol synthesis.

refers to the CO_2 emissions as a result of the process, *CO_2 indirect*, refers to the indirect CO_2 emissions because of electricity consumption and, in order to analyze the potential of net reduction of CO_2 emissions and the economic savings owed to tax saving between the different processes at different pressures, the Net CO_2 avoided and the saving CO_2 taxes have also been calculated.

Taking into account that the "Net CO_2 avoided" refers to the whole amount of CO_2 that is not emitted into the atmosphere, it is very interesting to observe how this value increases gradually with the pressure applied to the process, implying that, even if high pressure conditions are more energy demanding, it does not imply a higher environmental impact.

It is worth mentioning that comparing these results with the conventional process of methanol production via steam reforming with natural gas as a raw material, huge differences are evidenced. Jarvis et al. [56] report that the net CO_2 emissions for a plant producing methanol from the hydrogenation of captured CO_2 are 0.226 ton CO_2/ton methanol, compared to 0.768 ton CO_2/ton methanol for a plant using the conventional syngas route. Moreover, because the discussed hydrogenation process utilizes up to 1.4 ton CO_2/ton methanol [57], the net amount of CO_2 avoided is in the region of 2 ton CO_2/ton methanol [51]. Pérez-Fortes et al. [51] also point out that the CO_2 indicators underline the positive CO_2 balance for the CCU process. The direct and indirect emissions of the CCU process are around 0.1 Mton yr^{-1} while the conventional plant emits 1.17 Mton yr^{-1}. Even if differences are evidenced when comparing the energy consumption and costs between the conventional plant and the CCU plant, the authors indicate that those differences are mainly due to the relatively high price of H_2 and the variable price for electricity, which agrees with the actual analysis.

According to Bradford and Vannice [58], the syngas production step in a conventional plant, which normally uses natural gas as raw material (including oxygen production and compression), may account for at least 60% of the investment. This step in a conventional plant (steam reforming of natural gas or fuel oil) also increases the consumption of water. Therefore, lower capital costs and water consumption for the CCU plant are expected in a process lacking the syngas synthesis step. Although

Table 10.6 Environmental evaluation for the methanol CCU plant according to bibliographic studies.

	Environmental evaluation							
References	**Bellotin [48]**	**Savaete [47]**		**Nyary [50]**	**Ericcson[b)][49]**	**Perez-Fortes [51]**	**Simoes [54]**	**Xu [55]**
Process pressure	30 bar	30 bar		50 bar	76 bar	76 bar	78 bar	80 bar
$CO_{2\ inlet}$ (ton yr^{-1})	72 125	81 000	82 000	357 000	900 000	657 000	704 000	611 600
$CO_{2\ outlet}$ (ton yr^{-1})[a)]	—	—	15 267	—	—	—	—	—
$CO_{2\ indirect}$ (ton yr^{-1})[b)]	205	3.5E-06	0.256	984	2354	2040	0.0044	2315
Net $CO_{2,avoided}$ (ton yr^{-1})[c)]	71 920	81 000	66 733	356 016	897 646	654 960	704 000	609 285
Tax CO_2, $_{w/o\ CCS}$ (M€)	0.36	0.40	0.41	1.79	4.50	3.29	3.52	3.06
Tax CO_2, $_{methanol\ process}$ (€)	1.02E-03	1.7E-06	76.3E-03	4.9E-03	0.012E-03	10.2E-03	0	11.6E-03
(STC) Tax CO_2 saving (€)/ton methanol [d)]	7.2	4.9	5.6	7.1	8.4	7.28	7.2	6.96

a) CO_2 emissions as a result of the process
b) CO_2 emissions due to electricity consumption.
c) Is a CO_2 balance that is calculated as the difference between the CO_2 used for methanol production and the direct ($CO_{2,\ out}$) and indirect ($CO_{2,\ indirect}$) CO_2 emissions after the process [48].

NET $CO_{2\ avoided} = [CO_2\ inlet - CO_2\ outlet - CO_2\ indirect]_{process}$

d) STC is the economic incentive related to CO_2 taxes when the CO_2 is reused instead of being sent to the atmosphere.

$STC = Tax_{CO_2\ \ without\ \ CCS} - Tax_{CO_2\ \ Methanol\ \ process}$

the CCU plant consumes more electricity than the conventional plant, the final balance of CO_2 emissions and capital cost shows a clear advantage for the CCU plant.

Considering all of the above, the CCU technology for obtaining valuable products, such as methanol, should offer not only environmental advantages but also it should be attractive from a technical-economic point of view. This section has shown that the evaluation of any PtG technology depends on operating conditions, the origin of raw materials, and the price of electricity. In the discussed scenario, the most energy demanding part of the CCU plant is the hydrogen production system; therefore, in order to maintain a negative CO_2 balance for the process, it is imperative that a renewable source is used to supply electricity. Environmental metrics show that even if high-pressure conditions are more energy demanding, it does not directly imply a notorious environmental impact. The economic indicators reflect that even if the use of higher pressures involves a greater investment cost, the final investment strongly depends on the price of electricity. In this context, the best situation should present a low price of electricity to achieve a high benefit–cost ratio and low PBP period for methanol production.

10.5 Concluding Remarks

After analyzing the thermodynamic aspects of CO_2 hydrogenation toward methanol and methane, there are clear pieces of evidence of the several advantages of working under high-pressure condition. However, scaling-up such processes often involve a humongous raise on the prices, owing to the settling up of equipment suitable to work under high-pressure conditions. Consequently, the industries often look for novel catalytic processes that might allow reducing considerably the working pressure, decreasing the initial investment cost. Nevertheless, the techno-economic analysis of a proposed CCU methanol plant showed that even if high-pressure conditions are more energy and budget, demanding from an environmental and economic point of view, the approach is a feasible challenge that should not be overlooked for a sustainable future.

References

1 Sternberg, A., Jens, C.M., and Bardow, A. (2017). Life cycle assessment of CO_2-based C1-chemicals. *Green Chem.* 19: 2244–2259.

2 Artz, J., Müller, T.E., Thenert, K. et al. (2018). Sustainable conversion of carbon dioxide: an integrated review of catalysis and life cycle assessment. *Chem. Rev.* 118: 434–504.

3 Álvarez, A., Borges, M., Corral-Pérez, J.J. et al. (2017). CO_2 activation over catalytic surfaces. *ChemPhysChem* 18: 3135–3141.

4 Sternberg, A. and Bardow, A. (2016). Life cycle assessment of power-to-gas: syngas vs methane. *ACS Sustainable Chem. Eng.* 4: 4156–4165.

5 Prieto, G. (2017). Carbon dioxide hydrogenation into higher hydrocarbons and oxygenates: thermodynamic and kinetic bounds and progress with heterogeneous and homogeneous catalysis. *ChemSusChem* 10: 1056–1070.
6 Centi, G. and Perathoner, S. (2009). Opportunities and prospects in the chemical recycling of carbon dioxide to fuels. *Catal. Today* 148: 191–205.
7 Reymond, H., Corral-Pérez, J.J., Urakawa, A., and Rudolf von Rohr, P. (2018). Towards a continuous formic acid synthesis: a two-step carbon dioxide hydrogenation in flow. *React. Chem. Eng.* 3: 912–919.
8 Rumayor, M., Dominguez-Ramos, A., and Irabien, A. (2018). Formic acid manufacture: carbon dioxide utilization alternatives. *Appl. Sci.* 8: 9–14.
9 Daza, Y.A. and Kuhn, J.N. (2016). CO_2 conversion by reverse water gas shift catalysis: comparison of catalysts, mechanisms and their consequences for CO_2 conversion to liquid fuels. *RSC Adv.* 6: 49675–49691.
10 Mittal, C., Hadsbjerg, C., and Blennow, P. (2017). Small-scale CO from CO_2 using electrolysis. *Chem. Eng. World* 52: 44–46.
11 Álvarez, A., Bansode, A., Urakawa, A. et al. (2017). Challenges in the greener production of formates/formic acid, methanol, and DME by heterogeneously catalyzed CO_2 hydrogenation processes. *Chem. Rev.* 117: 9804–9838.
12 Bozzano, G. and Manenti, F. (2016). Efficient methanol synthesis: perspectives, technologies and optimization strategies. *Prog. Energy Combust. Sci.* 56: 71–105.
13 Navarro, C.J., Centeno, A.M., Laguna, H.O., and Odriozola, A.J. (2018). Policies and motivations for the CO_2 valorization through the Sabatier reaction using structured catalysts. A review of the most recent advances. *Catalysts* 8: 1–25.
14 Su, X., Xu, J., Liang, B. et al. (2016). Catalytic carbon dioxide hydrogenation to methane: a review of recent studies. *J. Energy Chem.* 25: 553–565.
15 Sordakis, K., Tang, C., Vogt, L.K. et al. (2018). Homogeneous catalysis for sustainable hydrogen storage in formic acid and alcohols. *Chem. Rev.* 118: 372–433.
16 Moret, S., Dyson, P.J., and Laurenczy, G. (2014). Direct synthesis of formic acid from carbon dioxide by hydrogenation in acidic media. *Nat. Commun.* 5: 4017.
17 H. Croezen and M. Afman. (2017). *Indicative GHG-balance for formic acid as a hydrogen carrier in transport*. https://www.cedelft.eu/publicatie/indicative_ghg_balance_for_formic_acid_as_a_hydrogen_carrier_in_transport/1949 (accessed 28 September 2020).
18 Bahmanpour, A.M., Hoadley, A., and Tanksale, A. (2015). Formaldehyde production via hydrogenation of carbon monoxide in the aqueous phase. *Green Chem.* 17: 3500–3507.
19 Bansode, A., Tidona, B., von Rohr, P.R., and Urakawa, A. (2013). Impact of K and Ba promoters on CO_2 hydrogenation over Cu/Al_2O_3 catalysts at high pressure. *Catal. Sci. Technol.* 3: 767–778.
20 Bertucco, A. and Vetter, G. (2001). *High Pressure Process Technology: Fundamentals and Applications*. Elsevier Science.
21 Jia, C., Gao, J., Dai, Y. et al. (2016). The thermodynamics analysis and experimental validation for complicated systems in CO_2 hydrogenation process. *J. Energy Chem.* 25: 1027–1037.

22 Gaikwad, R., Bansode, A., and Urakawa, A. (2016). High-pressure advantages in stoichiometric hydrogenation of carbon dioxide to methanol. *J. Catal.* 343: 127–132.

23 Stangeland, K., Li, H., and Yu, Z. (2018). Thermodynamic analysis of chemical and phase equilibria in CO_2 hydrogenation to methanol, dimethyl ether, and higher alcohols. *Ind. Eng. Chem. Res.* 57: 4081–4094.

24 Gao, J., Wang, Y., Ping, Y. et al. (2012). A thermodynamic analysis of methanation reactions of carbon oxides for the production of synthetic natural gas. *RSC Adv.* 2: 2358–2368.

25 Miguel, C.V., Soria, M.A., Mendes, A., and Madeira, L.M. (2015). Direct CO_2 hydrogenation to methane or methanol from post-combustion exhaust streams – A thermodynamic study. *J. Nat. Gas Sci. Eng.* 22: 1–8.

26 Iyer, S.S., Renganathan, T., Pushpavanam, S. et al. (2015). Generalized thermodynamic analysis of methanol synthesis: Effect of feed composition. *J. CO2 Util.* 10: 95–104.

27 Bennekom, J.G.v., Winkelman, J.G.M., Venderbosch, R.H. et al. (2012). Modeling and experimental studies on phase and chemical equilibria in high-pressure methanol synthesis. *Ind. Eng. Chem. Res.* 51: 12233–12243.

28 Torrente-Murciano, L., Mattia, D., Jones, M.D., and Plucinski, P.K. (2014). Formation of hydrocarbons via CO_2 hydrogenation – a thermodynamic study. *J. CO2 Util.* 6: 34–39.

29 Sandler, S.I. (2017). *Chemical, Biochemical, and Engineering Thermodynamics.* Wiley.

30 Atkins, P. and de Paula, J. (2010). *Atkins' Physical Chemistry*. Oxford: OUP.

31 Perry, R.H., Green, D.W., and Maloney, J.O. (1997). *Perry's Chemical Engineers' Handbook*. McGraw-Hill.

32 Sandler, S.I. (2015). *Using Aspen Plus in Thermodynamics Instruction: A Step-by-Step Guide*. Wiley.

33 Götz, M., Lefebvre, J., Mörs, F. et al. (2016). Renewable power-to-gas: a technological and economic review. *Renewable Energy* 85: 1371–1390.

34 Hobson, C. (2018). *Renewable Methanol Report* (ed. C. Márquez). Madrid: Methanol Institute.

35 Anttila, A.A.P. (2014). The realistic potential for forest biomass supply in the European Union. *Metlan Työraportteja* 289: 27–32.

36 C. R. International. (2019). *Carbon Recycling International*. www.carbonrecycling.is/ (accessed 28 September 2020).

37 ENERKEM. (2019). *ENERKEM*. https://enerkem.com/ (accessed 28 September 2020).

38 C. Böhme. (2019). *BASF develops process for climate-friendly methanol*. https://www.basf.com/global/en/media/news-releases/2019/05/p-19-218.html (accessed 28 September 2020).

39 Ancona, M.A., Antonucci, V., Branchini, L. et al. (2019). Thermal integration of a high-temperature co-electrolyzer and experimental methanator for power-to-gas energy storage system. *Energy Convers. Manage.* 186: 140–155.

40 Janke, C., Duyar, M.S., Hoskins, M., and Farrauto, R. (2014). Catalytic and adsorption studies for the hydrogenation of CO_2 to methane. *Appl. Catal., B* 152-153: 184–191.

41 Zheng, Q., Farrauto, R., and Chau Nguyen, A. (2016). Adsorption and methanation of flue gas CO_2 with dual functional catalytic materials: a parametric study. *Ind. Eng. Chem. Res.* 55: 6768–6776.

42 E. P. T. Gas. (2019). *European Power To Gas Platform.* https://www.afhypac.org/documents/European%20Power%20to%20Gas_White%20Paper.pdf.

43 Bailera, M., Lisbona, P., Romeo, L.M., and Espatolero, S. (2017). Power to gas projects review: lab, pilot and demo plants for storing renewable energy and CO_2. *Renewable Sustainable Energy Rev.* 69: 292–312.

44 Navarro-Jaén, S., Navarro, J.C., Bobadilla, L.F. et al. (2019). Size-tailored Ru nanoparticles deposited over γ-Al_2O_3 for the CO_2 methanation reaction. *Appl. Surf. Sci.* 483: 750–761.

45 Weatherbee, G.D. and Bartholomew, C.H. (1981). Hydrogenation of CO_2 on group VIII metals: I. Specific activity of $NiSiO_2$. *J. Catal.* 68: 67–76.

46 Weatherbee, G.D. and Bartholomew, C.H. (1984). Hydrogenation of CO_2 on group VIII metals: IV. Specific activities and selectivities of silica-supported Co, Fe, and Ru. *J. Catal.* 87: 352–362.

47 T. Savaete (2016). Catalytic CO_2 conversion: a techno-economic analysis and theoretical study, Master thesis. Universiteit Gen. https://lib.ugent.be/catalog/rug01:002300884.

48 Bellotti, D., Rivarolo, M., and Magistri, L. (2019). Economic feasibility of methanol synthesis as a method for CO_2 reduction and energy storage. *Energy Procedia* 158: 4721–4728.

49 K. Ericsson (2017). Biogenic carbon dioxide as feedstock for production of chemicals and fuels: A techno-economic assessment with a European perspective," Lund: Miljö- och energisystem, LTH, Lunds universitet. https://lup.lub.lu.se/search/publication/67d3a737-cf7c-4109-bc4f-a6346956d6a2.

50 J. Nyari, (2018) Techno-economic feasibility study of a methanol plant using carbon dioxide and hydrogen. 712 Student thesis, TRITA-ITM EX.

51 Pérez-Fortes, M., Schöneberger, J.C., Boulamanti, A., and Tzimas, E. (2016). Methanol synthesis using captured CO_2 as raw material: techno-economic and environmental assessment. *Appl. Energy* 161: 718–732.

52 Smejkal, Q., Rodemerck, U., Wagner, E., and Baerns, M. (2014). Economic assessment of the hydrogenation of CO_2 to liquid fuels and petrochemical feedstock. *Chem. Ing. Tech.* 86: 679–686.

53 T. Fleiter, A. Herbst, M. Rehfeldt, M. Arens (2019). Industrial Innovation: Pathways to deep decarbonisation of Industry. Part 2: Scenario analysis and pathways to deep decarbonisation. ICF Consulting Services Limited and Fraunhofer ISI to the European Commission, DG Climate Action, January.

54 Van-Dal, É.S. and Bouallou, C. (2013). Design and simulation of a methanol production plant from CO_2 hydrogenation. *J. Cleaner Prod.* 57: 38–45.

55 Xu, A., Indala, S., Hertwig, T.A. et al. (2005). Development and integration of new processes consuming carbon dioxide in multi-plant chemical production complexes. *Clean Technol. Environ. Policy* 7: 97–115.

56 Jarvis, S.M. and Samsatli, S. (2018). Technologies and infrastructures underpinning future CO_2 value chains: A comprehensive review and comparative analysis. *Renewable Sustainable Energy Rev.* 85: 46–68.

57 Kiss, A.A., Pragt, J.J., Vos, H.J. et al. (2016). Novel efficient process for methanol synthesis by CO_2 hydrogenation. *Chem. Eng. J.* 284: 260–269.

58 Bradford, M.C.J. and Vannice, M.A. (1999). CO_2 reforming of CH_4. *Catal. Rev.* 41: 1–42.

11

Sabatier-Based Direct Synthesis of Methane and Methanol Using CO_2 from Industrial Gas Mixtures

K. Müller, J. Israel, F. Rachow, and D. Schmeißer

BTU Cottbus-Senftenberg, Chair of Applied physics/sensor technology, K.-Wachsmann-Allee 17, Cottbus D-03046, Germany

11.1 Overview

Some major benefits of the idea of using industrial gases as the source for direct methanation are summarized here: such strategies must be based on existing technologies with a reasonable cost of investment. The ongoing research therefore has the challenge/duty to develop such concepts and bring them close to industrial applications. Equipment for methanation can be installed in running industrial plants, and the existing power plants can be easily fitted with this additional CO_2 reducing technology.

There is another advantage as lower investments are required for purifying the CO_2-containing industrial gas source, as there is no need for CO_2 capture techniques such as amine chemistry or else. Such refitted power plants may better be suited for flexible operation to compensate fluctuating demands. Furthermore, there are synergies by using the exothermal heat of reaction in the existing thermal budgets. These combined heat and power synergies, storage concepts, and efficient power plant regulations altogether will enable to reduce the anthropogenic CO_2 emissions [1]. In all applications, for each converted ton of CO_2, we obtain a true CO_2 reduction. Certainly, the product, methane (as well as methanol) opens new ideas for storage concepts.

There is still another aspect when the global challenge is considered. The high demands for electrical power make it necessary that the leading industrial nations develop technologies for refitting and modernizing CO_2-intensive power plants and/or industries in order to obtain the goals given by the international climate conferences. The latter aspect is important as it is independent from the actual power delivered from fluctuating nonfossil sources such as wind and photovoltaic (PV) [2].

A considerable increase in the development of such CO_2 reducing technologies will be expected when politics has understood that CO_2 definitely has to be paid – CO_2 taxes or other political regulations have to be raised – and the CO_2

Engineering Solutions for CO_2 Conversion, First Edition.
Edited by Tomas R. Reina, José A. Odriozola, and Harvey Arellano-Garcia.

reducing challenge is not a job for some coal power plants but requires a change in many aspects of our social systems [3]. This global challenge demands contributions from economics and from the society in order to establish innovative energy and storage concepts.

The Sabatier reaction is originally discovered by Paul Sabatier in 1902 [4]. It offers the most popular route for the conversion of CO_2 into methane.

CO_2 Methanation/Sabatier Reaction:

$$CO_2 + 4H_2 \rightarrow CH_4 + 2H_2O \quad \Delta H_{R,298\,K} = -165\,kJ\,mol^{-1} \tag{11.1a}$$

This chapter describes the Sabatier process for the direct conversion of flue gas emitted from a lignite coal power plant. In this context, the meaning of the term "direct" is that the flue gas is used without any separation/enrichment of CO_2. The concept was investigated under real conditions by installing a pilot methanation reactor (100 kW) at an operating power plant (1600 MW).

The direct CO_2 methanation from flue gas is envisaged as a post-combustion approach: even older power plants could be upgraded for methanation and cycling. A further process integration should be possible by using waste heat of the power plant for the catalytic reaction. Or as shown later, the heat production of the catalytic reactor is significant and could also be available for an improvement of the total process efficiency.

Of course, if the proof of principle for a direct methanation of flue gas without CO_2 enrichment is given, the possible field of applications for CO_2 methanation is enlarged: not only the emissions of power plants could be recycled but also the CO_2 content from flue gases of steel production, cement production, or refineries.

The hydrogen for the hydrogenation must be generated from renewable energy. In a "power-to-gas-to-power" approach (PTG), the hydrogen and methane store energy with the possibility to use it in times in the absence of regenerative power production. If methanation is integrated into plants for power or cement production, for example, it could be an important contribution for an improvement of the industrial CO_2 emission balance.

In the scope of the energy transition from fossil to renewable sources, we need new concepts for a sustainable energy supply and energy storage, preferably with a reduction of the greenhouse gas CO_2. The conversion of CO_2 into methane is a promising approach for a CO_2 neutral production circle. In contrast to the concept of carbon capture and storage, methanation opens an opportunity for a reintegration and reuse of CO_2 in a recycling process in the form of synthetic natural gas (SNG).

It should be mentioned that the following paragraphs will quote political statements such as "power-to-X" or others. However, in all of the experiments presented here, it is intended to give a feasibility study of how to integrate gas mixtures in realistic conversion projects. As a consequence, in the conclusions, there are no reliability issues or other commercial considerations of the individual processes.

In the last part of this chapter (Section 11.4), the concept of using the direct conversion of CO_2 to methanol is presented, which is closely related to the Sabatier reaction (11.1a).

Direct Methanol Synthesis:

$$CO_2 + 3H_2 \rightarrow CH_3OH + H_2O \quad \Delta H_{298\,K} = -49.5\,kJ\,mol^{-1} \tag{11.1b}$$

The direct methanol conversion is an exothermic reaction, too. This process has the advantage that again in the approach presented here, the conversion of the CO_2 content in industrial gas mixtures can be applied. Methanol as a product can be used as a fuel in conventional engines or as a precursor to produce higher hydrocarbons. Furthermore, the experimental setup, the catalysts used, and some optimization strategies for better conversion rates, including the implementation of a recirculation cycle, will be reported.

11.2 Methane Synthesis of Gas Mixtures

11.2.1 Thermodynamics of Methane Conversion

According to the Sabatier reaction (Eq. (11.1a)), the methanation is thermodynamically favored; however, the reaction is limited in kinetics, and catalysts are required to achieve an acceptable conversion rate from CO_2 to CH_4. Beside catalysts using ruthenium or rhodium on various oxide substrates (for example, SiO_2, Al_2O_3, TiO_2, CeO_2, and ZrO_2, [5]), the use of supported nickel also shows high selectivity to methane (close to 100%, [6]) and is cheaper in price.

The maximum conversion rate is limited by thermodynamics. If only the Sabatier reaction is considered, that means no side reactions taking place, the equilibrium constant can be easily calculated [7] as shown in Figure 11.1. The experimental data show that for the methanation reaction, temperatures around 350 °C are necessary to

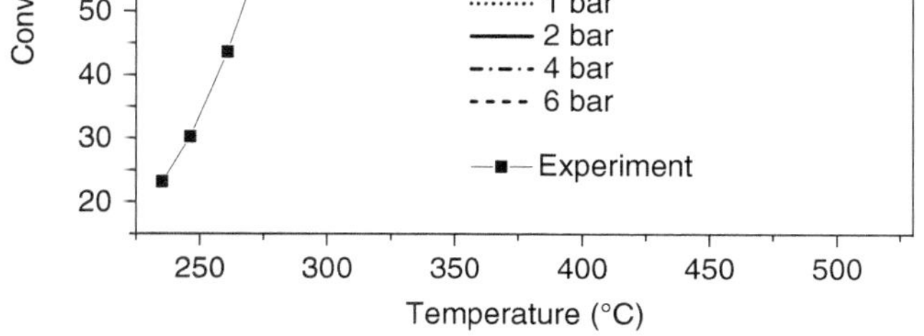

Figure 11.1 Calculation of the temperature and pressure dependency of the conversion rate of the Sabatier reaction at equilibrium and comparison with experimental data. As a catalyst we use NiO/SiO_2 (Cat. 1 in Table 11.1). A pressure drop of 1–2 bar along the reactor is due to the small grain size of the fixed bed reactor (0.2 mm).

come close to the thermodynamic values. We would like to emphasize that at atmospheric pressure, only a maximum conversion rate of around 90% can be achieved. Higher conversion rates are thermodynamically not possible.

For the exothermic methane synthesis, that takes place under a reduction of volume, lower temperatures and elevated pressure are favored following Le Chatelier's principle. This is also relevant because higher temperatures might promote unwanted side reactions (see below). Therefore, the Sabatier reaction itself is only a good description in a limited pressure and temperature window. For an accurate calculation of the equilibrium conversion and of the reaction kinetics in realistic conversion systems, several other reaction steps and possible reactions have to be taken into account [8].

11.2.2 Experimental Setup, General Definitions, and Catalysts

A schematic of the setup for the direct CO_2 conversion into methane is given in Figure 11.2. In all cases, for laboratory and technical scale, the experimental setup is built from five main units. Earlier experiments started at laboratory scale with relatively small flow rates of the reactants. Based on these experiments, a methanation reactor was up-scaled as a small pilot plant (technical scale) for the use in an operating coal power plant. For both cases, the experimental setup will be briefly described by five main blocks, also indicated in Figure 11.2.

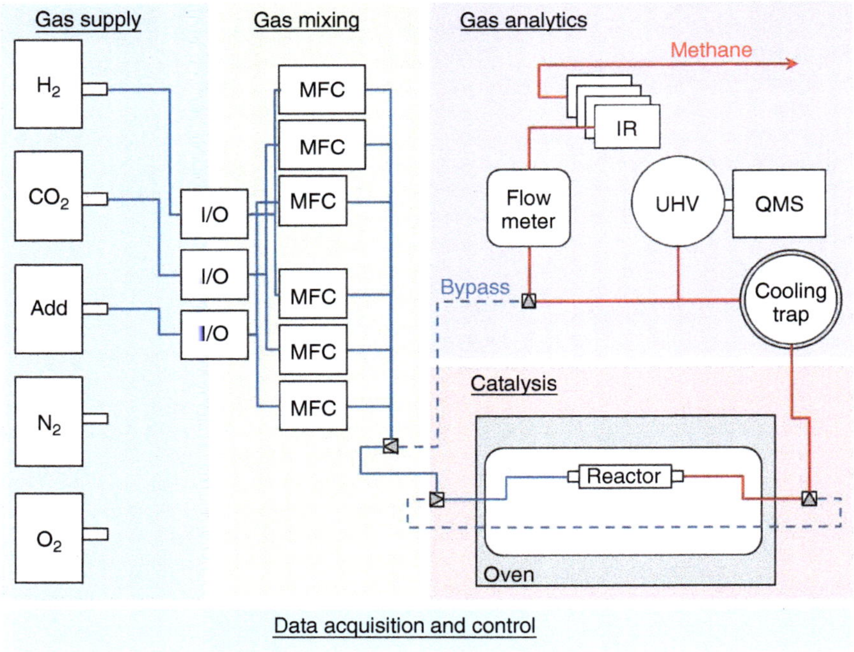

Figure 11.2 Schematic illustration of the setup for the direct methanation. The five individual blocks are described in the text.

The gas supply occurs from gas bottles for CO_2 and H_2 for the Sabatier reaction for the most experiments done in laboratory scale. For synthetic flue gas and also for N_2 and O_2, we also use special gas mixtures with defined compositions and contaminations (oxyfuel, coke oven gas, COG). The pressure of the educt gases is adjusted by standard valves.

Gas mixing and the flow of the process gas is adjusted by mass flow controllers (MFCs) for CO_2, H_2, O_2, and N_2 or flue gas. The maximal flow rate of each MFC is adapted to the different reactor sizes used during the experiments. The gases are mixed in stainless steel tubes and are guided to the reactor. During the power plant experiments, the flue gas is extracted from the emission stream, and only hydrogen is added for the Sabatier reaction of the CO_2 content [9].

The reactor with the catalyst is placed inside an oven for our experiments at laboratory scale. For methanation, temperatures above 300 °C are generally needed. In technical scale with higher flow rates and a larger reactor volume, heating tapes were used to preheat the gas and to achieve an initial reactor temperature of 350 °C. Because of the exothermic nature of the involved reactions, no further heating of the reactor was necessary (after the initial heating) in technical scale.

The reactor temperature was measured by a thin thermocouple placed inside the reactor. For the technical scale, the reactor design was modular, consisting of up to 10 segments (3 l volume) placed as two towers with 5 segments each. Temperatures were measured with three thermocouples at different position in each segment. Inlet pressures of up to 10 bar were necessary to overcome the flow resistance depending on the catalyst used.

The analytical system consists of gas sensors (AGM, Sensors Inc.) to record in the product gas the absolute quantities of CO_2, H_2, CH_4, and CO. Additionally, a quadrupole mass spectrometer (e-Vision2, mks) is installed. For the measurement of the mass flow after the catalytic reaction, we use a volumetric flow meter for the product gas stream ("Definer," Bios). With this information on chemical composition and flow rates of the product gas stream, a characterization of the catalytic performance in terms of conversion X, yield Y, and selectivity S, according to [10], is possible.

Control and data recording is done using a PC in a LabView environment. In our experiments, we regularly monitor the reactor line by recording the following data versus the reaction time: temperature, gas concentration, and input/output gas flow. In our system, we use the following definitions and conveniences: The catalytic performance is calculated according to standard definitions [10]. We calculate for the CO_2 conversion X_{CO_2}

$$X_{CO_2} = \frac{\dot{n}_{CO_2,in} - \dot{n}_{CO_2,out}}{\dot{n}_{CO_2,in}} \tag{11.2}$$

For the CH_4 yield Y_{CH_4}

$$Y_{CH_4} = \frac{\dot{n}_{CH_4,out}}{\dot{n}_{CO_2,in}} \tag{11.3}$$

Table 11.1 Overview of the catalysts employed in the methanation studies.

No., company	Metal oxide/wt%	Substrate	Grain size (mm)	Reactor type	Open volume (%)
Cat. 1, Aldrich	60, NiO	Silica gel	0.15–0.25	Fixed bed, SiO_2-sand: catalyst = 15 : 1	40
Cat. 2, Aldrich, 190 $m^2\ g^{-1}$ BET	65, Ni	Silica/ alumina	0.5–1.0	Fixed bed, SiO_2-sand: catalyst = 18–38 : 1	40
Cat. 3, METH 134, C&CS	20–25, NiO	Alumina	Spheres, diameter 3–6 mm	Packed bed of spheres	54

And for the selectivity S_{CH_4}

$$S_{CO_2}^{CH_4} = \frac{Y_{CH_4}}{X_{CO_2}} = \frac{\dot{n}_{CH_4,out}}{\dot{n}_{CO_2,in} - \dot{n}_{CO_2,out}} \tag{11.4}$$

The value "$\dot{n}$" is mass flow per unit of time, either as an educt ("in") or as a product ("out").

Nickel-based catalysts are used for methanation; some details are described in Table 11.1. As the Sabatier reaction is exothermic, the formation of hot spots during the Sabatier process is possible. Hot spots initiate a sintering process of the catalysts, resulting in a reduced surface area and a lower lifetime [11]. To prevent such effects, we distribute the NiO catalyst in a bed of quartz sand with an average grain size of 0.2 mm, and the weight ratio of quartz sand:catalyst varies around 15–38 : 1 (Cat. 1, 2).

The third catalyst used in our studies is a granulated material in the form of spheres with 3–6 mm in diameter. For this catalyst, an additional bed of catalytic inactive spheres (like quartz sand) is not necessary. The open volume between the spheres is approximately 54% [9].

All catalysts need an activation process in a reducing hydrogen atmosphere at temperatures around 400 °C. Afterward, the temperature is lowered to at 350 °C and the Sabatier reaction is started with a stoichiometric input composition of four parts of H_2 and one part of CO_2 in the gas flow.

In our experiments, we usually start analyzing the methanation process first for the pure gases. We produce "synthetic gas mixtures" by adding the respective contents to simulate the real gas mixtures. In addition, we need to check the influence of possible contaminations before approaching a real situation experiment.

11.2.3 Industrial Gas Mixtures

Carbon dioxide is the main agent for the Sabatier reaction, and it can be obtained at high purity from the air by adsorption systems [12] and can be enriched from the standard value of around 400 ppm to over 99%. Industrial gas mixtures contain a wide range of CO_2 concentrations. This is demonstrated with some values

Table 11.2 Overview of the CO_2 content of some gas mixtures in industrial applications (all values are given in vol%).

Gas	Application				
	Air	COG	Flue gas	Biogas	Oxyfuel
CO_2	<1	<2	15	50	99
N_2	78	<1	78	<11	<1
O_2	21	<1	6	<5	<1

For comparison, the corresponding values of nitrogen and oxygen are listed also. The detailed compositions of the individual systems will be given in the respective application (Sections 11.2 and 11.2.2)

in Table 11.2 for air, flue gas, COG, oxyfuel, and biogas. The values given are average values only. Usually, the gas composition of industrial mixtures is not fixed but varies in certain ranges, which will be given in the respective application in Section 11.3. Further, we will report on the content of contaminations or minor contributions which we have neglected here.

Other industrial sources for CO_2 are not reported here. It is just enough to mention that CO_2 is contained in most industrial waste gases, in variable concentrations. Some sources are cement production, breweries, bakeries, hospitals, air condition systems, to name a few. All of them could be addressed with our idea of reusing the produced CO_2 and convert it into methane or methanol. Some of these waste gases are delivered continuously; others appear only occasionally, most of them with a periodicity given by the industrial process.

Besides CO_2, the main content in most of the industrial gases is N_2; in some applications, there is also some remaining oxygen content. For methanation, the required hydrogen content has to be provided externally, except for the COG application.

The use of gas mixtures instead of pure gases for the Sabatier reaction implies the increasing influence of side reactions. For the methanation, these include the oxyhydrogen reaction, the reversed water gas shift, and the CO methanation reaction.

Oxyhydrogen Reaction:

$$2H_2 + O_2 \rightarrow 2H_2O \quad (\Delta_R H^0 = -571.6\,\text{kJ mol}^{-1}) \tag{11.1c}$$

Reverse Water Gas Shift Reaction (RWGS):

$$CO_2 + H_2 \leftrightarrow CO + H_2O \quad \Delta H^0_{298} = 41.2\,\text{kJ mol}^{-1} \tag{11.1d}$$

CO methanation:

$$CO + 3H_2 \leftrightarrow CH_4 + H_2O \quad \Delta H^0_{298} = -206.2\,\text{kJ mol}^{-1} \tag{11.1e}$$

Generally, these side reactions ((11.1c)–(11.1e)) have an influence on the overall performance as they reduce the selectivity. They also limit the temperature range for the methanation as they require that temperature must be in a range where side reactions become unlikely.

Another problem is in the influence of contaminations such as NO_x, H_2S, and SO_2. It is at least necessary to determine the critical concentration of these poisoning contaminations.

11.3 Applications

In this section, several applications for our Sabatier reaction system in different industrial processes applying industrial gas mixtures or their synthetic analogue. The section contains six items:

- APP-01: Combustion plant flue gas
- APP-02: Coke oven gas (COG)
- APP-03: Saline aquifer back-produced CO_2
- APP-04: Biogenic CO_2 sources
- APP-05: Oxyfuel operation in gas engines
- APP-06: Reusage of CH_4 product gas mixtures

The highlights of these selected applications demonstrate that industrial waste gases can be used as CO_2 sources without sophisticated cleaning procedures. The first four applications deal with systems with different CO_2 contents. In all the presented examples, we are able to give the proof of principle. For all investigated gas mixtures, we find high conversion rates similar to the pure systems. With these studies, we would like to emphasize that the use of industrial gas mixtures is a good opportunity to integrate PTG concepts in the existing industrial systems with high CO_2 emissions.

There are two more aspects that open further integration of PTG into more complex systems and allow to develop concepts that really contribute to reduce CO_2 emissions at global dimensions. Biogas as a CO_2 source for stable methanation (APP-04) can be achieved by avoiding sulfur-containing gas components. APP-05 deals with CO_2 reduction by employing oxyfuel concepts. The demonstration of closed cycle CO_2 systems or oxyfuel systems opens a new field for CO_2 reducing and avoiding technologies. Finally, in APP-06, we sketch the possible integration of the product gas in energy concepts.

11.3.1 APP-01: Combustion Plant Flue Gas

The typical ratio of N_2/CO_2 in flue gas from a conventional combustion process is around 5, with a residual amount of O_2, in the range of 5 vol%. This value is mainly due to an extra surplus of oxygen for an optimized combustion in the power plant [13]. We investigate a synthetic flue gas composition with N_2:CO_2 = 5 : 1 with four parts of H_2 as an additive for the Sabatier reaction. For this composition, we investigate the influence of oxygen as well as SO_2 and NO_2 contaminations.

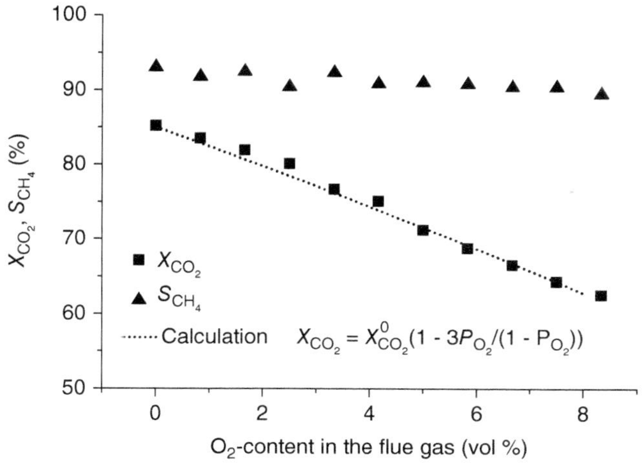

Figure 11.3 Performance as a function of the content of oxygen in synthetic flue gas (100 ml min^{-1} with N_2:CO_2:H_2 = 5 : 1 : 4 and additional oxygen). Dotted line: Calculation of CO_2 conversion including the oxyhydrogen reaction $H_2 + O_2 \Rightarrow 2H_2O$. The reactor temperature is stable at 350 °C.

The catalytic performance versus the oxygen content in the flue gas is shown in Figure 11.3. Under conservation of a high value for CH_4 selectivity, the conversion and yield drop down with increasing oxygen content. This is obviously due to the oxyhydrogen reaction (Eq. (11.1c)). For every oxygen molecule, two hydrogen molecules react into water and are not available for the Sabatier reaction. This assumption is confirmed by a calculation for an input gas composition of N_2:CO_2:H_2 = 5 : 1 : 4 [6, 14] and with the addition of oxygen. For 100% conversion of hydrogen by the oxyhydrogen reaction, we derive the conversion of CO_2 as a function of the oxygen content in the flue gas, as

$$X_{CO_2}(P_{O_2}) = X^0_{CO_2}(1 - 3P_{O_2}/(1 - P_{O_2})) \tag{11.5}$$

with $X^0_{CO_2}$ as the conversion without any oxygen content and P_{O_2} as content of oxygen (vol-%) in the flue gas. This line is also shown in Figure 11.3.

The calculation fits the measurement values; therefore, the oxyhydrogen reaction is responsible for the loss in catalytic performance. Because the oxyhydrogen reaction is strongly exothermic, an influence on the reactor temperature is expected. However the level of conversion and selectivity at an oxygen content of 5% remains constant over a period of at least 40 hours, as further measurements have shown.

The consideration of the temperature development shows that at a laboratory scale, no temperature gradient can be recorded despite the strongly exothermic oxyhydrogen and Sabatier reaction. In contrast, on a technical scale (Figure 11.4), a strong temperature increase is observed with increasing oxygen content in the educt gas. This can be attributed to the approximately 80 times larger surface area

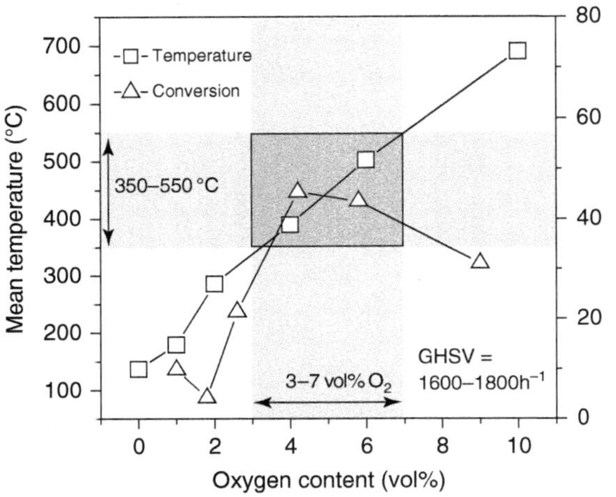

Figure 11.4 Synthetic flue gas: temperature and CO_2 conversion versus content of oxygen in the input gas stream. We use Cat. 2 (Table 11.1). The experiments are performed with one reactor segment. The gas hour space velocity (GHSV) is 1600 h^{-1} up to 1800 h^{-1}. The gray area between 350 and 550 °C marks the stable region for the Sabatier reaction without side reactions as revealed by experiments in laboratory scale. The typical range of oxygen residual is also shown. Source: Based on Israel; Rachow [15, 16].

to volume ratio of the laboratory reactor compared to the pilot plant reactor and the associated higher heat radiation and conduction.

The volume of the reactor used for the data shown in Figure 11.4 comprises 3000 cm^3 with the maximum space velocity (GHSV) of the synthetic flue gas including hydrogen of 1800 h^{-1}. The low conversion between 30% and 50% is due to the stoichiometric gas mixtures; they could be increased by additional H_2 content. Another reason can be found in the temperature gradient within the reactor, as the temperature given is only an average value over three measured values. A conversion maximum can be seen with an oxygen content of 3–7%; this is the typical range as it can be found in a conventional combustion process [13] and thus represents the ideal range for the Sabatier reaction to operate the direct methanation of exhaust gases from combustion processes without external heating.

As a next step, the influence of SO_2 and NO_2 contaminations is investigated, at laboratory scale. In Figure 11.5, we show a catalytic performance as a function of time, with SO_2 additive. We used a CO_2–SO_2 mixture with 516 ppm SO_2 during the experiments. This corresponds to $516/(5+1) = 86$ ppm SO_2 in the synthetic flue gas with N_2:CO_2 = 5 : 1. For dry flue gas after flue gas condensation and desulfurization, this is a typical value (Table 11.3).

The performance for the $Ni/SiO_2/Al_2O_3$ catalyst (Cat. 2, Table 11.1, Figure 11.5) remains stable for almost 2 days. After this time, the catalytic activity drops completely during a time of 9 hours. A possible reason for the stability range of the catalyst could be the high active surface area. A time of approximately 40 hours is necessary for a significant occupation of active sites. This deactivation caused by

Table 11.3 Composition of oxyfuel and conventional flue gas from lignite power plants, after condensation and desulfurization.

	$N_2 + H_2O$	CO_2 (vol%)	O_2 (vol%)	SO_2 (ppm)	NO_2 (ppm)
Oxyfuel [17]	<50 ppm	99.7	<0.3	2.5	<15
Conventional lignite p.p. [13]	78.5 vol%	15.5	6	63	46

Source: Modified from Schmeißer et al. [18].

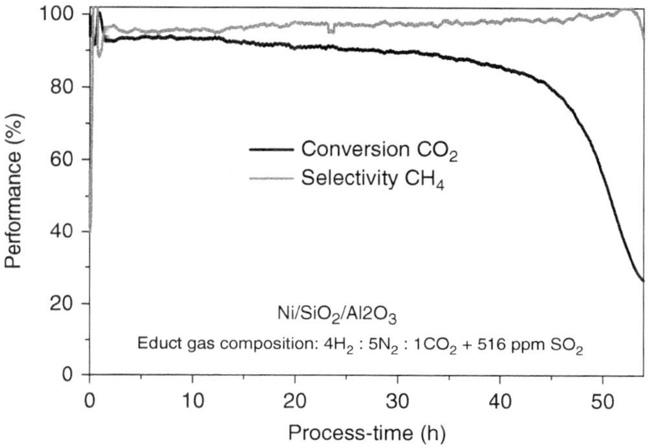

Figure 11.5 Catalytic performance versus time. CO_2 with 516 ppm SO_2 is used. This corresponds to a content of 516/6 = 86 ppm in the flue gas. We use a $Ni/SiO_2/Al_2O_3$ catalyst for the experiments (Cat. 2, Table 11.1).

the SO_2 content is lower than for a possible degradation caused by H_2S. The standard enthalpy of formation is $-21\ kJ\ mol^{-1}$ for H_2S [19] but $-297\ kJ\ mol^{-1}$ for SO_2 [20]. Because of this energy, it becomes again necessary to decompose the molecule, which indicates a much higher thermal stability of the SO_2 molecule, compared to H_2S. Similar tests with a NO_2 contamination of 100 ppm show no degradation of the catalyst activity after 100 hours of measurement. The conversion has a value of greater than 80%, with a constant selectivity near 100%.

With a conversion of 85% and around 100% selectivity, the resulting product gas composition from the direct conversion of CO_2 is 5 parts N_2, 0.15 parts CO_2, 0.85 parts CH_4, and 0.6 parts H_2. The produced water is removed in a cooling trap. The relative amount of CH_4 is here 12.9%, and the energy density is in the range of $5\ MJ\ m^{-3}$. This value belongs to the so-called "lean gas" quality [21]. Whether and in which form the use of lean gas can be realized is discussed in APP-07.

Furthermore, a test to investigate the stability of the Ni-based catalyst with original oxyfuel gas from a pilot plant (Vattenfall, Schwarze Pumpe) is performed at laboratory scale, with a gas flow of $100\ ml\ min^{-1}$. The composition of oxyfuel gas after flue gas condensation and desulfurization is given in Table 11.3 [18].

The catalytic performance for original oxyfuel CO_2 from the brown coal power plant is stable during the measurement time of 72 hours. In advance, we measured with a synthetic mixture of four parts of H_2 and one part of CO_2 with an additive of 3 ppm SO_2. No change in performance during the measurement time (17 hours) is found. Moreover, a test with 14 ppm NO_2 as an additive shows no drop-in performance during the 10 hours process time. In all these three cases, the conversion of CO_2 and the yield of CH_4 is high, both with a value in a range from 80% to 95%. This is the same performance as for a reference without SO_2 and NO_2 contaminations. The stability of the Ni-based catalyst against oxidized sulfur contaminations is a result of the relative stable molecule SO_2, if compared to H_2S, which is a typical sulfur contamination in biogas production. For new approaches to address this problem, see APP-04.

To summarize, we tested our system directly at a conventional combustion power plant. Without further cleaning of the flue gas, we established a stable conversion of CO_2 for a period of two weeks. This proof of principle with the up-scaled system is achieved with a reactor size of 3000 cm^3 and a GHSV of maximum 1800 h^{-1}. The energy density of the obtained product gas is 5 MJ m^{-3}, a typical value for a "lean gas" quality.

11.3.2 APP-02: Coke Oven Gas (COG)

The annual and total global COG production in 2007 is estimated to be 100 billion N m^3 [22]. The methanation of COG is a further opportunity in the context of a direct CO_2 conversion from industrial flue gas. The COG is a by-product of the carbon-rich coke production for steel industry. After several cleaning steps, it has a high amount of hydrogen (~60 vol%) and also contains CH_4 (~25 vol%), CO (5–8 vol%), and CO_2 (1–2 vol%).

In any power-to-gas concept, electrolysis for hydrogen production is a critical part, mainly because of the limited power of required electrolyzers. Commonly alkaline-based electrolyzers are used with the highest available value of hydrogen production with a power of up to 3.5 MW and a corresponding hydrogen flow of 760 N m^3 h^{-1} [23]. Compared to hydrogen production by water electrolysis, COG has an enormous potential as an alternative source of hydrogen or as a starting material for cheap mass production of e-fuels such as e-methane or e-methanol.

As a proof of principle, we investigate the direct catalytic conversion of the CO/CO_2 component of the COG into methane. In our experiments with COG, the nickel-based catalyst described as Cat. 2 in Table 11.1 is used. The reactor is a fixed bed reactor at laboratory scale, described above. In a first series of experiments, COG is guided directly in the reactor for catalysis, without any additional gases. Synthetic COG with 60% H_2, 25% CH_4, 7% CO, 4% CO_2, and 4% N_2 is used. It has been shown that the CO and CO_2 component of pure COG is fully converted into methane by a catalytic process with almost 100% selectivity. The CH_4 content of the resulting (product) gas is increased, thereby improving the heating value and reducing the CO component.

On the other hand, COG can be regarded as a hydrogen source in a power to gas process for instance. Within this approach, additional CO_2 is provided in order to obtain the optimum conversion of the hydrogen content in COG. Emissions from lignite power plants, refineries, or cement production could be used now, and this CO_2 could be involved in a recycling process into e-fuels such as e-methane, with COG as a hydrogen source.

Therefore, in a second series of experiments, additional CO_2 is mixed into the COG. As the ratio of H_2/CO_2 is adjusted as $H_2/CO_2 = 4/1$ in these experiments, this gas mixture is called "Sabatier mixture." Here, it is assumed that hydrogen preferably reacts according to the CO methanation reaction (Eq. (11.3)) and also according to the Sabatier reaction. Because of these two reactions, 3*3.5 ml min^{-1} CO + 4*2.0 ml min^{-1} CO_2 = 18.5 ml min^{-1} H_2 is consumed. For the Sabatier reaction with the residual hydrogen (11.5 ml min^{-1}), an additional amount of 2.9 ml min^{-1} CO_2 is needed. The high selectivity of 100% for methane is not changed, if additional CO_2 is mixed into the COG, up to the stoichiometric concentration in a "Sabatier mixture."

In Figure 11.6, we show in (a) the conversion of CO and CO_2 as a function of the temperature of the reactor. The full lines represent the direct conversion of COG, and the dashed lines are representative for the conversion of the COG with additional CO_2. As a reference, we use a measurement with pure gases in a stoichiometric feed ratio of $H_2/CO_2 = 4:1$. The total gas flow during those experiments is 50 ml min^{-1} COG. The conversion of CO is expressed by the flows. Regarding the CO conversion, both the pure COG and the Sabatier mixture show a maximum conversion of nearly 100% at 235 °C. For both, we find the CO component already consumed before the CO_2 conversion sets in. This indicates a preferred reaction of CO according to the CO methanation (Eq. (11.1e)) at temperatures below 235 °C and a subsequent Sabatier reaction at temperatures higher than 235 °C. In relation to the reference inlet gas composition, the CO_2 conversion of the "Sabatier mixture" shows almost similar performance: a CO_2 conversion of around 90% at 300 °C with an onset of CO conversion in the range of 230–240 °C.

Such performance is also confirmed by investigations of [24] for Ru, Rh, Pt, and Pd catalysts in $H_2/CO/CO_2$ mixtures. It was shown that hydrogenation of CO is the preferred reaction until a certain temperature is reached, depending on the catalyst. Also, a significant selectivity for ethylene (C_2H_4), ethane (C_2H_6), propene (C_3H_6), or propane (C_3H_8) is measured in the temperature range from 225 to 300 °C. In contrast to COG, the feed composition was 1% CO, 15% CO_2, and 50% H_2 in He [25]. These authors find the presence of higher hydrocarbons as deducted by a mass spectrometry analysis. For instance, with Ru as a catalyst, a selectivity of up to 10% for C_3H_8 at 250 °C is detected in this gas mixture.

In our studies, we also perform a mass spectrometry analysis of the product gas for COG and the Sabatier mixture. In both cases, the order of magnitude of the main peak around 16 amu (CH_4) in relation to the group at around 56/58 amu, which is attributed to butene/butane, is p16amu/p56/58amu = 104 even at 200 °C. A further signal around 44 amu is attributed to residuals of CO_2 with only minor parts of

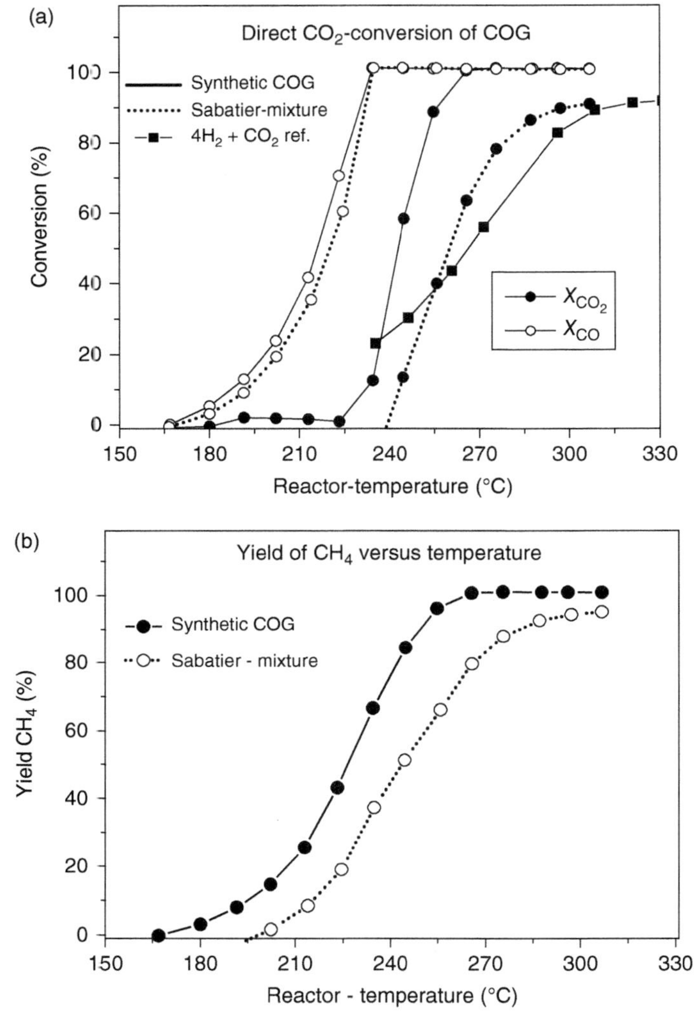

Figure 11.6 (a) Conversion of CO (empty circles) and CO_2 (black circles) versus the temperature of the reactor. The full lines represent the direct conversion of COG, the dashed lines are representative for the conversion of COG with additional CO_2 into Sabatier stoichiometry with H_2:CO_2 = 4 : 1. The black squares represent the reference measurement, described in text. (b) Yield of CH_4 versus temperature for COG and the Sabatier mixture.

propene/propane [26]. Thus, the selectivity for the formation of higher hydrocarbons is very low for both the pure COG and for the Sabatier mixture.

If we consider the selectivity for methane to be 100% over the temperature range, the yield Y of methane is equal to the conversion of CO/CO_2. We have a combined reaction of CO and CO_2 into CH_4 and use the modified gas flows

$$\dot{n}_{\mathrm{CO,CO_2,in}} = \dot{n}_{\mathrm{CO_2,in}} + \dot{n}_{\mathrm{CO,in}} \text{ and } \dot{n}_{\mathrm{CO,CO_2,out}} = \dot{n}_{\mathrm{CO_2,out}} + \dot{n}_{\mathrm{CO,out}}$$

With a selectivity of

$$S_{CH_4} = (\dot{n}_{CH_4,out} - \dot{n}_{CH_4,in})/(\dot{n}_{CO,CO_2,out} - \dot{n}_{CO,CO_2,in}) = 1 \tag{11.6}$$

The methane yield is given by

$$Y_{CH_4} = \frac{\dot{n}_{CO,CO_2,in} - \dot{n}_{CO,CO_2,out}}{\dot{n}_{CO,CO_2,in}} = X_{CO,CO_2} \tag{11.7}$$

The yield of CH_4 according to Eq. (11.7) for pure COG and for the Sabatier mixture is shown in Figure 11.6, (b): In COG, first, the CO component is converted into CH_4. In a second step, the residual CO_2 reacts into CH_4 in a Sabatier process. The maximum in yield is reached already at 260 °C. This is also due to a shift in the equilibrium for the surplus of H_2 in the COG. Furthermore, for the Sabatier mixture, the yield of CH_4 is higher than that for clean CO_2 and H_2 (reference composition, shown in Figure 11.5a). This high yield of CH_4 is of course due to the CO methanation reaction (Eq. (11.1e)) at temperatures below 260 °C.

In summary, we show the enormous potential for the use of COG either as a hydrogen source or in a process integration of CO_2 utilization in a much larger scale than for hydrogen production by electrolysis. In our studies, we used a Ni-based catalyst that shows very high selectivity (close to 100%) towards methane for synthetic COG and COG with an additional CO_2 input. The CO/CO_2 component of synthetic COG is completely converted into CH_4 at a temperature of 260 °C for this catalyst. Also, an additional CO_2 component is converted into methane, but according to the Sabatier reaction (Eq. (11.1a)) with lower equilibrium conversion and yield. As a further application of our concept for a direct use of industrial waste gases, the use of COG is an outstanding example.

11.3.3 APP-03: Saline Aquifer Back-Produced CO_2

As an additional example for our strategy of a "direct" CO_2 methanation, we investigate the catalytic performance of back-produced CO_2, stored in a saline aquifer, located in Ketzin near Berlin/Germany. As a pilot site for CO_2 storage in saline aquifers, the Ketzin project was initiated by the Germany Research Centre for Geosciences (GFZ) in 2004. Aim was a better understanding of onshore CO_2 deposition by different monitoring techniques [27]. The Ketzin saline aquifer is located in the triassic Stuttgart formation at a depth of 630–650 m [28]. The injection of CO_2 started in June 2008 and ended in August 2013. During this period, 67 000 t of clean CO_2 (with >99.9% purity) was stored. Further, 1500 t of captured CO_2 of an oxyfuel process from a pilot plant in Schwarze Pumpe was injected during May–June 2011. The purity of this CO_2 was higher than 99.7% [27].

In October 2014, in order to demonstrate a safe back-production, the stored CO_2 was partially recovered again from the aquifer. In total, 240 t of CO_2 was back-produced during this period [29]. Of important interest in the frame of a possible power-to-X-approach with the step of a CO_2 storage is an investigation of the utilization of recovered CO_2 for a conversion into methane.

As a special opportunity, in our laboratory, a small part of back-produced CO_2 from the Ketzin aquifer was available for a Sabatier experiment. Again, we use a nickel-based catalyst, described as "Cat. 1" in Table 11.1. We use the setup at laboratory scale (Figure 11.2). Four parts of hydrogen (5.0) are delivered as clean gas from gas bottles. One part of the recovered CO_2 is delivered from gas bags (10 l, Linde), flanged at our setup. We investigate the catalytic performance of three samples acquired at different dates and different extraction rates (800–1600 $kg\,h^{-1}$) (gas bag 1: back-producing rate 800 $kg\,h^{-1}$, 17.10.14; gas bag 2: 1600 $kg\,h^{-1}$, 22.10.14; gas bag 3: 800 $kg\,h^{-1}$, 27.10.14). As reference, we also use CO_2 with a purity of >99.9% as an educt gas for the Sabatier reaction. A mass spectrometry analysis of the back-produced CO_2 (from gas bag 1) reveals a content of 5% N_2, 1.25% O_2, 1% H_2O, and Ar (as mol%), obviously because of a contamination with air during handling of the gas samples. In relation to the CO_2 reference, no significant difference in methanation performance is found for the recovered CO_2 of all three samples. In all cases, we find a high conversion (>80%) of CO_2 into CH_4 [30].

To conclude, we demonstrated that the CO_2 back-produced after storage in a saline aquifer is available and we find no significant differences when compared to laboratory experiments using pure gases. This is a promising aspect for CO_2 recycling with a further utilization.

11.3.4 APP-04: Biogenic CO_2 Sources

Biogas is another important CO_2 source that can be implemented in PTG concepts. For this approach, the influence of typical contaminations on the catalyst stability, which limits the use of CO_2 from biogas plants, has to be investigated. A serious problem is the H_2S content of the biogas, which depends on the raw material for biogas production and which is in the range of 100–1000 ppm. Even after desulfurization, the H_2S content is in the critical range for the above-mentioned degradation of the catalysts [31].

Therefore, alternative techniques for the provision of CO_2 containing exhaust gases from biogenic products are to be investigated, which are suitable as a basis for catalytic methanation because of their low proportion of sulfur-containing components in the exhaust gas. Two promising approaches are presented here: The glycolate synthesis on microalgae basis according to the "New Green Chemistry" [32] and a high-performance fixed bed biogas process on the basis of a selective, membrane-supported biomass gasification [33] and [34].

Glycolate synthesis uses the efficient photosynthesis apparatus of the microalgae to bypass biomass growth. The organic molecule glycolate is excreted directly, with the microalgae acting as a biocatalyst. The cells of the green algae Chlamydomonas are used here, which favor glycolate production by adding oxygen in the photosynthesis process and thereby suppress further biomass production [35]. In this way, air CO_2, oxygen, water, and light can be turned into the basic material from which a variety of materials can be produced in further fermentation processes, such as methane [36], ethanediol, or polyglycolate. Compared to a conventional biomass gas plant, the land consumption is smaller by a factor of 10 [35], and in addition, the use of fertilizers can be dispensed with as the formation of biomass is avoided.

In the fermentation process to form photomethane, the CO_2 fixed in the glycolate synthesis is partially released again. The CO_2 obtained in this way as chemically pure "exhaust gas" can be fed into the catalytic methanation process in relatively high concentrations. The demand for oxygen for glycolate production could be met with oxygen from water electrolysis, which is a side product of hydrogen production for methanation. This would close another metabolic cycle and improve the efficiency of electrolysis.

Compared to conventional biogas plants, the high-performance fixed bed biogas process is characterized by two innovations that result in the obtained product gas having a proportion of sulfur compounds <1 ppm, and thus, the CO_2 proportion of the product gas can be used for catalytic methanation without further processing. These innovations include spatial separation by a membrane of the hydrolysis process, where the biomass is broken down into organic acids, and methanogenesis, where the organic acids are converted into methane and CO_2. The membrane process retains the solid biomass components in the hydrolysis zone, whereby the fermentation zone of methanogenesis is in a pure liquid phase and can therefore be subjected to continuous environmental treatment. This environmental impact leads to a homogeneous pH value >7.8 in the reactor, which suppresses sulfur gas compounds and keeps them as dissolved salts in the residual water of the reactor. This is also known as "New Green Chemistry."

11.3.5 APP-05: Oxyfuel Operation in Gas Engines

The oxyfuel process belongs to the techniques of CO_2 sequestration and offers the possibility to use the CO_2 emitted in high concentration with >99% directly for methanation. The basis of the concept presented here was created by the conversion experiments with real oxyfuel CO_2 from a pilot plant in Schwarze Pumpe, described in part APP-01 [37]. Conversion rates of >90% are achieved, a value that is close to the thermodynamic limit.

The aim of this oxyfuel system is to create a closed CO_2 cycle based on conventional system components, see Figure 11.7. On the energy demand side, a methane combustion system is proposed as an oxyfuel process, preferably in a combined cycle power plant or on a small scale in a combined heat and power (CHP) plant. The flame temperature is controlled by the recirculation of CO_2, which creates CO_2 with easily condensable water as the main component in the exhaust gas. As a by-product within a stoichiometric combustion, CO with a concentration of approximately 2000 ppm is formed [38] instead of NO_x and SO_x, as is the case of combustion with conventional oxygen from air. A CO content has a neutral or even a positive effect [39] on the methanation of the exhaust gas in the case of energy storage (Section 11.3.2). This depends on the reaction conditions, such as temperature and pressure, and on the catalyst, see also [40]. The hydrogen, which is necessary for the Sabatier reaction (Eq. (11.1a)), is obtained by water electrolysis in a previous step. This in turn produces the oxygen required for oxyfuel combustion in

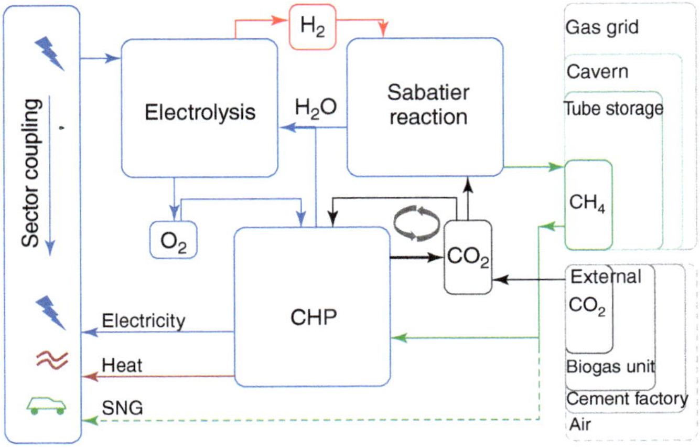

Figure 11.7 Diagram for an oxyfuel CO_2 circuit for an energy supply system with sector coupling. System components: electrolysis, methanation, and CHP, as well as an external CO_2 source and the gas storage facilities for oxygen, hydrogen, carbon dioxide, and methane.

a stoichiometric ratio. In addition, methanation and combustion together produce the necessary amount of water, which is split into hydrogen and oxygen during electrolysis.

If the mobility sector is also to be integrated into the system, external CO_2 must be fed back into the cycle in corresponding quantities. External CO_2 must also be taken into account if decentralized consumers are to be supplied with methane via the natural gas system. Accordingly, the electrolyzer must also be dimensioned larger. For reasons of sustainability, the CO_2 has to be extracted from air, but cement factories, breweries, or other CO_2-emitting industries can also be used [12].

It should be noted that compared to the conventional PTG methane system, additional tube storage tanks are required. These can be estimated with investment costs of approximately 3–5 € kWh^{-1} [41], which is negligible compared to the investment costs of electrolysis and methanation of approximately 2000 € kW^{-1} [42]. This application demonstrates that highly concentrated CO_2 flue gas mixture can be achieved by an oxyfuel combustion process. Therefore, a sustainable cycle process can be established with subsequent exhaust gas methanation with renewable hydrogen.

11.3.6 APP-06: Reusage of CH_4 Product Gas Mixtures

The direct methanation of industrial exhaust gases produces a gas mixture with a variable calorific value ranging from 5 MJ mol^{-1} (lean gas) to 36 MJ mol^{-1} (pure methane). For this range of variable calorific values, it is necessary to develop a reutilization concept in order to be able to represent a closed energy storage system. The possibilities presented here are intended to create the basis for such a cycle concept.

The concept of "direct power to gas process" is useful only if the circle with a process of "gas to power" can be closed again. In our case of the direct use of flue gas in general, we have to investigate the boundary condition for this strategy. With regard to this aspect, the use of flue gas of a conventional power plant turns out as a "limit of application" for possible realizations of the above-mentioned cycle. This of course in our scope, without any refinement or separation of CO_2. The dry product gas of direct methanation from the power plant with a conversion rate of 85% is composed of 12% CH_4, 78% N_2, 8% H_2, and 2% CO_2. The energy density (heating value) is near 5 MJ N^{-1} m^{-3}; this is in the range of a lean gas in a range of 5–7.5 MJ N^{-1} m^{-3}.

For the use of the lean gases from the direct methanation of CO_2 containing exhaust gases, two approaches are to be shown here. The first shows an integration of methanation and combustion of the lean gases into the water vapor cycle in the large power plant [43], and the second approach pursues decentralized recycling in the CHP, in the scope of a decentralized PTG to power approach, where we use a Wankel rotary engine driving an electrical generator.

The experiments with the Wankel engine are performed in cooperation with Wankel SuperTec GmbH, Cottbus. As fuel gas, we use various gas compositions, mixed from the components of product gas of the direct methanation at a brown coal power plant (see APP-01). In a mixing battery as in our further experiments (see Figure 11.2), we combine H_2, CH_4, N_2, and CO_2 in different ratios. Goal of the feedback between flow of fuel and the Wankel engine is a stable run at a constant frequency of 3000 min^{-1}, respectively, 50 Hz.

The range of the adjusted heating values is in a range of 5 up to 36 MJ N^{-1} m^{-3}. The lowest value is the minimal measured value of the product gas from the methanation at the power plant. The upper value is the energy density for pure methane. The resulting energy densities of the gas mixtures are plotted as x-axis in Figure 11.8. The demand of gas flow with this energy density, necessary for a stable engine run at 50 Hz, is plotted as y-axis.

In a first series (black squares), we only vary the ratio of H_2, CH_4, and CO_2, without addition of N_2. The gas ratios correspond to different CO_2 conversions, from 20% up to hypothetically 100%. We assume the pure gases CO_2 and H_2 in a ratio of $CO_2/H_2 = 1/4$ here. For 20% conversion, the product gas composition is therefore 4.76 vol% CH_4, 76.19 vol% H_2, and 19.05 vol% CO_2, with an energy density of 9.92 MJ N^{-1} m^{-3}. Also, pure methane with an energy density of 36 MJ N^{-1} m^{-3} is used as a fuel gas for Wankel engine (100% "conversion").

In a second series (blank squares), we assume a constant conversion of 80% for the Sabatier reaction, with 100% selectivity for methane. Now, the heating value is varied by stepwise addition of nitrogen into the gas. The starting composition for a conversion of 80% is 44.44% CH_4, 44.44% H_2, 11.11% CO_2, and 0% N_2, with an energy density of 20.72 MJ N^{-1} m^{-3}. The lowest value in energy density is 5 MJ N^{-1} m^{-3}, with a composition of 10.69% CH_4, 10.69% H_2, 2.67% CO_2, and 76% N_2. In a last step (star), we add 15 vol% CH_4 into 85 vol% N_2, with an energy density of 5.38 MJ N^{-1} m^{-3}. The multiplication of the energy density in MJ N^{-1} m^{-3} with the demand of gas flow in N m^3 h^{-1} represents the power supply of the Wankel engine. For 50 Hz, this is the constant value of 36 kW.

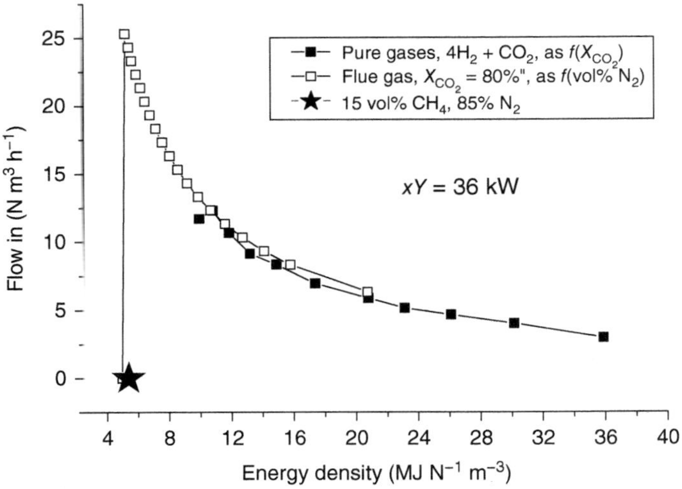

Figure 11.8 Demand of gas flow of the Wankel engine, stable at 50 Hz, with respect to the energy density of the injected gas. Black squares: variation of CO_2 conversion without N_2 addition, blank squares: variation of nitrogen content, for a Sabatier reaction with 80% CO_2 conversion, star: only CH_4 with 85% nitrogen.

In the critical energy density for a lean gas and also for our product gas from direct methanation, the Wankel engine works in the relevant frequency for a power supply into the electrical grid. For a stable operation at such low energy density, a certain amount of hydrogen is necessary. In the case of only CH_4 with a major part of nitrogen (85%), no power is delivered by the Wankel engine (indicated as a star in Figure 11.8).

To summarize here, the lean gas of a direct CO_2 conversion from a conventional brown coal power plant is usable to close the desired circle of energy storage as chemical energy and back electrification in a conventional CHP.

Another concept is the reuse of the methanation lean gas in conventional power plants. The concept sketch shown in Figure 11.9 describes the system integration of the elements' electrolysis, methanization, and pressure storage in a coal-fired power plant. The CO_2 in the flue gas is converted to methane directly after flue gas desulfurization, stored in the pressure reservoir, and fed back to the power plant as a fuel if required. The calorific value of the lean gas of 5–6 MJ N^{-1} m^{-3} is high enough for direct recirculation as fuel gas into the power plant boiler. Energy storage takes place both in the form of chemical energy (lean gas) and compression energy (gas reservoir). The stored energy is recovered by means of direct re-electricity generation (expansion energy from the pressure reservoir for the lean catalysis gas) and by combustion of the lean catalysis gas (caloric value). These two contributions are intended to enable the controllability in the case of the reduced load of the power plant. The oxygen obtained during electrolysis replaces the air in minimum load operation and thus increases the flame stability and optimizes (reduces) the required dry lignite use. This system integration thus shows, on the one hand, a probable stabilizing

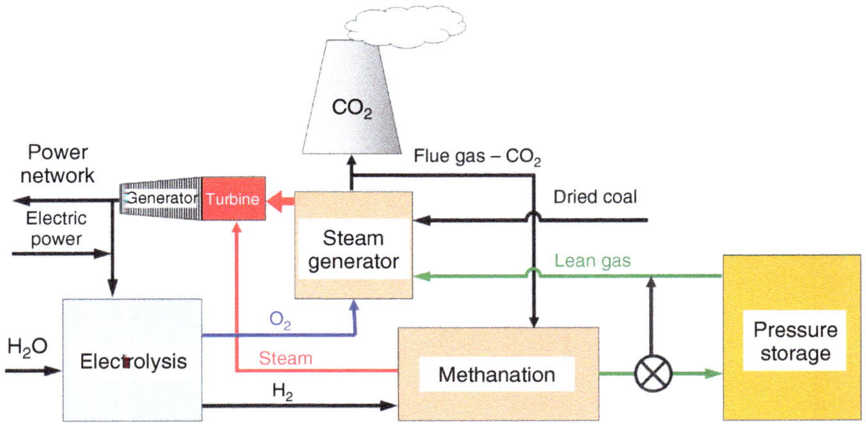

Figure 11.9 Sketch of an integrated system applying electrolysis, direct flue gas methanation, and pressure storage as well as the feedback loop of the lean gas into the reactor. The oxygen from the electrolyzer (blue) and the lean gas from the methanation are used within the reactor. The thermal balance from the exothermal methanation is used in the reactor's steam circulation.

effect of the operation of the power plant especially in medium and minimum load operation and, on the other hand, it represents a controllable power module with adapted load even below the normal minimum load operation.

The heat recuperation of the reaction heat from the exothermic Sabatier reaction and the hydrogen–oxygen reaction into the power plant process ensures a more constant temperature profile in the process steam of the steam turbine. This allows an increase in efficiency to be achieved at a low power plant-rated output. In addition to increasing power plant efficiency, the thermodynamic equilibrium of the Sabatier reaction can also be optimized by heat dissipation. As shown in Section 11.2.1 ("Thermodynamics"), a minimum in reactor temperature is desired for the CO_2 conversion in a range of 90%.

With the help of this system integration, the efficiency of the power plant can be improved by three contributions. Firstly, by recirculating the lean gas from the catalysis for partial or full load operation, secondly, by including the oxygen from the electrolysis process, especially in minimum load operation of the power plant, and thirdly, by using the exothermic reaction heat in the steam cycle.

The reusability of methane converted from waste gases in the form of a lean gas is demonstrated and this option gives another key argument for direct methanation. We sketched here two possible scenarios: decentralized power solutions can be realized based on our successful reconversion experiments at the Wankel CHP for wide ranges of fuel gas compositions. Further, the implementation of the lean gas in the large power plants opens chances for easy refitting and improved flexibility in the operation to react on fluctuating wind and PV power. This can be realized by storage of the reused methane in a pressurized storage tank with demand-oriented removal and subsequent combustion in the steam cycle.

11.4 Methanol Synthesis

Methanol synthesis is another option for a chemical storage process useful in many energy concepts. Methanol as a product may be used in conventional automotive systems as a fuel or as a precursor for further chemical synthesis of higher quality fuels such as (dimethylether and oxymethylether) [44–47]. Indeed, the application of industrial CO_2 sources demands for a methanol synthesis, which does not depend on synthesis gas formation. We follow the direct methanol synthesis as the modified Sabatier reaction (Eq. (11.1b).

The main aspect is that only this direct path for methanol synthesis enables the use of CO_2-containing gas mixtures – the focus of our studies. Another advantage is – when compared to the methanation process – that the N_2 components can be easily separated as the methanol product is liquid. This direct synthesis route differs from the more common reaction path via synthesis gas, which offers higher conversion rates but has the disadvantage of lower selectivity, i.e. formation of unintended higher hydrocarbons causing also a low lifetime of the catalysts [48].

The direct synthesis of methanol is supposed to occur at temperatures around 250 °C. It requires a high pressure of 50–80 bar. The temperature window for the methanol synthesis is limited by the CO reaction (Eq. (11.1e)) and by the reverse water gas shift (Eq. (11.1c)); the pressure range is limited by the supercritical behavior of CO_2 that starts at above 50 bar. In order to ensure a gaseous educt phase, we limit the pressure at 45 bar.

Very common are catalysts based on Cu–ZnO, which provide a yield of about 20% [5, 49]. The thermodynamic limit is around 30% at 250 °C, 50 bar, and a stoichiometric ratio of H_2:CO_2 = 3 : 1 (https://portal.uni-freiburg.de/molchem/research/encon). The search for catalysts with higher yield and better selectivity is a challenge in ongoing research projects (see e.g. [44]).

Our approach is a completely new approach in concepts for methanol synthesis that benefits from the advantages already described in Section 11.1. The main aspect again is to avoid cumbersome CO_2 capture techniques [17, 50] and to use the industrial flue gas directly.

The experimental setup for the direct methanol synthesis is depicted in Figure 11.10; it has the following parameters: It is built to carry out experiments up to 200 bar pressure. The reactor tube can be filled with about 25 g of catalysts and can be exchanged easily. The temperature is controlled in a tube furnace and can be adjusted up to 500 °C. The gas flow enables values of GHSV up to 5000 h^{-1}. Product gases can be analyzed by infrared detectors and mass spectroscopy. An external characterization of catalyst composition and morphology is possible by X-ray photoelectron spectroscopy (XPS), thermal desorption spectroscopy (TDS), and energy-dispersive X-ray (EDX).

Our setup uses a similar gas mixture supply as described in Section 11.2.2. We determine the methanol yield by the amount of condensate obtained in the cold trap. In a later state, a distillation setup will be implemented, too. We use a single compressor stage to feed the gas into the reactor. The setup can also be used by adjusting the pressure via the gas bottles without any compressor. Such experiments

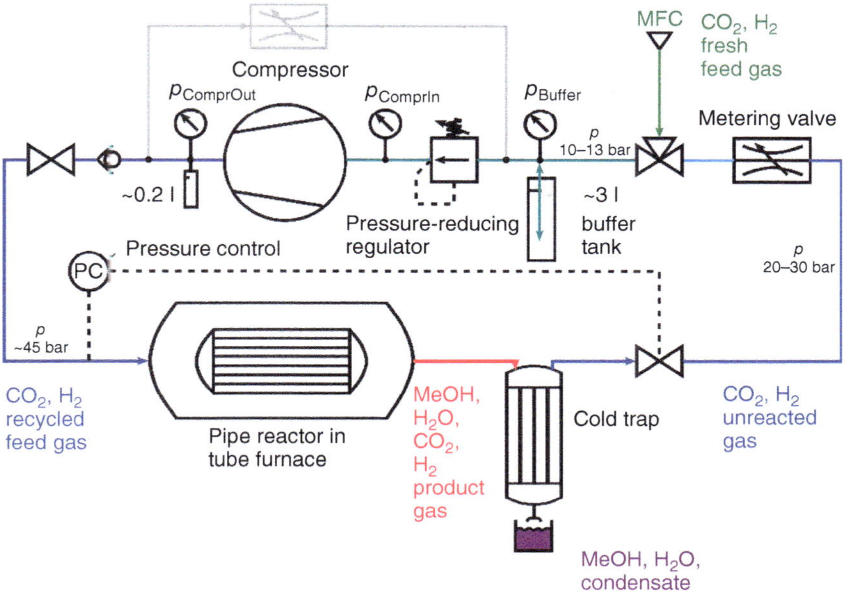

Figure 11.10 Setup of our high-pressure methanol reactor with an integrated recycling route. Source: Modified from Hagendorf [51].

are the basis for studies with gas mixtures. We started adding several concentrations of N_2 to the gas mixtures and found that the reaction is still going on near the thermodynamic limits.

The performance of different catalysts for the direct methanol synthesis is a main focus of our studies, and our experimental setup is suited for this issue, in particular. We started using a home-made catalyst based on a porous ZrO_2 ceramic, impregnated with the nitrates of Cu and Zn in a ratio of 80 : 20. This catalyst showed a high selectivity around 60% and a low conversion rate of about 7% [51]. In order to perform the tests with different N_2 contents, we use conventional Cu/Zn catalysts that are readily available. For pure CO_2 gases, we find again good results as these catalysts have a rather high conversion rate of about 20% with a reasonable selectivity of around 60%.

The influence of N_2 admixture is studied, and in Figure 11.11, we show some results with the Cu/Zn catalysts upon varying the N_2 content. Here, the reaction is performed at a pressure 45 bar, the reactor temperature is set to 250 °C, and the total gas flow is adjusted to give a GHSV value of around 1560 h^{-1}. As evident in Figure 11.11, up to admixtures of 40%, there is no change in the obtained conversion rate and only small deviations occur at higher N_2 contents. However, the selectivity is decreasing continuously. Currently, we further improve the experimental setup, in particular, the temperature gradients within the reactor tube, and a preheating of the input gases. In our ongoing studies, this is a necessary step toward a better overall performance [51]. It will allow us to optimize the parameters such as GHSV

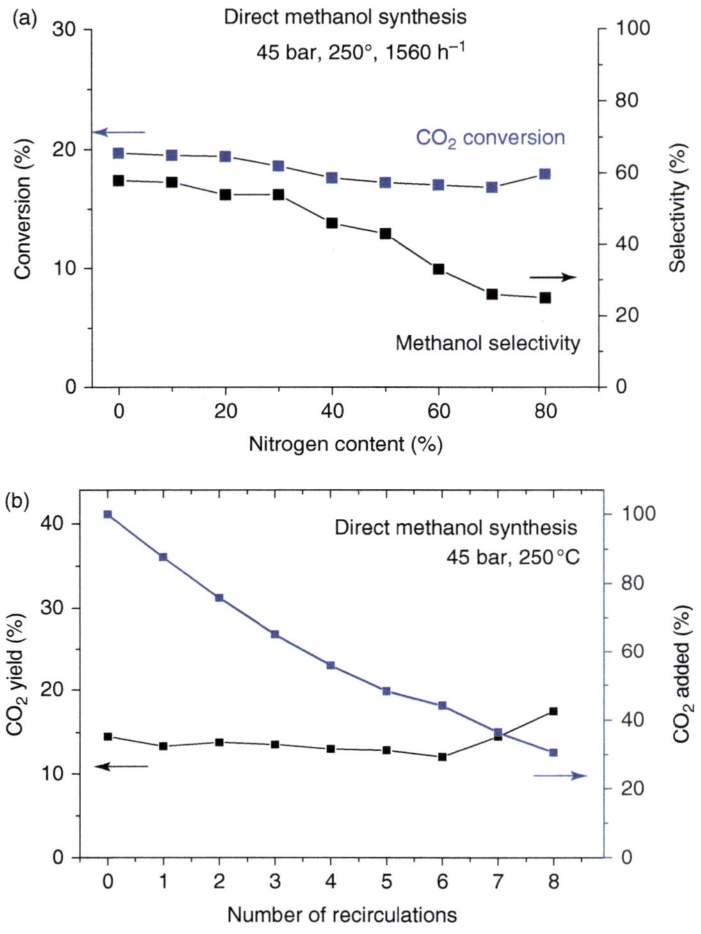

Figure 11.11 (a) The influence of N_2 on the CO_2 conversion rate (left scale) and selectivity (right scale) of the direct methanol synthesis. (b) Methanol yield per cycle and amount of added CO_2 in a recirculation experiment (no additional nitrogen content).

and reactor volume to enhance the conversion rates and selectivity also for higher N_2 contents.

The implementation of a recirculation loop as shown in Figure 11.10 gives another opportunity for our high-pressure system. We aim to improve the obtained methanol amount and to reduce the remaining reactants in the residual gas. Our first results are shown in Figure 11.11b, whereupon by increasing the number of recirculation, the CO_2 conversion remains constant while the amount of refilled CO_2 is successively decreasing ("fresh feed gas" in Figure 11.10). These data are taken using pure CO_2 gas; yet, they indicate that we actually can expect conversion rates of 90% when the recirculation is performed about 10 times [52].

Concerning the recirculation loop, any inert gas component requires a critical check of the provided gas flows, reaction rates, and volume and dimension of the buffer and the recirculation loop. Our tests aim to learn the parameters needed for an up-scaling for a pilot plant, in particular, to learn how to handle the extra gas volume caused by the inert gas content.

We believe that with our setup, we can deliver enough reliable data for a modeling procedure that will enable us to optimize the layout of a direct methanol synthesis adapted for the individual application.

Acknowledgments

We would like to acknowledge the various funding sources of our studies by BMBF (GeoEn), BMWI, ZIM, and EFRE. We had several industrial partners that helped us to discuss possible applications of our ideas and studies: Mitsubishi, Vattenfall, PantaRhei, Wankel SuperTec GmbH, Logesoft, and GFZ Potsdam. We would like to thank K. Henkel, V. Günther, and F. Mauß for stimulating discussions. We appreciate the technical support by G. Beuckert, C. Schwiertz, J. Paloumpa, E. Charlafti, as well as the experimental assistance in parts of this chapter by our master students: D. Hoffmannbeck, N. Beinlich, G. Kirner, M. Fleige, M. Hagendorf, Jan Ullrich, and Robin Herrmann.

References

1 Socolow, R.H. and Pacala, S.W. (2007). Spektrum der Wissenschaft-Spezial. *Energie und Klima* 1 (7): 14–21.

2 Schmitt, D. https://braunkohle.de/media/pressemitteilungen/

3 von Weizäcker, E.U., Hargroves, K., and Smith, M. (2010). *Faktor Fünf: Die Formel für nachhaltiges Wachstum*. München: Droemer Knaur GmbH & Co. KG. ISBN: 978-3426274866.

4 Sabatier, P. and Senderens, J.B. (1902). *Comptes Rendus Des Séances DeL'Académie Des Sciences, Section VI – Chimie*. Paris: Imprimerie Gauthier-Villars.

5 Wang, W. and Gong, J. (2011). Methanation of carbon dioxide: an overview. *Front. Chem. Sci. Eng.* 5 (1): 2–10.

6 Müller K., Fleige, M., Rachow, F., and Schmeißer, D. (2013). Sabatier based CO_2-methanation of flue gas emitted by conventional power plants. *Energy Procedia* 40: 240–248.

7 Rachow, F. (2017). *Prozessoptimierung für die Methanisierung von CO_2: vom Labor zum Technikum*. Brandenburgische Technische Universität Cottbus-Senftenberg.

8 Gao, J., Wang, Y., Ping, Y. et al. (2012). A thermodynamic analysis of methanation reactions of carbon oxides for the production of synthetic natural gas. *RSC Adv.* 2 (6): 2358.

9 Müller, K., Rachow, F., Israel, J. et al. (2017). Direct methanation of flue gas at a lignite power plant. *Int. J. Environ. Sci.* 2: 425–432.

10 Dittmeyer, R. (2006). Kapitel 5, Katalyse. In: *Chemische Technik. Prozesse und Produkte*, 5e, vol. 1 (eds. R. Dittmeyer, W. Keim, G. Kreysa and A. Oberholz). Weinheim: Wiley-VCH Verlag.

11 Rostrup-Nielsen, J.R., Pedersen, K., and Sehested, J. (2007). High temperature methanation sintering and structure sensitivity. *Appl. Catal., A* 330: 134–138.

12 W. Leitner, R. Eichel, K. Wagemann; *Erneuerbare Energien vielfältig nutzen - Öffentlicher Statusbericht des Kopernikus-Projektes P2X: Erforschung*, Validierung und Implementierung von Power-to-X-Konzepten; 2018.

13 Bernstein, W. (2007). Energietechnik. In: *Grundwissen des Ingenieurs* (ed. E. Hering), 581–618. München: Carl Hanser.

14 Fleige, M. (2015). *Direkte Methanisierung von CO_2 aus dem Rauchgas konventioneller Kraftwerke -Experimentelle Untersuchung und Verfahrens-aspekte.* Springer Spektrum https://doi.org/10.1007/978-3-658-09225-2.

15 Israel, J. (2016). Sabatierbasierte autothermale katalytische Rauchgasmethanisierung im Technikumsmaßstab und Rückverstromung im BHKW. Dissertation BTU Cottbus. BTU Cottbus-Senftenberg, Cottbus, Germany.

16 Rachow, F. (2017). Prozessoptimierung für die Methanisierung von CO_2 – Vom Labor zum Technikum. Dissertation BTU Cottbus. BTU Cottbus-Senftenberg, Cottbus, Germany.

17 Wall, T.F. (2007). Combustion processes for carbon capture. *Proc. Combust. Inst.* 31: 31–47.

18 Schmeißer, D., Müller, K., and Rademacher, B. (2015). CO2-Methanisierung aus dem Rauchgas: Rauchgasmethanisierung: gemeinsamer Schlussbericht im BMWi-Projekt: Cottbus. https://www.tib.eu/de/suchen/id/TIBKAT%3A866683100/CO2-Methanisierung-aus-dem-Rauchgas-Rauchgasmethanisierung/?tx_tibsearch_search%5Bsearchspace%5D=tibub (accessed 08 October 2020).

19 Zumdahl, S. *Chemical Principles* 6th ed. Boston: Houghton Mifflin Company; 2009.

20 Eckman, J.R. and Rossini, F. (1929). The heat of formation of sulfur dioxide. *Bur. Stand. J. Res.* 3: 597.

21 Cerbe, G., Carlowitz, O., Hölzel, G. et al. (1992). *Grundlagen der Gastechnik*, 4e. München: Hanser.

22 Razzaq, R., Li, C., and Zhang, S. (2013). Coke oven gas: availability, properties, purification, and utilization in China. *Fuel* 113: 287–299.

23 Ursua, A., Gandia, L.M., and Sanchis, P. (2012). Hydrogen production from water electrolysis: current status and future trends. *Proc. IEEE* 100 (2): 410.

24 Panagiotopoulou, P., Kondarides, D.I., and Verykios, X.E. (2008). Selective methanation of CO over supported noble metal catalysts: effects of the nature of the metallic phase on catalytic performance. *Appl. Catal., A* 344: 45–54.

25 Panagiotopoulou, P., Kondarides, D.I., and Verykios, X.E. (2009). Selective methanation of CO over supported Ru catalysts. *Appl. Catal., B* 88: 470–478.

26 Müller, K., Rachow, F., Günther, V., and Schmeißer, D. (2019). Methanation of coke oven gas with nickel-based catalysts. *Int. J. Environ. Sci.* 4: 73–79.

27 Martens, S., Möller, F., Streibel, M., and Liebscher, A., and the Ketzin Group (2014). Completion of five years of safe CO_2 injection and transition to the post-closure phase at the Ketzin pilot site. *Energy Procedia* 59: 190–197. https://doi.org/10.1016/j.egypro.2014.10.366.

28 Kempka, T. and Norden, B. (2017). Inverse modelling of hydraulic testing to revise the static reservoir model of the Stuttgart formation at the Ketzin pilot site. *Energy Procedia* 125: 640–649. https://doi.org/10.1016/j.egypro.2017.08.264.

29 Martens, S., Kempka, T., Liebscher, A. et al. (2015). Field experiment on CO_2 back-production at the Ketzin pilot site. *Energy Procedia* 76: 519–527.

30 Beinlich, N. (2015). *Katalytische Methanisierung von rückgefördertem CO_2 vom Pilotstandort zur CO_2-Speicherung in Ketzin.* Projektarbeit BTU Cottbus-Senftenberg.

31 Mescia, D. (2011). MSW landfill biogas desulfurization. *Fuel Energy Abstr.* 36 (13): 7884–7890.

32 Wilhelm, C. (2012). The biological perspective, ideas from new green chemistry concepts for improving the performance of microalgae. *TATuP* 21 (1): 46–53.

33 Voigt, H.R., Kühnert, J.-Th., Weyd, M., Tusel, E., Brüschke, H.E.A.. Sustainable, energy efficient and economic biorefineries by membrane enhanced processes, *ICOM International Congress on Membranes and Membrane Processes*, 23-29 July, 2011, Amsterdam, Netherlands.

34 Weyd, M. (2010). Ceramic membranes for the separation of glucose and the dewatering of ethanol. *18th European Biomass Conference & Exhibition*, Lyon (5 May 2010).

35 Taubert, A., Jakob, T., Wilhelm, C. et al. (2019). Glycolate from microalgae: an efficient carbon source for biotechnological applications. *Plant Biotechnol. J.*: 1–9. https://doi.org/10.1111/pbi.13078.

36 Günther, A., Jakob, T., Goss, R. et al. (2012). Methane production from glycolate excreting algae as a new concept in the production of biofuels. *Bioresour. Technol.* 121: 454–457: „Photomethan" Deutsche Patentanmeldung Nr. 10 2010 040 440.3-41, Patenteigner Universität Leipzig, KIT, IUV Bremen.

37 Müller, K., Beinlich, N., Rachow, F. et al. (2015). Methanation of recovered oxyfuel-CO_2 from Ketzin and of flue gas emitted by conventional power plants. *Symposium "ERE – Energy, Resources and the Environment", Session "ERE1 – Integrated studies"*, Wien, Österreich (12–17 April 2015). European Geosciences Union General Assembly.

38 Glarborg, P. and Bentzen, L.L.B. (2008). Chemical effects of a high CO_2 concentration in oxy-fuel combustion of methane. *Energy Fuels* 22: 291–296.

39 Monachino, E., Greiner, M., Knop-Gericke, A. et al. (2014). Reactivity of carbon dioxide on nickel: role of CO in the competing interplay between oxygen and graphene. *J. Phys. Chem. Lett.* 5: 1929–1934.

40 Gao, J., Liu, Q., Gu, F. et al. (2015). Recent advances in methanation catalysts for the production of synthetic natural gas. *RSC Adv.* 5: 22759.

41 Zloklikovits, M. (2012). Erdgasröhrenspeicher- Der österreichische Speichermarkt im Wandel der Zeit, Diplomarbeit, Hochschule Mittweida, University of Applied Science, Mittweida.

42 Sterner, M. and Stadler, I. (2019). *Energiespeicher – Bedarf, Technologien, Integration*. Springer Verlag.

43 Technische Universität Wien (März 2013). Überblick über die Schwachgasnachbehandlung bei Biogasaufbereitungsanlagen. Forschungsbereich Thermische Verfahrenstechnik und Simulation. IEE/10/130, Task 3.1.1.

44 Wang, W., Wang, S., Ma, X., and Gong, J. (2011). Recent advances in catalytic hydrogenation of carbon dioxide. *Chem. Soc. Rev.* 40: 3703–3727.

45 Semelsberger, T.A., Ott, K.C., Borup, R.L., and Greene, H.L. (2006). Generating hydrogen-rich fuel-cell feeds from dimethyl ether (DME) using Cu/Zn supported on various solid-acid substrates. *Appl. Catal.* 309: 210–223.

46 Martanez-Suarez, L., Siemer, N., Frenzel, J., and Marx, D. (2015). Reaction network of methanol synthesis over Cu/ZnO nanocatalysts. *ACS Catal.* 5 (7): 4201–4218. https://doi.org/10.1021/acscatal.5b00442.

47 George, A.O. (2005). Jenseits von Öl und Gas: die Methanolwirtschaft. *Angew. Chem. Int. Ed.* 117 (18): 2692–2696. https://doi.org/10.1002/ange.200462121.

48 Tsakoumis, N.E., Rønning, M., Borg, Ø. et al. (2010). Deactivation of cobalt based Fischer–Tropsch catalysts. A review. *Catal. Today* 154: 162–182.

49 Le Valant, A., Comminges, C., Tisseraud, C. et al. (2015). The Cu/ZnO synergy in methanol synthesis from CO_2. Part 1: Origin of active site explained by experimental studies and a sphere contact quantification model on Cu + ZnO mechanical mixtures. *J. Catal.* 324: 41–49. https://doi.org/10.1016/j.jcat.2015.01.021.

50 Kohl, A.B. and Nielsen, R.B. (1997). Chapter 2: Alkanolamines for hydrogen sulfide and carbon dioxide removal. In: *Gas Purification*, 40–186. Gulf Publishing Company.

51 Hagendorf, M. (2016). Heterokatalytische Synthese von Methanol aus CO_2 und H_2 an ZrO_2 geträgerten Cu/ZnO Katalysatoren. Masterarbeit. BTU Cottbus-Senftenberg.

52 Ullrich, J. (Juli 2019). Process optimization of the methanol synthesis of flue gas. Masterarbeit. BTU Cottbus-Senftenberg.

12

Survey of Heterogeneous Catalysts for the CO_2 Reduction to CO via Reverse Water Gas Shift

Thomas Mathew[1], *Simi Saju*[1], *and Shiju N. Raveendran*[2]

[1] *St. John's College, Department of Chemistry, Anchal, Kerala, India*
[2] *University of Amsterdam, Van't Hoff Institute for Molecular Sciences, Science Park 904, Amsterdam 1090GD, The Netherlands*

12.1 Introduction

Increasing amounts of CO_2 in our atmosphere is a major concern for the entire planet, which motivated significant research efforts in recent times to capture and/or utilize it. There are many prospective ways to utilize CO_2 including mineralization and conversion to chemicals and fuels. Indeed, CO_2 is a promising C1 feedstock of chemicals and fuels because it is renewable and cheap and also the conversion processes will be overall carbon neutral. One of the options to utilize CO_2 as a chemical feedstock is to convert it to CO because CO or syngas is already the basis of production of many chemicals. Conversion of CO_2 to CO using H_2, thus producing H_2O as a by-product, is known as the reverse water gas shift (RWGS) reaction. If economical and stable conditions for RWGS can be developed, it can be coupled with the existing CO-based industry, thus making the existing processes overall carbon neutral. The RWGS reaction is endothermic and requires high reaction temperature for obtaining significant conversions. Thus, the catalysts should be highly thermally stable. This review chapter summarizes the performance of different catalysts reported so far in the literature for RWGS.

12.2 RWGS Catalysts

Studies are still ongoing to find a high-performance catalyst, which is highly active and selective for the CO formation and also stable for long periods of continuous operation. Because methane and methanol are possible products at high $H_2 : CO_2$ ratios, the catalysts should be able to produce CO selectively. Structural and chemical properties of active metal species and supports, activity toward CO_2 activation, hydrogen dissociation, and relatively moderate strength with the adsorption of reaction intermediates are decisive factors for CO formation. According to the type

Engineering Solutions for CO_2 Conversion, First Edition.
Edited by Tomas R. Reina, José A. Odriozola, and Harvey Arellano-Garcia.

of active species, the catalysts are divided into noble metal catalysts and non-noble metal catalysts. During the past decades, a number of research works for CO_2 hydrogenation to CO over heterogeneous catalysts have been reported. In the following, the progress of RWGS reaction using different categories of catalysts is comprehensively reviewed. A brief section on mechanism is also included in the end of this chapter.

12.2.1 Supported Metal Catalysts

Supported metal catalysts with dual functionalities from the support and the metal sites that can activate CO_2 and dissociate H_2 in a synergistic manner are extensively studied for RWGS reaction. The structural and electronic characteristics of both the metal and support influence the catalytic performance. Particle size as well as the dispersion of the active metal component, metal loading, oxidation resistance, corrosion resistance, thermal stability of the metal, support characteristics such as its reducibility, thermal and mechanical stability, metal–support interaction, and nature of active species formed at the metal–support interfacial sites are all major factors that influence the catalytic activity and long-term stability for RWGS reaction. Supported noble metal catalysts containing one of the precious metals such as Au, Pd, Pt, Rh, and Ru are the most active catalysts [1–4]. These catalysts can work at relatively lower temperature for RWGS reaction. Noble metals, particularly Pt, Rh, and Ru, are resistant to coking and corrosion as well as display high activity toward H_2 dissociation and relatively moderate strength with the adsorption of reaction intermediates [1–8]. They show CO_2 conversions close to equilibrium and the main side product CH_4 does not occur above 500 °C under the RWGS conditions. Despite the fact that noble metal catalysts exhibit high activity, stability, and good resistance to carbon deposits, their cost and limited availability necessitates the development of cheaper and more abundant non-noble metal catalysts, particularly 3d transition elements [1–8]. To better understand the developments, we divided this section into various subsections (Sections 12.2.1.1–12.2.1.7) according to the metal used (in the following order: Au, Pt, Rh, Ru, Pd, Cu, Ni, Fe, and Co).

12.2.1.1 Au-Based Catalysts

Compared to Pt and Rh, Au with filled d-orbital electrons shows relatively low activity toward H_2 dissociation and hence lower hydrogenation ability [4]. However, it is known that Au nanoparticles (NPs) when finely dispersed on a suitable oxide support show high activity and selectivity for a variety of reactions [9, 10]. Though the superior catalytic activity of supported Au nanoparticles have been realized and extensively studied for reactions such as WGS, CO oxidation, and CO_2 hydrogenation reactions [9–13], they were studied for RWGS reaction only recently [14–19]. A range of supports such as Al_2O_3 [15, 16], TiO_2 [15, 17], CeO_2 [18, 19], MoO_x-promoted SiO_2 [20, 21], transition metal carbide (TMC) [22], and metal–organic framework (MOF) [23] were used to stabilize Au NPs with high dispersion aiming to enhance their catalytic performance for RWGS reaction with maximum CO selectivity.

The performance of Au catalysts depends strongly on the nature of the support [16]. In general, basic supports give low conversion in CO_2 hydrogenation reaction because of the strong interaction of CO_2 with basic sites. Acidic supports predominantly give CO and only small amounts of CH_4. When Au nanoparticles were supported on TiO_2, ZnO, CeO_2, ZrO_2, and Fe_2O_3, CO_2 hydrogenation to CH_4 or CH_3OH was negligible and CO formation was predominant [11, 16].

The mechanism of CO_2 hydrogenation to CO over supported Au catalysts involves the formate species as a key intermediate [16, 24]. Its decomposition is also strongly influenced by the support characteristics. Ishito et al. studied the decomposition of formate species on oxide supports such as SiO_2, γ-Al_2O_3, and CeO_2 [16]. Among these, the amphoteric Al_2O_3 converted formate species into CO with higher selectivity.

Au nanoparticles when deposited on non-reducible supports such as Al_2O_3 and SiO_2 exhibit only Au sites for RWGS. However, addition of electronic modifiers such as Mo to Au/SiO_2 occurs on undercoordinated sites of Au nanoparticles to form new active AuMoO_x interfacial sites and $Au^{\delta+}$ sites that are active for the RWGS reaction [20, 21].

The high oxygen storage capacity of CeO_2 makes it one of the prime choices as a support for RWGS reaction. Thus, as shown in Figure 12.1, a high CO_2 conversion, approaching the thermodynamic equilibrium value, under various conditions and nearly 100% CO selectivity were achieved over 3%Au/CeO_2 [18]. Au facilitates the reduction of surface oxygen species because of its strong interaction with CeO_2 support [18, 25], which is evidenced from the lower reduction temperature observed in H_2 temperature-programmed reduction (TPR) profile of Au/CeO_2 [18, 25].

Kyriakou et al. investigated the influence of supports on the catalytic activity of Au-based catalysts in RWGS [17]. According to their results, as illustrated in Figure 12.2, Au nanoparticles (2.2 nm) deposited on TiO_2 by the deposition–precipitation method gave CO_2 conversion as high as 50% (close to thermodynamic equilibrium values) and 100% CO selectivity at 400 °C for at least 60 hours. Bare TiO_2 and Au/Al_2O_3 showed less than 5% CO_2 conversion at

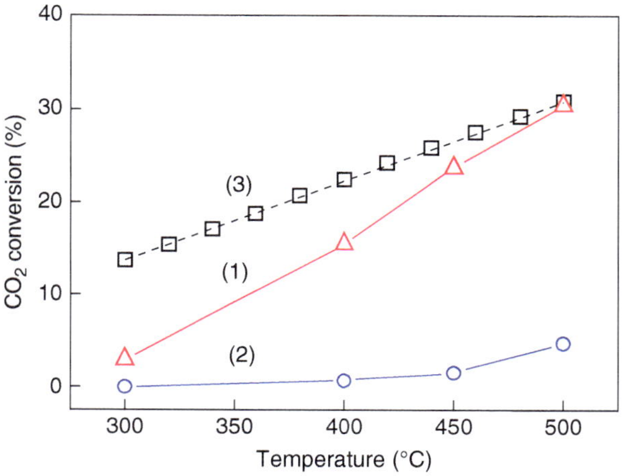

Figure 12.1 CO_2 conversion as a function of temperature over (1) Au/CeO_2 catalyst and (2) CeO_2 support. Reaction conditions H_2/CO_2 molar ratio = 1 and GHSV = 12 000 ml h^{-1} g^{-1}. Dashed line (3) represents thermodynamic equilibrium values for reaction under same conditions. Source: Zhu et al. [18]. Copyright 2016 Elsevier.

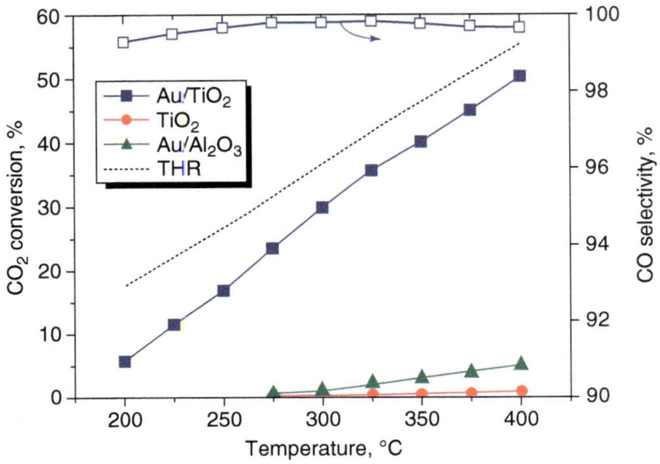

Figure 12.2 Dependence of CO_2 conversion on temperature for Au/TiO_2, TiO_2, and Au/Al_2O_3 and CO selectivity (empty symbols) for Au/TiO_2. The thermodynamic equilibrium curve of CO_2 conversion (THR) is plotted as a dotted line. Reaction conditions: catalyst weight = 0.4 g; flow rate = 100 cm^3 min^{-1}; P_{CO2} = 10 kPa, P_{H2} = 90 kPa; GHSV = 20 000 h^{-1} Source: Kyriakou et al. [17]. Copyright 2017 Elsevier.

400 °C [17]. Despite having lower Brunauer–Emmett–Teller (BET) surface area (51 m^2 g^{-1}) compared to Au/Al_2O_3 (210 m^2 g^{-1}), the high RWGS activity of Au/TiO_2 indicates a synergistic interaction between Au and TiO_2. The X-ray photoelectron spectroscopy (XPS) and TPR results revealed the existence of partially oxidized gold species (Au^+), in contrast to the pure metallic state (Au^0) on Al_2O_3. These results suggest that there is a charge flow between Au nanoparticles and TiO_2 support and favorable active sites may be formed at the Au–TiO_2 interface, which enhances CO_2 hydrogenation reaction to CO.

Recently, Bobadilla et al. [15] compared the catalytic performance of 1%Au/Al_2O_3 and 1%Au/TiO_2 for RWGS reaction and found that 1%Au/TiO_2 exhibited higher RWGS activity, reaching values close to the thermodynamic limit under the optimized reaction conditions as illustrated in Figure 12.3. Their results confirm that depositing Au on TiO_2 is beneficial for RWGS, as observed by Kyriakou et al. (see previous paragraph). Despite the Au loading and the average particle size, consequently the Au dispersion, for both catalysts are similar, the activation energy for RWGS reaction estimated for Au/TiO_2 (30 kJ mol^{-1} – 0.30 eV) was much lower than that of Au/Al_2O_3 (79 kJ mol^{-1} – 0.79 eV) as shown in Figure 12.4. Therefore, it is reasonable to assume that the support participates very actively in the reaction [15]. The superior performance of Au/TiO_2 over Au/Al_2O_3 is due to the presence of Ti^{3+} and an increased amount of oxygen vacancies in the former catalyst as evidenced by *operando* UV–vis spectroscopy.

Metal carbides can be excellent supports for enhancing the ability of transition metals to adsorb and activate CO_2 toward the synthesis of methanol or methane. Au NPs dispersed on TMC surface such as TiC(001) show high activity for CO_2

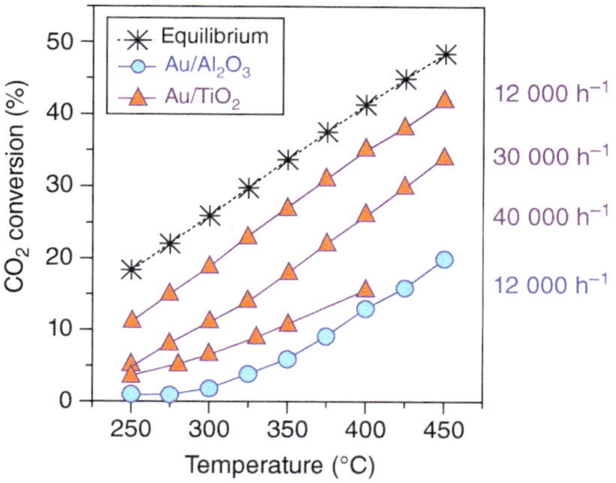

Figure 12.3 Catalytic performance of Au/TiO_2 and Au/Al_2O_3 catalysts in RWGS reaction (H_2/CO_2 molar ratio = 4) Source: Bobadilla et al. [15]. Copyright 2018 American Chemical Society.

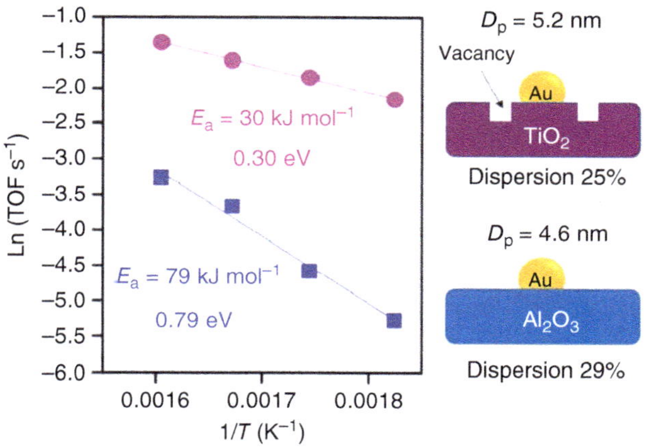

Figure 12.4 Arrhenius plots for the Au/TiO_2 and Au/Al_2O_3 catalysts in RWGS reaction. The Au particle size and dispersion are also shown for both catalysts. Source: Bobadilla et al. 15. Copyright 2018 American Chemical Society.

reduction to CO alongside a substantial amount of methanol compared to the low reactivity of Au(100), suggesting that the catalytic property of Au/TMC composites can be tuned in favor of RWGS reaction by optimizing the catalyst composition and reaction conditions [22].

Sintering and fast deactivation are serious problems for Au-based catalytic systems because of the endothermic nature of this reaction [4, 23]. Xu et al. explored the use of thermally stable and high surface area Zr(IV)-based porous MOFs (UIO-67) as supports to obtain monodispersed and isolated metal nanoparticles (NPs) for avoiding sintering and aggregation of metal nanoparticles and thereby enhancing the RWGS performance [23]. Such an Au-supported UIO-67 material showed a CO_2 conversion of 31% at a reaction temperature of 408 °C for RWGS reaction, which is higher than CO_2 conversion observed over Au/TiO_2 and Au/MgO. The high surface area and porosity of UIO-67 are favorable for Au nanoparticles to be adsorbed in the exterior pores or cavities of UIO-67 leading to isolated nanoparticles. Compared with UIO-67, common metal oxide supports such as TiO_2 or MgO have limited pores and surface area, making the adsorption of Au^{3+} not easy on them under the same conditions. Nevertheless, Au NPs deposited on UIO-67 tend to migrate and sinter, reducing the number of active sites at high temperature. When the NP loading is high, overcrowding of AuNPs increases the potential for migration and sintering, decreasing CO_2 conversion.

12.2.1.2 Pt-Based Catalysts

Supported Pt catalysts have been studied for CO_2 reduction for decades because of their high activity at low reaction temperatures, high hydrogen dissociation ability, relatively moderate strength with the adsorption of reaction intermediates, as well as excellent activity and stability [5, 6, 26–39]. As for gold, the support influences the catalytic activity of Pt in CO_2 hydrogenation to CO. The support materials employed in dispersing Pt nanoparticles for RWGS reaction include Al_2O_3 [6, 26–28], SiO_2- and Mo-promoted SiO_2 [29, 30], CeO_2 [5, 27, 31–33], TiO_2 [6, 30, 34–36], K-promoted Mullite [37], K-promoted zeolite [38], and La-doped ZrO_2 [39].

In general, reducible supports such as TiO_2 and CeO_2 give higher CO_2 conversion in RWGS than nonreducible supports such as Al_2O_3 and SiO_2. This effect can be assigned to the strong metal–support interaction and higher reducibility of the support, leading to the formation of new active interfacial sites such as Pt–O–M^{n+} at the Pt–oxide interface [6]. These Pt–O–M interfacial sites are selective toward CO formation. For example, the Pt–Ov–Ti^{3+}sites (O_v = oxygen vacancy) generated from the interaction between Pt and reducible TiO_2 support are the active sites for the formation of CO, while large Pt particles facilitate the hydrogenation of CO to produce CH_4 at higher temperature [34–36]. These Pt–O–M interfacial sites are also formed with the addition of promoters as in K-promoted zeolite [38] and Mo-promoted SiO_2 [29].

Controlled addition of potassium promoter by impregnation of KOH to Pt/L catalyst generates new active sites Pt–O(OH)–K, enhancing the catalytic activity and selectivity toward CO in RWGS reaction, as illustrated in Figure 12.5 [38]. K promoter enhances the adsorption of CO_2 and the interface between Pt particles and K promoter serves as the main active site, facilitating the formation of formate intermediates. Another advantage of such a catalyst system is that zeolite pores encapsulate Pt, enhancing Pt dispersion and stabilizing Pt nanoparticles even at severe reaction conditions [38]. The presence of potassium also decreased the strength of adsorption of CO on Pt particles, which hindered further hydrogenation of CO into CH_4 through CO bond dissociation [37].

12.2.1.3 Rh-Based Catalysts

Like Pt catalysts, supported Rh catalysts also show high activity toward H_2 dissociation and relatively moderate strength with the adsorption of reaction intermediates. The catalytic behavior of Rh catalysts varies depending on the support material, the Rh particle size, and the reaction conditions such as CO_2/H_2 ratio, temperature, and pressure [7, 40–49]. Different support materials have been used to disperse Rh nanoparticles for RWGS reaction, including TiO_2 [7, 40, 41], Al_2O_3 [42], SiO_2 [43, 44], Zeolite Y [45, 46], and $SrTiO_3$ [47].

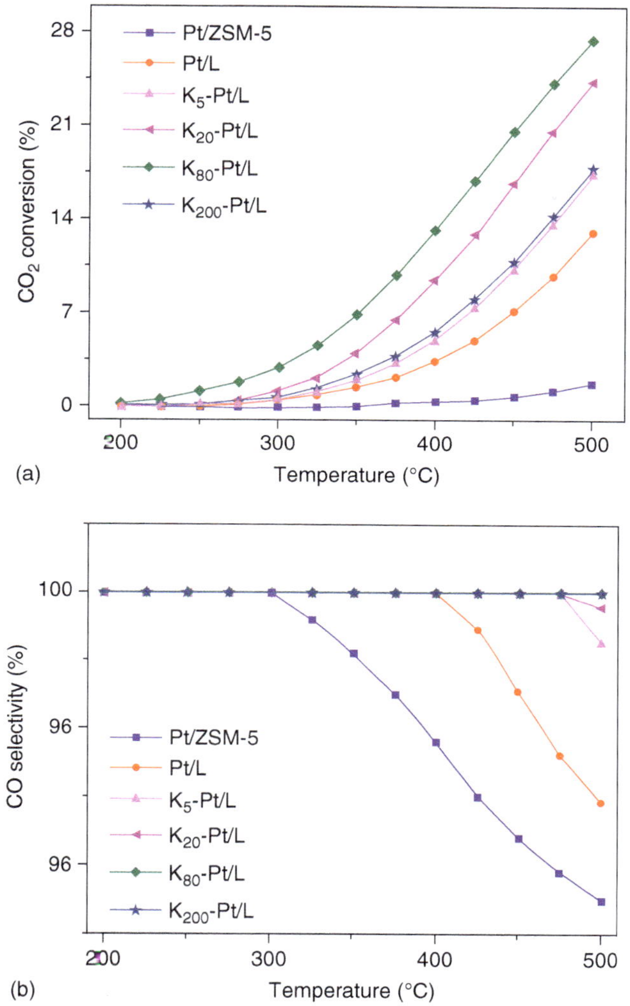

Figure 12.5 (a) CO_2 conversion and (b) CO selectivity as a function of reaction temperature for Pt/zeolite catalysts. Reaction conditions: 0.1 MPa, $H_2/CO_2/N_2$ = 45/45/10 (v/v/v), GHSV is 30 000 m.$^{-1}$ g^{-1} h^{-1} Source: Yang et al. 38. Copyright 2017 Elsevier.

Rh catalysts are largely applied for hydrogenation of CO_2 to CH_4 [48, 49]. However, activity and selectivity to CO through RWGS reaction can be enhanced if oxide supports such as $SrTiO_3$ that contain surface defects are used [47]. The highly dispersed Rh clusters efficiently dissociate H_2 and the oxygen vacancies of the oxide support activate CO_2 molecule. The stronger interactions between Rh atoms and the support weakens the binding of CO on Rh resulting its fast desorption [47].

The CO/CH_4 selectivity strongly depends on the particle size as well as on the amount of Rh loaded on the support. The methane production rate increases as metal Rh particle size increases up to c. 7 nm over Rh/TiO_2 [48]. In another work,

CO_2 hydrogenation was studied over silica-supported Rh catalysts (Rh/SiO_2) with Rh loading ranging from 1 to 10 wt%. CO was the main product for the low loaded catalyst, whereas CH_4 was dominantly produced on 10 wt% Rh/SiO_2 [44]. Compared to 10 wt% Rh/SiO_2, relatively large amount of surface hydroxyl groups present in 1 wt% Rh/SiO_2 is favorable to produce fine Rh carbonyl clusters and subsequently the descrption of CO.

K and Li are excellent promoters of Rh-based catalysts for CO_2 hydrogenation to CO [42, 45]. Remarkable differences in the catalytic behavior were observed over Ba- and K-containing Rh/Al_2O_3 and the pure Rh/Al_2O_3 catalysts in RWGS reaction in the temperature range 300–700 °C [42].

CO_2 was selectively converted to CO without CH_4 formation on K-containing Rh/Al_2O_3 in the entire temperature range, whereas both the Ba-containing and especially the pure Rh/Al_2O_3 catalysts showed high selectivity to CH_4 below 500 °C [42]. The addition of K weakens the CO interaction with the catalyst, leading to the release of CO before further reduction to elementary C and CH_4.

Bando et al. observed that the CO/CH_4 ratio increases with increasing Li content in CO_2 hydrogenation on Li-promoted Rh ion-exchanged zeolites (Li/RhY). For the undoped catalyst (RhY), the selectivity to methane was almost 100%, exemplifying the promoting effect of Li toward RWGS reaction [45]. The microporous RhY catalyst showed 10 times higher activity than a SiO_2-supported Rh catalyst (Rh/SiO_2) because of enhanced CO_2 adsorption and its diffusion to active Rh sites. However, the material deactivated fast because of the accumulation of the water produced inside the zeolite cage and consequent suppression of the further adsorption of CO_2 [46].

12.2.1.4 Ru-Based Catalysts

Supported ruthenium catalysts are generally treated as methanation catalysts because of their ability to dissociate hydrogen and to bind CO [50–57]. However, the properties of Ru such as its similar catalytic activity as that of Pt, large number of oxidation states, and interesting electronic properties in combination with other oxide supports or transition metals prompted to test them in RWGS reaction [8, 57–59]. Ru is the most economical among the noble metals [8]. Thus, there were efforts to support Ru nanoparticles either as such or in bimetallic form on oxide materials such as CeO_2 [8, 57] and Al_2O_3 [59] in an aim to produce CO selectively via RWGS reaction.

Aitbekova et al., in a recent study, demonstrated the importance of pretreatment conditions on tuning the structures of supported Ru NPs toward selective CO_2 reduction [57]. Ceria-supported Ru NPs act as selective methanation catalysts for the CO_2 reduction reaction when Ru is in metallic state after reductive pretreatment in H_2 (Figure 12.6). However, after oxidative pretreatment at a relatively low temperature, Ru/CeO_2 gave above 90% selectivity to CO via RWGS. In situ and ex situ XAS analysis revealed that this oxidative treatment induced the redispersion of initial Ru NPs into atomically dispersed RuO_x species, which substantially changed the reaction pathway from CO_2 methanation into RWGS. However, the pretreatment-induced product selectivity depends on the metal–support combination. For example, the

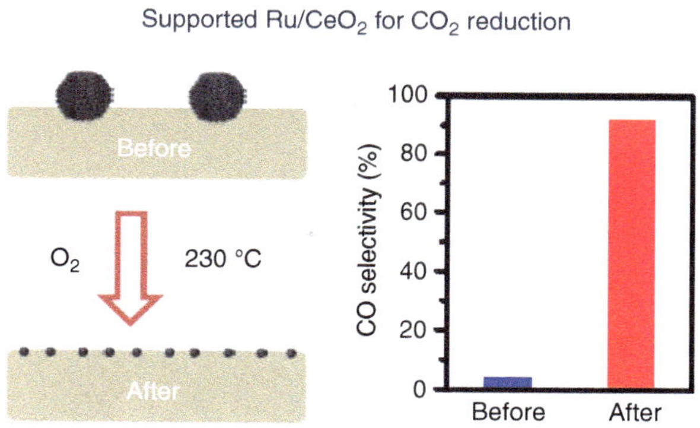

Figure 12.6 Effect of reduction and oxygen pretreatment of Ru/CeO_2 catalyst in CO selectivity in the CO_2 hydrogenation reaction. Source: Aitbekova et al. 57. Copyright 2018 American Chemical Society.

redispersion of Ru is complete on CeO_2 but only partial on TiO_2 and Al_2O_3 because of the strong stabilization of RuO_x species on the CeO_2 surface [57].

Panaritis et al. applied Ru–Fe nanoparticles with different Ru_xFe_{100-x} compositions ($x = 100$, 80, 45, 20, and 0 atomic%) supported on samarium-doped ceria (SDC) for RWGS in the temperature range of 300–800 °C [8]. Among all the compositions, $Ru_{45}Fe_{55}$/SDC (2 wt%) was the most promising, which showed 100% CO selectivity above 500 °C. A stability test of 100 hours at 650 °C confirmed the excellent stability of the catalyst. The high thermal resistance of Fe and the high thermal stability of SDC enhanced the stability of this catalyst [8].

Kwak and coworkers reported that CO was the major product on Ru/Al_2O_3 catalysts in the Ru loading range of ≤0.5 wt% in which Ru is in a highly dispersed state or exist as single atoms on the alumina support, whereas CH_4 increases proportionally with Ru loadings and is prevalent on large (three-dimensional [3D]) Ru clusters [59]. Scire et al. reported that the electronic state of the active metal (Ru), which varies with particle size, largely influences the CO selectivity in the CO_2 hydrogenation process and proposed that a more oxidized Ru surface favors methanation activity because of higher H_2 and lower CO coverage [60]. Regardless of Ru loading, the apparent activation energies of 82 and 62 kJ mol^{-1} were determined for CO and CH_4 formation over Ru/Al_2O_3, suggesting that the formation of CO and CH_4 follows different reaction pathways or proceed on active centers of different nature. Thus, the above results show that Ru-based catalysts can be made selective to form CO instead of CH_4 by carefully manipulating the geometrical and electronic properties.

12.2.1.5 Pd- and Ir-Based Catalysts

Supported Pd and Ir catalysts are known for its high activity in hydrogenation reactions. Both metals show high selectivity to CH_4 formation in CO_2 reduction [61–66], especially at higher metal loading. Studies using two Pd/γ-Al_2O_3 catalysts with different Pd loadings (5% and 0.5%) revealed that the selectivity to methane (22–40%) on 5% Pd/Al_2O_3 was 2–3 times higher than that on 0.5% Pd/Al_2O_3 [62].

Similar to Ru, the selectivity of Pd catalysts in CO_2 hydrogenation depends on several factors. Kwak and collaborators reported different reactivity patterns for atomically dispersed Pd and traditional 3D clusters [63]. They also substantiated the requirement of two different catalyst functionalities in the CO_2 hydrogenation reaction on Pd/Al_2O_3 and Pd/multi-walled carbon nanotube (MWCNT) catalysts [63]. These include a component that can activate CO_2, here either the support oxide (Al_2O_3) or an oxide promoter (here La_2O_3), and a metallic component that can dissociate H_2 (here Pd). When both of these functionalities are present, the CO and CH_4 selectivities seem to be determined by the sizes of the metal particles [63].

Combining Pd with synergistically interacting elements to form bimetallic nanoparticles can improve the selectivity toward CO formation while using less amount of expensive noble metal Pd [64]. Thus, in comparison with Pd/SiO_2, Pd–In/SiO_2 showed 100% CO selectivity though the latter catalyst resulted less CO yield [64]. The combined experimental and density functional theory (DFT) studies revealed that the lower activity and 100% CO selectivity on Pd–In/SiO_2 are due to (i) PdIn is less active than Pd for H_2 dissociation, (ii) CO adsorption on PdIn surface is much weaker than on Pd surface, and (iii) CO hydrogenation on PdIn surface is energetically less favored than that on the Pd surface [64].

RWGS over CeO_2-supported Ir nanoparticles indicates that the selectivity of CO_2 hydrogenation reaction can be tuned by simply changing the Ir loading [66]. A dramatic change in CH_4/CO selectivity pattern was observed from 88% CH_4 selectivity on 20% Ir/CeO_2 to an almost 100% CO selectivity when the Ir loading was decreased to 5%. The chemical state of Ir species induced by strong metal–support interaction has a major impact on the reaction selectivity. Ir in metallic state shows selectivity to CH_4, while partially oxidized Ir species is favorable for CO formation [66].

12.2.1.6 Cu-Based Catalysts

Cu-based catalysts have been extensively studied and commercially used for water gas shift reaction and CO_2 hydrogenaton to methanol in which RWGS reaction is an undesirable side reaction [4]. The excellent catalytic hydrogenation activity and selectivity at relatively low temperature makes Cu-based oxide catalysts a promising candidate for RWGS reaction also. The activity and selectivity are comparable to those of noble metal catalysts at low reaction temperature [67]. Nonetheless, Cu nanoparticles have poor thermal stability and strong tendency to sinter at reaction temperatures above 300 °C [68], while the endothermic nature of the RWGS reaction demands higher reaction temperatures for maximum catalytic performance. Also, CO_2 leads to the oxidation of part of the metallic Cu to its higher oxidation state in several supported Cu catalyst systems during reaction. This oxidation process of metallic Cu results in Cu agglomerates, leading to a decrease in the metal surface area and, hence, a lower catalyst activity and stability in due course of time [4, 68–70].

As sintering is a major concern for the efficiency and lifetime of Cu-based RWGS catalysts, efforts have been made to develop highly stabilized Cu nanoparticles through strong Cu–support interaction that can withstand sintering even at high temperature. Different strategies have been adopted to improve the activity and stability of Cu-based catalysts, which include dispersion of the active copper components to different kinds of supports [67–88], alloying Cu with other metals (Ru, Zn, Fe, etc.) [78, 80, 82], controlling the morphology of the catalyst support [89, 90], Cu loading, surface area, and the oxidation state of Cu [78, 86], controlling

Cu–support interaction, using basic metal promoters such as Cs, K [69, 78], and electronic promoters such as Fe incorporated in supports such as Al_2O_3 and SiO_2 [76, 78, 86, 87]. A variety of supports including SiO_2 [69–72, 86, 87]; ZnO [67, 88]; CeO_2 [74, 75]; Al_2O_3 [74, 76, 78]; ZrO_2 [74, 79]; TiO_2 [74]; MOF-derived porous C [80]; perovskites [81]; TMCs [68]; mixed oxide supports such as ZnO–Al_2O_3 [82–84]; supports modified with ZnO, ZrO_2, and MgO over Al_2O_3 [77]; mesoporous ceria [85] etc., have been used to disperse the active Cu species.

Cu^0 nanoparticles are generally the active phase and copper–oxide interface are the active sites for CO_2 hydrogenation, suggesting that the extent of copper dispersion and the oxidation state of Cu play a decisive role in the activity and long-term stability of the catalyst [88, 91]. Stabilization of Cu nanoparticles and their oxidation states are crucial to achieve high RWGS activity and CO selectivity as RWGS reaction is favored on metallic copper as opposed to partially oxidized copper. For example, a direct correlation between the Cu^+/Cu^0 ratio and methanol selectivity was observed in CO_2 hydrogenation over Cu/ZnO/Al_2O_3 catalysts at the conventional industrial operating conditions [92]. RWGS reaction over Cu supported on non-reducible supports such as Al_2O_3 and SiO_2 shows fast decay of catalytic activity because of the decrease of copper surface area and the oxidation of copper to Cu^{2+}in due course of time [70–74, 86, 87].

Addition of oxide components such as Fe_2O_3 or alloying copper with suitable elements such as Fe and Zn [76, 83, 93] are effective ways to inhibit sintering and oxidation of Cu and obtain a highly dispersed copper-based catalyst with improved microstructural stability while maintaining stable catalytic activity for high-temperature (>600 °C) RWGS applications [76, 86]. For example, at the gas space velocity of 60 000 h^{-1} and at a temperature of 700 °C, the conversion of CO_2 was improved from 10% (blank test) to 42% over the CuFe/Al_2O_3 catalyst by applying in solid oxide electrolysis cell (SOEC) single cells for the integrated SOEC-RWGS system [76]. Different arguments were made on the reasons for the enhanced stability and oxidation state of Cu in Fe-incorporated Cu catalysts in RWGS. One of the arguments is that the Cu–Fe interaction can also stabilize the Fe active phase because of the partial or complete oxidation of Cu during the reaction, leading to a higher catalytic performance [76, 78, 86, 87]. A different opinion is that the synergetic interaction between iron oxide species and Cu particles in catalyst systems such as iron-promoted Cu/SiO_2, Cu/Al_2O_3, and commercial Cu/ZnO/Al_2O_3 prevents sintering of Cu at high temperatures. The oxygen released from the dissociative adsorption of CO_2 on Cu can rapidly diffuse to Fe surfaces forming Fe oxides through a spillover process while keeping Cu in its reduced state [86]. Chen et al. reported that iron additives in Cu/SiO_2 catalyst significantly improve the Cu surface area and inhibit the sintering of Cu particles and oxidation of Cu even at high reaction temperature [87].

ZnO is a widely used support for stabilizing Cu nanoparticles for WGS reaction, CO_2 hydrogenation reaction to methanol or CO [88, 94]. As shown in Figure 12.7, RWGS activity results on Cu/ZnO catalysts with different Cu/Zn atomic ratio showing that catalysts derived from ZnO-enriched surface (Zn content, 50–70% atom) provides an optimized Cu/ZnO interface for the activation and dissociation

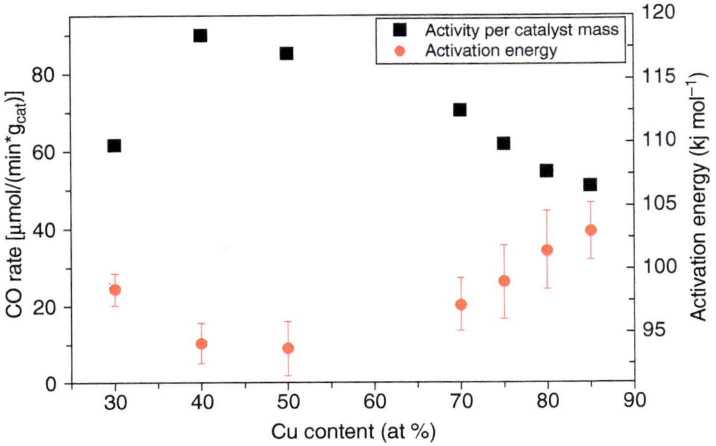

Figure 12.7 CO formation rate per catalyst weight and activation energy as a function of Cu content for RWGS reaction at 503 K. Source: Galván et al. 88. Copyright 2016 Elsevier.

of CO_2 and hence more active in RWGS with lower apparent activation energies than Cu-enriched catalyst surface (Zn content, 15–30% atom) [88].

ZnO is considered to play multiple roles in the above reactions such as by providing active sites for H_2 spillover or as a structure-directing support in controlling dispersion of metallic Cu particles through synergetic interactions of Cu and ZnO [77, 82–84, 94, 95]. However, ZnO is more efficient for RWGS reaction when combined with other oxide materials such as Al_2O_3 and ZrO_2 [77, 82–84]. When high surface area supports such as Al_2O_3 is used, the crystallite sizes of both CuO and ZnO are reduced, and moreover, the amount of Cu content can be increased at the expense of the high surface area of the support [83].

CeO_2 with good oxygen storage capacity and excellent redox properties a good support for Cu for RWGS reaction. Being a reducible support, CeO_2 is reduced by H_2 to form surface oxygen vacancies, whereas CuO is reduced to form active Cu species. The surface oxygen vacancies and the stabilized active Cu species can act as active centers for RWGS reaction through the formation of Ce^{3+}–□–Cu^0 structure [75]. The presence of a high amount of lattice oxygen surface species in ceria promotes CO_2 activation, resulting in high CO_2 conversion and CO selectivity. However, an optimum loading of Cu is important to achieve maximum catalytic performance [75].

Zhang et al. reported that β-Mo_2C, a TMC, is an efficient support capable of dispersing and stabilizing Cu particles and prevents the aggregation of Cu particles during the RWGS reaction [68]. As evidenced in Figure 12.8, under relatively high weight hourly space velocity (WHSV = 300 000 ml g^{-1} h^{-1}), the optimized 1 wt% Cu/β-Mo_2C exhibited superior activity over traditional oxide-supported Pt- and Cu-based catalysts. The activity was well maintained in a 40 hours stability test conducted at 600 °C when compared to Cu/ZnO/Al_2O_3 catalyst. The Cu–β–Mo_2C interface tunes the electronic structure of loaded Cu species and the strong interaction between copper and hexagonal β-Mo_2C substrate greatly enhances the metal

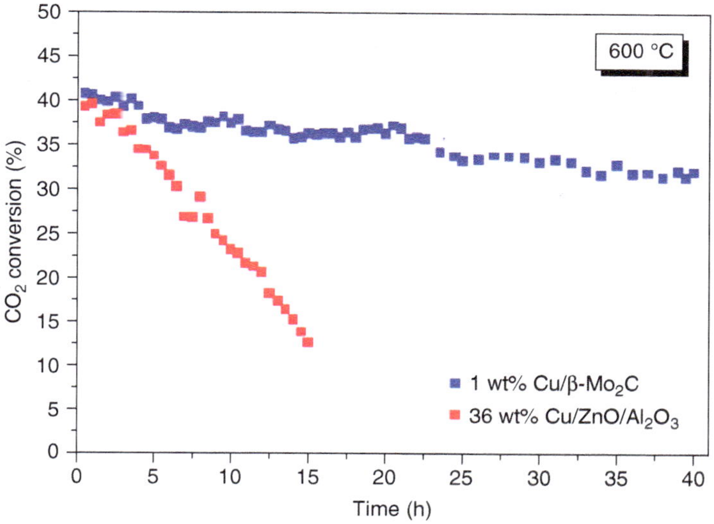

Figure 12.8 Stability of 1 wt% Cu/β-Mo_2C catalyst and the commercial 36 wt% Cu/ZnO/Al_2O_3 catalyst in RWGS reaction (reaction conditions: atmospheric pressure, 600 °C, CO_2:H_2 = 1:2, WHSV = 300 000 ml g^{-1} h^{-1}. Source: Zhang et al. 68. Copyright 2017 American Chemical Society.

dispersion and prevents the agglomeration of supported Cu particles under the reaction conditions, which accounts for the good catalytic performance.

The use of multicomponent oxides as supports for Cu nanoparticles in general improves the RWGS activity, CO selectivity, and stability [1–4, 79, 82, 83]. The structural characterization of these catalysts shows that the enhancement in the catalytic performance with respect to single metal oxide supported Cu catalyst is due to one or several of the reasons such as higher concentration of active sites at the Cu–metal oxide interface, high reducibility, and increased adsorption capacity of CO_2, H_2, etc. [76, 85].

Another way to improve the catalytic stability of Cu nanoparticles is to encapsulate them in a porous carbon matrix derived via pyrolysis of Cu–BTC MOFs [80]. The porous structure of the carbon matrix protects the NPs from sintering, leading to stable catalytic activity at 500 °C under which RWGS reaction is favored. The catalytic activity of this MOF-derived Cu catalyst could be further improved by adding other elements such as Zn because of the synergistic effects between the well-mixed Cu and Zn from pyrolysis [80]. Most importantly, the porous architecture of the carbon matrix prevents the phase separation of Cu/Zn bimetallic catalysts, which is a major concern for traditional Cu/ZnO/Al_2O_3 catalysts at high temperatures.

The addition of an optimum amount of basic species such as MgO, K [69, 77] or Cs [78] makes the supported Cu catalysts comparatively more reducible in nature with respect to the parent sample with improved dispersion of Cu species. Alkali promoters or MgO species are easily adsorbed on the acid surface of supports such as Al_2O_3. This helps the dispersion of Cu species, enhancing CO_2 adsorption and

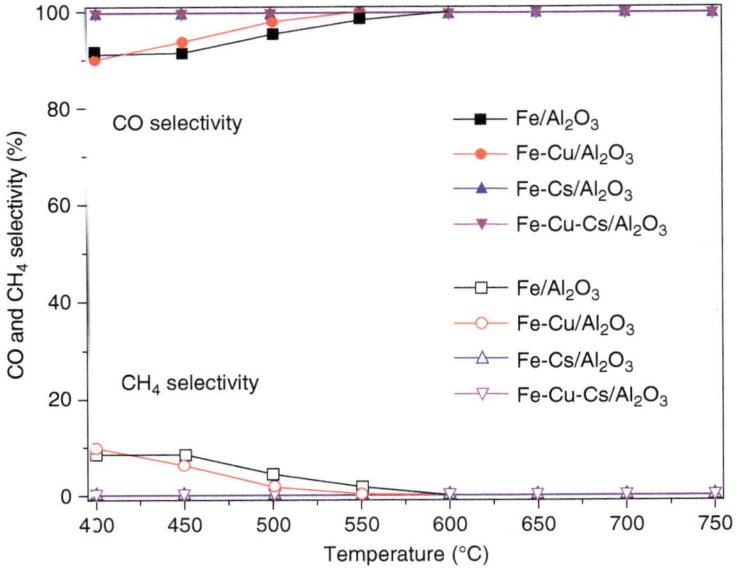

Figure 12.9 CO selectivity and CH_4 selectivity for Fe/Al_2O_3, Fe–Cu/Al_2O_3, Fe–Cs/Al_2O_3, and Fe–Cu–Cs/Al_2O_3. Cs-doped catalysts maintain a CO selectivity of ≥99% over a wide temperature range. Source: Pastor-Pérez et al. 78. Copyright 2017 Elsevier.

facilitating the formation of intermediate compounds such as formate species and its eventual decomposition to CO [69, 77, 78]. The presence of alkali promoters also prevents undesirable side reactions that lead to CH_4 in the relatively higher temperature regime of RWGS reaction at which CO formation favors. However, an optimum loading of promoters is essential for achieving maximum CO_2 conversion and desired product selectivity in CO_2 hydrogenation reaction. Ren and coworkers observed that too much MgO species (Mg/(Mg + Zr) molar ratio > 0.1) in Cu–ZnO–ZrO_2–MgO/Al_2O_3 catalyst makes the reduction of Cu^{2+} very difficult and disadvantageous to the formation of highly dispersed active Cu^0 [77].

As illustrated in Figure 12.9, Cs-doped Fe–Cu/Al_2O_3 and Fe/Al_2O_3 catalysts maintain a CO selectivity of ≥99% over a wide temperature range of 400–750 °C compared to catalysts without any Cs doping [78]. The higher activity and CO selectivity of the Cs doped sample indicate that Cs enhances the electron transfer with the CO_2 molecules even at low temperature and inhibits the methanation reaction occurring on the iron and copper active phases at lower temperatures [78].

Addition of small amounts of Ru to Cu can enhance the activity and selectivity to CO, although Ru is known for its high activity in methylation reactions. Thus, the 0.5 wt% Ru–Cu/ZnO/Al_2O_3 catalyst gave a two-fold increase in CO_2 conversion with respect to Ru-free Cu/ZnO/Al_2O_3 catalyst while maintaining complete selectivity to CO formation [82]. The formation of the Ru–Cu core–shell nanoparticles could be an explanation in this case.

12.2.1.7 Ni-Based Catalysts

Nickel-based catalysts are well known for their high activity in various catalytic hydrogenation reactions and are considered to be an alternative to precious metals because of their lower cost and higher availability [96]. However, Ni-based catalysts in RWGS reaction have shown relatively low CO selectivity because of the favorable methanation reaction during the hydrogenation of CO_2 [1, 97]. Efforts have been made to improve the activity and CO selectivity over Ni-based catalysts, such as by dispersing the active nickel components on different kinds of supports [96–109], alloying Ni with other metals (Fe and Cu) [110, 111], using promoters such as K [109], controlling the Ni particle size, Ni loading, and the oxidation state of Ni [96, 112]. A variety of supports including Al_2O_3 [98, 99], SiO_2 [96, 112], MgO [100], CeO_2 [101–105], perovskites [106], SBA-15 [107], Ce–Zr mixed oxides [97], different transition metal-promoted CeO_2–Al_2O_3 [108], Ni–Fe bimetallic/CeO_2 [110], and Ni–Cu bimetallic/γ-Al_2O_3 [111] have been used to disperse the active Ni species.

The studies using Ni/Al_2O_3 [98], Ni/SiO_2 [96], Ni/MgO [100], and Ni/CeO_2 [101, 102, 110] catalysts show that lower nickel loading promoted RWGS reaction while methanation of CO_2 dominated at higher nickel loading. It has been reported that the selectivity of CO start decreases and the side product CH_4 formed via Sabatier reaction starts increases when the Ni content increases [98, 100]. For example, CO_2 hydrogenation was in favor of CH_4 formation with the best activity for methanation was achieved at around 25% and 7% Ni loading over Ni/Al_2O_3 and Ni/MgO catalysts, respectively [98, 100]. The reducible nature of cerium oxide supports with lattice oxygen vacancies and highly dispersed Ni nanoparticles in the ceria matrix play important roles in the activation of CO_2 and selective formation of CO, whereas bulk nickel favors methane formation [101]. The surface-sensitive in situ ambient pressure XPS studies [101] and X-ray absorption near-edge structure (XANES) analysis [110] on CeO_2-supported Ni at different Ni loadings revealed that Ni at the surface of Ni/CeO_2 catalysts remains metallic in both CO-selective and CH_4-selective Ni/CeO_2 catalysts and the selectivity trend to CO/CH_4 formation is due to structural rather than oxidation state effects.

A direct correlation between the nickel particle size and the CO/CH_4 product selectivity is established for the RWGS reaction. CO selectivity can be improved by decreasing the Ni particle size, and regardless of the reaction temperature, 100% CO selectivity can be achieved over monodispersed Ni particles supported on mesoporous CeO_2 [102] and SBA-15 [107]. Besides the reaction selectivity, difference in Ni particle size also affects the kinetic parameters of CO_2 hydrogenation, the formation pathways of CO and CH_4, and stability of the catalyst [96].

Ni particle size can be tuned by controlled loading of Ni on the support. Tiny particles, similar to atomically dispersed Ni, were observed on 0.5 wt% Ni/SiO_2 while large nonuniform Ni particles with an average particle size 9 nm were observed when the Ni loading increased to 10 wt%. As shown in Figure 12.10, 0.5 wt% Ni/SiO_2 shows a comparatively higher catalytic activity for CO_2 hydrogenation with high CO selectivity and the selectivity for CH_4 increases on increasing

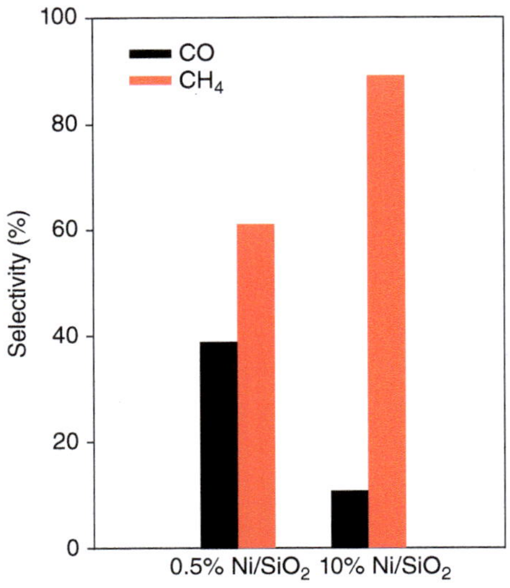

Figure 12.10 Comparison of CO and CH_4 selectivity on 0.5% and 10% Ni/SiO_2 catalysts at CO_2 conversion ~10%. Temperature = 623 K. Source: Wu et al. 96. Copyright 2015 Royal Society of Chemistry.

the Ni loading to 10 wt% [96]. The apparent activation energy estimated in the temperature range between 573 and 673 K for the overall CO_2 reduction increases from 57 to 73 kJ mol^{-1} when the Ni loading increases from 0.5 to 10 wt% over SiO_2 support [96].

Use of basic supports such as MgO or promoters such as potassium increases the basicity of catalyst and forms new active sites on it, which increases the adsorption of CO_2, facilitating the formation and decomposition of formate intermediate as well as hindering CH_4 formation [100, 109].

Without significantly compromising the CO_2 hydrogenation activity, CO selectivity can be improved by adding another transition metal component such as Fe to form Ni–Fe bimetallic species on CeO_2 oxide support as demonstrated in Figure 12.11 [110]. In the bimetallic catalysts, Ni enhances H_2 dissociation, whereas the presence of Fe oxides weakens the adsorption strength of CO with the catalyst and thereby decreases the extent to which the adsorbed CO is hydrogenated to CH_4 [110].

However, a major drawback of ceria-supported catalysts is its fast deactivation under the reaction conditions of CO_2 hydrogenation [101]. The deactivation of ceria catalysts can be decreased by adding other oxide components such as Zr, Al, and Fe to the CeO_2 to form a multicomponent solid oxide solution [97, 108, 110]. For example, Ni nanoparticles dispersed on Ce–Zr-mixed oxide via an impregnation method performed well in RWGS [97]. As shown in Figure 12.12, the stability study shows that CO_2 conversion reached 50% and remained stable after 80 hours of reaction over 3 wt% Ni/Ce–Zr–O_x [97]. The oxygen storage capability and thermal stability can be improved over Ce–Zr–O_x solid solution compared with pure oxides, which enables it to mitigate the carbon deposition through carbon gasification and to maintain the catalyst stability for a long period of time [97, 113, 114]. Besides CeO_2, the addition of elements such as iron also remarkably boosts the activity, selectivity, and

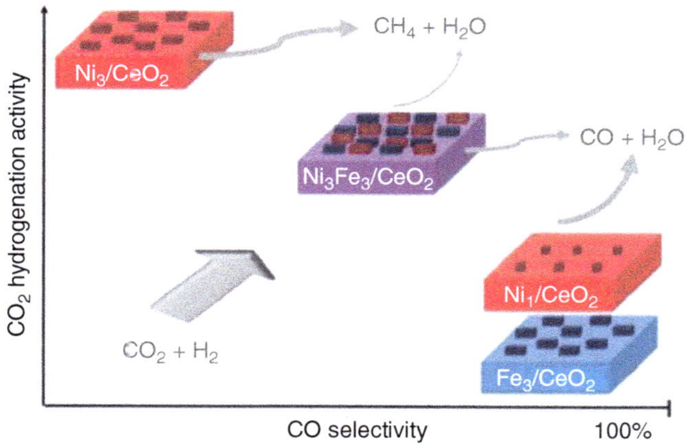

Figure 12.11 Relative activity of Ni/CeO_2, Fe/CeO_2, and Fe-incorporated Ni/CeO_2 catalysts in CO_2 conversion. Source: Winter et al. 110. Copyright 2018 Elsevier.

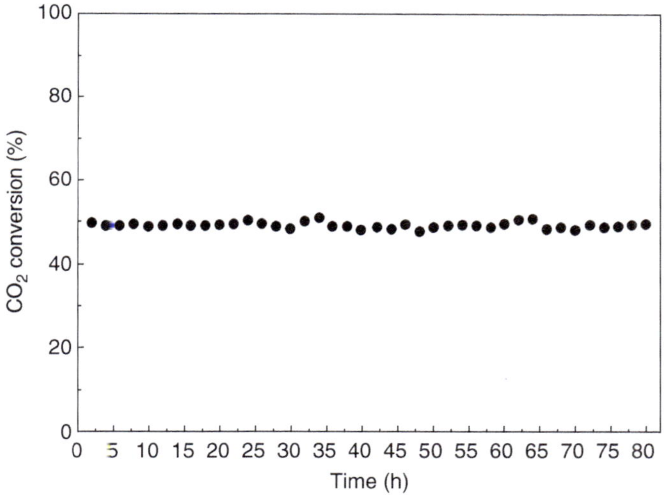

Figure 12.12 Stable CO_2 conversion was obtained over 3 wt% Ni/Ce–Zr–Ox catalyst for 80 hours at 750 °C. Source: Sun et al. 97. Copyright 2015 Elsevier.

stability of the Ni/Al_2O_3catalysts [108]. The high activity/selectivity and exceptional stability exhibited by the multicomponent Ni–FeO_x/CeO_2–Al_2O_3 for the RWGS reaction is attributed to an electronic enrichment of the surface Ni atoms because of the FeO_x–Ni interaction [108].

CH_4 is considered to form through CO over Ni-based catalysts [109, 112, 115, 116]. Although under the hydrogenation reaction conditions nickel is more likely to be reduced to Ni^0, the selectivity of CO_2 hydrogenation can be significantly tuned by controlling the valence state of nickel using a suitable support such as lanthanum–iron–nickel perovskites [106]. The enhanced Ni–support interaction in

$LaFe_{0.5}Ni_{0.5}O_3$ maintains the higher valance states of Ni species, leading to a higher CO selectivity, whereas metallic Ni is formed over $LaNiO_3$ under reaction conditions [106]. Goncalves et al. used the magnetron sputtering deposition technique to obtain silica-supported Ni nanoparticles with small metal particle size (2.3 nm) and controlled Ni loading and composition (77% Ni and 23% Ni oxides) and such a catalyst displayed an enhanced selectivity for the RWGS reaction [112].

12.2.2 Oxide Systems

Metal oxides are an attractive option for RWGS reaction because of their low cost, resistance to sintering when compared to noble metals, and metal-supported catalysts [76, 83, 117–130]. Particularly, materials such as perovskites with ABO_3 structure (where A is a large-sized A-site cation and B is a medium-sized B-site cation), which shows structural flexibility, thermal stability, and stability under nonstoichiometric conditions and harsh reaction conditions, are promising alternatives to precious metal-based catalysts [117, 118]. Besides its structural stability, the possibility to tune their composition with multiple cation combinations on each site to obtain desired properties such as high oxygen mobility make them valuable for specific catalytic applications such as RWGS [118, 119].

Among a series of Y-, Zn-, and Ce-doped barium zirconate-based perovskite-type catalysts with the general formula $Ba–Ce_xZr_{0.8-x}Y_{0.16}Zn_{0.04}O_3$ ($x = 0, 0.2, 0.5,$ and 0.7), $BaZr_{0.8}Y_{0.16}Zn_{0.04}O_3$ gave the best catalytic performance for the RWGS reaction with an average CO_2 conversion of 38% and a CO selectivity of 97% [119]. Although the formation of oxygen vacancies enhances the proton conductivity and RWGS activity, an increase in oxygen vacancies could decrease the chemical stability of catalysts, suggesting that the interaction between the elements in the catalyst during a chemical reaction should be carefully considered along with optimization of elemental composition of the catalyst [119]. DFT investigations based on $La_{(1-x)}Sr_xFe_{(1-y)}Co_yO_{(3-\delta)}$ perovskite demonstrated that carbon dioxide adsorption strength increases with increasing oxygen vacancies in the material, showing the importance of oxygen vacancies [120]. The bulk oxygen vacancy formation energies for these mixed perovskite oxides are found to increase with increasing lanthanum and iron contents in the "A" site and "B" site, respectively [120]. By carefully choosing different elements in suitable combinations in the A and B sites, perovskite materials can be fine-tuned to achieve favorable properties and optimum catalytic performance in RWGS.

The BaFe–hexaaluminate catalysts with and without K loading in RWGS reaction displayed a strong promotion effect of K on the activity of Ba–Fe–hexaaluminates [121]. At an optimal K content of 6 wt%, the RWGS rates increased by a factor of 12–15 without changing the turnover frequency number and was attributed to increasing concentration of active sites because of the reduction of Fe^{3+} ions to Fe^{2+} ions facilitated by K [121]. In another variation, Utsis and coworkers reported that K-promoted Fe-substituted Ba–hexaaluminate after partial reduction of iron showed high activity and selectivity in low-temperature CO_2 hydrogenation to C_5^+ hydrocarbons via an integrated RWGS-FTS process [122]. The pure

BaFe–hexaaluminate itself can act as an active RWGS catalyst with negligible methanation at 320 °C [122, 123].

However, perovskite oxides need higher reaction temperature compared to metal-incorporated oxide catalysts. The reaction temperature can be decreased by adding an active metal component such as Cu to the oxide system [81]. The effect of Cu doping in Fe-based perovskites having the general formula $La_{0.75}Sr_{0.25}Fe_{1-Y}Cu_YO_3$ with different compositions Y = 0, 0.1, 0.25, 0.5, 0.75, and 1 on the conversion of CO_2 to CO was investigated. The incorporation of Cu promotes the formation of oxygen vacancies at lower temperatures but also increased the instability of the perovskite. Only the perovskites with lowest Cu substitution (Y = 0.1) and no Cu (Y = 0) were able to produce CO [81]. Cu incorporation into the perovskite structure induced a loss of oxygen affinity and decreased the ability of the perovskite to reoxidize.

Catalysts containing iron are active and largely applied for reaction that requires higher temperature because of their high stability and high oxygen mobility [78, 86, 124]. The endothermic nature of the RWGS s demands high reaction temperature to achieve maximum CO_2 conversion and CO yield, and hence, iron is usually a choice to use either as a promoter or as one of the oxide components in combination with other RWGS active metals. Especially, in combination with relatively low-temperature active RWGS catalysts such as Cu, iron catalysts show several advantages such as improved mechanical properties (i.e. sintering tolerance), synergistic electronic effects, and stability [78, 86, 124].

Co- and Ni-incorporated iron oxide catalyst supported on Al_2O_3 showed a significant enhancement in the RWGS activity compared to the undoped Fe/Al_2O_3 catalyst [125]. Among the doped catalysts, $Co–Fe/Al_2O_3$ showed highest CO yield (48%) and stable time-on-stream performance for 40 hours at 650 °C and atmospheric pressure. Addition of Co and Ni increases the reducibility of iron oxide and resulted in the formation of new mixed oxide phase, which is nonselective for methane and stable to the RWGS reaction conditions [125].

Studies of $Fe/\gamma\text{-}Al_2O_3$ and $Fe–K/\gamma\text{-}Al_2O_3$ catalysts at temperatures between 723 and 753 K and atmospheric pressure revealed that both catalysts have high CO selectivity (>99%). However, the incorporation of potassium into Fe/Al_2O_3 results in a significant increase (~3 times) in CO formation rates [126]. Gas-switching experiments (CO_2 or H_2 along) and in situ diffuse reflectance Fourier transform infrared spectroscopy (DRIFTS) analysis showed that stable intermediates were formed on $Fe–K/\gamma\text{-}Al_2O_3$ but not on Fe/Al_2O_3, suggesting that redox mechanism is the only reaction pathway on the Fe/Al_2O_3 catalyst, whereas associative mechanism also operates on $Fe–K/Al_2O_3$ along with the dominant redox mechanistic pathway [126]. Similarly, addition of 2 wt% K significantly enhanced the rate of RWGS reaction tenfold on Fe–Al–O oxide spinel phase compared with the unpromoted catalyst [127]. This was attributed to the partial reduction of Fe^{3+} to Fe^{2+} in the spinel structure because of interaction with K and formation of oxygen vacancies, which act as active sites for RWGS [121, 127].

Besides the above elements, Mo is also reported to enhance the catalytic activity and stability of Fe/Al_2O_3 compared to Mo-free Fe/Al_2O_3 counterpart. Mo increases

Fe dispersion and produces iron species of smaller particle size as well as a new $Fe_2(MoO_4)_3$ phase. Existence of Fe–O–Mo structure in Fe-Mo/Al_2O_3 catalyst increases electron-deficient state of Fe on the surface of catalyst, which is suggested as the reason for the enhanced catalytic activity and CO selectivity [128].

Co is mainly used in combination with other active metals either as a promoter or in the form of mixed oxide for RWGS reaction. The Co–CeO_2 catalysts with different Co contents prepared by a coprecipitation method were explored for RWGS reaction [129]. The selectivity was sensitive to Co particle size, Co loading, and its dispersion. Highly dispersed Co oxide nanoparticles with smaller particle size were essential for the RWGS catalyst with high activity and selectivity, whereas large metal particle was more favorable to methanation [129, 130]. The 2% Co–CeO_2 catalyst showed excellent catalytic performance in terms of activity and selectivity, which is related to the small particle size and highly dispersed Co_3O_4 on CeO_2. Increased Co content results in larger Co particle size, leading to increased carbon deposition and CH_4 formation [129]. Co/CeO_2 catalyst with Co particles larger than 10 nm prepared by impregnation method were reported to be very active for methanation reaction [131].

However, the stability can be an issue with highly dispersed Co catalyst prepared by conventional methods such as coprecipitation [129, 130]. Incorporating Co oxide nanoparticles in mesoporous CeO_2 framework was attempted as a method to solve this issue [130]. Wang and Liu synthesized a series of mesoporous Co–CeO_2 catalysts with different Co contents by a colloidal solution combustion method [130]. The mesoporous 5%Co–CeO_2 catalyst showed high activity and selectivity in RWGS and also showed a good stability during 10 hours at 600 °C. The excellent performance of the mesoporous 5%Co–CeO_2 catalyst was assigned to the small Co particles intimately interacted with CeO_2 in the pore walls.

12.2.3 Transition Metal Carbides

As described before, metals such as Cu, Au, and Pt are highly selective to CO but require high temperatures (>500 °C) for a sufficient CO_2 conversion. TMCs, on the other hand, are active at low temperatures with relatively lower CO selectivity [68, 132]. The high catalytic activity together with the other advantageous aspects of TMCs such as noble metal-like properties [133], low cost, and the ability to catalyze H_2 dissociation as well as C=O bond scission make them a promising support for dispersing active metal species for RWGS [22, 68, 132, 134–141].

Porosoff and coworkers have demonstrated that Mo_2C and cobalt-modified Mo_2C exhibit better performance in terms of both activity and selectivity than the traditional CeO_2-supported bimetallic RWGS catalysts such as Pt–Co, Pt–Ni, Pd–Co, Pd–Ni, and monometallic Ni and Co catalysts [132]. However, the long-term stability of TMCs is a major concern because of the tendency of TMCs to form oxycarbides in the presence of oxygen-rich molecules [133]. Porosoff et al. suggested that Mo_2C can undergo an oxidation-carburization cycle and sustain the activity in RWGS reaction as demonstrated by in situ XANES studies [132]. RWGS reaction on TMCs may originate from its direct participation in the reaction through facile oxygen transfer [132]. The RWGS activities of Mo_2C (high activity) and TiC (low activity) were also found to correlate with the ease for removal of surface oxygen to complete the catalytic cycle, which was confirmed by in situ XANES measurements [134].

Gao et al. reported that α–MoC_{1-x} and β–Mo_2C nanowires offer reasonably high CO_2 conversion (~27%) with 100% CO selectivity at 600 °C and at a CO_2/H_2 ratio of 1/1 [135]. An enhanced CO_2 conversion of >60% together with nearly 100% CO selectivity was achieved at 600 °C even at a higher H_2/CO_2 molar ratio of 4/1. A formate decomposition mechanism for the RWGS reaction was proposed based on the in situ DRIFTS results [135]. The high CO selectivity on molybdenum carbide nanowire is most likely due to weak adsorption of CO on molybdenum carbides, which prevents further hydrogenation to CH_4. Studies on 3d transition elements (Co, Cu, and Ni) supported over Mo_2C revealed that the best catalyst for the production of CO is Cu/Mo_2C at 200 or 300 °C and CO increases in the order $Co/Mo_2C < Ni/Mo_2C < \beta\text{-}Mo_2C < Cu/Mo_2C$ [136]. Small Au, Cu, and Ni particles supported on TiC(001) showed a high activity for the catalytic hydrogenation of CO_2[22]. However, the product distribution changed with respect to the active metal component: Ni/TiC(001) produced a mixture of CO, methanol, and methane, whereas on Au/TiC(001) and Cu/TiC(001) methanol was the predominant side product and no methane was observed [22].

The low surface area and inaccessible sites due to the particle growth and coarsening of metal carbide crystallites [137] are limiting factors for TMCs. As described above, incorporation of metal nanoparticles is a way to modify the electronic properties of TMCs, thereby regulating the activity and selectivity of the CO_2 hydrogenation reaction [137–141]. Recently, Xiong et al. demonstrated that the catalytic activity and stability of Mo_2C can be improved by incorporating N-functional groups on the interface of Mo_2C–C [137]. The incorporation of N-functional groups can also enhance the affinity of supports for metal nanoparticles such as Cu and prevent aggregations during the reaction. The N-functional groups serve as basic sites to further improve the adsorption and activation ability for CO_2 molecules [137]. Use of promoters such as K and/or dispersing TMCs over high surface area materials such as alumina are other possible ways to improve the catalytic performance of TMCs [139]. By modifying $Mo_2C/\gamma\text{-}Al_2O_3$ with a K promoter, the CO selectivity and yield increased significantly, which approached the maximum thermodynamic yield for RWGS under the appropriate reaction conditions. Addition of K also improved the catalyst stability, with only 7% deactivation after 68 hours on stream (Table 12.1) [139].

Table 12.1 Selection of some recent work in CO_2 hydrogenation to CO.

		Selectivity, %			Reaction conditions				
Catalyst	CO_2 Conv., %	CO	CH_4	CH_3OH	Temperature, °C	H_2/CO_2 ratio	Pressure	Space velocity, ml g_{cat}^{-1} h^{-1}	Ref.
2.4%Au@UIO-67 MOF	30.5	96.5	3.5		408	3 : 1	2 MPa	12 000	[23]
3%Au@TiO_2	1.9				408	3 : 1	2 MPa	12 000	[23]
3%Au@MgO	1.7				408	3 : 1	2 MPa	12 000	[23]
3%Au/CeO_2	27.6	100			400	3 : 1		12 000	[18]
1%Au/TiO_2	12.2	100			300	1 : 1	$P_{CO2} = 10$ kPa	15 000	[17]
1%Au/TiO_2	24.5	100			300	5 : 1	$P_{CO2} = 10$ kPa	15 000	[17]
1%Au/TiO_2	29.9	100			300	9 : 1	$P_{CO2} = 10$ kPa	15 000	[17]
1%Au/TiO_2	49.8	100			400	9 : 1	$P_{CO2} = 10$ kPa	15 000	[17]
Au/ZnO	1	30		70	240	3 : 1	50 bar	NA	[21]
Pt/CeO_2	13.7	100			225	4 : 1	0.1 MPa	75 000	[33]
Pt/TiO_2	4.8	98			250	1:1	NA	6000	[34]
Pt/TiO_2	21	65			400	1 : 1	NA	6000	[34]
PtK/Mullite	5	100			300	1 : 1	0.1 MPa	30 000	[37]
PtK/Mullite	10	100			350	1 : 1	0.1 MPa	30 000	[37]
PtK/Mullite	30.9	99.2			550	1 : 1	0.1 MPa	30 000	[37]
Pt/L	13.1	100			500	1 : 1	0.1 MPa	30 000	[38]
K_{80}-Pt/L	27.4	100			500	1 : 1	0.1 MPa	30 000	[38]
Pt/TiO_2	4.5	99.1			300	1 : 1	0.1 MPa	119.7 h^{-1}	[29]
Pt/SiO_2	3.3	100			300	1 : 1	0.1 MPa	24.7 h^{-1}	[29]

Rh-$SrTiO_3$	7.9	95.4			300	1 : 1	NA	12 000	[47]
Li/RhY	13.1	86.6			250	3 : 1	30 bar		[45]
Fe	26.7	84.1	15.9		540	4 : 1		1166 h^{-1}	[58]
Co	71.2	1.4	98.6		388	4 : 1		965 h^{-1}	[58]
Ni	54.7	20.1	79.9		519	4 : 1		956 h^{-1}	[58]
Cu	58.9	98.4	1.6		532	4 : 1		1200 h^{-1}	[58]
Ru/Al_2O_3	74.7	1.2	98.8		379	4 : 1		5305 h^{-1}	[58]
Cu/β-Mo_2C	6	96.5			300	2 : 1		300 000	[68]
Cu/β-Mo_2C	16	97.6			400	2 : 1		300 000	[68]
Cu/β-Mo_2C	28	99			500	2 : 1		300 000	[68]
Cu/β-Mo_2C	39	99.2			600	2 : 1		300 000	[68]
CuFe/Al2O3	42				700	1 : 1		60 000 h^{-1}	[76]
Cu/Zn@C-Submm MOF	5	100			500	3 : 1	1 bar		[80]
Cu/Zn@C-Submm MOF	7.5	11.3	87.9	0.8	300	3 : 1	1 bar		[80]
Cu/Zn@C-Submm MOF	17	88.5	0	11.5	300	3 : 1	40 bar		[80]
Cu/Zn@C-Submm MOF	12.8	58.3	0	41.7	260	3 : 1	40 bar		[80]

(Continued)

Table 12.1 (Continued)

Catalyst	CO_2 Conv., %	Selectivity, %			Reaction conditions			Space velocity, $ml\,g_{cat}^{-1}\,h^{-1}$	Ref.
		CO	CH_4	CH_3OH	Temperature, °C	H_2/CO_2 ratio	Pressure		
Cu@C-Submm MOF	1.9	90.4	9.6	0	500	3 : 1	1 bar		[80]
Mesoporous Cu/CeO_2	31.34	100			400	4 : 1		60 000	[85]
Mesoporous Mn/CeO_2	8.12	100			400	4 : 1		60 000	[85]
Mesoporous Fe/CeO_2	8.69	100			400	4 : 1		60 000	[85]
Mesoporous Co/CeO_2	34.91	63			400	4 : 1		60 000	[85]
Mesoporous Ni/CeO_2	69.21	28			400	4 : 1		60 000	[85]
Ni/CeO_2	5	90			400	1 : 1		12 000	[104]
Ni/CeO_2	24	100			550	1 : 1		12 000	[104]
Ni/CeO_2	42	100			750	1 : 1		12 000	[104]
Ni/CeO_2	5	100			400	1 : 1		7200	[102]
Ni/CeO	47	100			750	1 : 1		7200	[102]
Ni/SiO_2	64	100			660	4 : 1		400 000	[112]
$LaNiO_3$	39.7	8	92		400	2 : 1		40 ml min^{-1}	[106]
$LaFe_{0.5}Ni_{0.5}O_3$	16.3	96.6	3.2		400	2 : 1		40 ml min^{-1}	[106]
2%Co/CeO_2	6	97	3		400	1 : 1		300 000	[129]
2%Co/CeO_2	16	100	0		450	1 : 1		300 000	[129]

Co/β-Mo_2C	9.5	98.1		300	2 : 1		36 000	[132]
PtCo/CeO_2	6.6	81.8		300	2 : 1		36 000	[132]
β-Mo_2C	28	98	2	400	4 : 1		36 000	[135]
α–MoC_{1-x}	25	100		400	4 : 1		36 000	[135]
α–MoC_{1-x}	60	100		600	4 : 1		36 000	[135]
β-Mo_2C	24	28	45	300	5 : 1			[136]
Cu/Mo_2C	19	38	32	300	5 : 1			[136]
Ni/Mo_2C	29	22	64	300	5 : 1			[136]
Co/β-Mo_2C	31	19	26	300	5 : 1			[136]
Mo_2C/γ-Al_2O_3	19.9	73.5	19.7	300	3 : 1	2.1	0.000 841 g^{-1} s^{-1}	[139]
1%K-Mo_2C/γ-Al_2O_3	24.3	81	12.4	300	3 : 1	2.1	0.000 841 g^{-1} s^{-1}	[139]
2%K-Mo_2C/γ-Al_2O_3	18.1	95.9	2.7	300	3 : 1	2.1	0.000 841 g^{-1} s^{-1}	[139]
3%K-Mo_2C/γ-Al_2O_3	7.5	98.5	1.5	300	3 : 1	2.1	0.000 841 g^{-1} s^{-1}	[139]
Mo/γ-Al_2O_3	7.1	82.5	15.3	300	3 : 1	2.1	0.000 841 g^{-1} s^{-1}	[139]
2%K-Mo/γ-Al_2O_3	6.5	96.9	3.1	300	3 : 1	2.1	0.000 841 g^{-1} s^{-1}	[139]
$BaZr_{0.8}Y_{0.16}Zn_{0.04}O_3$	37.5	97		600	1 : 1		2400	[119]
La-Ni-Fe	16.3	96.6		350	2 : 1		24 000	[106]
Fe-Mo-$A_{12}O_3$	~45	~100		600	1 : 1	1	30 000	[128]

12.3 Mechanism of RWGS Reaction

The nature and the reaction mechanism of the RWGS reaction over oxides and metal-supported catalysts are elusive and still under debate. In general, two different mechanisms have been postulated for RWGS reaction, namely dissociative (also known as regenerative or redox mechanism) and associative [1–4]. In the redox (dissociative) mechanism, the reactants are oxidized or reduced separately on the catalyst surface as simply modeled by Eqs. (12.1, 12.2):

$$CO_2 + * \rightleftharpoons CO + O - * \tag{12.1}$$

$$H_2 + O \rightleftharpoons H_2O + * \tag{12.2}$$

where * denotes a surface site such as a metal surface site. The redox reaction mechanism, as for example discussed over Cu-based catalysts, is illustrated by the following equations [2]:

$$CO_2 + 2Cu^0 \rightarrow Cu_2O + CO \tag{12.3}$$

$$H_2 + Cu_2O \rightarrow 2Cu^0 + H_2O \tag{12.4}$$

The dissociative reaction mechanism is generally postulated over Cu-based catalysts, wherein Cu^0 atoms facilitate the dissociation of CO_2 and the oxidized Cu species are subsequently reduced by H_2 [67]. The redox mechanism for the RWGS reaction is possible only if two fundamental conditions are fulfilled: if (i) the surface of the catalyst can be reduced by H_2 and (ii) if CO_2 can efficiently reoxidize the partially reduced catalyst surface at the RWGS reaction conditions [19]. Wang et al. proposed a redox mechanism over Au/CeO_2 catalyst with the aid of quantitative temporal analysis of products (TAP) measurements [19]. In the first step of their mechanism, surface oxygen is removed by reaction with H_2 as $H_2 + O_{Ce_{O_2}} \rightarrow H_2O + {}_{Ce_{O_2}}$ where ${}_{Ce_{O_2}}$ and $O_{Ce_{O_2}}$ denote oxygen vacancies and oxygen atoms at the surface of the CeO_2 support. Subsequently, CO_2 would act as an oxidant for the partly reduced Au/CeO_2 catalyst surface, according to

$$CO_2 + {}_{Ce_{O_2}} \rightarrow CO + O_{Ce_{O_2}} +$$

In the associative mechanism, the reaction occurs via a surface intermediate as described below:

$$CO_2 + H_2 \rightleftharpoons \text{Intermediate}$$
$$\text{Intermediate} \rightleftharpoons CO + H_2O$$

This mechanism proposes that H_2 is dissociatively adsorbed on the metal surface followed by the spillover of H atoms to metal–oxide interface sites at which CO_2 activation takes place. This association of H_2 and CO_2 leads to a variety of intermediate species, the nature, and amount of these intermediates depend on the type of metal–support characteristics. This mechanism involves the formation of a bicarbonate species from adsorbed CO_2, followed by its conversion to formate species in the presence of hydrogen. The formate species is proposed as the key intermediate for CO formation in RWGS reaction over K-promoted Cu/SiO_2 catalysts and several

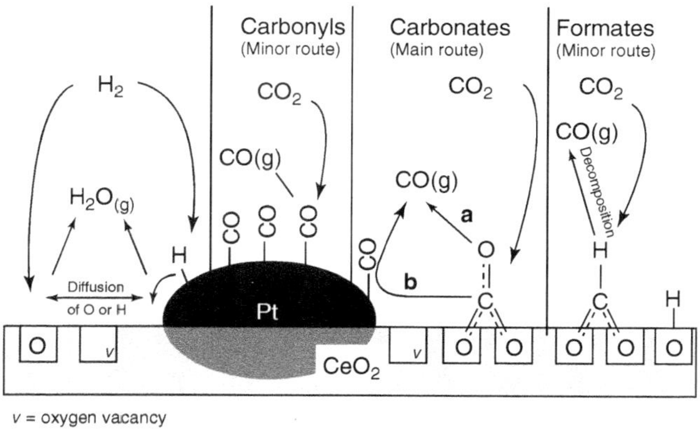

Figure 12.13 Proposed reaction mechanism of the RWGS reaction over Pt/CeO_2 catalyst. Source: Goguet et al. [33]. Copyright 2017 American Chemical Society.

other supported Cu and metal catalysts [69, 72, 73, 84]. The formate intermediate (HCOO*) is then decomposed to CO at the metal–oxide interface, which is promoted by hydrogen. A variety of in situ and ex situ characterization techniques have been employed to prove the existence of intermediates and CO formation mechanism [15, 31, 33, 38, 135]. Goguet et al. used a combination of steady-state isotopic transient kinetic analysis (SSITKA) and DRIFTS analysis to dynamically detect the surface species and proposed a model for the reaction mechanism of the RWGS reaction over Pt/CeO_2 catalyst as shown in Figure 12.13 [33]. This proposes that the main route involves the formation of carbonate species. The formation of carbonyls and formates are likely but they are minor species.

References

1 Wang, W., Wang, S., Ma, X., and Gong, J. (2011). Recent advances in catalytic hydrogenation of carbon dioxide. *Chem. Soc. Rev.* 40: 3703–3727.

2 Daza, Y.A. and Kuhn, J.N. (2016). CO_2 conversion by reverse water gas shift catalysis: comparison of catalysts, mechanisms and their consequences for CO_2 conversion to liquid fuels. *RSC Adv.* 6: 49675–49691.

3 Porosoff, M.D., Yan, B., and Chen, J.G. (2016). Catalytic reduction of CO_2 by H_2 for synthesis of CO, methanol and hydrocarbons: challenges and opportunities. *Energy Environ. Sci.* 9: 62–73.

4 Su, X., Yang, X., Zhao, B., and Huang, Y. (2017). Designing of highly selective and high-temperature endurable RWGS heterogeneous catalysts: recent advances and the future directions. *J. Energy Chem.* 26: 854–867.

5 Goguet, A., Meunier, F., Breen, J.P. et al. (2004). Study of the origin of the deactivation of a Pt/CeO_2 catalyst during reverse water gas shift (RWGS) reaction. *J. Catal.* 226: 382–392.

6 Kim, S.S., Lee, H.H., and Hong, S.C. (2012). A study on the effect of support's reducibility on the reverse water-gas shift reaction over Pt catalysts. *Appl. Catal., A* 423–424: 100–107.

7 Matsubu, J.C., Yang, V.N., and Christopher, P. (2015). Isolated metal active site concentration and stability controls catalytic CO_2 reduction selectivity. *J. Am. Chem. Soc.* 137: 3076–3084.

8 Panaritis, C., Edake, M., Couillard, M. et al. (2018). Insight towards the role of ceria-based supports for reverse water gas shift reaction over RuFe nanoparticles. *J. CO_2 Util.* 26: 350–358.

9 Carabineiro, S.A.C. and Thompson, D.T. (2007). Catalytic applications for gold nanotechnology. In: *Nanocatalysis* (eds. U. Heiz and U. Landman), 377–489. Berlin, Heidelberg: Springer.

10 Hashmi, A.S.K. and Hutchings, G.J. (2006). Gold catalysis. *Angew. Chem. Int. Ed.* 45: 7896–7936.

11 Sakurai, H., Tsubota, S., and Haruta, M. (1993). Hydrogenation of CO_2, over gold supported on metal oxides. *Appl. Catal., A.* 102: 125–136.

12 Burch, R. (2006). Gold catalysts for pure hydrogen production in the water-gas shift reaction: activity, structure and reaction mechanism. *Phys. Chem. Chem. Phys.* 8: 5483–5500.

13 Hartadi, Y., Widmann, D., and Behm, R.J. (2015). CO_2 hydrogenation to methanol on supported Au catalysts under moderate reaction conditions: support and particle size effects. *ChemSusChem* 8: 456–465.

14 Upadhye, A.A., Ro, I., Zeng, X. et al. (2015). Plasmon-enhanced reverse water gas shift reaction over oxide supported Au catalysts. *Catal. Sci. Technol.* 5: 2590–2601.

15 Bobadilla, L.F., Santos, J.L., Ivanova, S. et al. (2018). Unravelling the role of oxygen vacancies in the mechanism of the reverse water–gas shift reaction by *Operando* DRIFTS and ultraviolet–visible spectroscopy. *ACS Catal.* 8 (8): 7455–7467.

16 Ishito, N., Hara, K., Nakajima, K., and Fukuoka, A. (2016). Selective synthesis of carbon monoxide via formates in reverse water–gas shift reaction over alumina-supported gold catalyst. *J. Energy Chem.* 25: 306–310.

17 Kyriakou, V., Vourros, A., Garagounis, I. et al. (2017). Highly active and stable TiO_2-supported Au nanoparticles for CO_2 reduction. *Catal. Commun.* 98: 52–56.

18 Zhu, X., Qu, X., Li, X. et al. (2016). Selective reduction of carbon dioxide to carbon monoxide over Au/CeO_2 catalyst and identification of reaction intermediate. *Chin. J. Catal.* 37: 2053–2058.

19 Wang, L.C., Tahvildar Khazaneh, M., Widmann, D., and Behm, R.J. (2013). TAP reactor studies of the oxidizing capability of CO_2 on a Au/CeO_2 catalyst – a first step toward identifying a redox mechanism in the reverse water–gas shift reaction. *J. Catal.* 302: 20–30.

20 Carrasquillo-Flores, R., Ro, I., Kumbhalkar, M.D. et al. (2015). Reverse water–gas shift on interfacial sites formed by deposition of oxidized molybdenum moieties onto gold nanoparticles. *J. Am. Chem. Soc.* 137: 10317–10325.

21 Ro, I., Carrasquillo-Flores, R., Dumesic, J.A., and Huber, G.W. (2016). Intrinsic kinetics of plasmon-enhanced reverse water gas shift on Au and Au–Mo interfacial sites supported on silica. *Appl. Catal., A* 521: 182–189.

22 Rodriguez, J.A., Evans, J., Feria, L. et al. (2013). CO_2 hydrogenation on Au/TiC, Cu/TiC, and Ni/TiC catalysts: Production of CO, methanol, and methane. *J. Catal.* 307: 162–169.

23 Xu, H., Li, Y., Luo, X. et al. (2017). Monodispersed gold nanoparticles supported on a zirconium-based porous metal–organic framework and their high catalytic ability for the reverse water–gas shift reaction. *Chem. Commun.* 53: 7953–7956.

24 Koeppel, R.A., Baiker, A., Schild, C., and Wokaun, A. (1991). Carbon dioxide hydrogenation over Au/ZrO, catalysts from amorphous precursors: catalytic reaction mechanism. *J. Chem. Soc., Faraday Trans.* 87: 2821–2828.

25 Zhang, S., Li, X.S., Chen, B.B. et al. (2014). CO oxidation activity at room temperature over Au/CeO_2 catalysts: disclosure of induction period and humidity effect. *ACS Catal.* 4: 3481–3489.

26 Jadhav, S.G., Vaidya, P.D., Bhanage, B.M., and Joshi, J.B. (2016). Kinetics of reverse water–gas shift reaction over Pt/Al_2O_3 catalyst. *Can. J. Chem. Eng.* 94: 101–106.

27 Porosoff, M.D. and Chen, J.G.G. (2013). Trends in the catalytic reduction of CO_2 by hydrogen over supported monometallic and bimetallic catalysts. *J. Catal.* 301: 30–37.

28 Ferri, D., Burgi, T., and Baiker, A. (2002). Probing boundary sites on a Pt/Al_2O_3 model catalyst by CO_2 hydrogenation and in situ ATR-IR spectroscopy of catalytic solid–liquid interfaces. *Phys. Chem. Chem. Phys.* 4: 2667.

29 Ro, I., Sener, C., Stadelman, T.M. et al. (2016). Measurement of intrinsic catalytic activity of Pt monometallic and Pt-MoO_x interfacial sites over visible light enhanced $PtMoO_x/SiO_2$ catalyst in reverse water gas shift reaction. *J. Catal.* 344: 784–794.

30 Kattel, S., Yan, B., Chen, J.G., and Liu, P. (2016). CO_2 hydrogenation on Pt, Pt/SiO_2 and Pt/TiO_2: importance of synergy between Pt and oxide support. *J. Catal.* 343: 115–126.

31 Wang, Y., Arandiyan, H., Scott, J. et al. (2018). Single-atom and nano-clustered Pt catalysts for selective CO_2 reduction. *ACS Appl. Energy Mater.* 1 (12): 6781–6789.

32 Jacobs, G. and Davis, B.H. (2005). Reverse water-gas shift reaction: steady state isotope switching study of the reverse water-gas shift reaction using in situ DRIFTS and a Pt/ceria catalyst. *Appl. Catal., A* 284: 31.

33 Goguet, A., Meunier, F.C., Tibiletti, D. et al. (2004). Spectrokinetic investigation of reverse water–gas-shift reaction intermediates over a Pt/CeO_2 catalyst. *J. Phys. Chem. B* 108: 20240–20246.

34 Chen, X., Su, X., Duan, H. et al. (2017). Catalytic performance of the Pt/TiO_2 catalysts in reverse water gas shift reaction: controlled product selectivity and a mechanism study. *Catal. Today* 281: 312–318.

35 Kim, S.S., Park, K.H., and Hong, S.C. (2013). A study of the selectivity of the reverse water–gas-shift reaction over Pt/TiO_2 catalysts. *Fuel Process. Technol.* 108: 47–54.

36 Kim, S.S., Lee, H.H., and Hong, S.C. (2012). The effect of the morphological characteristics of TiO_2 supports on the reverse water–gas shift reaction over Pt/TiO_2 catalysts. *Appl. Catal., B* 119–120: 100–108.

37 Liang, B., Duan, H., Su, X. et al. (2017). Promoting role of potassium in the reverse water gas shift reaction on Pt/mullite catalyst. *Catal. Today* 281: 319–326.

38 Yang, X., Su, X., Chen, X. et al. (2017). Promotion effects of potassium on the activity and selectivity of Pt/zeolite catalysts for reverse water gas shift reaction. *Appl. Catal., B* 216: 95–105.

39 Oshima, K., Shinagawa, T., Nogami, Y. et al. (2014). Low temperature catalytic reverse water gas shift reaction assisted Byan electric field. *Catal. Today* 232: 27–32.

40 Gogate, M.R. and Davis, R.J. (2010). Comparative study of CO and CO_2 hydrogenation over supported Rh–Fe catalysts. *Catal. Commun.* 11: 901–906.

41 Zhang, Z.L., Kladi, A., and Verykios, X.E. (1994). Effects of carrier doping on kinetic parameters of CO_2 hydrogenation on supported rhodium catalysts. *J. Catal.* 148: 737–747.

42 Büchel, R., Baiker, A., and Pratsinis, S.E. (2014). Effect of Ba and K addition and controlled spatial deposition of Rh in Rh/Al_2O_3catalysts for CO_2 hydrogenation. *Appl. Catal., A* 477: 93–101.

43 Kusama, H., Bando, K.K., Okabe, K., and Arakawa, H. (2001). CO_2 hydrogenation reactivity and structure of Rh/SiO_2 catalysts prepared from acetate, chloride and nitrate precursors. *Appl. Catal., A* 205: 285–294.

44 Kusama, H., Bando, K.K., Okabe, K., and Arakawa, H. (2000). Effect of metal loading on CO_2 hydrogenation reactivity over Rh/SiO_2 catalysts. *Appl. Catal., A* 197: 255–268.

45 Bando, K.K., Soga, K., Kunimori, K., and Arakawa, H. (1998). Effect of Li additive on CO_2 hydrogenation reactivity of zeolite supported Rh catalysts. *Appl. Catal., A* 175: 67–81.

46 Bando, K.K., Soga, K., Kunimori, K. et al. (1998). CO_2 hydrogenation activity and surface structure of zeolite-supported Rh catalysts. *Appl. Catal., A* 173: 47–60.

47 Yan, B., Wu, Q., Cen, J. et al. (2018). Highly active subnanometer Rh clusters derived from Rh-doped $SrTiO_3$ for CO_2 reduction. *Appl. Catal., B* 237: 1003–1011.

48 Karelovic, A. and Ruiz, P. (2013). Mechanistic study of low temperature CO_2 methanation over Rh/TiO_2 catalysts. *J. Catal.* 301: 141–153.

49 Beuls, A., Swalus, C., Jacquemin, M. et al. (2012). Methanation of CO_2: further insight into the mechanism over Rh/γ-Al_2O_3 catalyst. *Appl. Catal., B* 113: 2–10.

50 Upham, D.C., Derk, A.R., Sharma, S. et al. (2015). CO_2 methanation by Ru-doped ceria: the role of the oxidation state of the surface. *Catal. Sci. Technol.* 5: 1783–1791.

51 Sharma, S., Hu, Z., Zhang, P. et al. (2011). CO_2 methanation on Ru-doped ceria. *J. Catal.* 278: 297–309.

52 Guo, Y., Mei, S., Yuan, K. et al. (2018). Low-temperature CO_2 methanation over CeO_2-Supported Ru single atoms, nanoclusters, and nanoparticles competitively tuned by strong metal–support interactions and H-spillover effect. *ACS Catal.* 8: 6203–6215.

53 Dreyer, J.A.H., Li, P., Zhang, L. et al. (2017). Influence of the oxide support reducibility on the CO_2 methanation over Ru-based catalysts. *Appl. Catal., B* 219: 715–726.

54 Wang, F., He, S., Chen, H. et al. (2016). Active site dependent reaction mechanism over Ru/CeO_2 catalyst toward CO_2 methanation. *J. Am. Chem. Soc.* 138: 6298–6305.

55 Wang, F., Li, C., Zhang, X. et al. (2015). Catalytic behavior of supported Ru nanoparticles on the {100}, {110}, and {111} facet of CeO_2. *J. Catal.* 329: 177–186.

56 Tada, S., Ochieng, O.J., Kikuchi, R. et al. (2014). Promotion of CO_2 methanation activity and CH_4 selectivity at low temperatures over $Ru/CeO_2/Al_2O_3$ catalysts. *Int. J. Hydrogen Energy* 39: 10090–10100.

57 Aitbekova, A., Wu, L., Wrasman, C.J. et al. (2018). Low-temperature restructuring of CeO_2-supported Ru nanoparticles determines selectivity in CO_2 catalytic reduction. *J. Am. Chem. Soc.* 140 (42): 13736–13745.

58 Mutschler, R., Moioli, E., Luo, W. et al. (2018). CO_2 hydrogenation reaction over pristine Fe, Co, Ni, Cu and Al_2O_3 supported Ru: Comparison and determination of the activation energies. *J. Catal.* 366: 139–149.

59 Kwak, J.H., Kovarik, L., and Szanyi, J. (2013). CO_2 reduction on supported Ru/Al_2O_3 catalysts: cluster size dependence of product selectivity. *ACS Catal.* 3 (11): 2449–2455.

60 Scire, S., Crisafulli, C., Maggiore, R. et al. (1998). Influence of the support on CO_2 methanation over Ru catalysts: an FT-IR study. *Catal. Lett.* 51: 41–45.

61 Lebarbier, V., Dagle, R., Datye, A., and Wang, Y. (2010). The effect of PdZn particle size on reverse-water–gas-shift reaction. *Appl. Catal., A* 379: 3–6.

62 Wang, X., Shi, H., Kwak, J.H., and Szanyi, J. (2015). Mechanism of CO_2 hydrogenation on Pd/Al_2O_3 catalysts: kinetics and transient DRIFTS-MS studies. *ACS Catal.* 5: 6337–6349.

63 Kwak, J.H., Kovarik, L., and Szanyi, J. (2013). Heterogeneous catalysis on atomically dispersed supported metals: CO_2 reduction on multifunctional Pd catalysts. *ACS Catal.* 3: 2094–2100.

64 Ye, J.Y., Ge, Q.F., and Liu, C.J. (2015). Effect of PdIn bimetallic particle formation on CO_2 reduction over the Pd–In/SiO_2 catalyst. *Chem. Eng. Sci.* 135: 193–201.

65 Erdohelyi, A., Pasztor, M., and Solymosi, F. (1986). Catalytic hydrogenation of CO_2 over supported palladium. *J. Catal.* 98: 166–177.

66 Li, S., Xu, Y., Chen, Y. et al. (2017). Tuning the selectivity of catalytic carbon dioxide hydrogenation over iridium/cerium oxide catalysts with a strong metal-support interaction. *Angew. Chem. Int. Ed.* 56: 10761–10765.

67 Fujita, S.-I., Usui, M., and Takezawa, N. (1992). Mechanism of the reverse water gas shift reaction over Cu/ZnO catalyst. *J. Catal.* 134: 220.

68 Zhang, X., Zhu, X., Lin, L. et al. (2017). Highly dispersed copper over β–Mo_2C as efficient and stable catalysts for RWGS reaction. *ACS Catal.* 7: 912–918.

69 Chen, C.-S., Cheng, W.-H., and Lin, S.-S. (2003). Study of reverse water gas shift reaction by TPD, TPR and CO_2 hydrogenation over potassium-promoted Cu/SiO_2 catalyst. *Appl. Catal., A* 238: 55–67.

70 Chen, C.S., Wu, J.H., and Lai, T.W. (2010). Carbon dioxide hydrogenation on Cu nanoparticles. *J. Phys. Chem. C* 114: 15021–15028.

71 Chen, C.-S., Lin, J.H., You, J.H., and Chen, C.R. (2006). Properties of $Cu(thd)_2$ as a precursor to prepare Cu/SiO_2 catalyst using the atomic layer epitaxy technique. *J. Am. Chem. Soc.* 128: 15950–15951.

72 Chen, C.S. and Cheng, W.H. (2002). Study on the mechanism of CO formation in reverse Water gas shift reaction over Cu/SiO_2 catalyst by pulse reaction, TPD and TPR. *Catal. Lett.* 83: 121–126.

73 Chen, C.-S., Cheng, W.-H., and Lin, S.-S. (2000). Mechanism of CO formation in reverse water–gas shift reaction over Cu/Al_2O_3 catalyst. *Catal. Lett.* 68: 45–48.

74 Jurković, D.L., Pohar, A., Dasireddy, V.D., and Likozar, B. (2017). Effect of copper-based catalyst support on reverse water-gas shift reaction (RWGS) activity for CO_2 reduction. *Chem. Eng. Technol.* 40: 973–980.

75 Zhou, G., Dai, B., Xie, H. et al. (2017). CeCu composite catalyst for CO synthesis by reverse water–gas shift reaction: effect of Ce/Cu mole ratio. *J. CO_2 Util.* 21: 292–301.

76 Zhao, K., Bkour, Q., Hou, X. et al. (2018). Reverse water gas shift reaction over CuFe/Al_2O_3 catalyst in solid oxide electrolysis cell. *Chem. Eng. J.* 336: 20–27.

77 Ren, H., Xu, C.-H., Zhao, H.-Y. et al. (2015). Methanol synthesis from CO_2 hydrogenation over Cu/γ-Al_2O_3 catalysts modified by ZnO, ZrO_2 and MgO. *J. Ind. Eng. Chem.* 28: 261–267.

78 Pastor-Pérez, L., Baibars, F., Sache, E.L. et al. (2017). CO_2 valorisation via reverse water-gas shift reaction using advanced Cs doped Fe-Cu/Al_2O_3 catalysts. *J. CO_2 Util.* 21: 423–428.

79 Fornero, E.L., Chiavassa, D.L., Bonivardi, A.L., and Baltanás, M.A. (2017). Transient analysis of the reverse water gas shift reaction on Cu/ZrO_2 and Ga_2O_3/Cu/ZrO_2 catalysts. *J. CO_2 Util.* 22: 289–298.

80 Zhang, J., An, B., Hong, Y. et al. (2017). Pyrolysis of metal–organic frameworks to hierarchical porous Cu/Zn-nanoparticle@carbon materials for efficient CO_2 hydrogenation. *Mater. Chem. Front.* 1: 2405–2409.

81 Daza, Y.A., Maiti, D., Hare, B.J. et al. (2016). More Cu, more problems: decreased CO_2 conversion ability by Cu-doped $La_{0.75}Sr_{0.25}FeO_3$ perovskite oxides. *Surf. Sci.* 648: 92–99.

82 Zhuang, Y., Currie, R., McAuley, K., and Simakov, D.S.A. (2019). Highly-selective CO_2 conversion via reverse water gas shift reaction over the 0.5wt% Ru-promoted Cu/ZnO/Al_2O_3 catalyst. *Appl. Catal., A* 575: 74–86.

83 Stone, F.S. and Waller, D. (2003). Cu–ZnO and Cu–ZnO/Al_2O_3; catalysts for the reverse water gas shift reaction. The effect of the Cu/Zn ratio on precursor characteristics and on the activity of the derived catalysts. *Top. Catal.* 22: 305–318.

84 Gines, M.J.L., Marchi, A.J., and Apestegufa, C.R. (1997). Kinetic study of the reverse water-gas shift reaction over CuO/ZnO/$A1_2O_3$ catalysts. *Appl. Catal., A* 154: 155–171.

85 Dai, B., Zhou, G., Ge, S. et al. (2017). CO_2 reverse water-gas shift reaction on mesoporous M-CeO_2 catalysts. *Can. J. Chem. Eng.* 95: 634–642.

86 Chen, C.-S., Cheng, W.-H., and Lin, S.-S. (2004). Study of iron-promoted Cu/SiO_2 catalyst on high temperature reverse water gas shift reaction. *Appl. Catal., A* 257: 97–106.

87 Chen, C.-S., Cheng, W.-H., and Lin, S.-S. (2001). Enhanced activity and stability of a Cu/SiO_2 catalyst for the reverse water gas shift reaction by an iron promoter. *Chem. Commun.*: 1770–1771.

88 Galván, C.Á., Schumann, J., Behrens, M. et al. (2016). Reverse water–gas shift reaction at the Cu/ZnO interface: Influence ofthe Cu/Zn ratio on structure-activity correlations. *Appl. Catal., B* 195: 104–111.

89 Meunier, F.C. (2011). Mixing copper nanoparticles and ZnO nanocrystals: a Route towards understanding the hydrogenation of CO_2 to methanol? *Angew. Chem. Int. Ed.* 50 (18): 4053–4054.

90 Liao, F., Huang, Y., Ge, J. et al. (2011). Morphology-dependent interactions of ZnO with Cu nanoparticles at the materials' interface in selective hydrogenation of CO_2 to CH_3OH. *Angew. Chem. Int. Ed.* 50 (9): 2162–2165.

91 Saeidia, S., Najarib, S., Fazlollahic, F. et al. (2017). Mechanisms and kinetics of CO_2 hydrogenation to value-added products: a detailed review on current status and future trends. *Renewable Sustainable Energy Rev.* 80: 1292–1311.

92 Dasireddy, V.D.B.C. and Likozar, B. (2019). The role of copper oxidation state in Cu/ZnO/Al_2O_3 catalysts in CO_2 hydrogenation and methanol productivity. *Renewable Energy* 140: 452–460.

93 Choi, Y., Futagami, K., Fujitani, T., and Nakamura, J. (2001). The role of ZnO in Cu/ZnO methanol synthesis catalysts – morphology effect or active site model? *Appl. Catal., A* 208: 163–167.

94 Topsøe, N.-Y. and Topsøe, H. (1999). On the nature of surface structural changes in Cu ZnO methanol synthesis catalysts. *Top. Catal.* 8: 267–270.

95 Baltes, C., Vukojevic, S., and Schuth, F. (2008). Correlations between synthesis, precursor, and catalyst structure and activity of a large set of CuO/ZnO/Al_2O_3 catalysts for methanol synthesis. *J. Catal.* 258: 334–344.

96 Wu, H.C., Chang, Y.C., Wu, J.H. et al. (2015). Methanation of CO_2 and reverse water gas shift reactions on Ni/SiO_2 catalysts: the influence of particle size on selectivity and reaction pathway. *Catal. Sci. Technol.* 5: 4154–4163.

97 Sun, F.-M., Yan, C.-F., Wang, Z.-D. et al. (2015). Ni/Ce-Zr-O catalyst for high CO_2 conversion during reverse water gas shift reaction (RWGS). *Int. J. Hydrogen Energy* 40: 15985–15993.

98 Zhang, Z., Tian, Y., Zhang, L. et al. (2019). Impacts of nickel loading on properties, catalytic behaviors of Ni/γ–Al_2O_3 catalysts and the reaction intermediates formed in methanation of CO_2. *Int. J. Hydrogen Energy* 44: 9291–9306.

99 Wolf, A., Jess, A., and Kern, C. (2016). Syngas production via reverse water–gas shift reaction over a Ni-Al_2O_3 catalyst: catalyst stability, reaction kinetics, and modeling. *Chem. Eng. Technol.* 39: 1040–1048.

100 Ranjbar, A., Irankhah, A., and Aghamiri, S.F. (2018). Reverse water gas shift reaction and CO_2 mitigation: nanocrystalline MgO as a support for nickel based catalysts. *J. Environ. Chem. Eng.* 6: 4945–4952.

101 Winter, L.R., Chen, R., Chen, X. et al. (2019). Elucidating the roles of metallic Ni and oxygen vacancies in CO_2 hydrogenation over Ni/CeO_2 using isotope exchange and in situ measurements. *Appl. Catal., B* 245: 360–366.

102 Lu, B. and Kawamoto, K. (2014). Preparation of mesoporous CeO_2 and monodispersed NiO particles in CeO_2, and enhanced selectivity of NiO/CeO_2 for reverse water gas shift reaction. *Mater. Res. Bull.* 53: 70–78.

103 Wang, L., Liu, H., Liu, Y. et al. (2013). Effect of precipitants on Ni–CeO_2 catalysts prepared by a co-precipitation method for the reverse water–gas shift reaction. *J. Rare Earths* 31: 969–974.

104 Luhui, W., Hui, L., Yuan, L. et al. (2013). Influence of preparation method on performance of Ni-CeO_2 catalysts for reverse water-gas shift reaction. *J. Rare Earths* 31: 559–564.

105 Wang, L.H., Zhang, S.X., and Liu, Y. (2008). Reverse water gas shift reaction over Co-precipitated Ni-CeO_2 catalysts. *J. Rare Earths* 26: 66–70.

106 Zhao, B., Yan, B., Jiang, Z. et al. (2018). High selectivity of CO_2 hydrogenation to CO by controlling the valence state of nickel using perovskite. *Chem. Commun.* 54: 7354–7357.

107 Lu, B. and Kawamoto, K. (2013). Preparation of monodispersed NiO particles in SBA-15, and its enhanced selectivity for reverse water gas shift reaction. *J. Environ. Chem. Eng.* 1: 300–309.

108 Yang, L., Pastor-Pérez, L., Gu, S. et al. (2018). Highly efficient Ni/CeO_2-Al_2O_3 catalysts for CO_2 upgrading via reverse water–gas shift: effect of selected transition metal promoters. *Appl. Catal., B* 232: 464–471.

109 Campbell, T.K. and Falconer, J.L. (1989). Carbon dioxide hydrogenation on potassium-promoted nickel catalysts. *Appl. Catal.* 50: 189–198.

110 Winter, L.R., Gomez, E., Yan, B. et al. (2018). Tuning Ni-catalyzed CO_2 hydrogenation selectivity via Ni-ceria support interactions and Ni-Fe bimetallic formation. *Appl. Catal., B* 224: 442–450.

111 Liu, Y. and Liu, D. (1999). Study of bimetallic Cu–Ni/γ-Al_2O_3 catalysts for carbon dioxide hydrogenation. *Int. J. Hydrogen Energy* 24: 351–354.

112 Gonclaves, R.V., Vono, L.L.R., Wojcieszak, R. et al. (2017). Selective hydrogenation of CO_2 into CO on a highly dispersed nickel catalyst obtained by magnetron sputtering deposition: a step towards liquid fuels. *Appl. Catal., B* 209: 240–246.

113 Laosiripojana, N. and Assabumrungrat, S. (2005). Methane steam reforming over Ni/Ce-ZrO_2 catalyst: influences of Ce-ZrO_2 support on reactivity, resistance toward carbon formation, and intrinsic reaction kinetics. *Appl. Catal., A* 290: 200–211.

114 Chen, X.Y., Yik, E., Butler, J., and Schwank, J.W. (2014). Gasification characteristics of carbon species derived from model reforming compound over Ni/Ce-Zr-O catalysts. *Catal. Today* 233: 14–20.

115 Lin, W., Stocker, K.M., and Schatz, G.C. (2017). Mechanisms of hydrogen-assisted CO_2 reduction on nickel. *J. Am. Chem. Soc.* 139 (13): 4663–4666.

116 Heine, C., Lechner, B.A., Bluhm, H., and Salmeron, M. (2016). Recycling of CO_2: probing the chemical state of the Ni(111) surface during the methanation reaction with ambient-pressure X-ray photoelectron spectroscopy, X-ray photoelectron spectroscopy. *J. Am. Chem. Soc.* 138: 13246–13252.

117 Zhang, C., Sunarso, J., and Liu, S. (2017). Designing CO_2-resistant oxygen-selective mixed ionic–electronic conducting membranes: guidelines, recent advances, and forward directions. *Chem. Soc. Rev.* 46: 2941.

118 Peña, M.A. and Fierro, J.L.G. (2001). Chemical structures and performance of perovskite oxides. *Chem. Rev.* 101: 1981–2017.

119 Kim, D.H., Park, J.L., Park, E.J. et al. (2014). Dopant effect of barium zirconate-based perovskite-type catalysts for the intermediate-temperature reverse water gas shift reaction. *ACS Catal.* 4: 3117–3122.

120 Maiti, D., Daza, Y.A., Yung, M.M. et al. (2016). Oxygen vacancy formation characteristics in the bulk and across different surface terminations of $La_{(1-x)}Sr_xFe_{(1-y)}Co_yO_{(3-\delta)}$ perovskite oxides for CO_2 conversion. *J. Mater. Chem. A.* 4: 5137–5148.

121 Utsis, N., Landau, M.V., Erenburg, A. et al. (2018). Performance of reverse water gas shift on co-precipitated and carbon-templated BaFe-hexaaluminate. *ChemCatChem* 10: 3795–3805.

122 Utsis, N., Vidruk-Nehemya, R., Landau, M., and Herskowitz, M. (2016). Novel bifunctional catalysts based on crystalline multi-oxide matrices containing iron ions for CO_2 hydrogenation to liquid fuels and chemicals. *Faraday Discuss.* 188: 545–563.

123 Landau, M.V., Meiri, N., Utsis, N. et al. (2017). Conversion of CO_2, CO and H_2 in CO_2 hydrogenation to fungible liquid fuels on Fe-based catalysts. *Ind. Eng. Chem. Res.* 56: 13334–13355.

124 Wenzel, M., Dharanipragada, N.A., Galvita, V.V. et al. (2017). CO production from CO_2 via reverse water–gas shift reaction performed in a chemical looping mode: kinetics on modified iron oxide. *J. CO_2 Util.* 17: 60–68.

125 Sengupta, S., Jha, A., Shende, P. et al. (2019). Catalytic performance of Co and Ni doped Fe-based catalysts for the hydrogenation of CO_2 to CO via reverse water-gas shift reaction. *J. Env. Chem. Eng.* 7: 102911.

126 Loiland, J.A., Wulfers, M.J., Marinkovic, N.S., and Lobo, R.F. (2016). Fe/γ-Al_2O_3 and Fe–K/γ- Al_2O_3 as reverse water gas shift catalysts. *Catal. Sci. Technol.* 6: 5267–5279.

127 Amoyal, M., Vidruk-Nehemya, R., Landau, M.V., and Herskowitz, M. (2017). Effect of potassium on the active phases of Fe catalysts for carbon dioxide conversion to liquid fuels through hydrogenation. *J. Catal.* 348: 29–39.

128 Kharaji, A.G., Shariati, A., and Takassi, M.A. (2013). A novel-alumina supported Fe-Mo bimetallic catalyst for reverse water gas shift reaction. *Chin. J. Chem. Eng.* 21: 1007–1014.

129 Wang, L., Liu, H., Chen, Y., and Yang, S. (2017). Reverse water gas shift reaction over co-precipitated Co-CeO_2 catalysts: effect of Co content on selectivity and carbon formation. *Int. J. Hydrogen Energy* 42: 3682–3689.

130 Wang, L. and Liu, H. (2018). Mesoporous Co-CeO_2 catalyst prepared by colloidal solution combustion method for reverse water-gas shift reaction. *Catal. Today* 316: 155–161.

131 Díez-Ramírez, J., Sánchez, P., Kyriakou, V. et al. (2017). Effect of support nature on the cobalt-catalyzed CO_2 hydrogenation. *J. CO_2 Util.* 21: 562–571.

132 Porosoff, M.D., Yang, X., Boscoboinik, J.A., and Chen, J.G. (2014). Molybdenum carbide as alternative catalysts to precious metals for highly selective reduction of CO_2 to CO. *Angew. Chem. Int. Ed.* 53 (26): 6705–6709.

133 Levy, R.B. and Boudart, M. (1973). Platinum-like behavior of tungsten carbide in surface catalysis. *Science* 181 (4099): 547–549.

134 Porosoff, M.D., Kattel, S., Li, W. et al. (2015). Identifying trends and descriptors for selective CO_2 conversion to CO over transition metal carbides. *Chem. Commun.* 51 (32): 6988–6991.

135 Gao, J., Wu, Y., Jia, C. et al. (2016). Controllable synthesis of α–MoC_{1-x} and β–Mo_2C nanowires for highly selective CO_2 reduction to CO. *Catal. Commun.* 84: 147–150.

136 Xu, W., Ramirez, P.J., Stacchiola, D. et al. (2015). The carburization of transition metal molybdates (MxMoO4, M = Cu, Ni or Co) and the generation of highly active metal/carbide catalysts for CO_2 hydrogenation. *Catal. Lett.* 145: 1365–1373.

137 Xiong, K., Zhou, G., Zhang, H. et al. (2018). Bridging Mo_2C–C and highly dispersed copper by incorporating N functional groups to greatly enhance catalytic activity and durability for the carbon dioxide hydrogenation. *J. Mater. Chem. A* 6: 15510–15516.

138 Liu, X., Kunkel, C.R., de la Piscina, P.R. et al. (2017). Effective and highly selective CO generation from CO_2 using a polycrystalline α-Mo_2C catalyst. *ACS Catal.* 7: 4323–4335.

139 Porosoff, M.D., Baldwin, J.W., Peng, X. et al. (2017). Potassium-promoted molybdenum carbide as a highly active and selective catalyst for CO_2 conversion to CO. *ChemSusChem* 10: 2408–2415.

140 Posada-Pérez, S., Ramirez, P.J., Evans, J. et al. (2016). Highly active Au/δ-MoC and Cu/δ-MoC catalysts for the conversion of CO_2: the metal/C ratio as a key factor defining activity, selectivity, and stability. *J. Am. Chem. Soc.* 138 (26): 8269–8278.

141 Posada-Pérez, S., Ramírez, P.J., Gutiérrez, R.A. et al. (2016). The conversion of CO_2 to methanol on orthorhombic β-Mo_2C and Cu/β-Mo_2C catalysts: mechanism for admetal induced change in the selectivity and activity. *Catal. Sci. Technol.* 6: 6766–6777.

13

Electrocatalytic Conversion of CO_2 to Syngas

Manuel Antonio Díaz-Pérez[1], A. de Lucas Consuegra[2], and Juan Carlos Serrano-Ruiz[1]

[1] *Universidad Loyola Andalucía, Materials and Sustainability Group, Department of Engineering, Avda. de las Universidades s/n, Dos Hermanas, Seville 41704, Spain*
[2] *University of Castilla-La Mancha, School of Chemical Sciences and Technology, Department of Chemical Engineering, Avda. Camilo José Cela 12, Ciudad Real 13005, Spain*

13.1 Introduction

Fossil fuels currently supply most of the energy consumed worldwide. Combustion of fossil fuels releases large amounts of CO_2 into the atmosphere. Because the demand of fossil fuels is increasing steadily over the years as a result of strong population growth and economic development, anthropogenic emissions of CO_2 are also expected to increase continuously in the near future [1]. CO_2 is a well-known powerful greenhouse gas that has a large impact on global warming. Thus, the reduction of CO_2 emissions can mitigate this challenging environmental problem. Several approaches are proposed in this regard: (i) the partial replacement of fossil fuels with renewable sources of energy, (ii) increasing the energy efficiency of conventional fossil routes, (iii) the development of technologies for the capture, sequestration, and storage (CSS) of CO_2, and finally, (iv) the development of technologies for the conversion of CO_2 into value-added chemicals and fuels. The last approach is particularly interesting and a good number of technologies have been proposed involving catalytic, biological, photocatalytic, and electrochemical routes [2]. While catalytic approaches are effective in reducing CO_2 into a plethora of interesting products (e.g. CH_3, CH_3OH, and CO), they typically require harsh temperature and pressure conditions to overcome the high stability of the CO_2 molecule. Another drawback is that the reduction is carried out with large amounts of external (and typically fossil fuelderived) H_2. On the other hand, biological and photocatalytic routes operate under significantly milder conditions compared to catalytic processes, but they suffer from low efficiencies and sluggish reaction kinetics, which hinders practical applications.

Electrochemical reduction is particularly suitable for converting CO_2 into useful products. This route combines efficient conversion into valuable products with mild operation conditions. In addition, because the electrochemical reduction of CO_2 can

Engineering Solutions for CO_2 Conversion, First Edition.
Edited by Tomas R. Reina, José A. Odriozola, and Harvey Arellano-Garcia.

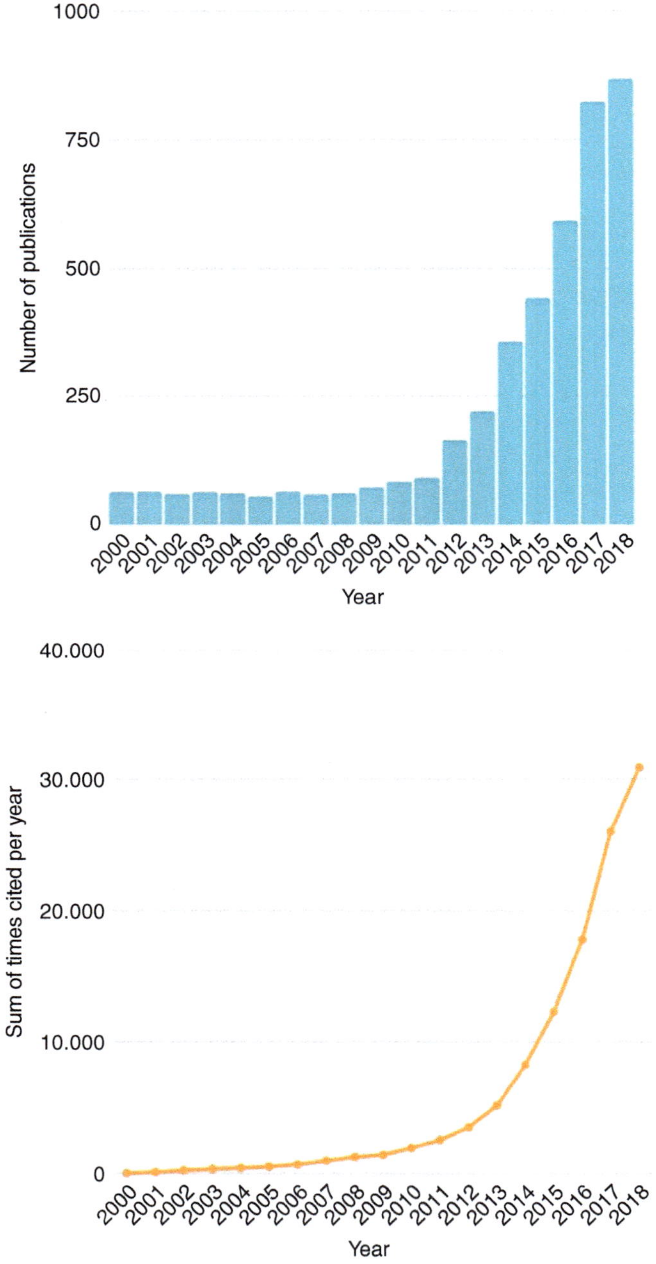

Figure 13.1 Number of scientific publications containing the topic *"electrochemical reduction of CO_2"* and total number of citations of these publications. Source: Web of Science (WOS).

be potentially carried out with renewable electricity, this technology can be used to store energy and therefore mitigate the inherent demand/production mismatch of renewables energies. Electroreduction of CO_2 allows versatile production of a variety of interesting products (e.g. CO, CH_4, syngas, formaldehyde, methanol, and ethylene) by simply controlling the reaction conditions (e.g. potential, intensity, and electrolyte) and the composition of the electrodes, among other parameters. These products are generated by means of reduction processes involving two, four, six, and eight electrons, depending on the extent of the reduction process. Furthermore, the compact and modular nature of electrochemical reaction systems favors potential scaling up and implementation in a wide variety of industrial applications.

Accordingly, the electroreduction of CO_2 has received a great deal of research attention in recent years. As detailed in Figure 13.1, the number and relevance (i.e. number of citations) of papers devoted to this field have increased dramatically within the past five years. The electroreduction of CO_2 to high value liquid products has been widely studied in the literature under both liquid- [3, 4] and gas-phase [5–7] conditions. A comprehensive review on the main CO_2 electroreduction technologies is out of the scope of the present chapter, but the reader can find excellent reviews on this topic elsewhere [8–10].

In this chapter, we will review the most relevant electroreduction technologies available to this date for the conversion of CO_2/water mixtures into syngas (i.e. H_2/CO mixtures). As detailed further on, the production of syngas by this route is particularly relevant because of (i) the industrial interest of this mixture for a wide variety of downstream processes and (ii) the advantages of this route compared to conventional fossil fuel-derived approaches.

13.2 Production of Syngas

Syngas (abbreviation of synthesis gas) is a mixture of CO and H_2 that can be used to produce electricity and a wide variety of chemical products depending on the composition (i.e. H_2/CO molar ratio) [11]. These products may include synthetic natural gas (SNG) and different oxygenated fuels and chemicals (e.g. bioethanol, dimethyl ether, and methanol). Liquid hydrocarbon fuels (i.e. gasoline, diesel, and jet fuels) can be also produced from syngas by a well-developed catalytic process denoted as Fischer–Tropsch synthesis [12]. Syngas can be virtually produced from any carbon-containing material (e.g. natural gas, coal, biomass, and sewage sludge) by means of a highly endothermic process conducted under a well-controlled oxidizing atmosphere. High temperatures and pressures are necessary to obtain significant yields of syngas. Syngas is industrially produced by steam reforming of natural gas at high temperatures (800–1000 °C) and pressures of 20–40 atm over a metal-based catalyst, according to the following reaction [13]:

$$CH_4 + H_2O \leftrightarrow CO + 3H_2 \qquad \Delta H_{298\,K} = 49.1\ \text{kcal mol}^{-1} \tag{13.1}$$

The composition of the syngas product depends on a number of factors including the oxidizing power of the atmosphere, the design of the gasifier, the texture and morphology of the carbon source, and the residence time, among others. Thus, pure oxygen atmospheres, small particle sizes, and high temperatures favor the production of syngas versus a more complex and less-valuable mixture denoted as producer gas [14]. Apart from the harsh temperature and pressure requirements, industrial syngas production also requires additional and complex cleaning steps with the aim of removing impurities such as H_2S, COS, and mercaptans derived from the initial carbon feedstocks. This issue is particularly important when biomass is used as a carbon source and impurities such as tars (condensable high-molecular-weight hydrocarbons), HCl, and NH_3 being typically found in the gas stream accompanying H_2 and CO. Because the catalysts used downstream to convert syngas into fuels and chemicals are highly sensitive to these impurities, deep cleaning is required, which complicates the process and increases costs.

Electrocatalysis provides an alternative route for the generation of syngas by simultaneous reduction of CO_2 and water to CO and H_2, respectively. Unlike the thermochemical process described above, the electroreduction route can be carried out at significantly milder conditions (room temperature) and without complex downstream cleaning steps. As will be detailed in Section 13.3, CO_2 electroreduction also allows fine-tuning of the syngas composition, providing control over the H_2/CO ratio by proper selection of the reaction conditions (e.g. potential, intensity, and temperature) and electrode (cathode) composition. This is valuable in that it provides flexibility for downstream processes requiring different H_2/CO ratios (e.g. 2 for Fischer–Tropsch and methanol synthesis and 0.3–1 for syngas fermentation) and facilitates the potential integration in current chemical infrastructures.

The present chapter summarizes the most relevant syngas production via electrochemical technologies developed to date. Particular emphasis will be given to the composition of the cathodes (over which reduction of CO_2 is carried out), the reaction conditions (and how these can be used to control the composition of the syngas), and the different designs developed for the electrochemical cells.

13.3 Electroreduction of CO_2/Water Mixtures to Syngas

The electrosynthesis of syngas from CO_2/H_2O mixtures involves two reduction processes, namely, the electroreduction of CO_2 to CO (Eq. (13.2)) and the hydrogen evolution reaction (HER, Eq. (13.3)).

$$CO_2 + 2H^+ + 2e^- \rightarrow CO + H_2O \qquad (13.2)$$

$$2H^+ + 2e^- \rightarrow H_2 \qquad (13.3)$$

The most common experimental approach utilizes membrane–electrode assemblies (MEAs) composed of cathodic and anodic electrodes separated by a proton-conducting membrane (PCM, Figure 13.2).

In this approach, CO_2 and H^+ are reduced simultaneously on the cathode, while the electrolytic dissociation of water or oxygen evolution reaction (OER, Eq. (13.4))

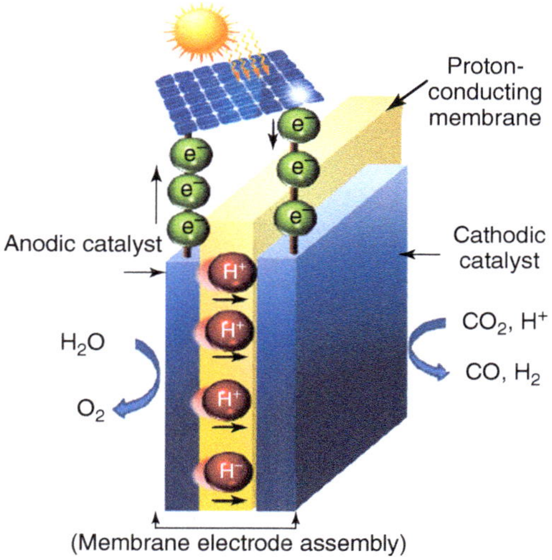

Figure 13.2 Scheme of a typical membrane electrode assembly used to carry out the electrosynthesis of syngas from CO_2/water mixtures.

takes place on the anode, which then provides the protons and electrons required to conduct processes (13.2) and (13.3):

$$2H_2O \rightarrow O_2 + 4H^+ + 4e^- \quad (13.4)$$

These protons are transported from the anode to the cathode through the PCM. The reduction potential required to transform CO_2 into CO ($E°$ versus the reversible hydrogen electrode (RHE) = –0.104 V) only differs slightly from that of the HER ($E°$ versus RHE = 0 V), which results in both reactions competing for the active sites on the cathode and consuming the electrical energy supplied to the cell. According to the above analysis, the composition of the syngas mixture could be adjusted by controlling the reaction rates of the competing processes (13.2) and (13.3), which can be done by selecting proper reaction conditions (e.g. potential, intensity, pH of the solution, and temperature) and the cathode composition. For instance, syngas with low H_2/CO ratios can be produced by combining metals with high activity and selectivity toward the reduction of CO_2 to CO (e.g. Ag and Au) [15] with metals known for hindering the HER (e.g. In, Sn, and Pb) [8]. In this section, we will cover the most relevant and recent technologies for the electroreduction of CO_2/water mixtures into syngas. These technologies and their most relevant experimental aspects (e.g. operation temperature, cathode composition, composition of the syngas, and electrolyte) are summarized in Table 13.1. The effect of the electrochemical cell configuration, the cathode composition, and the reaction parameters on the electroreduction process is also discussed in detail in this section.

13.3.1 Effect of Cell Configuration and Chemical Environment

The design of the electrochemical cell can alter the performance of the cathode significantly during the electroreduction of CO_2/water mixtures to syngas. One paradigmatic example in this sense was reported by Delacourt et al. [17]. These

Table 13.1 Summary of some of the most relevant studies dealing with the production of syngas by electroreduction of CO_2/water mixtures.

Metal	Support	Electrolyte	Temperature (°C)	Potential (V) (versus RHE)	H_2/CO ratio	Total current density (mA cm^{-2})	Ref.
Cu-enriched Au	Unsupported	$KHCO_3$ (0.5 M)	RT	−0.35 to −0.65	0.6–2	>20	[11]
Nanoporous Ag	Unsupported	$KHCO_3$ (0.5 M)	RT	−0.6	Not reported	~20	[16]
Ag	Unsupported	Ion-exchange membrane + buffer ($KHCO_3$-supported on glass fiber)	RT	−2	2 : 1	80	[17]
Ag ultrathin nanowires (25 nm)	Carbon black	$KHCO_3$ (0.1 M)	RT	−0.46 to −0.66	Variable	0.4 –1	[18]
Ag nanoparticles	g-C_3N_4	KH_2PO_4/K_2HPO_4 (0.1 M)	RT	−1.05 to −1.15	2 : 1	c. 5–12[a)]	[19]
Zn[b)]	Cu foam	$KHCO_3$ (0.1 M)	40	−0.9	0.35–5	12	[20]
Ag nanowires	Macroporous carbon sheet	$KHCO_3$ (0.2 M)	Not reported	c. −1.2	2.15	8.72[c)]	[21]
Pd	Vulcan carbon	$NaHCO_3$ (0.5 M)	RT	−0.5 to −0.6	2 : 1–1 : 1	~0.5	[22]
Cu/Au core–shell nanowire	Unsupported	$KHCO_3$ (0.5 M)	RT	−0.65	2 : 1	14	[23]
Cu/In alloy	Unsupported	$KHCO_3$ (0.1 M)	RT	−1	2.6 : 1	1.5–10	[24]
MoSeS	Unsupported	KCl	Not reported	−1.15	1 : 1	43	[25]

RT, room temperature.

a) Current densities obtained for 20 wt% Ag-supported electrode at a potential between −1.05 and −1.15 V (versus RHE).

b) Data corresponding to the so-called Zn-3 (1 mM cetrimonium bromide).

c) Total current density obtained with 0.5 Ω Ag nanowires for a H_2/CO ratio of 2.15.

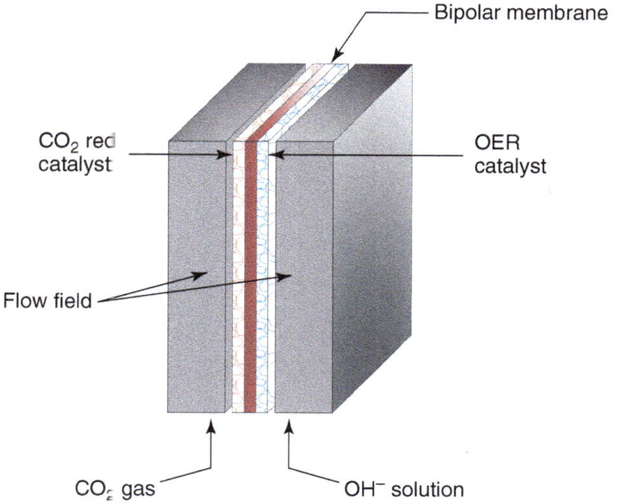

Figure 13.3 Scheme of a bipolar membrane-based electrochemical cell used for the electroproduction of syngas. Source: Adapted from Li et al. [26].

authors carried out the electroreduction of CO_2/water mixtures over an Ag cathode with several cell configurations. The utilization of a regular proton-conducting fuel cell configuration resulted in pure H_2 streams, while the insertion of a pH buffer layer of $KHCO_3$ between the cathode and the PCM generated syngas mixtures with a H_2/CO ratio suitable for methanol synthesis. The $KHCO_3$ buffer layer helped control the HER rate by preventing high concentrations of protons from reaching the cathode. This configuration is interesting although, as pointed out by the authors, the thick buffer layer increased the resistance of the cell significantly.

Li et al. [26] built an electrochemical cell for the production of syngas by combining the CO_2 reduction with the OER (Figure 13.3). The most relevant aspect of this technology lies in the utilization of a bipolar membrane (i.e. a membrane composed of a cation exchange layer and an anion exchange layer) that allowed the transport of H^+ and OH^- (formed by dissociation of water) to the cathode and anode, respectively. CO_2 was continuously fed to the cathode chamber, while a basic aqueous solution served as a feed for the anode. This cell configuration generated syngas mixtures of different H_2/CO ratios by changing the potential of the cell from −1.9 to −1.3 V (versus Ag/AgCl electrode). Remarkably, the overall Faradaic efficiency for H_2 and CO was close to 100% over the entire potential range, indicating that no reduction products other than H_2 or CO were produced over the Ag cathode. The cell was tested for eight hours under stable conditions (i.e. 50 mA cm^{-2}) with a slight decay of the Faradaic efficiency for CO.

The chemical environment in which the electroreduction of CO_2 takes place has a strong influence on the reaction performance. It has been well documented that the presence of certain ionic components in the electrolyte can modify the intrinsic activity of metals for the electroreduction of CO_2 [27]. Rosen et al. exploited this phenomenon by carrying out the electroreduction of CO_2 in the presence of ionic liquids [28]. The presence of the ionic compound improved the electrochemical performance of Ag for the production of CO. This improvement was explained by complexation of the reaction intermediates with the ionic compound, which reduced the energy barrier for CO production. As a result, outstanding Faradaic efficiencies as high as 96% at very low overpotentials (0.2 V) were obtained. This pioneer work paved the way for the utilization of ionic liquids as cocatalysts of transition metals for the electroreduction of CO_2 to CO.

13.3.2 Effect of the Cathode Composition and Structure

The vast majority of cathodes used for the electrosynthesis of syngas are metal based. As pointed out before, in order to produce pure syngas streams, these metals are required to simultaneously promote reactions (13.2) and (13.3) with high selectivity and Faradaic efficiency. Concerning the reduction of CO_2 to CO, the performance of transition metals in this reaction has been associated with their bonding strength toward CO. Thus, metals adsorbing CO strongly (e.g. Pt, Rh, Pd, and Ni) show poor performance and deactivate quickly by CO poisoning. However, metals with low CO adsorption strength (e.g. Ag and Au) tend to release this gas once formed, avoiding over-protonation of the adsorbed CO to more reduced products [29]. This trend has been widely confirmed by the literature. Au is particularly suitable for syngas production because it combines high activity for the production of CO at very low overpotentials with moderate HER activities over a wide range of operation conditions [30–33]. High Faradaic efficiencies toward syngas were reported several decades ago over polycrystalline Au nanoparticles [4]. Ross et al. [11] modified the intrinsic activity of Au by performing a careful and well-controlled Cu-enrichment process. Cu modified the electronic structure of the Au atoms, which in turn altered the CO binding strength on these surfaces. This approach allowed to control the relative rates of the two competing reduction processes (i.e. CO_2 to CO and HER) to provide syngas mixtures with a variety of compositions and with high production rates (Table 13.1). Moderate activity losses (30%) were reported after eight hours of reaction. The H_2/CO ratio of the syngas increased with the amount of Cu, in line with differential functional theory (DFT) calculations performed over Cu-enriched Au model surfaces. This technology is relevant in that it achieved syngas with H_2/CO ratios suitable for Fischer–Tropsch, fermentation, and methanol synthesis (0.6–2) by simply controlling the extent of the Cu enrichment process.

The high cost of Au has motivated researchers to find cheaper alternatives to perform the electroreduction of CO_2/water mixtures into syngas, with Ag being mostly preferred. However, polycrystalline Ag electrocatalysts present high overpotential for the electroreduction of CO_2 to CO. However, most of the reported works have focused on overcoming these kinetic barriers of Ag. With this aim, Lu et al. [16] developed a nanoporous Ag electrocatalyst that demonstrated high rates (3000 times higher than polycrystalline Ag) and Faradaic efficiencies toward CO (92%) at overpotentials as low as 0.5 V. The higher electrochemical surface area of the nanoporous material compared to its polycrystalline counterpart was invoked to explain this outstanding performance. The rate of the HER was found to increase

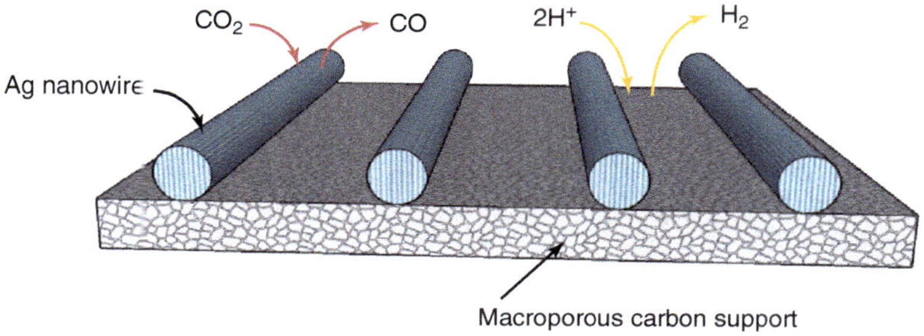

Figure 13.4 Scheme of the electrocatalyst developed by Cho and coworkers to produce syngas from CO_2/water mixtures. Source: Adapted from Cho et al. [21].

with the voltage of the cell at the expense of CO. Ag supported on graphitic carbon nitride (g-C_3N_4) was reported to be effective in producing syngas mixtures from CO_2 using a phosphate buffer as electrolyte [19]. The composition of the syngas mixture was controlled by both the Ag loading and the reduction potential. Thus, syngas streams with H_2/CO ratios close to 2 (suitable for methanol synthesis) were obtained for Ag loadings ranging from 20 to 40 wt% over a narrow potential window (−1.05 to −1.15 V versus RHE). Apart from nanoporous Ag, Ag nanowires have also been used to transform CO_2/water mixtures into syngas. Liu et al. [18] compared the performance of Ag nanoparticles and nanowires toward the electrosynthesis of CO. The diameter of the nanowire was important in determining the performance of these materials. Syngas with controlled H_2/CO ratios was obtained by decreasing the potential below −0.456 V. The outstanding performance of Ag nanowires for producing CO was exploited by Cho et al. [21] to develop a catalyst with high syngas production activity. With this aim, Ag nanowires were deposited over a macroporous carbon sheet with high ability to promote the HER reaction (Figure 13.4). The electrocatalyst yielded syngas mixtures with varying H_2/CO ratios, which were controlled by the loading of Ag nanowires on the carbon sheet.

The electroreduction of CO_2 has shown to be a structure-sensitive reaction. Back et al. [34] conducted first-principles calculations to identify Au and Ag active sites for the electroreduction of CO_2. These authors found that the size of the Au and Ag nanoparticles was key in controlling the relative rates of the CO production and HER competing processes. Three different catalytic sites were considered, namely, low-index planes, edges, and corners. Corner and edge sites were found to be the most active for Au and Ag, respectively. These results were not in line with those published by Zhu et al. [31]. According to these authors, Au edges were the most active sites for the electroreduction CO_2 proven both experimentally and theoretically (DFT)

Metals other than Au and Ag have been employed as cathodes for the electroreduction of CO_2 into syngas. Thus, syngas stream with a H_2/CO ratio between 1 and 2 were obtained over a Pd/C electrocatalyst at a potential of −0.6 V (versus RHE) having formic acid as the only liquid product [22]. As confirmed by X-ray diffraction (XRD), β Pd hydride (PdH) was the active phase of the electrocatalyst.

This phase showed lower binding energy for both CO and H_2, thereby favoring syngas production and avoiding poisoning issues. Chen and coworkers [22] conducted electroreduction of CO_2 on three types of Cu-based electrocatalysts (i.e. Cu/Au core–shell nanowires, Cu nanowires, and polycrystalline Cu). The Cu/Au catalysts demonstrated significantly higher Faradaic efficiencies to CO than the polycrystalline monometallic material. A maximum CO Faradaic efficiency of 33% was obtained at a potential of −0.65 V (versus RHE). These conditions resulted in a syngas stream with a H_2/CO ratio of 2. A mixture of syngas and formate was obtained over In-modified Cu electrocatalysts [24]. The product distribution was found to depend highly on both the In loading and the overpotential. Thus, pure Cu electrocatalysts showed high HER activity, which was moderated with the In loading in the electrocatalyst. A syngas stream with a H_2/CO ratio of c. 2.6 was produced along with formate (50% selectivity) at −1 V (versus RHE) over a 40 at% In–Cu electrocatalyst. Zn electrocatalysts with different crystallographic orientations ((101) and (002)) were electrodeposited over a Cu foam and subsequently used to produce syngas by electroreduction of CO_2/water mixtures [20]. The Zn(101)/Cu electrocatalyst was found to be highly effective in producing syngas (Faradaic efficiency of 83%) at −0.9 V (versus RHE) with a wide range of H_2/CO ratios (0.35–5).

p-Block metals (e.g. Pb, In, Bi, and Sn) are well known for preferentially producing formate versus CO by electroreduction of CO_2 [35, 36]. In addition, the HER is hindered over these metals, which typically present high overpotential and poor kinetics. Therefore, p-block metals are apparently not suitable for generating syngas mixtures by electroreduction of CO_2/water mixtures. Despite these limitations, Li et al. [37] demonstrated that the intrinsic activity of these metals can be significantly modified toward the production of syngas by alloying them with a second metal. These authors prepared core–shell particles by coating thin layers of SnO_2 on Cu nanoparticles. The selectivity toward formate, CO, and H_2 was modulated by controlling the thickness of the SnO_2 shell and the voltage of the cell. Thus, particles with thin SnO_2 shells (0.8 nm) displayed high Faradaic efficiencies toward CO over a wide range of voltages. Syngas mixtures with a total Faradaic efficiency for H_2 + CO close to 85% were obtained over this cathode at −0.5 V (versus RHE). This radical change in the electrochemical performance of Sn was ascribed to geometrical effects of SnO_2 over Cu and to partial decoration of SnO_2 with Cu, as suggested by DFT studies. This work opens the possibility of using nonexpensive metals for generating syngas mixtures in replacement of more commonly used noble metals such as Au and Pd.

In an interesting work, MoSeS alloy monolayers were employed for the electrochemical conversion of CO_2 to syngas [25]. The MoSeS alloyed material presented higher activities than the MoS_2 and $MoSe_2$ monolayer materials. The MoSeS alloyed material showed a Faradaic efficiency for CO of 45.2% and generated a syngas stream with a H_2/CO ratio of 1 at −1.15 V (versus RHE). The superior performance of MoSeS was ascribed to its enhanced CO_2 adsorption capacity (revealed by density of states [DOS] calculations) and its higher electrical conductivity compared to MoS_2 and $MoSe_2$. Moreover, temperature-programmed desorption (TPD) revealed that CO was weakly adsorbed on these MoSeS alloyed monolayers.

Carbon-based materials have also demonstrated good performance for the reduction of CO_2 to CO. N-doping has been proposed as an effective approach to improve the performance of carbon-based materials in this reaction. N-doping modifies the electronic structure of carbon, which is believed to stabilize certain reaction intermediates on graphitic or pyridinic N sites [38]. This strategy was used by Kumar et al. [39] to prepare N-doped carbon nanofibers displaying outstanding activity for the reduction of CO_2 to CO (one order of magnitude higher than that of Ag). The outstanding performance of this material was ascribed to the high binding energy of key intermediates on positively charged carbon atoms, which were generated by the presence of electronegative nitrogen atoms in their vicinities. More recently, Fe-doping has also been proposed as a suitable strategy to drive the electroreduction of CO_2/water toward syngas production. Huan et al. [40] prepared Fe- and N-doped carbon materials by pyrolysis of precursors containing these elements. They stated that the electroreduction of CO_2 over these materials is controlled by the ratio of isolated FeN_4 sites versus isolated Fe nanoparticles. Thus, FeN_4 sites catalyzed the CO_2 to CO process preferentially, while isolated Fe or iron carbide sites promoted the HER under similar conditions. Syngas with H_2/CO ratios ranging from 0.2 to 4 were obtained with this technology, depending on the amount of FeN_4 sites in the final material as revealed by Mössbauer spectroscopy. This can be experimentally controlled by the iron content of the precursor material and the impregnation methodology. Varela et al. [41] functionalized a number of commercial carbon blacks with N, Fe, and Mn by pyrolysis and acid leaching. The as-obtained functionalized solids showed outstanding activities and selectivities to syngas, which was preferentially produced on N sites. Fe sites were found to promote CO protonation to hydrocarbons (e.g. methane), most likely as a result of the strong adsorption of CO on this metal. This hypothesis was supported by temperature-programmed desorption of CO (CO-TDP) experiments. The functionalized materials showed higher activities than a low-area polycrystalline Au catalyst used as a benchmark.

Regarding the HER, noble metals such as Pt supported on carbon materials are particularly active and selective to H_2 [42]. Transition metals (e.g. Co, Ni, Fe, and Mo) and related compound such as MoS_2 and WC have also been well described in the literature as HER catalysts [43–46], despite undergoing corrosion under acidic conditions. In recent years, carbon-based materials such as carbon nitrides have also been reported to present HER activities comparable to those of noble metals [47]. This work is particularly relevant in that it provides a robust and corrosion-resistant alternative to expensive noble metals to conduct the HER under acidic conditions.

13.3.3 Effect of the Reaction Parameters

The reaction parameters (e.g. current, electrical potential, and temperature) can be adjusted conveniently to control the relative rates of the HER and CO_2 to CO processes and therefore to drive the electroreduction process to syngas mixtures of different compositions. For example, Kumar et al. [39] carried out the electroreduction of CO_2/water mixtures over N-doped carbon nanofibers. Different syngas mixtures were obtained by varying the applied potential from −0.53 to −1.14 V (versus RHE).

In a subsequent work, the same authors designed a methodology to produce syngas of varying compositions over a Cu cathode [48]. By applying square voltage pulses, these researchers were able to control the H_2/CO ratio (from 32 to near 0.5) by simply varying the pulse period within the 10–50 ms range. Thus, the obtained syngas could be potentially used for a variety of chemical processes, including Fischer–Tropsch and hydroformylation of alkenes to aldehydes. As a result of the voltage pulses, the surface of the Cu cathode underwent repetitive oxidation and reduction processes, altering the structure of the nanoparticles by the continuous generation of Cu oxide and metallic Cu phases. This continuous restructuration resulted in the preferential formation of CO versus hydrocarbons such as methane and ethane, which are more commonly produced over this metal under regular conditions.

13.3.4 Electrochemical Promotion of Catalyst (EPOC) for CO_2 Hydrogenation

Over the past years, a new electrocatalytic concept based on the use of solid electrolytes as electroactive supports has been developed for producing syngas by CO_2 hydrogenation. This phenomenon, known as electrochemical promotion of catalysis (EPOC) or non-Faradaic electrochemical modification of the catalytic activity (NEMCA), was discovered by Stoukides and Vayenas in 1981 [49]. Since then, EPOC has been used to promote and enhance catalytic reaction rates by coupling catalysis and electrochemistry. EPOC achieves catalyst activation by electrochemically supplying promoter ions from an electroactive support (e.g. a solid electrolyte) through applying electrical polarization on a solid electrolyte cell. This electrochemical promotion allows the control of in situ addition or removal of promoters to the catalyst under reaction conditions in a continuous and reversible manner. This effect makes a fast and precise evaluation and optimization of the coverage of the promoter on the catalyst surface possible [50]. The electrochemical promotion of CO_2 hydrogenation has been studied in the past over a wide variety of metal catalysts including Cu [51], Rh [52], Pt [53], Pd [54], and Ni [55] film electrodes deposited on different ceramic electrolytes such as yttria-stabilized zirconia (YSZ) [51] and K-β''-Al_2O_3 [56]. Although methanation was promoted under the application of positive potentials on the electrochemical cell, the reverse water gas shift (RWGS) reaction leading to syngas mixtures was catalyzed at negative potentials. The EPOC approach provides higher Faradaic efficiencies toward CO and CH_4 compared to conventional electrocatalysis.

We have exploited the EPOC approach to produce syngas mixtures of different compositions. Ni deposited on K-βAl_2O_3 showed an outstanding electrochemical activation performance under a wide variety of reaction conditions [57]. As shown in Figure 13.5, a cathodic (negative) polarization favored CO production (electrophilic behavior) while an anodic (positive) polarization promoted methanation (electrophobic behavior). The CO production rate was more sensitive to potential variations compared to CH_4. This behavior was explained by the different reaction orders of CO and H_2 and the electronic effect of K^+ ions supplied or removed under polarization conditions. One of the main benefits of the EPOC approach is that it allows

Figure 13.5 Effect of the CO_2 concentration on the reaction rates of (a) CO_2 consumption, (b) CO production, and (c) CH_4 production under different potentials (U_{WR}). Reaction conditions: $H_2 = 30\%$ (N_2 balance), flow = 100 N ml min^{-1}, $T = 240\,°C$. Cathode: Ni/K-βAl_2O_3. Source: Gutiérrez-Guerra et al. [57].

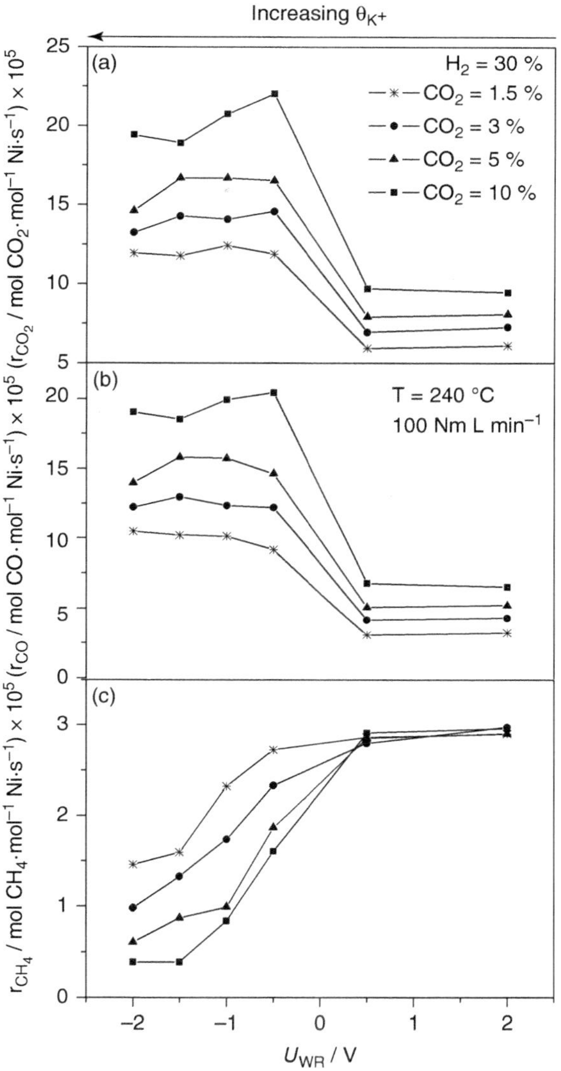

control over the competing chemisorption of the different adsorbates on the catalyst surface, which in turn provides control over the selectivity to the different products (CO, H_2, and CH_4) and the composition of the resulting syngas mixtures.

13.4 Conclusions

The electroreduction of CO_2/water mixtures to syngas is a paradigmatic process allowing for the production of a wide variety of fuels and chemicals. This reaction requires the cathode material to simultaneously promote two competing reduction processes, namely, the conversion of CO_2 into CO and the HER. The composition of the cathode is therefore key for driving the reaction selectivity toward the selected products. Regarding the reduction of CO_2 into CO, metals such as Au and Ag on which CO is weakly adsorbed are preferred versus other metals such as Pt, Rh, Pd, and Ni with higher adsorption heats that result in higher tendency to undergo CO poisoning. Furthermore, Ag and Au possess moderate activity toward the HER, making them excellent candidates for syngas production. More recently, carbon-based materials have also been applied to this reaction upon doping with N and Fe. N and Fe doping modifies the electronic structure of the carbon atoms and generates new active sites for the generation of CO and H_2. Some N- and Fe-doped carbon materials have outperformed Au and Ag, thereby opening the possibility of replacing these expensive metals. The configuration of the electrochemical cell and the reaction conditions (mostly the voltage of the cell) can be used to drive the electroreduction process toward the generation of syngas mixtures with varying compositions, as some examples were provided herein. The potential of the EPOC effect in enhancing the hydrogenation of CO_2 and in driving the electroreduction process to CO and H_2 was also stressed in this work.

Acknowledgments

JCSR would like to thank the Spanish Ministry of Economy, Industry and Competitiveness for financial support through the Ramón y Cajal Program, Grant: RYC-2015-19230.

References

1 International Energy Agency, Global Energy & CO_2 Status Report 2017. (2018). https://www.iea.org/geco/.

2 Centi, G. and Perathoner, S. (2009). Opportunities and prospects in the chemical recycling of carbon dioxide to fuels. *Catal. Today* 148: 191–205. https://doi.org/10.1016/J.CATTOD.2009.07.075.

3 Ma, J., Sun, N., Zhang, X. et al. (2009). A short review of catalysis for CO_2 conversion. *Catal. Today* 148: 221–231. https://doi.org/10.1016/J.CATTOD.2009.08.015.

4 Hori, Y., Takahashi, I., Koga, O., and Hoshi, N. (2003). Electrochemical reduction of carbon dioxide at various series of copper single crystal electrodes. *J. Mol. Catal. A: Chem.* 199: 39–47. https://doi.org/10.1016/S1381-1169(03)00016-5.

5 Genovese, C., Ampelli, C., Perathoner, S., and Centi, G. (2013). Electrocatalytic conversion of CO_2 on carbon nanotube-based electrodes for producing solar fuels. *J. Catal.* 308: 237–249. https://doi.org/10.1016/J.JCAT.2013.08.026.

6 Genovese, C., Ampelli, C., Perathoner, S., and Centi, G. (2013). Electrocatalytic conversion of CO_2 to liquid fuels using nanocarbon-based electrodes. *J. Energy Chem.* 22: 202–213. https://doi.org/10.1016/S2095-4956(13)60026-1.

7 Centi, G., Perathoner, S., Win, G., and Gangeri, M. (2007). Electrocatalytic conversion of CO_2 to long carbon-chain hydrocarbons. *Green Chem.* 9: 671. https://doi.org/10.1039/b615275a.
8 Larrazábal, G.O., Martín, A.J., and Pérez-Ramírez, J. (2017). Building blocks for high performance in electrocatalytic CO_2 reduction: materials, optimization strategies, and device engineering. *J. Phys. Chem. Lett.* 8: 3933–3944. https://doi.org/10.1021/acs.jpclett.7b01380.
9 Marken, F. and Fermin, D. (2018). *Electrochemical Reduction of Carbon Dioxide*. Cambridge: Royal Society of Chemistry https://doi.org/10.1039/9781782623809.
10 Kuhl, K.P., Cave, E.R., Abram, D.N., and Jaramillo, T.F. (2012). New insights into the electrochemical reduction of carbon dioxide on metallic copper surfaces. *Energy Environ. Sci.* 5: 7050. https://doi.org/10.1039/c2ee21234j.
11 Ross, M.B., Dinh, C.T., Li, Y. et al. (2017). Tunable Cu enrichment enables designer syngas electrosynthesis from CO_2. *J. Am. Chem. Soc.* 139: 9359–9363. https://doi.org/10.1021/jacs.7b04892.
12 Jacobs, G., Das, T.K., Zhang, Y. et al. (2002). Fischer–Tropsch synthesis: support, loading, and promoter effects on the reducibility of cobalt catalysts. *Appl. Catal., A* 233: 263–281. https://doi.org/10.1016/S0926-860X(02)00195-3.
13 Subramani, V., Sharma, P., Zhang, L., and Liu, K. (2009). Catalytic steam reforming technology for the production of hydrogen and syngas. In: *Hydrogen and Syngas Production and Purification Technologies*, 14–126. Hoboken, NJ: Wiley https://doi.org/10.1002/9780470561256.ch2.
14 Serrano-Ruiz, J.C. and Dumesic, J.A. (2011). Catalytic routes for the conversion of biomass into liquid hydrocarbon transportation fuels. *Energy Environ. Sci.* 4: 83–99. https://doi.org/10.1039/C0EE00436G.
15 Hori, Y., Wakebe, H., Tsukamoto, T., and Koga, O. (1994). Electrocatalytic process of CO selectivity in electrochemical reduction of CO_2 at metal electrodes in aqueous media. *Electrochim. Acta* 39: 1833–1839. https://doi.org/10.1016/0013-4686(94)85172-7.
16 Lu, Q., Rosen, J., Zhou, Y. et al. (2014). A selective and efficient electrocatalyst for carbon dioxide reduction. *Nat. Commun.* 5: 3242. https://doi.org/10.1038/ncomms4242.
17 Delacourt, C., Ridgway, P.L., Kerr, J.B., and Newman, J. (2008). Design of an electrochemical cell making syngas ($CO+H_2$) from CO_2 and H_2O reduction at room temperature. *J. Electrochem. Soc.* 155: B42. https://doi.org/10.1149/1.2801871.
18 Liu, S., Wang, X.-Z., Tao, H. et al. (2018). Ultrathin 5-fold twinned sub-25 nm silver nanowires enable highly selective electroreduction of CO_2 to CO. *Nano Energy* 45: 456–462. https://doi.org/10.1016/J.NANOEN.2018.01.016.
19 Sastre, F., Muñoz-Batista, M.J., Kubacka, A. et al. (2016). Efficient electrochemical production of syngas from CO_2 and H_2O by using a nanostructured Ag/g-C_3N_4 catalyst. *ChemElectroChem* 3: 1497–1502. https://doi.org/10.1002/celc.201600392.
20 Qin, B., Li, Y., Fu, H. et al. (2018). Electrochemical reduction of CO_2 into tunable syngas production by regulating the crystal facets of earthabundant Zn

catalyst. *ACS Appl. Mater. Interfaces* 10: 20530–20539. https://doi.org/10.1021/acsami.8b04809.

21 Cho, M., Seo, J.-W., Song, J.T. et al. (2017). Silver nanowire/carbon sheet composites for electrochemical syngas generation with tunable H_2/CO ratios. *ACS Omega* 2: 3441–3446. https://doi.org/10.1021/acsomega.7b00846.

22 Sheng, W., Kattel, S., Yao, S. et al. (2017). Electrochemical reduction of CO_2 to synthesis gas with controlled CO/H_2 ratios. *Energy Environ. Sci.* 10: 1180–1185. https://doi.org/10.1039/C7EE00071E.

23 Chen, K., Zhang, X., Williams, T. et al. (2017). Electrochemical reduction of CO_2 on core-shell Cu/Au nanostructure arrays for syngas production. *Electrochim. Acta* 239: 84–89. https://doi.org/10.1016/J.ELECTACTA.2017.04.019.

24 Hoffman, Z.B., Gray, T.S., Moraveck, K.B. et al. (2017). Electrochemical reduction of carbon dioxide to syngas and formate at dendritic copper–indium electrocatalysts. *ACS Catal.* 7: 5381–5390. https://doi.org/10.1021/acscatal.7b01161.

25 Xu, J., Li, X., Liu, W. et al. (2017). Carbon dioxide electroreduction into syngas boosted by a partially delocalized charge in molybdenum sulfide selenide alloy monolayers. *Angew. Chemie Int. Ed.* 56: 9121–9125. https://doi.org/10.1002/anie.201704928.

26 Li, Y.C., Zhou, D., Yan, Z. et al. (2016). Electrolysis of CO_2 to syngas in bipolar membrane-based electrochemical cells. *ACS Energy Lett.* 1: 1149–1153. https://doi.org/10.1021/acsenergylett.6b00475.

27 Murata, A. and Hori, Y. (1991). Product selectivity affected by cationic species in electrochemical reduction of CO_2 and CO at a Cu electrode. *Bull. Chem. Soc. Jpn.* 64: 123–127. https://doi.org/10.1246/bcsj.64.123.

28 Rosen, B.A., Salehi-Khojin, A., Thorson, M.R. et al. (2011). Ionic liquid-mediated selective conversion of CO_2 to CO at low overpotentials. *Science* 334: 643–644. https://doi.org/10.1126/science.1209786.

29 Peterson, A.A. and Nørskov, J.K. (2012). Activity descriptors for CO_2 electroreduction to methane on transition-metal catalysts. *J. Phys. Chem. Lett.* 3: 251–258. https://doi.org/10.1021/jz201461p.

30 Chen, Y., Li, C.W., and Kanan, M.W. (2012). Aqueous CO_2 reduction at very low overpotential on oxide-derived Au nanoparticles. *J. Am. Chem. Soc.* 134: 19969–19972. https://doi.org/10.1021/ja309317u.

31 Zhu, W., Michalsky, R., Metin, Ö. et al. (2013). Monodisperse Au nanoparticles for selective electrocatalytic reduction of CO_2 to CO. *J. Am. Chem. Soc.* 135: 16833–16836. https://doi.org/10.1021/ja409445p.

32 Liu, M., Pang, Y., Zhang, B. et al. (2016). Enhanced electrocatalytic CO_2 reduction via field-induced reagent concentration. *Nature* 537: 382–386. https://doi.org/10.1038/nature19060.

33 Khanova, L.A. and Krishtalik, L.I. (2011). Kinetics of the hydrogen evolution reaction on gold electrode. A new case of the barrierless discharge. *J. Electroanal. Chem.* 660: 224–229. https://doi.org/10.1016/j.jelechem.2011.01.016.

34 Back, S., Yeom, M.S., and Jung, Y. (2015). Active sites of Au and Ag nanoparticle catalysts for CO_2 electroreduction to CO. *ACS Catal.* 5: 5089–5096. https://doi.org/10.1021/acscatal.5b00462.

35 Zhang, S., Kang, P., and Meyer, T.J. (2014). Nanostructured tin catalysts for selective electrochemical reduction of carbon dioxide to formate. *J. Am. Chem. Soc.* 136: 1734–1737. https://doi.org/10.1021/ja4113885.

36 Zhang, Z., Chi, M., Veith, G.M. et al. (2016). Rational design of Bi nanoparticles for efficient electrochemical CO_2 reduction: the elucidation of size and surface condition effects. *ACS Catal.* 6: 6255–6264. https://doi.org/10.1021/acscatal.6b01297.

37 Li, Q., Fu, J., Zhu, W. et al. (2017). Tuning Sn-catalysis for electrochemical reduction of CO_2 to CO via the core/shell Cu/SnO_2 structure. *J. Am. Chem. Soc.* 139: 4290–4293. https://doi.org/10.1021/jacs.7b00261.

38 Sharma, P.P., Wu, J., Yadav, R.M. et al. (2015). Nitrogen-doped carbon nanotube arrays for high-efficiency electrochemical reduction of CO_2: on the understanding of defects, defect density, and selectivity. *Angew. Chemie Int. Ed.* 54: 13701–13705. https://doi.org/10.1002/anie.201506062.

39 Kumar, B., Asadi, M., Pisasale, D. et al. (2013). Renewable and metal-free carbon nanofibre catalysts for carbon dioxide reduction. *Nat. Commun.* 4: 2819. https://doi.org/10.1038/ncomms3819.

40 Huan, T.N., Ranjbar, N., Rousse, G. et al. (2017). Electrochemical reduction of CO_2 catalyzed by Fe–N–C materials: a structure–selectivity study. *ACS Catal.* 7: 1520–1525. https://doi.org/10.1021/acscatal.6b03353.

41 Varela, A.S., Ranjbar Sahraie, N., Steinberg, J. et al. (2015). Metal-doped nitrogenated carbon as an efficient catalyst for direct CO_2 electroreduction to CO and hydrocarbons. *Angew. Chemie Int. Ed.* 54: 10758–10762. https://doi.org/10.1002/anie.201502099.

42 Conway, B.E. and Tilak, B.V. (2002). Interfacial processes involving electrocatalytic evolution and oxidation of H_2, and the role of chemisorbed H. *Electrochim. Acta* 47 3571–3594. https://doi.org/10.1016/S0013-4686(02)00329-8.

43 Artero, V., Chavarot-Kerlidou, M., and Fontecave, M. (2011). Splitting water with cobalt. *Angew. Chemie Int. Ed.* 50: 7238–7266. https://doi.org/10.1002/anie.201007987.

44 Du, P. and Eisenberg, R. (2012). Catalysts made of earth-abundant elements (Co, Ni, Fe) for water splitting: Recent progress and future challenges. *Energy Environ Sci.* 5: 6012. https://doi.org/10.1039/c2ee03250c.

45 Popczun, E.J., McKone, J.R., Read, C.G. et al. (2013). Nanostructured nickel phosphide as an electrocatalyst for the hydrogen evolution reaction. *J. Am. Chem. Soc.* 135: 9267–9270. https://doi.org/10.1021/ja403440e.

46 Eftekhari, A. (2017). Electrocatalysts for hydrogen evolution reaction. *Int. J. Hydrogen Energy* 42: 11053–11077. https://doi.org/10.1016/J.IJHYDENE.2017.02.125.

47 Zheng, Y., Jiao, Y., Zhu, Y. et al. (2014). Hydrogen evolution by a metal-free electrocatalyst. *Nat. Commun.* 5: 3783. https://doi.org/10.1038/ncomms4783.

48 Kumar, B., Brian, J.P., Atla, V. et al. (2016). Controlling the product syngas H_2:CO ratio through pulsed-bias electrochemical reduction of CO_2 on copper. *ACS Catal.* 6: 4739–4745. https://doi.org/10.1021/acscatal.6b00857.

49 Stoukides, M. and Vayenas, C.G. (1981). The effect of electrochemical oxygen pumping on the rate and selectivity of ethylene oxidation on polycrystalline silver. *J. Catal.* 70: 137–146. https://doi.org/10.1016/0021-9517(81)90323-7.

50 Yentekakis, I., Konsolakis, M., Lambert, R. et al. (2000). Successful application of electrochemical promotion to the design of effective conventional catalyst formulations. *Solid State Ionics* 136–137: 783–790. https://doi.org/10.1016/S0167-2738(00)00547-6.

51 Papaioannou, E.I., Souentie, S., Hammad, A., and Vayenas, C.G. (2009). Electrochemical promotion of the CO_2 hydrogenation reaction using thin Rh, Pt and Cu films in a monolithic reactor at atmospheric pressure. *Catal. Today* 146: 336–344. https://doi.org/10.1016/J.CATTOD.2009.06.008.

52 Bebelis, S., Karasali, H., and Vayenas, C.G. (2008). Electrochemical promotion of CO_2 hydrogenation on Rh/YSZ electrodes. *J. Appl. Electrochem.* 38: 1127–1133. https://doi.org/10.1007/s10800-008-9574-7.

53 Pekridis, G., Kalimeri, K., Kaklidis, N. et al. (2007). Study of the reverse water gas shift (RWGS) reaction over Pt in a solid oxide fuel cell (SOFC) operating under open and closed-circuit conditions. *Catal. Today* 127: 337–346. https://doi.org/10.1016/J.CATTOD.2007.05.026.

54 Bebelis, S., Karasali, H., and Vayenas, C.G. (2008). Electrochemical promotion of the CO_2 hydrogenation on Pd/YSZ and Pd/β″-Al_2O_3 catalyst-electrodes. *Solid State Ionics* 179: 1391–1395. https://doi.org/10.1016/J.SSI.2008.02.043.

55 Jiménez, V., Jiménez-Borja, C., Sánchez, P. et al. (2011). Electrochemical promotion of the CO_2 hydrogenation reaction on composite Ni or Ru impregnated carbon nanofiber catalyst-electrodes deposited on YSZ. *Appl. Catal., B* 107: 210–220. https://doi.org/10.1016/J.APCATB.2011.07.016.

56 Ruiz, E., Cillero, D., Martínez, P.J. et al. (2013). Bench scale study of electrochemically promoted catalytic CO_2 hydrogenation to renewable fuels. *Catal. Today* 210: 55–66. https://doi.org/10.1016/J.CATTOD.2012.10.025.

57 Gutiérrez-Guerra, N., González-Cobos, J., Serrano-Ruiz, J.C. et al. (2015). Electrochemical activation of Ni catalysts with potassium ionic conductors for CO_2 hydrogenation. *Top. Catal.* 58: 1256–1269. https://doi.org/10.1007/s11244-015-0488-4.

14

Recent Progress on Catalyst Development for CO_2 Conversion into Value-Added Chemicals by Photo- and Electroreduction

Luqman Atanda[1,2], Mohammad A. Wahab[3], and Jorge Beltramini[1,4]

[1]*Queensland University of Technology, Institute for Future Environments, Centre for Agriculture and the Bioeconomy, 2, George Street, Brisbane, Queensland 4000, Australia*
[2]*Queensland University of Technology, School of Mechanical, Medical and Process Engineering, Science and Engineering Faculty, 2, George Street, Brisbane, Queensland 4000, Australia*
[3]*Institute for Advanced Study, Chengdu University, No 2025, Chengluo Road, Longquanyi District, Chengdu City, Sichuan Province 610106, China*
[4]*Kumamoto University, IROAST, Department of Chemistry, Faculty of Advanced Science and Technology, 2-39-1 Kurokami, Chuo-ku, Kumamoto 860-8555, Japan*

14.1 Introduction

Carbon dioxide (CO_2) is a nontoxic, nonflammable gas that its availability in high amount in the atmosphere makes it a potential carbon resource. Importantly, the increasing level of CO_2 emission arising from the burning of fossil fuels for energy generation has allegedly contributed significantly to the greenhouse effect – global warming of the environment [1]. An Intergovernmental Panel for Climate Change (IPCC) predicts that a rise in the level of CO_2 in the atmosphere to a concentration of about 600 ppm will lead to increased mean global temperature by 1.9 °C by the year 2100, which would lead to a devastating effect on humans and the environment [2]. In view of this, the global climate action was taken by nearly 190 countries in 2015 at the Conference of Parties – COP 21 to limit the CO_2 concentration in the atmosphere and to prevent average temperature rise by the end of this century [3]. Based on the COP 21, some strategies such as energy conservation and efficiency, employing renewable and low carbon fuels, adapting geoengineering strategies such as afforestation, and most importantly developing CO_2 capture and storage (CCS) techniques were recommended [4]. In spite of the proven success of the CCS technology in preventing the release of CO_2 into the atmosphere, a more sustainable and effective reduction of CO_2 emissions can be achieved by subsequently utilizing the stored CO_2 via chemical conversion techniques to produce useful chemicals that may serve as potential energy carriers. With global CO_2 emissions around 30–40 gigatons per annum and deployment of CO_2 capture facilities on the rise, it is expected that concentrated sources of CO_2 will become readily available at a competitive cost, making it a nondepleting renewable chemical feedstock [5]. Within this context, the integration of CO_2 capture and catalytic-driven conversion strategies

Engineering Solutions for CO_2 Conversion, First Edition.
Edited by Tomas R. Reina, José A. Odriozola, and Harvey Arellano-Garcia.

represent an important route for carbon recycling. An approach that has consistently drawn attention in recent decade is the production of hydrocarbons and chemicals such as methanol and formic acid. However, the catalytic conversion of CO_2 was found to be extremely challenging because of its thermodynamic stability and chemical inertness at relatively mild reaction conditions because of a high C=O energy bond of 750 kJ mol^{-1} [6]. In spite of these encountered challenges, the catalytic conversion of CO_2 has remained an intriguing research problem for future generations. Hence, the quest for a sustainable energy economy via novel catalytic methods initiated tremendous activity in fundamental research, especially in the concept of utilizing solar energy for CO_2 conversion techniques. Solar-driven conversion of CO_2 directly into hydrocarbons via photocatalytic reduction of CO_2 is an attractive technique to overcome the high energy requirement for such reaction. Alternatively, the low-temperature electrocatalytic strategy is also an interesting route of CO_2 conversion, allowing the reactions to be conducted under relatively mild conditions such as ambient temperatures and pressures. Therefore, this review focuses on the most recent reports on material design for the photo- and electroreduction processes of CO_2 conversion in generating energy carrier chemicals and renewable fuels.

14.2 CO_2 Catalytic Conversion by Photoreduction

Photocatalytic reaction is a catalytic process in which the catalysts are activated by photons or light radiation (such as UV light or solar radiation) of a suitable wavelength for generating electron–hole pairs that initiate chemical reactions. CO_2 activation via photocatalytic reduction proceeds via multiple electron generation and when coupled with a proton transfer, an array of C_1 to C_4 species of chemicals and fuels can be produced as shown in Eqs. (14.1)–(14.5).

$$CO_2 + 2H^+ + 2e^- \rightarrow HCOOH \tag{14.1}$$

$$CO_2 + 2H^+ + 2e^- \rightarrow CO + H_2O \tag{14.2}$$

$$CO_2 + 4H^+ + 4e^- \rightarrow HCHO + H_2O \tag{14.3}$$

$$CO_2 + 6H^+ + 6e^- \rightarrow CH_3OH + H_2O \tag{14.4}$$

$$CO_2 + 8H^+ + 8e^- \rightarrow CH_4 + H_2O \tag{14.5}$$

However, the chemical activation of CO_2 by splitting the structure through chemical reactions is a significant limiting factor because of the requirement of high energy input. Given that the hourly supply of energy from the sun is far more than that consumed in a year [7], the broad spectral distribution of infrared, visible, and ultraviolet (IR–Vis–UV) radiation from solar energy can be used not only for thermal power generation but also for chemical manufacturing. For the latter, the unique properties of the sun's spectrum permits the Vis–UV to be used for catalytic reaction activation, while the IR can supply needed endothermic energy to produce lower cost higher energy products. Therefore, as an alternative to the conventional process heat derived from burning fossil fuels, effective utilization of the thermal energy of the solar irradiation is an attractive technique for exploiting renewable energy resources for CO_2-to-fuel conversions via a photothermal catalytic reduction [8].

The photothermal CO_2 conversion is thus a process where solar radiation is used as the energy source of high-temperature process heat for driving the catalytic reduction of CO_2 to "solar fuels." Photothermal CO_2 conversion via catalytic reduction is closely related to photocatalysis but distinguishes itself by obtaining the heat to drive the reaction directly from or initiated by the absorbed light to produce energy directly and hence may be referred to as a process of artificial photosynthesis. It combines light-driven reaction (photocatalysis) with the thermal energy of the sun as the energy source for the reaction while attaining effective catalytic conversions typically obtained for thermal catalysis. The photothermal conversion of CO_2 provides a straightforward and effective method for the highly efficient production of solar fuels with high solar light utilization efficiency. A broad range of the solar spectrum can be effectively utilized during photothermal CO_2 conversion, thus achieving reaction rates (mol h^{-1} g^{-1}) that are several orders of magnitude larger than those obtained (μmol h^{-1} g^{-1}) with photocatalytic methods [9]. Methane and methanol are examples of solar fuels, which may be produced by the direct photothermal reduction of CO_2 with hydrogen. CO is another product of CO_2 photoreduction with hydrogen. Water may also be used as a reductant for the CO_2 photoreduction reaction; yet, it is thermodynamically a less preferred route compared to hydrogen as shown in the following equations:

$$CO_{2(g)} + 3H_{2(g)} \rightarrow CH_3OH_{(g)} + H_2O_{2(g)} \quad \Delta G^{\circ}_{r} = +2.9\ \text{KJ mol}^{-1} \tag{14.6}$$

$$CO_{2(g)} + 4H_{2(g)} \rightarrow CH_{4(g)} + 2H_2O_{(g)} \quad \Delta G^{\circ}_{r} = -113.6\ \text{KJ mol}^{-1} \tag{14.7}$$

$$CO_{2(g)} + 2H_2O_{(g)} \rightarrow CH_3OH_{(g)} + 1.5\ H_2O_{(g)} \quad \Delta G^{\circ}_{r} = +689\ \text{KJ mol}^{-1} \tag{14.8}$$

Nonetheless, a more feasible approach proposed for the overall water-based CO_2 photoreduction conversion process is to integrate CO_2 and water-splitting processes together, whereby the H_2 required for the photothermal CO_2 conversion is derived from sunlight-driven water electrolysis as shown in Figure 14.1. Novel nanomaterials as catalysts have made significant improvements to water-splitting efficiencies and solar-to-energy chemical conversions. Herein, the focus will be on metal nanoparticles (NPs) that possess photothermal effect, thereby having the ability to extend the photocatalytic reduction of CO_2 from visible light to near-infrared (NIR) region in order to boost the utilization efficiency of solar energy.

14.2.1 Principle of CO_2 Photothermal Reduction

Chemical reactions driven by photothermal catalysis exploit broad absorption of the solar spectrum to stimulate a combination of thermochemical and photochemical processes, which contribute synergistically to driving catalytic reactions. The process relies on photon-induced transitions between the electronic states of

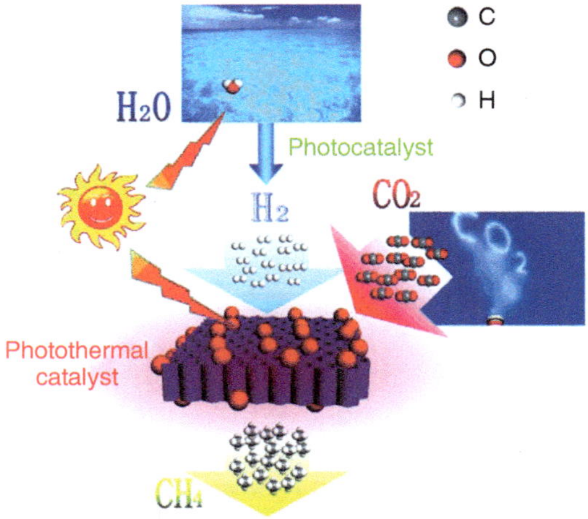

Figure 14.1 A schematic view of solar-driven CO_2-to-fuel strategy; a two-step CO_2 conversion by photocatalytic water splitting and photothermal CO_2 hydrogenation. Source: Meng et al. [9] Copyright (2014) Wiley-VCH.

the catalyst and consequent interaction with the molecular orbitals of adsorbed reactant species. The first step in any photothermal conversion process is the near-instantaneous absorption of solar radiation by the catalyst material. Photons interacting with a light-absorbing semiconductor material lead to the formation of electron–hole pairs that are able to participate in the surface chemistry, whereas heating of the system via the localized surface plasmon resonance (LSPR) effect or carrier relaxation contributes to raising the temperature at the active site [10]. The heating of a catalyst material by solar irradiation involves the transfer of energy from photon-induced electron excitations to lattice phonons via processes that are non-radiative such as hot carrier relaxation, Auger recombination, and Shockley–Read–Hall recombination. The energy transfer efficiency, together with the thermal properties (i.e. heat capacity and thermal conductivity) of the catalyst and the surrounding media, determines the resulting temperature experience by the surface reaction. Photothermal catalysis ultimately demands good thermal conductivity between the nanoheater component of the system and the catalyst, while simultaneously favoring poor thermal conductivity of the catalyst itself, in order to focus thermal energy at the reaction site.

Therefore, the application of a photothermal strategy to gas-phase heterogeneous catalytic systems, i.e. CO_2 reduction, enables high production rates while operating under relatively mild conditions because of localized heating of the catalyst material. One of the key advantages of photothermal effect is that it offers more effective harvesting of the solar spectrum, including absorption of lower energy infrared light that would be insufficient to excite purely photochemical reactions. Because CO_2 chemical conversion is an endothermic reaction that requires high temperature, the generation of high local temperatures at the surface of the catalyst through the combined contributions of plasmonic and nanoscale thermal transport effects make the heat-related processes particularly efficient. For this reason, the photothermal process offers a promising approach for converting gaseous CO_2 to chemical and fuels as it allows for greater exploitation of the solar spectrum as compared to purely photocatalytic systems, which are limited by their semiconductor bandgap, and eliminates the need for solar concentrators to produce the exceedingly high temperature needed for purely thermochemical conversion.

Photothermal reduction of CO_2 is a heterogeneous catalytic reaction that requires the initial adsorption step of CO_2 onto the reaction sites of the catalyst surface, followed by activation, reaction, and then desorption of product molecules. On this note, both the photochemical and thermochemical processes of the photothermally enabled catalytic reaction can contribute to the overall reaction rate. In this context, the photothermal effect can facilitate both the reaction kinetics and energy efficiency in terms of increasing the rate of reaction by enhancing the temperature experienced at the active site of the catalyst (thermochemical component), especially by increasing the formation of products or, alternatively, enabling the desorption of unwanted species from the active surface sites. Furthermore, the enhancement of the reaction kinetics could also be achieved by lowering the activation barrier of one or more reaction steps that arise from charge transfer to the catalyst surface causing an increase in acidity and/or basicity of the active site, or from energy transfer to a vibrational mode of the transition state effectuating bond elongation (photochemical component) [10].

14.2.2 Catalyst Development for CO_2 Photothermal Reduction

The pioneering work of Fujishima and coworkers [11] on the photoreduction of carbon dioxide to formic acid and methanol utilizing semiconductors has generated a growing interest in developing photocatalysts with remarkable carbon dioxide conversion efficiencies under solar irradiation for the production of solar fuels [12]. The key factor in the effectiveness of such a semiconducting material is its capability to capture sunlight to generate the requisite energy to activate the CO_2 reduction reaction. Since then, photoreduction of CO_2 using TiO_2 as the photocatalyst has been extensively investigated because of its remarkable properties including excellent electronic conductivity, good optical transmittance, and superior chemical stability. Recently, for the emerging direct gas-phase catalytic CO_2 reduction using sunlight directly as the energy source, nanostructuring strategies in developing hybrid photocatalysts that can combine both photothermal and photochemical processes are being explored. These new catalytic materials are envisioned to effectively harvest a broader solar spectrum and able to generate a localized heating (photothermal) effect to drive the direct conversion of CO_2 and H_2O or indirectly producing H_2 from H_2O and then converting CO_2 to fuels. For instance, the CO_2 photoconversion efficiency of TiO_2 photocatalysts was enhanced by fabricating TiO_2 photonic crystals (TiO_2 PCs) with particular photon energy/heat radiation capture ability. Low et al. [13] reported that these TiO_2 PCs exhibited slow photon and localized surface photothermal (LSPT) effect, thereby enhancing light utilization and catalytic reaction rate, respectively. As a result, TiO_2 PCs exceeded TiO_2 nanotube arrays (TiO_2 NTAs) greatly in photocatalytic CO_2 reduction performance.

The photothermal catalysts based on metal nanocrystals may transform light into high local temperatures that drive CO_2 reductions through the surface plasmon resonance (SPR) effect. Exploiting the photothermal effect or rapid heating of nanomaterials induced by SPR absorption, Meng et al. [9] described the use of certain group VIII metals as nanocatalysts that can utilize solar energy for the selective production of solar chemicals directly from CO_2. Photo-irradiation of these nanocatalysts resulted in intense absorption, which implies that high utilization efficiency of solar energy would be feasible. The photothermal effect of the group VIII catalysts enabled effective harvesting of a broad range of radiations including ultraviolet light, visible light, and infrared radiation. Aside the excellent photothermal properties of the group VIII nanocatalysts, they can effectively activate hydrogen to create active H atoms, which are the key intermediates for CO_2 hydrogenation at low-onset temperature. These two predominant features played a crucial role in the highly efficient photoinduced conversion of CO_2, resulting in reaction rates in the magnitude of mol h^{-1} g^{-1} that are several orders of magnitude greater than those observed with traditional semiconductor photocatalysts. To rationalize this, the photothermal CO_2 conversion was done in the absence of H_2 and a reduced photoreduction activity was observed. A control reaction in the dark or with a monochromatic light greatly suppressed light-induced thermal effect with no obvious temperature increase, while CH_4 production was at very low rates of ca. 0.02 mmol/100 minutes. This result reflects the fact that the photoinduced hydrogen-assisted CO_2 activation and methanation over the group VIII catalysts was possible even with the low-energy photons of visible light and infrared radiation. Loading plasmonic metals as cocatalysts can simultaneously lead to enhanced visible light harvesting, charge separation, and accelerated surface reaction kinetics. More importantly, cocatalysts can also greatly improve the charge separation performance and photostability of semiconductors [14]. In this context, the plasmon exciton coupling of Au and Pt nanoparticles co-decorated on TiO_2 nanofiber was demonstrated by Zhang et al. [15] to greatly enhance the solar harvesting capability of TiO_2. Electrospun Au/Pt/TiO_2 nanofiber displayed strong visible light absorption caused by the SPR effect of the embedded Au nanoparticle. On the contrary, Pt nanoparticle on the composite nanofiber acts as active sites for H_2 production. For instance, pure TiO_2 NF showed poor H_2 production rate of 0.009 μmol h^{-1}. In comparison, Pt-decorated TiO_2 (Pt_1/TiO_2) displayed more than 900 times higher H_2 production rate (8.289 μmol h^{-1}) because Pt served as an electron sink to suppress the recombination of photogenerated charge carriers and reduce the overpotential for H_2 evolution. Partial replacement of Pt with Au in the composite nanofiber is marked by a further increase in H_2 production rate to 11.658 μmol h^{-1}, suggesting a photocatalytic enhancement through the plasmonic excitation of Au. Simultaneously, photoreduction of CO_2 to CH_4 was significantly improved on $Au_{0.25}/Pt_{0.75}/TiO_2$ because of enhanced charge separation of the photoexcited TiO_2, resulting in 1.35 times higher activity in CH_4 production than Pt_1/TiO_2 (0.57 versus 0.42 μmol h^{-1}). The plasmon-enhanced photocatalytic activity over the Au/Pt/TiO_2 NFs is illustrated as Figure 14.2. The solar irradiation causes SPR excitation of the nanofiber to induce hot electrons of high energy levels to be transferred into the conduction band of TiO_2.

To avoid the high cost and the low abundance of noble plasmonic materials, noble-metal-free plasmonic metals, such as nickel, have been explored for assisting

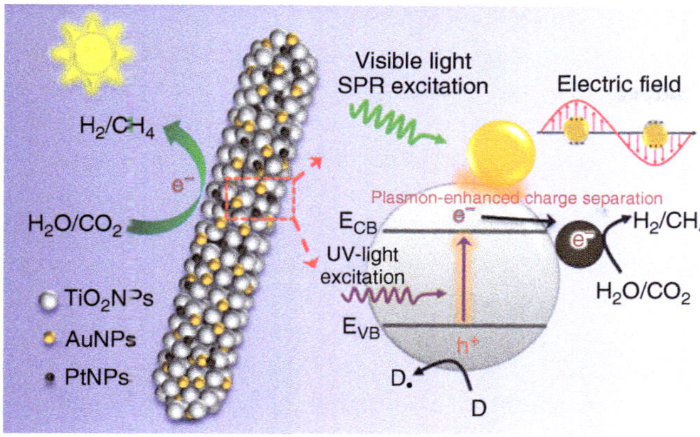

Figure 14.2 Schematic diagram showing the photocatalytic process for H_2 production and CO_2 reduction on the Au/Pt/TiO_2 NFs. Source: Zhang et al. [15] Copyright (2013) American Chemical Society.

the photothermal reduction of CO_2 [16]. Interestingly, Ni/Al_2O_3 catalyst, under light irradiation, was active for CO_2 photoreduction with methane (CRM). Ni plasmonic effect, which is controlled by the optical property of the Ni particles, was primarily responsible for the photothermal catalytic performance of the catalyst. Meanwhile, the optical property of Ni exhibited strong size-dependent effects. Ni particle size correlation with optical property showed that Ni LSPR dominates the optical property of Ni particles of small sizes (<17.2 nm), interband transition accounts for higher percentage of the optical property of Ni particles with increased Ni particle sizes (>33.5 nm) while bulk Ni particle with extremely large Ni particles (approx. 200 nm) displayed weak and negligible optical property with no obvious enhanced activity for the CRM reaction. By using the Mie theory for theoretical simulation to calculate the number of photons and apparent quantum efficiency (AQE), a wavelength-dependent performance was obtained and it was proposed that the enhanced activity of Ni/Al_2O_3 was related to the properties, i.e. size, of Ni under light irradiation. Thus, the size-dependent optical property of Ni as it is related to its photothermal efficiency offers new insight into the application of Ni-based catalysts for photothermal catalytic processes.

Besides the SPR effect, catalysts with suitable oxygen vacancies are also vital for the photothermal reduction of CO_2. For example, Wang et al. [17] demonstrated that the photothermal reduction of CO_2 with water into CH_4 and CH_3OH over mesoporous WO_3 is greatly influenced by its concentration of active oxygen vacancies. The reaction was found to be initiated by the decomposition of CO_2 into its elements, C or CO, which was enabled by electron transfer by the surface oxygen vacancies of WO_3. With the reduction of protons by the formed electrons to produce hydrogen, CH_4 and CH_3OH are formed when the generated carbon or CO reacts with hydrogen. Increased oxygen vacancies by hydrogen treatment remarkably improved the photothermal catalytic efficiency of WO_3 for solar-to-methane (or methanol), STM, and energy. Hydrogen-treated m-WO_3 thus has stronger catalytic performance and selectivity toward CH_4 because of bandgap narrowing and a much improved visible light absorption, which results in more electron excitation and relaxation, thereby favoring the energetics of the CO_2 reduction with H_2O selectively toward CH_4 evolution under photothermal condition. Similarly, Hang et al. [18] explored the oxygen vacancies of double perovskites $LaSrCoFeO_{6-\delta}$ (LSCF) and a three dimensionally ordered macroporous $LaSrCoFeO_{6-\delta}$ (3DOM-LSCF) for solar-to-methane (STM) conversion. Both LSCF and 3DOM-LSCF catalysts have shown good catalytic performance and high selectivity for the reduction of CO_2 with H_2O vapor to CH_4 associated with suitable structural geometry of the catalysts that are energetically favorable for the adsorption/dissociation of CO_2 and H_2O on the oxygen defect sites present on the catalyst surface. Particularly under photothermal reaction conditions, both catalysts displayed reaction rates five times better than that under thermal-only conditions, thus suggesting that the photothermal effect plays a critical role in improving conversion of CO_2 into CH_4 over active oxygen vacancies. Moreover, with the reduction of CO_2 to CH_4 under photothermal reaction conditions (350 °C + Vis-light) in eight hours, 3DOM-LSCF catalysts showed a better reaction rate than that of LSCF (557.88 versus 351.32 $\mu mol\, g^{-1}$) because of the self-formed heterostructures and oxygen vacancies of the 3DOM structure that enhanced the absorption efficiency of solar irradiation and the separation of photoinduced electron–hole pairs. These findings may further provide new strategies for fabricating catalysts with more oxygen vacancies that can efficiently catalyze the photothermal conversion of CO_2 into solar fuels, consequently offering new insights into the prominent role of self-formed oxygen vacancies in improving the CO_2 and H_2O absorption and reaction.

The design of bifunctional cocatalysts with a proper structure is also vital for enhancing the photoreduction rate of CO_2. The photothermal efficiency of layered double hydroxide (LDH) in the production of CH_4 by the hydrogenation of CO_2 under light irradiation was greatly enhanced by the exfoliation of LDHs to produce ultrathin nanosheets, as demonstrated by Ren et al. [19] The nanohybrid ultrathin LDH nanosheets and Ru nanoparticles (denoted as Ru@FL-LDHs) were designed to realize targeting CO_2 activation by the surface OH groups generated in the nanosheets combined with the H_2 activation function of Ru metal. To explore the validity of the catalyst design strategy, the photothermal conversion of CO_2 was carried out in a flow-type reaction system, wherein a 277 $mmol\, h^{-1}\, g^{-1}$ (catalyst) of maximum yield rate for CH_4 generation could be attained over Ru@FL-LDHs. More than 96% CO_2 can be photothermally converted to CH_4 with a selectivity of about 99% after 60 minutes irradiation. The highest photothermal CO_2 conversion over Ru@FL-LDHs catalyst could be partially ascribed to its strong light absorption capacity. Monitoring the reaction temperature of the catalysts showed that the Ru@FL-LDHs underwent rapid heating and maintained a reaction temperature of approx. 350 °C, a temperature range sufficient for Ru nanoparticles to activate the H_2 molecule. Compared with FL-LDHs, very low temperature only about 67 °C was measured, indicating that the Ru nanoparticles played a key role in generating the photothermal effect. In synergy with the photothermal effect created by Ru, the chemical compositions and the ultrathin 2D nanostructure of Ru@FL-LDHs also played a critical role in the photothermal efficiency of the catalyst to reduce CO_2. Even though the ultrathin FL-LDH matrix possesses no capability to activate H_2 and barely exhibits any activity in the photothermal CO_2 hydrogenation, it considerably provided high-density basic active sites for the chemical adsorption and activation of CO_2. Meanwhile, the Ru nanoparticles not only dramatically increase local temperature under light irradiation but also activate H_2 to initiate CO_2 (or CO_3^{2-}) hydrogenation, thus revealing the second function of Ru in the catalytic system, namely, the activation agent of H_2.

Inspired by the successful application of ultrathin LDH nanosheets in constructing efficient photothermal catalysts for the methanation of CO_2, Chen et al. [20] demonstrated that solar-driven CO_2 hydrogenation can be tuned toward higher value C_{2+} hydrocarbons such as ethane, propane, and butane on a CoFe-based catalyst formed by the hydrogen reduction of a CoFeAl-LDH nanosheet precursor. The surface chemical composition of the CoFe-x catalysts (where x is the reduction temperature) was facilely manipulated by changing the reduction temperature of CoFeAl-LDH nanosheet precursor. The H_2-reduction-induced topological transformation of the CoFeAl-LDH nanosheets resulted in three distinct CoFe-x catalysts that showed markedly different activity and product selectivities in CO_2 hydrogenation under UV–Vis excitation. It was observed that for CoFe-300, the uniformly dispersed Fe_2O_3 nanoparticles that appeared on a CoAl-mixed metal oxide nanosheet support served as catalytic active species to promote the CO_2 hydrogenation to CO. For reduction temperatures between 450 and 550 °C, both CoO_x and FeO_x were identified as the active species on an amorphous alumina nanosheet support. The formed FeO_x/CoO_x nanoparticles were predominantly active for the production of CH_4 and trace amount of CO, with the CoO_x domains responsible for high selectivity to CH_4 and the FeO_x domains favoring CO production. With increased LDH-nanosheet reduction temperature to above 600 °C, alumina-supported CoFe-alloy nanoparticles were formed, which demonstrated high CO_2 conversions (78.6%) and remarkable selectivity (≈95%) toward hydrocarbons (>35% selectivity to C_{2+} hydrocarbons). density functional theory (DFT) calculations confirmed the key role of the CoFe alloy structure in promoting C–C coupling reactions during CO_2 hydrogenation similar to Fischer–Tropsch-like carbon chain growth essential for higher hydrocarbon synthesis. Lastly, it was concluded that the catalytic activity of the CoFe-x catalysts for the CO_2 hydrogenation under a simulated solar excitation originated from the photothermal effects that caused catalyst heating, however, tuning the selectivity of the CO_2 hydrogenation progressively from CO to CH_4 and eventually high-value C_{2+} hydrocarbons was found to be largely dependent on the reduction temperature of the LDH precursor (Figure 14.3).

Heterostructures, especially metal–metal oxide hetero-nanostructures, are a class of materials with optical properties that make them compelling choices as photothermal CO_2 reduction catalysts. The photothermal effect in these hetero-nanostructure materials arises from broadband optical absorption and non-radiative relaxation of photoexcited surface plasmons or intra-/interband electrons in the metal nanocomponents [21]. For instance, nanostructured Pd@Nb_2O_5 was shown to photothermally activate the hydrogenation of gaseous CO_2 using visible and NIR light

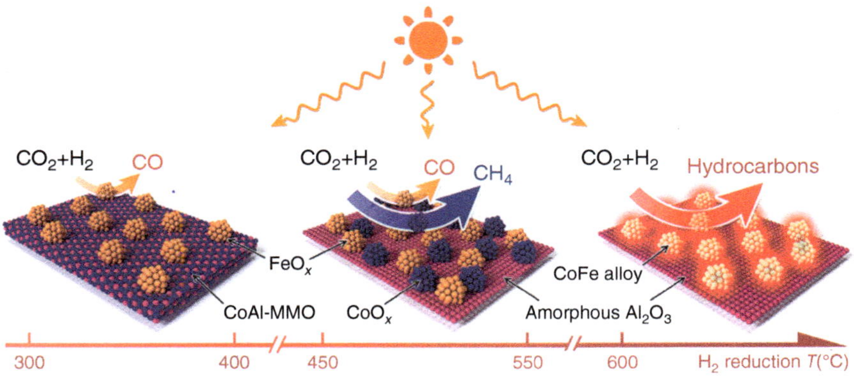

Figure 14.3 Tuning the selectivity of solar-driven CO_2 hydrogenation over CoFe-*x* catalysts via hydrogen reduction of a CoFeAl-LDH nanosheet precursor at different temperatures. Source: Chen et al. [20] Copyright (2018) Wiley-VCH.

[22]. The catalyst photothermally activated the reduction of CO_2 via a reverse water gas shift (RWGS) reaction to produce CO. The RWGS reaction was initiated by heat generated from the thermalization of charge carriers in the Pd nanocrystals that are excited by interband and intraband absorption of the visible and NIR light, thus achieving CO_2 conversion rates as high as 1.8 mmol $gcat^{-1}$ h^{-1}. Jia et al. [22] proposed the mechanism of the photothermal reaction was initiated by H_2 dissociation on Pd nanocrystals and subsequent spillover of H to the Nb_2O_5 nanorods, whereupon adsorbed CO_2 is hydrogenated to CO. The synergy of Pd nanocrystals acting as local nanoheaters and the existence of oxygen vacancies and reduced niobium oxide surface states that provided an additional driving force to promote the adsorption and dissociation of CO_2. In addition, with Pd enabling the photothermal effect on Nb_2O_5 nanorods to occur, the formation of reaction intermediates and subsequent hydrogenation are favored, which significantly increased the CO production rate and enhancement of the observed RWGS reaction rate by a factor of 2.6.

As an extension of the work on Pd@Nb_2O_5 hetero–nanostructured catalyst for CO_2 reduction, the catalytic activity and selectivity of the catalyst were systematically tailored to either CO or CH_4 products by varying the size of the Pd nanocrystals supported on the Nb_2O_5 nanorods [23]. The size-controlled Pd nanocrystals interfaced with Nb_2O_5 nanorods could achieve a remarkable CO production rate of 18.8 mol h^{-1} $g_{Pd}{}^{-1}$ with 99.5% CO selectivity for a small-sized Pd nanocrystal, whereas CO and CH_4 production rates as high as 0.75 and 0.11 mol h^{-1} $g_{Pd}{}^{-1}$, respectively, were observed when large Pd nanocrystals are used. Higher CH_4 selectivity, which rose with increase in the size of Pd nanocrystals, originated from an elevated temperature generated by larger population of Pd nanoheaters, thus indicating that methanation of CO_2 is more effectively influenced by photothermal-driven kinetics than the RWGS process. Furthermore, study of the surface chemistry of the Pd@Nb_2O_5 catalyst also provided more insight into the effectiveness of the catalyst for the CO_2 photoreduction reaction. X-ray photoelectron spectroscopy (XPS) and ultra-violet photoemission spectroscopy (UPS) measurement showed a shift in the binding energy of Pd, which caused electrons at the interface between the Pd and the Nb_2O_5 to transfer from the initially *d* electron-rich metal (Pd) to the metal oxide (Nb_2O_5). As a result, the Pd nanocrystals develop a positive surface charge that is more pronounced for the smaller Pd nanocrystals, a key factor that can affect the strength of binding of adsorbed reactants, intermediates, and products [24], thus controlling the reactivity and selectivity of the CO_2 hydrogenation reaction. With the aid of an *in situ* diffuse reflectance infrared Fourier transform spectroscopy (DRIFTS) measurement and DFT calculations, the mechanism governing the CO_2 photoreduction reaction was determined to be via the formate pathway. It was rationalized that formate and monodentate surface carbonate species that are intermediates in CO_2-to-CO hydrogenation were observed in smaller Pd nanocrystals or samples with low Pd loading. When the loadings and sizes of Pd nanocrystals are small, surface local positive charge are more pronounced, which facilitate the desorption of CO but raises the activation barrier for methanation, thereupon favoring highly efficient and selective production of CO. Contrastingly, samples with large loadings and sizes of Pd nanocrystals possess a lower surface positive charge, thus creating more methanation active sites and local heating sources that not only hydrogenate CO_2 to formate and then to CO but can efficiently retain CO intermediate species on their surfaces, where it can be further hydrogenated into CH_4. The remarkable control over the catalytic performance of Pd@Nb_2O_5 is demonstrated to stem from a combination of photothermal, electronic, and size effects, which is rationally tunable. Hence, through a creative catalyst design and photoreactor engineering, a remarkable energy conversion efficiency of the gas-phase CO_2 reduction was demonstrated to be feasible in a flow-type photoreactor, simulating the possibility of an energetically viable solar refinery in terms of industrial application. A photothermally driven CO_2-to-CO conversion rate of 0.91 $mol\,h^{-1}\,g_{Pd}{}^{-1}$ was obtained with an unprecedented turnover frequency (TOF) as high as 0.61 s^{-1} and an energy conversion efficiency exceeding 0.23%.

In summary, photothermal catalysis is an emergent and promising strategy for the conversion of carbon dioxide into synthetic fuels using sunlight. Photothermal catalysis is an effective means of improving the gas-phase CO_2 photocatalytic conversion rates using both photogenerated charge carriers and thermal energy originating from sunlight. Noble metals are predominantly used as photothermal catalysts because of their efficient photoinduced thermal effect and LSPR, leading to the enhanced generation of energetic hot electrons for CO_2 activation. Nanostructuring strategies have largely dominated the research and development of photothermal catalysts because it allows the ease of tuning the bandgap and surface properties of the materials. Other techniques that have been explored such as hierarchical architecturing or metal–metal oxide heterostructure design offer increased dispersion, stability, and photoresponse of the photothermal catalysts, thereby resulting in their improved efficiency and durability. In spite of these breakthroughs in materials synthesis, the principle challenge of limited absorption and utilization of the solar spectrum still remains because photothermal catalysis is spectrum selective. With materials and process engineering optimization, the performance metrics of photothermal catalysts may likely be further enhanced and the cost of the process can be reduced. To accomplish this realization of cost-effective photothermal CO_2 catalysis, rational design of noble metal-free catalysts that can facilitate optimization of the solar heat generation and catalytic conversion capabilities is highly desired. One strategy is to develop a composite plasmonic transition metal (TM) as the nanoheater with an effective heat trapping material that has high solar absorption and low thermal emissivity. Complementary to material design is to explore the development of hybridized photoreactors with systems such as photovoltaics (PV) that can utilize the full spectrum of concentrated sunlight to generate heat and electricity to convert both CO_2 and H_2O into solar fuels. Also crucial to successfully implementing photothermal catalysis is to seek a better understanding of CO_2 conversion pathways through experimental and computational techniques.

14.3 CO_2 Catalytic Conversion by Electroreduction

The electrochemical conversion of CO_2 (EC-CO_2) to produce value-added chemicals is a promising route for reducing CO_2 emission as well as meeting the future energy demand [3, 4]. Figure 14.4 shows the schematic presentation of a sustainable energy economy based on electrocatalysis [1d]. In this scheme, CO_2 can be converted back to fuels and chemical feedstock. To facilitate the catalytic cleavage of C=O bonds in CO_2 molecules, low-temperature EC-CO_2 conversions (≤200 °C)

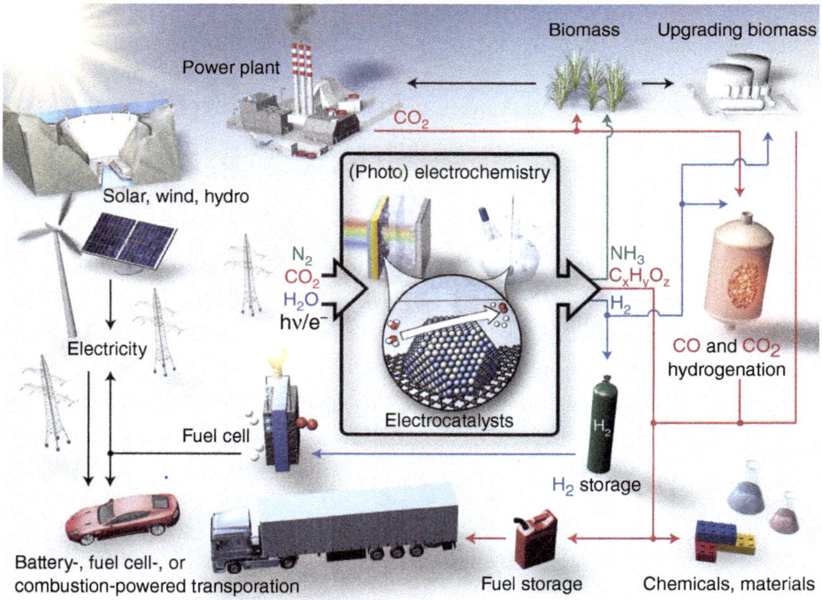

Figure 14.4 An overall schematic presentation of the sustainable energy economy based on electrocatalysis. Source: Seh et al. [1d] Copyright (2017) American Association for the Advancement of Science (AAAS).

based on the transition metal electrodes in liquid electrolytes have shown high selectivity and accelerated reaction rates than that of high-temperature (≥800 °C) conversion process such as in solid oxide cells [25]. It is found that the electrochemical conversion of CO_2 has mostly employed suitable transition metals as the catalysts either as pure or co-doped with noble metal to activate CO_2 by the direct input of electrical energy [2b, 3, 4]. The activated CO_2 and subsequent conversions proceed under ambient experimental conditions, making it possible to stabilize a catalyst and to identify the selective active sites on the catalyst surface for further optimization [4, 26]. Highly efficient catalysts for practical electrocatalytic CO_2 reduction (ECR) still remain a challenge; however, numerous approaches have been reported to rationally design and develop active catalysts that are more active, selective, and stable during the electrochemical conversion reactions [1d, 4a, b]. Nanostructured metals have shown more surface-active sites as well as enhanced large surface areas than that of bulk materials. The nanostructured catalysts have the large portion of edge/low-coordinated sites at their surfaces, which will show different catalytic activity during EC-CO_2 process than bulk materials with flat surface/fully coordinated sites [4b, c, 25–27]. Among the numerous transition metal catalytic systems employed for the electrochemical CO_2 conversion, copper (Cu)-based catalysts have produced chemicals such as CH_4, CH_3OH, C_2H_5OH, C_3H_5OH, and C_2H_4, while other metals have only produced products that involve two-electron transfer such as $HCOO^-$, CO, or H_2 [4c, 25, 26].

14.3.1 Principle of CO_2 Electrocatalytic Reduction

Electrocatalysis can be defined as the heterogeneous catalysis of electrochemical reactions, which occur at the electrode–electrolyte interface and where the electrode plays both the role of electron donor/acceptor and of the catalyst. This catalytic process is controlled by the elementary reactions that occur at the interface between the electrode and the solution phase along with the electrochemical potential that results across this interface. An electrocatalyst can accomplish the surface catalytic reaction through different reaction pathways with different thermodynamic potentials. Elucidating the electrochemical behavior at this interface, however, presents a considerable challenge because of the complexity of the surface chemistry, the explicit atomic and molecular structure of the solution phase at the interface, the presence and formation of different ions and their specific location in solution or on the surface, and the surface potentials and electric fields that results as a function of the surface reactivity.

The electrocatalytic reduction of CO_2 is appealing because of its multiple merits. It operates under ambient temperature and pressure conditions, and when it is interfaced with a solar input (a photo-electrochemical [PEC] cell), the energy for CO_2 reduction can come from sunlight. Aqueous electrocatalytic reduction involves multi-electron/proton transfer processes to convert CO_2 into value-added useful chemicals [1]. The equations below outline the standard electrode potentials for CO_2 reduction reactions that were established at 25 °C with respect to a standard hydrogen electrode (SHE) [26a]. The standard potentials are given at a neutral pH, where most of actual CO_2 reductions are measured. As can be seen, CO_2 undergoes different types of electrochemical reduction, either a $2e^-$ reduction into HCOOH or CO (Eqs. (14.9)–(14.10)) or a multiple electron process into CH_3OH, CH_4, C_2H_4, C_2H_5OH, or C_3H_7OH (Eqs. (14.11)–(14.15)). In most cases, ECR generates mixed products containing several species. The pathways of CO_2 reduction are strongly dependent on the electrocatalysts applied. Besides, electrolyte acts as a medium to transfer electrons and protons and can greatly influence the performance of electrocatalysts. Thus, rational design of the electrocatalytic interface, including selection of electrode and electrolyte compositions and their optimal structure, requires establishing composition–structure–function relationships.

$$CO_2 + H_2O + 2e^- \leftrightarrow CO + 2OH^- \quad E^\circ = -0.52\,V \tag{14.9}$$

$$CO_2 + H_2O + 2e^- \leftrightarrow HCOO^- + OH^- \quad E^\circ = -0.43\,V \tag{14.10}$$

$$CO_2 + 5H_2O + 6e^- \leftrightarrow CH_3OH + 6OH^- \quad E^\circ = -0.39\,V \tag{14.11}$$

$$CO_2 + 6H_2O + 8e^- \leftrightarrow CH_4 + 8OH^- \quad E^\circ = -0.25\,V \tag{14.12}$$

$$CO_2 + 8H_2O + 12e^- \leftrightarrow C_2H_4 + 12OH^- \quad E^\circ = -0.34\,V \tag{14.13}$$

$$CO_2 + 9H_2O + 12e^- \leftrightarrow C_2H_5OH + 12OH^- \quad E^\circ = -0.33\,V \tag{14.14}$$

$$CO_2 + 13H_2O + 18e^- \leftrightarrow C_3H_7OH + 18OH^- \quad E^\circ = -0.32\,V \tag{14.15}$$

$$H_2O + 2e^- \leftrightarrow H_2 + 2OH^- \quad E^\circ = -0.31\,V \tag{14.16}$$

The CO_2 ECR process on a catalyst surface consists mainly of the following three steps: (i) CO_2 adsorption and interaction with atoms on the employed catalyst surface, (ii) CO_2 activation and reduction via catalyst-initiated electron/proton transfers, and (iii) product desorption and recovery of the catalyst surface for next round of reaction [25, 28]. To enhance the reaction rate toward the target reaction, activation energy must be decreased by minimizing the electrode potential that in turn depends on the electrode material, reactants, products, and intermediates at the interface of electrode–electrolyte. Therefore, a good catalyst should minimize the activation barrier for CO_2 reduction in relation to proton reduction, driving CO_2 reduction selectively at low overpotential (i.e. high energy efficiency) with satisfactory reaction rates (i.e. high turnover number).

Parameters that have been used to evaluate the performance of an electrocatalyst include the following: (i) onset potential or overpotential (h, i.e. the difference between the thermodynamic and actual electrode reduction voltages), (ii) current density (current divided by the geometric surface area of the working electrode) – reflects the rate of CO_2 reaction, (iii) faradaic efficiency – describes the product selectivity in the reduction reaction, (iv) energetic efficiency, (v) TOF – a measure of per-site activity of catalysts, and (vi) Tafel slope – indicates the reaction pathways and rate-determining step (RDS) [29].

14.3.2 Catalysts Development for CO_2 Electroreduction

The PEC conversion of CO_2 (PEC-CO_2) into formic acids, formaldehyde, and methanol using nonporous metal phosphides (MP) such as p-GaP without any cocatalyst on the surface was reported [11, 30]. The results from these catalysts were found to show very poor efficiency [30]. Apart from the complex junctions, the catalyst was also responsible for the limited efficiency. It should be noted that high surface area is important for making good dispersion of the targeted catalyst particles on the surface of the supports. Therefore, based on these initial important reports, a suitable catalyst is required with favorable characteristic sites that allow the efficient conversion of the CO_2 into high value-added chemical products. Thereafter, there have been many reports on developing efficient electrocatalysts for the conversion of CO_2 to ethylene, ethanol, and propanol [27b–f]. In order to produce hydrocarbons and alcohols, to date, among the employed catalysts, copper (Cu) is found to be more promising because of its low cost, availability, and easy processing and catalyzes to produce a high quantity of hydrocarbons by applying a moderate to high overpotential [4, 25, 26, 31]. The possible reaction pathway that seems most consistent with experimental data for the electrocatalytic reduction of CO_2 to hydrocarbons is shown in Figure 14.5 [32]. The overall RDS in the reduction of CO_2 is the first electron transfer to form $CO\cdot^{-}_{2,ads}$. This intermediate then reacts to form CO, which can then be reduced to hydrocarbons *via* two pathways, a C_1 and a C_2 pathway, with the overall product distribution being determined by the relative importance of the local RDS within each pathway [32].

Nanostructured Cu can enhance the reaction rate for CH_4 and C_2H_4 [4, 25, 26, 31]. Carbon dioxide reduction using a Cu_2O-derived copper working electrode gave ethylene (C_2H_4) and ethanol as main C_2 products, with optimized Faradaic efficiencies (FEs) of 32.1 and 16.4% at −1.0 V versus RHE (reversible hydrogen electrode) [27a, 31f, g]. It was reported that the nature of the catalyst and its structures largely affect the electrochemical CO_2 conversion process for producing chemicals and also for selective chemicals [4, 25, 26, 31]. Moreover, we note that Cu catalysts have their own limit, with at least ~30% of the Faradaic current accounts for the formation of unwanted CO and H_2. It is obvious that the improvement in CO_2 conversion efficiency reaction rate is highly desired for producing selective chemicals. Therefore, in order to advance the development of catalytically active electrodes for the reduction of CO_2 and improving the conversion efficiency, tuning the morphology (including shape, surface facets, and size), and nanostructuring of metal catalysts and employing electrolytes would be potential factors for determining the CO_2 reduction catalyst.

To achieve the aforementioned goals, alloying Cu with metals such as Ag, Zn, or Sn have proven to modify the electronic structure of Cu, thereby introducing active sites for producing intermediates such as CO for sequential catalysis [11, 25, 27, 30, 31d–g]. As Cu is a very attractive catalyst for ECR, various methods have been adopted to modify the structure of Cu (such as doping, nanostructuring by mixing or heating at various temperatures, introducing active sites, crystal planes, and surface) for developing an efficient ECR catalysts but particles are found to be

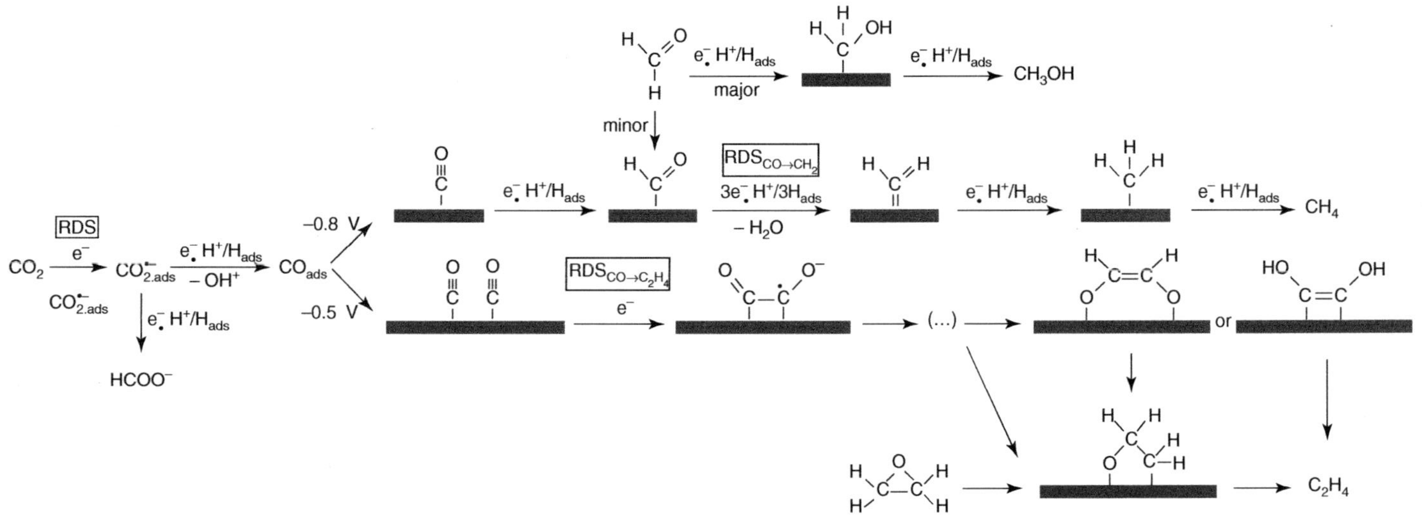

Figure 14.5 Proposed reaction pathway for CO_2 electrocatalytic reduction on Cu. Source: Schouten et al. [32] Copyright (2011) Royal Society of Chemistry.

aggregated with low surface area and no defined surface facets and no porosity in the framework that limits the diffusion of electrolytes within the electrodes during electrochemical reactions [11, 25, 27, 30, 31d–g]. As recently, the facet-dependent activity of Cu in CO_2 reduction has been reported for energetic efficiency and selectivity toward the reduction of carbon dioxide to commercially vital C_2 products such as ethylene (C_2H_4) [27a, b, 31g]. The studies on single-crystal Cu electrodes demonstrated that Cu crystal facets in catalyzing ECR selectively, which is very crucial in the field of CO_2 conversion into selective product formation [31e, 33]. Among the various facets, Cu(111) facet was found to be important for the formation of CH_4 (FE ~50.2% at −1.52 V versus SHE) while Cu(100) and some high index planes ((911), (711), (511)) were more selective to C_2 products at −1.3 V (versus SHE), giving C_2H_4 (FE ~37.7–58.5%) and C_2H_5OH (FE ~8.8–17.9%) [31e, 33]. This has motivated the synthesis of shape-controlled Cu NPs, which allows the studies of crystal facet effects on the ECR selectivity and to develop new NP catalysts with optimum efficiency for a specific reduction product [31e, 33]. Nanosizing catalysts can increase the active sites for ECR without the mass transport limit issue. By controlling the surface structure and size of NPs, the surface of an NP could exhibit varying corners and edges and therefore particles with different coordination numbers and chemical interaction energy are expected [11, 27d–f, 30, 31, 34]. For example, a study of particle size effects during the catalytic CO_2 electroreduction on size-controlled Cu nanoparticles (NPs) was prepared to investigate their catalytic activity and selectivity during CO_2 electroreduction. This study has prepared nanoparticles with six sizes, establishing the dramatic particle size effect on the overall catalytic CO_2 electroreduction activity of Cu NPs. Controlling the size of the particles accounts for the increase in the Faradaic activity observed on the Cu NPs, which led to the enhancement of CO and H_2 formation. In addition, shape-oriented product formation from ECR process using a core–shell particle has been reported to play a significant role in selective CO_2 electroreduction. Cu and Cu/alloy core–shell particle and their use to achieve selective electrochemical reduction of CO_2 in 0.5 M $KHCO_3$ [35]. It is clearly found that the Cu/SnO_2 NPs with the different thicknesses of SnO_2 can produce different products. For instance, the 1.8 nm SnO_2 shell favored the formation of formate (85% FE at −0.9 V versus RHE), whereas the 0.8 nm SnO_2 shell was selective to CO with FE reaching 93% at −0.7 V versus RHE. From DFT calculations, it was indicative that trace amounts of Cu are able to alloy with a thinner SnO_2 shell to compress the Sn–O lattice, hence favoring CO over formic acid. This study also suggested that the active Cu/SnO_2 NP catalyst could replace expensive noble metal catalysts (Au) for the electrochemical reduction of CO_2 to CO.

In spite of the notable progress in the electrocatalytic property of Cu NPs, single-atom catalysts (SACs) are emerging as intriguing catalytic materials in CO_2 electrocatalysis owing to their high catalytic activity, stability, selectivity, and maximum atom utilization [36]. Wang et al. [27e] rationally designed a Cu-doped CeO_2 (Cu-CeO_2) electrocatalyst that possessed highly dispersed Cu metal catalytic centers at a single atomic level for the selective reduction of CO_2 to CH_4. Cu-CeO_2-4% (4% Cu loading) nanorods exhibited high efficiency for CH_4 production in a CO_2-saturated 0.1 M $KHCO_3$, attaining an unprecedented ~58% FE of methane at

−1.8 V versus RHE associated with a clearly reduced amount of H_2 side product from the water reduction. The Cu-CeO_2-4% also presents an excellent ECR stability, as more than 40% CH_4 FE was retained after >8000 seconds of continuous test at −1.8 V versus RHE. The high ECR activity and CH_4 selectivity on Cu-substituted CeO_2 is enabled by the highly cooperative effect between the single-atomic Cu site and the multiple oxygen vacancies in its neighboring positions, thus yielding a catalytic active site, where a single CO_2 molecule is strongly adsorbed and activated, and then stepwise reduction to different ECR products. Giving that the catalyst is strongly oxophilic because of the measured binding energy of −4.89 eV for the O_{ads} [37], CH_3OH could not be coproduced with CH_4. Furthermore, as Cu mainly exists as single-atomically dispersed catalytic center, the C–C coupling pathway between two C_1 species into one C_2 product was substantially inhibited and thus dramatically enhancing CH_4 formation. Contrastingly, Jiao et al. [38] reported that C_2 hydrocarbon/alcohol can be produced on single-atom Cu when molecularly scaffolded with graphitic carbon nitride (g-C_3N_4). The study revealed that g-C_3N_4 allowed for an appropriate modification of the electronic structure of Cu by efficiently shifting higher the d-orbital position of Cu toward the Fermi level in the resultant Cu–C_3N_4 complex. As a result, stronger adsorption behavior of some key reaction intermediates was facilitated on the Cu–C_3N_4 surface, which greatly benefitted the activation of CO_2 and led to a more facile CO_2 reduction to desired products. The unique structure of Cu metal coordinated by the g-C_3N_4 framework led to a synergistic reaction between Cu and C induced by the migration of the active center, wherein the Cu atom serves as the active center for carbon (and C-containing groups, e.g. *COOH, *CO, and *CHO) adsorption, while the adjacent carbon atom on g-C_3N_4 serves as the active center for reaction intermediates having oxygen as the connecting atom (e.g. *OCH_2, *OCH_3, *O, and *OH). Thus, the Cu–C_3N_4 electrocatalyst showed a wider variety of C_2 products (C_2H_5OH, C_2H_6, and C_2H_4), as well as higher selectivity toward deep reduction products (CH_4 and CH_3OH) because of its unique dual active center catalysis that facilitated low free energy pathway for CO_2 hydrogenation.

Typically, Cu-based materials remain the most efficient and selective CO_2 reduction reaction electrocatalysts owing to their appropriate adsorption strength of key reaction intermediates as delineated by DFT computational studies [39]. Hence, as recent advances in SACs are promising in optimizing selectivity and activity for various reactions, both theoretical and experimental investigations are proving important to develop uniform single-atom dispersion with well-defined configuration as high-performance electrocatalysts for the ECR to multi-electron transfer products [40]. DFT calculation is a widely used method in ECR mechanism research, which also provides perspectives for the design of new catalysts. To this end, DFT calculations have been performed to systematically study the possibility of a series of single transition metal (TM) atoms (Sc to Zn, Mo, Rh, Ru, Pd, Ag, Pt, and Au) supported on the experimentally available defective boron nitride monolayers with a boron monovacancy (TM/defective BN) as electrocatalysts for CO_2 reduction to CH_4 [41]. The results revealed that Fe/defective BN, Co/defective BN, and Pt/defective BN nanosheets possessed outstanding ECR activities with quite low (less negative) onset potentials of −0.52, −0.68, and −0.60 V, respectively. The high ECR activities of these nanosheets were correlated to be a function of the electrocatalysts' affinity for C and O, given that the minimum energy pathways for ECR to CH_4 on these nanosheets are the same as those on the (111), (100), and (211) facets of Cu crystal [42]. Moreover, the comparison of the free energy pathway of CO_2 electroreduction reaction toward HCOOH, CO, CH_3OH, and CH_4 on the TM/defective BN showed that Fe/defective BN and Pt/defective BN displayed higher selectivity of ECR to CH_4 than Co/defective BN.

Another class of materials gaining attention for electroreduction of CO_2 is the two-dimensional (2D) nanosheets featuring abundant surface atoms of low coordination and large specific surface area [29]. These materials appear to be beneficial for fast interfacial charge transfer and facile electrochemical catalysis. Moreover, the properties of these nanosheets can be readily tuned by doping with heteroatoms, alloying, chemical functionalization, or heterostructuring, providing new routes for engineering 2D materials for CO_2 electrocatalysis. As an example, 2D mesoporous bismuth (Bi) nanosheets were proposed to be an efficient electrocatalyst for the reduction of CO_2 with great formate selectively and satisfactory stability [43]. The electrocatalyst was fabricated by the cathodic conversion of atomic thick Bi oxycarbonate ($Bi_2O_2CO_3$) nanosheets. When assessed for CO_2 ECR in 0.5 M $NaHCO_3$, it enabled a robust CO_2 reduction to formate with a large cathodic current density (≈18 mA cm^{-2} at −1.1 V) and an excellent FE (≈100%) over a broad potential window. Additionally, it demonstrated satisfactory stability, showing no activity or selectivity loss for 12 hours. Underscoring the superior ECR performance of the mesoporous Bi nanosheets to most Bi-based materials as well as other formate producing CO_2 reduction electrocatalyst (N-doped carbon) [44], its great performance was attributed to its unique 2D mesoporous nanosheet morphology with enlarged surface areas, abundant undercoordinated Bi sites, and structural robustness. The mesoporosity of the nanosheets also facilitated a good electrochemically active surface area (ECSA), with 13.4% of total Bi sites electrochemically accessible to the liquid electrolyte.

Cathodic exfoliation was demonstrated as a facile and efficient route for producing 2D antimony nanosheets (SbNS) that are active for CO_2 ECR from inactive bulk Sb crystals [45]. Exfoliation of the bulk Sb crystals successfully produced a few-layer SbNSs with a dramatically greater number of exposed edge sites. In contrast to the bulk counterpart, SbNSs were found to selectively and efficiently catalyze the reduction of CO_2 to formate, thereby ascribing the catalytic activity to the increased number of active edge sites in the SbNSs. Moreover, this cathodic exfoliation of Sb can be coupled with the anodic exfoliation of graphite in a single-compartment cell for *in situ* production of a few-layer Sb nanosheets and graphene composite SbNS-G. CO_2 ECR to formate on SbNS-G was enhanced because of the synergetic effect between SbNSs and graphene, arising from a strong electronic coupling between SbNSs and graphene. The SbNS-G electrocatalyst presented a higher selectivity toward formate at lower overpotentials compared to bulk Sb and SbNSs. At a potential of −0.96 V (overpotential of 0.87 V), it delivered the maximum FE for formate of 88.5%, whereas SbNS gave its maximum FE of 84% at −1.06 V. The electronic interaction-induced electrocatalytic performance enhancement of the SbNS-G can be explained by the increased electron density at the surface of the SbNS because of electron migration from graphene (at a lower work function, 4.43 eV) across the Sb-G boundary to SbNS (at higher work function, 4.55–4.70 eV). Furthermore, a Tafel slope of 110 mV dec^{-1} was observed for SbNS-G, which is in an excellent agreement with the expected value of 117 mV dec^{-1}. This indicated enhanced CO_2 adsorption ability, a factor that also contributed to the improved catalytic activity of SbSN-G.

Aside the edge sites of ultrathin metal layers that enhance the ECR performance of such electrocatalysts, the resultant abundant surface atoms can also serve as active sites for CO_2 adsorption. Ultrathin metal layers can be highly active for CO_2 reduction relative to bulk metal crystal because of abundant surface atoms that act as the active sites for CO_2 adsorption [46]. However, for some metals with higher chemical reactivity, they tend to be very unstable at ambient conditions and can be prone to oxidation [47], thereby leading to loss of electron and consequently causing a rapid decay in cyclability. Graphene-confined ultrathin layers of highly reactive metals provide a promising technique to designing efficient and robust catalysts for ECR [48]. As shown in Figure 14.6, graphene not only confines and protects the reactive metals from oxidation but it also affords fast electron penetration to the metal ultrathin layers because of its high conductivity. Moreover, the metal ultrathin layers could serve as a "spacer" to favor electrolyte diffusion into the matrix of graphene [49], thus substantially promoting CO_2 reduction performances. For instance, a spatially reduction strategy of preparing metallic tin (Sn) quantum sheets confined in graphene was reported by Lei et al. [48]. A sandwich-like 2D structure formed between the ultrathin Sn and graphene layers prevented the aggregation of the quantum-sized Sn sheets. Moreover, thermogravimetric analysis in air showed that the encapsulated Sn-quantum sheets in the graphene retained its stability up to 570 °C, whereas metallic Sn nanoparticles (15 nm size) without the protection of a graphene interlayer started to be oxidized at 200 °C. ECR activity of the graphene-confined Sn quantum sheets was examined in a CO_2-saturated 0.1 M $NaHCO_3$ solution. At −1.8 V versus SCE (saturated calomel electrode), a current density of 21.1 mA cm^{-2} and an onset potential of −0.85 V were measured. In addition, it

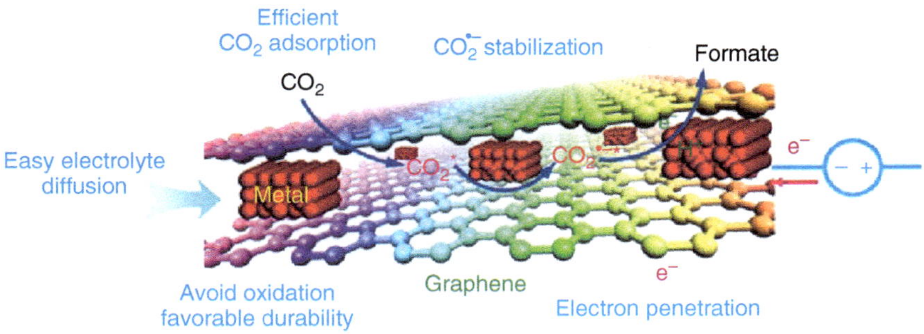

Figure 14.6 Schematic illustration depicts the several advantages of ultrathin metal layers confined in graphene for CO_2 electroreduction into hydrocarbon fuels. Source: Lei et al. [48] Copyright (2016) Nature.

attained a maximum FE 89% for formate production, a performance that was about two times higher relative to bulk Sn. The Sn quantum sheets confined in graphene also showed a reduced Tafel slope of 83 mV dec^{-1} relative to 176 mV dec^{-1} of the bulk Sn. The authors concluded that the high intrinsic catalytic activity conferred on the Sn quantum sheets confined in graphene is a combination of (i) high electrochemical surface area favoring CO_2 adsorption, (ii) rapid rate-limiting electron transfer from CO_2 to $CO_2{\cdot}^-$ benefiting from conductive graphene, and (iii) low Sn–Sn coordination numbers that enabled stabilization of the $CO_2{\cdot}^-$.

In recent years, metal–organic frameworks (MOFs) are an interesting category of materials that are finding significant applications in the electrocatalytic reduction of CO_2 because of their unique structural features, large surface area, and porosity [50]. Owing to their tunable porous structure and properties, strategies that involved fabrication of MOF-derived nanomaterials including metal nanoparticles or single-atom encapsulated *in situ* formed carbon hybrid, and metal-free nanocarbon are being explored [51]. These strategies have been adopted to improve the conductivity of the pristine MOF, increase the number of exposed surface active sites, and enhance the stability. The strong interactions between MOFs and substrates contribute to increasing the number of exposed surface-active sites and facilitate the charge transfer from the electrode to catalysts. As a result, the composites exhibit high activity and current density for electrocatalytic reduction of CO_2. Regarding MOF-derived nanomaterials for CO_2 ECR, Zhao et al. [52] reported a strategy to fabricate oxide-derived Cu/carbon (OD Cu/C) electrocatalysts by facile carbonization of a Cu-based MOF (HKUST-1) at different temperatures. The obtained OD Cu/C materials were hierarchically porous with confined Cu nanoparticles and Cu_2O. The CO_2 electroreduction activities of the OD Cu/C-x (where x is the carbonization temperature at 900, 1000, and 1100 °C) showed that CO_2 can be selectively reduced to alcohol compounds with a total FE of 45.2–71.2% at −0.1 to −0.7 V versus RHE. In addition, all the OD Cu/C catalysts exhibited the same onset potential for methanol and ethanol formation of −0.1 V versus RHE, which corresponded to 120 and 190 mV of overpotential for generation of methanol and ethanol from CO_2, respectively. The high alcohol production performance of the OD Cu/C materials was ascribed to the synergistic effect between the highly dispersed Cu and the matrix of porous carbon. The embedded Cu nanoparticles in the carbon matrix served as the active sites to catalytically activate and dissociate CO_2 to CO and then allowed the subsequent protonation of the adsorbed CO to generate alcohol products. In tandem, the porous carbon structure maintained high pressure of CO_2 on the electrode solution interface that facilitated C–C coupling, which is beneficial for the formation of ethanol. Additionally, the porous carbon structure minimized the diffusion resistance for mass transfer, thereby favoring the mass transport of CO_2 to the active sites and the fast emission of the alcohol products. More importantly, the carbon matrix protected the incorporated Cu from deactivation, thus prolonging the catalyst life to maintain a stable and similar production rate of alcohol for up to five successive CO_2 electroreduction tests. Notably, the electroreduction behavior of the OD Cu/C materials was dependent on factors such as the pore structure, specific surface area, particle diameters, and metal content, which can be easily controlled by the carbonization temperature [53]. The highest production rates of 5.1–12.4 and 3.7–13.4 mg l^{-1} h^{-1} for methanol and ethanol, respectively, were obtained on OD Cu/C-1000 with an FE of 13.8–43.2% for methanol and FE of 24–34.8% for ethanol at potential between −0.1 and −0.7 V. In addition, FE of 24% for ethanol production was obtained at −0.1 V versus RHE, corresponding to 190 mV of overpotential for these products. It was concluded that the superior electroreduction activity of OD Cu/C-1000 relative to others was due to the high Cu_2O content, correct Cu particle size, and well-dispersed 3D porous structure that enabled the charge transfer properties of OD Cu/C-1000 toward a lower charge transfer and mass transfer resistance during CO_2 reduction. In another report, HKUST-1-derived electrocatalyst can also be applied for selective reduction of CO_2 to hydrocarbons [54]. This was achieved by structural distortion of the paddle wheel Cu dimers of HKUST-1 to undercoordinated site-promoted Cu clusters by thermal activation. The formation of Cu clusters afforded control over the Cu electronic configuration, local atomic structure, and bonding strain toward an asymmetric Cu dimer motif with a low Cu–Cu coordination number. The Cu dimer distortion strategy appreciably enabled an increased FE for C_2H_4 from 10% to 45% FE.

Benefitting from the good electron transfer and mass transport of species within the porous structure of MOFs, strategies to fabricate MOF-derived SACs are being explored in order to achieve optimum exposure of active sites for the electrocatalytic reduction of CO_2 by. As an example, a ZIF-assisted strategy was adopted to rationally design a distributed Ni single atom in nitrogen-doped porous carbon (Ni SAs/N-C) [55]. This strategy involved the initial partial ionic exchange between Zn nodes and adsorbed Ni ions within the cavities of MOF, ZIF-8, followed by pyrolysis. The Ni SAs/N-C exhibited an excellent electrocatalytic performance with a current density of 10.48 mA cm^{-2} at an overpotential of 0.89 V and the FE of over 71.9% for CO production. The Ni SAC possessed large number of exposed reactive active sites, as well as a lower interfacial charge transfer resistance, which enabled faster electron transfer from the electrodes to CO_2 and easier formation of the $CO_2{\cdot}^-$ radical anion intermediate. Hence, the interplay of increased number of surface active sites, lower adsorption energy of CO over single Ni sites, and better electronic conductivity of Ni SAs/N-C accounted for its remarkable performance for CO_2 electroreduction. Wang et al. [56] prepared a series of atomically dispersed Co with different nitrogen coordination numbers using bimetallic Co/Zn zeolitic imidazolate frameworks (ZIFs) as the precursor. They found out that the surrounding N numbers of central Co can be modulated by controlling the volatile C–N fragments, which is dependent on the pyrolysis temperature. As a result, three atomically dispersed Co catalysts with different Co–N numbers were selectively prepared at 800, 900, and 1000 °C and designated as Co–N_4, Co–N_3, and Co–N_2, respectively. The electrochemical measurements of these Co–N catalysts clearly demonstrated that remarkable activity and selectivity toward CO_2 reduction was obtained at a lower Co–N coordination number of 2. The decreased coordinating N resulted in more unoccupied 3d orbitals of Co atoms, which benefitted the CO_2 reduction rate in terms of reduced charge transfer resistances, low activation energy to form $CO_2{\cdot}^-$ intermediates, and the strong adsorption capacity of $CO_2{\cdot}^-$ on active sites. As a result, the single-atom cobalt with two coordinate nitrogen atoms (Co–N_2) exhibited higher activity and selectivity than a nearly inert catalyst mainly comprising Co–N_4 sites, with current densities of about 18.1 mA cm^{-2} and a CO FE of 94% at a low overpotential of 520 mV. Moreover, the CO formation TOF catalyzed by Co–N_2 sites reached a record value of 18 200 h^{-1}.

In summary, the recent strategies of material design being employed have shown great prospect of fabricating catalysts with high activity and selectivity toward electrochemical reduction of CO_2. However, great challenges still remain to develop electrocatalysts with both high efficiency and good selectivity toward liquid fuels or other valuable chemicals at low overpotential. Furthermore, particularly challenging to achieve in some cases is catalyst selectivity to produce desired products. Thus, further efforts to design and optimize new catalyst structures with respect to improving catalytic activity, product selectivity, and catalytic stability are of great importance. A combination of theoretical calculations and in situ experimental technologies is an effective approach to exploring the intermediate species formed during electrocatalysis. Better fundamental understanding of the mechanism of catalyst degradation through both experiments and theoretical modeling is also necessary to mitigate the challenge of catalyst instability. Besides improving the catalyst material's stability, improving the electrode/reactor design to optimize the operating conditions is also an important performance indicator as it contributes to mass transport issues, i.e. controlling the rate of CO_2 transfer to the electrode surface. By altering the structure and composition of the electrode material, reaction selectivity can be tuned to the desired product.

References

1 (a) Appel, A.M., Bercaw, J.E., Bocarsly, A.B. et al. (2013). *Chem. Rev.* 113: 6621–6658. (b) Foit, S.R., Vinke, I.C., de Haart, L.G.J., and Eichel, R.-A. (2017). *Angew. Chem. Int. Ed.* 56: 5402–5411. (c) Duan, X., Xu, J., Wei, Z. et al. (2017). *Adv. Mater.* 29: 1701784. (d) Seh, Z.W., Kibsgaard, J., Dickens, C.F. et al. (2017). *Science* 355: eaad4998.

2 (a) Anwar, M.N., Fayyaz, A., Sohail, N.F. et al. (2018). *J. Environ. Manage.* 226: 131–144. (b) Leung, D.Y.C., Caramanna, G., and Maroto-Valer, M.M. (2014). *Renewable Sustainable Energy Rev.* 39: 426–443.

3 Kumar, M., Sundaram, S., Gnansounou, E. et al. (2018). *Bioresour. Technol.* 247: 1059–1068.

4 (a) Chalmers, H. and Gibbins, J. (2007). *Fuel* 86: 2109–2123. (b) Odenberger, M. and Johnsson, F. (2010). *Int. J. Greenhouse Gas Control* 4: 327–340. (c) Wang, J., Huang, L., Yang, R. et al. (2014). *Energy Environ. Sci.* 7: 3478–3518.

5 Ozin, G.A. (2015). *Adv. Mater.* 27: 1957–1963.

6 Fang, Y. and Wang, X. (2018). *Chem. Commun.* 54: 5674–5687.

7 Lewis, N.S. and Nocera, D.G. (2006). *Proc. Natl. Acad. Sci. U.S.A.* 103: 15729–15735.

8 Kho, E.T., Tan, T.H., Lovell, E. et al. (2017). *Green Energy Environ.* 2: 204–217.

9 Meng, X., Wang, T., Liu, L. et al. (2014). *Angew. Chem. Int. Ed.* 53: 11478–11482.

10 Ghoussoub, M., Xia, M., Duchesne, P.N. et al. (2019). *Energy Environ. Sci.* 12: 1122–1142.

11 Inoue, T., Fujishima, A., Konishi, S., and Honda, K. (1979). *Nature* 277: 637–638.

12 (a) Chueh, W.C., Falter, C., Abbott, M. et al. (2010). *Science* 330: 1797–1801. (b) Yui, T., Kan, A., Saitoh, C. et al. (2011). *ACS Appl. Mater. Interfaces* 3: 2594–2600. (c) Dimitrijevic, N.M., Vijayan, B.K., Poluektov, O.G. et al. (2011). *J. Am. Chem. Soc.* 133: 3964–3971. (d) Liu, Y., Huang, B., Dai, Y. et al. (2009). *Catal. Commun.* 11: 210–213.

13 Low, J., Zhang, L., Zhu, B. et al. (2018). *ACS Sustainable Chem. Eng.* 6: 15653–15661.

14 (a) Ran, J., Zhang, J., Yu, J. et al. (2014). *Chem. Soc. Rev.* 43: 7787–7812. (b) Li, X., Yu, J., Jaroniec, M., and Chen, X. (2019). *Chem. Rev.* 119: 3962–4179.

15 Zhang, Z., Wang, Z., Cao, S.-W., and Xue, C. (2013). *J. Phys. Chem. C* 117: 25939–25947.

16 Liu, H., Song, H., Meng, X. et al. (2019). *Catal. Today* 335: 187–192.

17 Wang, L., Wang, Y., Cheng, Y. et al. (2016). *J. Mater. Chem. A* 4: 5314–5322.

18 Ha, M.N., Lu, G., Liu, Z. et al. (2016). *J. Mater. Chem. A* 4: 13155–13165.

19 Ren, J., Ouyang, S., Xu, H. et al. (2017). *Adv. Energy Mater.* 7: 1601657.

20 Chen, G., Gao, R., Zhao, Y. et al. (2018). *Adv. Mater.* 30: 1704663.

21 (a) Sarina, S., Zhu, H.-Y., Xiao, Q. et al. (2014). *Angew. Chem. Int. Ed.* 53: 2935–2940. (b) Sarina, S., Zhu, H., Jaatinen, E. et al. (2013). *J. Am. Chem. Soc.* 135: 5793–5801.

22 Jia, J., O'Brien, P.G., He, L. et al. (2016). *Adv. Sci.* 3: 1600189.

23 Jia, J., Wang, H., Lu, Z. et al. (2017). *Adv. Sci.* 4: 1700252.

24 (a) Nørskov, J.K., Abild-Pedersen, F., Studt, F., and Bligaard, T. (2011). *Proc. Natl. Acad. Sci. U.S.A.* 108: 937–943. (b) Nilsson, A., Pettersson, L.G.M., Hammer, B. et al. (2005). *Catal. Lett.* 100: 111–114. (c) Zhang, W., Wang, L., Liu, H. et al. (2017). *Nano Lett.* 17: 788–793. (d) Sharif, M.J., Kitano, M., Inoue, Y. et al. (2015). *J. Phys. Chem. C* 119: 11725–11731. (e) Lin, X., Yang, B., Benia, H.-M. et al. (2010). *J. Am. Chem. Soc.* 132: 7745–7749.

25 (a) Zheng, Y., Zhang, W., Li, Y. et al. (2017). *Nano Energy* 40: 512–539. (b) Zheng, Y., Wang, J., Yu, B. et al. (2017). *Chem. Soc. Rev.* 46: 1427–1463.

26 (a) Hori, Y. (2008). Electrochemical CO_2 reduction on metal electrodes. In: *Modern Aspects of Electrochemistry* (eds. C.G. Vayenas, R.E. White and M.E. Gamboa-Aldeco), 89–189. New York, NY: Springer New York. (b) Costentin, C., Drouet, S., Robert, M., and Savéant, J.-M. (2012). *Science* 338: 90–94.

27 (a) Chen, C.S., Wan, J.H., and Yeo, B.S. (2015). *J. Phys. Chem. C* 119: 26875–26882. (b) Ren, D., Fong, J., and Yeo, B.S. (2018). *Nat. Commun.* 9: 925. (c) Huang, Y., Handoko, A.D., Hirunsit, P., and Yeo, B.S. (2017). *ACS Catal.* 7: 1749–1756. (d) Ma, S., Sadakiyo, M., Luo, R. et al. (2016). *J. Power Sources* 301: 219–228. (e) Wang, Y., Chen, Z., Han, P. et al. (2018). *ACS Catalysis* 8: 7113–7119. (f) Min, S., Yang, X., Lu, A.-Y. et al. *Nano Energy* 2016 (27): 121–129.

28 (a) Zhang, L., Zhao, Z.-J., and Gong, J. (2017). *Angew. Chem. Int. Ed.* 56: 11326–11353. (b) Zhu, D.D., Liu, J.L., and Qiao, S.Z. (2016). *Adv. Mater.* 28: 3423–3452.

29 Sun, Z., Ma, T., Tao, H. et al. (2017). *Chem* 3: 560–587.

30 Halmann, M. (1978). *Nature* 275: 115–116.

31 (a) Benson, E.E., Kubiak, C.P., Sathrum, A.J., and Smieja, J.M. (2009). *Chem. Soc. Rev.* 38: 89–99. (b) Kuhl, K.P., Cave, E.R., Abram, D.N., and Jaramillo, T.F. (2012). *Energy Environ. Sci.* 5: 7050–7059. (c) Schreier, M., Héroguel, F., Steier, L. et al. (2017). *Nat. Energy* 2: 17087. (d) Ren, D., Ang, B.S.-H., and Yeo, B.S. (2016). *ACS Catal.* 6: 8239–8247. (e) Xie, H., Wang, T., Liang, J. et al. (2018). *Nano Today* 21: 41–54. (f) Hao, J. and Shi, W. (2018). *Chin. J. Catal.* 39: 1157–1166. (g) Gao, J., Ren, D., Guo, X. et al. (2019). *Faraday Discuss.* 215: 282–296.

32 Schouten, K.J.P., Kwon, Y., van der Ham, C.J.M. et al. (2011). *Chem. Sci.* 2: 1902–1909.

33 (a) Hori, Y., Takahashi, I., Koga, O., and Hoshi, N. (2002). *J. Phys. Chem. B* 106: 15–17. (b) Loiudice, A., Lobaccaro, P., Kamali, E.A. et al. (2016). *Angew. Chem. Int. Ed.* 55: 5789–5792. (c) Gao, D., Zegkinoglou, I., Divins, N.J. et al. (2017). *ACS Nano* 11: 4825–4831. (d) Kim, D., Kley, C.S., Li, Y., and Yang, P. (2017). *Proc. Natl. Acad. Sci. U.S.A.* 114: 10560–10565. (e) Wang, Z., Yang, G., Zhang, Z. et al. (2016) *ACS Nano* 10: 4559–4564.

34 Reske, R., Mistry, H., Behafarid, F. et al. (2014). *J. Am. Chem. Soc.* 136: 6978–6986.

35 Li, Q., Fu, J., Zhu, W. et al. (2017). *J. Am. Chem. Soc.* 139: 4290–4293.

36 Jia, M., Fan, Q., Liu, S. et al. (2019). *Curr. Opin. Green Sustainable Chem.* 16: 1–6.

37 Kuhl, K.P., Hatsukade, T., Cave, E.R. et al. (2014). *J. Am. Chem. Soc.* 136: 14107–14113.

38 Jiao, Y., Zheng, Y., Chen, P. et al. (2017). *J. Am. Chem. Soc.* 139: 18093–18100.

39 (a) Nie, X., Luo, W., Janik, M.J., and Asthagiri, A. (2014). *J. Catal.* 312: 108–122. (b) Li, Y., Chan, S.H., and Sun, Q. (2015). *Nanoscale* 7: 8663–8683. (c) Xiao, H., Cheng, T., and Goddard, W.A. (2017). *J. Am. Chem. Soc.* 139: 130–136. (d) Cheng, T., Xiao, H., and Goddard, W.A. (2016). *J. Am. Chem. Soc.* 138: 13802–13805. (e) Peterson, A.A., Abild-Pedersen, F., Studt, F. et al. (2010). *Energy Environ. Sci.* 3: 1311–1315. (f) Liu, X., Xiao, J., Peng, H. et al. (2017). *Nat. Commun.* 8: 15438.

40 Yu, F., Wei, P., Yang, Y. et al. (2019). *Nano Mater. Sci.* 1: 60–69.

41 Tan, X., Tahini, H.A., Arandiyan, H., and Smith, S.C. (2019). *Adv. Theor. Simul.* 2: 1800094.

42 (a) Durand, W.J., Peterson, A.A., Studt, F. et al. (2011). *Surf. Sci.* 605: 1354–1359. (b) Kim, S.K., Zhang, Y.-J., Bergstrom, H. et al. (2016). *ACS Catal.* 6: 2003–2013.

43 Yang, H., Han, N., Deng, J. et al. (2018). *Adv. Energy Mater.* 8: 1801536.

44 (a) Kim, S., Dong, W.J., Gim, S. et al. (2017). *Nano Energy* 39: 44–52. (b) Koh, J.H., Won, D.H., Eom, T. et al. (2017). *ACS Catal.* 7: 5071–5077. (c) Wang, H., Jia, J., Song, P. et al. (2017). *Angew. Chem. Int. Ed.* 56: 7847–7852. (d) Zhang, S., Kang, P., Ubnoske, S. et al. (2014). *J. Am. Chem. Soc.* 136: 7845–7848.

45 Li, F., Xue, M., Li, J. et al. (2017). *Angew. Chem. Int. Ed.* 56: 14718–14722.

46 (a) Gao, S., Jiao, X., Sun, Z. et al. (2016). *Angew. Chem. Int. Ed.* 55: 698–702. (b) Liang, L., Lei, F., Gao, S. et al. (2015). *Angew. Chem. Int. Ed.* 54: 13971–13974.

47 (a) Li, C.W., Ciston, J., and Kanan, M.W. (2014). *Nature* 508: 504. (b) Zhang, S., Kang, P., and Meyer, T.J. (2014). *J. Am. Chem. Soc.* 136: 1734–1737.

48 Lei, F., Liu, W., Sun, Y. et al. (2016). *Nat. Commun.* 7: 12697.

49 Chen, S., Duan, J., Ran, J. et al. (2013). *Energy Environ. Sci.* 6: 3693–3699.

50 (a) Kornienko, N., Zhao, Y., Kley, C.S. et al. (2015). *J. Am. Chem. Soc.* 137: 14129–14135. (b) Lei, Z., Xue, Y., Chen, W. et al. (2018). *Adv. Energy Mater.* 8: 1801587.

51 Shao, P., Yi, L., Chen, S. et al. (2020). *J. Energy Chem.* 40: 156–170.

52 Zhao, K., Liu, Y., Quan, X. et al. (2017). *ACS Appl. Mater. Interfaces* 9: 5302–5311.

53 Yang, L., Gao, M., Dai, B. et al. (2016). *Electrochim. Acta* 191: 813–820.

54 Nam, D.-H., Bushuyev, O.S., Li, J. et al. (2018). *J. Am. Chem. Soc.* 140: 11378–11386.

55 Zhao, C., Dai, X., Yao, T. et al. (2017). *J. Am. Chem. Soc.* 139: 8078–8081.

56 Wang, X., Chen, Z., Zhao, X. et al. (2018). *Angew. Chem. Int. Ed.* 57: 1944–1948.

15

Yolk@Shell Materials for CO_2 Conversion: Chemical and Photochemical Applications

Cameron Alexander Hurd Price[1,2], *Laura Pastor-Pérez*[1], *Tomas Ramirez-Reina*[1], *and Jian Liu*[1,2]

[1] *University of Surrey, Department of Chemical and Process Engineering, Guildford, GU27HX, UK*
[2] *University of Surrey, DICP-Surrey Joint Centre for Future Materials, Department of Chemical and Process Engineering, Guildford, GU2 7XH, UK*

15.1 Overview

Encapsulated materials fall into two distinct types: core–shell (CS) and yolk–shell (YS). These two categories hold the same basic structure and nomenclature, i.e. core@shell.

CS is the commonest form of encapsulated materials where the core is in contact with the shell. A YS material on the other hand has a void space between the core and the shell, which is sometimes emphasized in its nomenclature as core@void@shell. Figure 15.1 provides a visual aid to understand these variants of encapsulated materials. The nomenclature usually remains the same, however, and the extended core@void@shell terminology is reserved for more descriptive purposes. The void space causes profound improvements in the properties of the encapsulated material, not least because it provides a homogeneous reaction environment around the core particle(s), and a means to maintain activity through the confinement effect [1]. In both the CS and the YS variants, the core is only accessible through the shell, making porosity a key parameter in controlling the effectiveness of these materials. The unique hierarchical structure offers considerable variation and tailoring toward the intended application. YS and CS particles represent extremely versatile solutions for current problems regarding deactivation, as they offer a high degree of thermal and chemical stability that does not impact on the selectivity or conversion but rather enhances these performance characteristics.

The development of encapsulated materials was sparked with the production of hollow silica spheres from silica alkoxides. This was achieved through a process widely termed "the Stöber process" [2]. Since this initial work, several variants of the CS morphology have been produced for catalytic purposes, most commonly involving inert shell media like SiO_2 [3–8], and more recently, Al_2O_3–SiO_2 [4]. The use of active instead of inert shell media has been a recent trend to enhance the properties of hollow nanospheres. Materials such as Pd–Pt [9] or TiO_2 [10] hollow spheres are

Engineering Solutions for CO_2 Conversion, First Edition.
Edited by Tomas R. Reina, José A. Odriozola, and Harvey Arellano-Garcia.

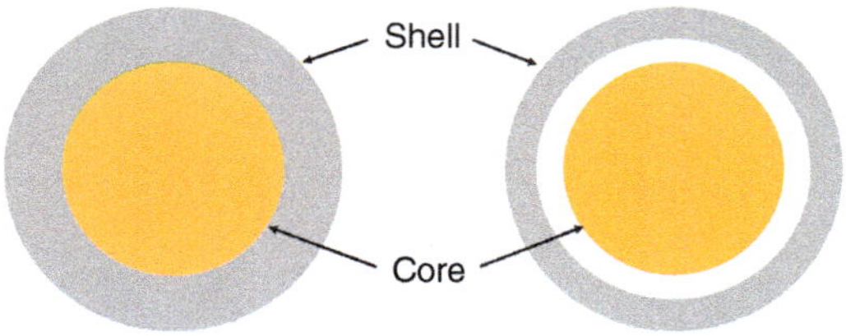

Figure 15.1 Graphic representation of core–shell and yolk–shell particles.

used for their photocatalytic activity or N-doped CoO shells for reverse water gas shift (RWGS) activity [11].

The developing field of the carbon Stöber process, which produces highly stable porous activated carbon spheres with favorable CO_2 adsorption capabilities, falls under the umbrella term of active shell media hollow spheres [12].

Given the stability of the CO_2 molecule and the predominant endothermic nature of the majority of CO_2 conversion reactions, higher temperatures are required in order to achieve high levels of conversion. This requirement also carries significant drawbacks in the form of deactivation mechanisms, most commonly coking and sintering. Encapsulated materials could hold an answer to these problems as they have proven highly resistant to deactivation.

Most current literature regarding the use of CS/YS morphology catalysts for chemical CO_2 reforming through reactions such as the dry reforming of methane (DRM), the RWGS, and methanation references the use of Ni. This offers high CO_2 conversion activity, while being substantially cheaper than comparable noble metal-based options. Owing to the chemical and physical robustness of this structure, these materials offer a potential solution to common modes of deactivation in CO_2 recycling. As can be seen in Figure 15.2, research trends concerning the prevention of sintering and coking within CO_2 recycling applications are still a key focal point of the literature that details a year-on-year increase in the number of publications. Meanwhile,

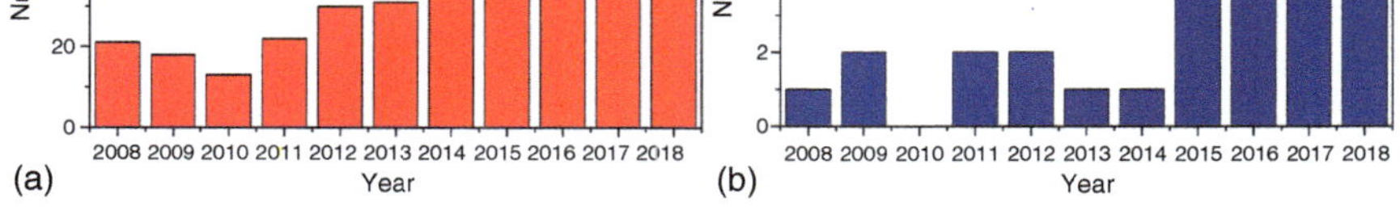

Figure 15.2 (a) Publication metrics for papers published concerning sintering and coking solutions within CO_2 reforming. (b) Publication metrics for the use of yolk–shell catalysts in CO_2 recycling reactions. Both sets of data were obtained from Web of Science for the 2008–2018 period using the search terms (a) "sintering resistance OR coke resistance AND CO_2 reduction," (b) "yolk shell AND CO_2 reduction."

Figure 15.2b shows that the use of YS catalysts, which will be shown in this chapter to be effective methods of deactivation prevention in CO_2 recycling reactions, is a developing field that is ripe for advancement.

Within this chapter, the major benefits of the hierarchical yolk@shell morphology will be discussed. Namely, the forms of spatial confinement effects at play as these are believed to be largely responsible for the impressive catalytic performance these materials report. Following this, a number of the most popular chemical and photochemical applications of these materials will be considered, with a particular emphasis on the synergy between the core and its shell. Finally, the most significant techniques toward synthesizing yolk–shell catalysts will be presented.

15.2 Key Benefits of Hierarchical Morphology

When designing materials toward the recycling of CO_2, it is important to consider the ability of the material to resist the prevalent mechanisms of deactivation within the reaction. Chemically and physically tough materials are required to provide significant improvement over existing catalysts [13]. Where CO_2 reduction reactions or processes are involved, the most prevalent forms of deactivation are sintering and coking [14, 15]. Although both of these phenomena can occur individually, they have been linked; sintering increases the active metal particle size and reduces the active surface area. Research has shown that larger particles (certainly for Ni) increase the amount of coke deposition found on the spent catalyst [16]. It is worth noting that this is not the only mechanism of coke formation that contributes to the overall "deactivation via coking" problem. Coking can also occur through the dissociation of carbon monoxide or decomposition of methane (depending on the reaction/conditions) to form C_α (surface-absorbed carbon), which in turn can react to form C_β (polymeric carbon) [17]. A catalyst that is resistant to both of the major routes of deactivation must be able to eliminate as many potential routes for these undesirable occurrences as possible.

This section will attempt to examine how the hierarchal structure of yolk–shell materials protects the overarching catalytic performance, under conditions that promote both sintering and coking.

15.2.1 Confinement Effects

A number of phenomena are attributed to the confinement effect caused by the YS morphology, ranging from mass transfer driving forces to catalytic efficiency [18]. The confinement effect is wielded as both statement and explanation for the improved performance exhibited by encapsulated materials, although the actual definition is rarely described in the literature.

The importance of catalytically active sites is well known and has been long held to be among the most important principles of catalysis. However, the confined microenvironment has been hitherto overlooked. Research into this has led to the field of confined catalysis [19, 20].

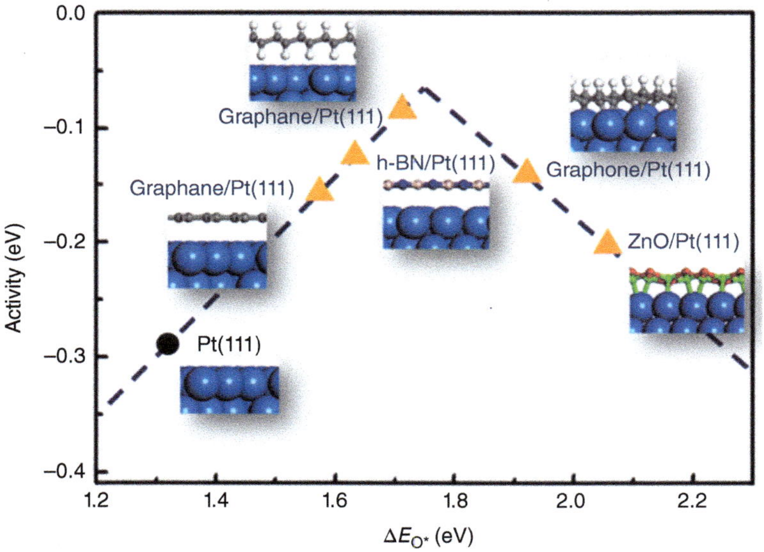

Figure 15.3 Volcano plot displaying the activity against binding energies of the various Pt catalysts. The peak illustrated by the dashed line shows the ideal binding energy and activity for the oxygen reduction reaction. Source: From Li et al. [25].

Confined catalysis is thought to be caused by the quantum confinement and nest effects. Some first-principles investigations have revealed that a spatially confined environment increases molecular orbital energies, such that it changes the activity of the confined molecule(s) [21, 22]. This is seen experimentally in microporous and mesoporous cavities in materials such as zeolites and molecular organic frameworks (MOFs) that demonstrate increased catalytic activities [22–24]. Some recent work proposed a two-dimensional (2D) model of graphene/Pt(111) to best describe these phenomena, owing to the relatively well-defined structure of the overall model as well as ease of modeling [25]. This study demonstrated how different forms of graphene-based "covers" (hexagonal boron nitride (h-BN), graphane, graphene, graphone, and ZnO/Pt) altered the activity of Pt(111) in the oxygen reduction reactions (ORRs), finding that a graphane "cover" offered the best improvement for this reaction over plain Pt(111). It was seen to produce the best activity with favorable surface binding energies (ΔE_O) for O atoms demonstrated on a Volcano Plot as being close to the apex. This can be interpreted as a "Goldilocks Zone" of activity where conditions are "just right" (Figure 15.3). The study found significant weakening of adsorption in atoms in the void between Pt(111) and graphene.

To summarize the work and apply its findings more generally, attractive van der Waals forces between the core and shell materials establish a stable distance (d_{M-S}) (Figure 15.4) that both surfaces attempt to maintain. This manner of reciprocal interaction and stabilization is known as the "nest effect" [26]. Once a molecule adsorbs to the surface of the core, the system responds by attempting to increase the distance between the highest points of the core–molecule surface but is unable to do so. This

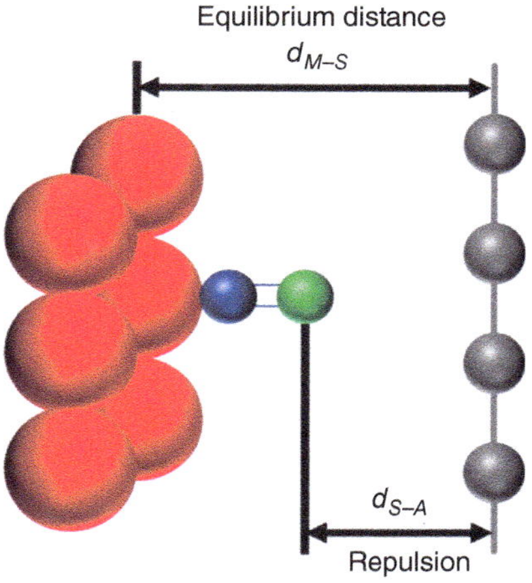

Figure 15.4 Demonstration of the established equilibrium distance (d_{M-S}) between a confined surface (red spheres) and an outer layer or shell (gray spheres). The distance d_{S-A} details how once a molecule adsorbs to the confined surface, the new distance is too close to the outer layer, causing a repulsion effect between the adsorbed molecule and the outer layer.

new distance (d_{S-A}) between the shell material and the adsorbent molecule is much smaller than the equilibrium value, sharply increasing the potential energy of the overall system and thus reducing stability. In a free system, an adsorbed molecule would relax sufficiently away from the surface to achieve its own most stable equilibrium distance. However, a confined system is, by definition, not free. Within the void space between the core and shell, there is competition between the attractive van der Waals forces attempting to maintain d_{M-S}, while the repulsive van der Waals forces attempt to expand d_{S-A} to stabilize the system. Ultimately, this leads to a significant weakening of the adsorption bond, as the system is not stable enough for the adsorbent to remain at the interface. This effect then permeates layers of adsorbed molecules, potentially causing the van der Waals forces present to generate significant pressure (>1 GPa) because of the intense geometric confinement [27, 28]. All these factors contribute to the reduction in desorption energy required for molecules within the void space, provided that there is no strong interaction between the adsorbate and the shell that would otherwise cause the adsorbate to remain within the void. In terms of the quantum confinement effect, the shell material acts as constraining boundary on the molecular orbitals of the adsorbed material, increasing energy within these orbitals, and causing the adsorption to become less stable. As can be expected, the confinement effect response increases as the distance between the core and shell increases as a result of molecule adsorption.

This explains why there is no coke deposition within a YS material and demonstrates that the use of a shell serves to weaken the adsorption between molecules and core within the void space, as well as making plain the efficacy of encapsulation as a method to prevent deactivation.

The porous shell is the key to the success of the morphology, as it produces both spatial and quantum confinement effects. It also acts as a physical barrier between the inner particle(s) and other catalyst particles and to any other material present because of the synthetic procedure. This, then, is the first way in which the morphology prevents sintering of the active phase. If the active material cannot agglomerate, then sintering has been prevented, allowing for the active surface area of the material to remain unchanged [29]. The spatial confinement effect is also the accepted reason that the materials are resistant to coking [29].

15.3 Materials for Chemical CO_2 Recycling Reactions

As with any catalytic process or reaction, the catalytic material must be selected with consideration of the desired outcome. Those catalysts tailored toward the intended purpose, benefit from simplistic synthetic methodologies, and offer compositional variation and greater advantages to the field. As will be seen in Section 15.4, YS particles can be produced using many different methods, all of which offer different levels of control over the physical properties of the final material. However, these methods are more arduous and time consuming than those for supported metal nanoparticles (SMPs), generally the current standard for catalysis. SMPs can be produced cheaply, quickly, and with a high degree of control over dispersion and metallic loading. However, despite their more arduous production, YS catalysts demonstrate themselves to be a significant enhancement in several sub-fields of low cost, long life catalysts. [7, 30–34] While both CS and YS particles benefit from encapsulation, and the potential functionalization of both the core and shell materials, it is the YS variant alone that pushes the possibilities further. The void space not only increases the surface area for catalytic reaction and functionalization of the core and shell media but it also provides a homogeneous reaction environment within the shell. The shell also provides the benefit of both spatial and quantum confinement effect weakening adsorption to the inner core particle(s), although this is discussed in Section 15.2.1.

15.3.1 CO_2 Utilization Reactions

Energy production and consumption is an inescapable necessity of human industry and development. The overwhelming majority of past and current human energy production has come from combustion of fossil fuels or biomass to release heat, which is in turn converted largely to electricity to meet our growing demand [35, 36]. More recently, however, with concerns rising over the release of CO_2 by these processes, there has been a focus upon the requirement for "cleaner" sources of generation [37]. Target technologies must be both ecologically sustainable and economically viable. As either as a bridging technology or as a solution, recycling or sequestering the products of power generation, such as carbon dioxide and methane, represents a substantial contribution to the goal of reducing pollution and undesirable emissions. Of the number of greenhouse gases, it is carbon dioxide that has captured the public imagination. So much so, that the reduction in CO_2 emissions has become the rhetoric for global agreements toward climate change (The Paris Agreement) [38] and scientific publication [13, 39–41]. In response, there has been a huge increase in research toward both capture [42] and recycling reactions [43], although the use of YS materials toward these goals is still a relatively young field, as can be seen in Figure 15.2b.

The most common reactions found in the literature and industry that utilize CO_2 are DRM, bi-reforming of methane (BRM), RWGS, and the Sabatier reaction. These reactions are largely catalyzed by Ni-based materials, owing to a number of reasons: Ni (Nickel) has been long established as a highly active metal toward CO_2 activation and a practical alternative to noble metals [44–47]; it is cheap, abundant, and can be made to produce comparable, if not better, conversion and selectivity [48]. These benefits come at a cost, however. Nickel is prone to coke formation and sintering [48, 49], which drastically affects its long-term efficacy within a CO_2 recycling process [48, 50, 51].

There is also an interest in the sequestration of CO_2, which has seen a number of recent developments including underground storage, compressing it into rock [52], or using the gas as a monomer in biodegradable polymers [53, 54]. Additional carbon capture technologies include recent work that concern YS materials [54]. This work details the use of $CaCO_3$@C YS particles that make use of the reported excellence of $CaCO_3$ toward high-temperature adsorption of CO_2 [55]. This material demonstrated a significant enhancement in the amount of CO_2 adsorbed compared to solely $CaCO_3$. Furthermore, the shell allowed for the capture and storage of CO_2 within the void space, which served to enhance the CO_2 adsorption to higher levels. Additionally, the $CaCO_3$@C YS particles were found to be easily recycled through a facile etching step using NaOH [54].

The use of YS catalysts in CO_2 recycling reactions is an increasingly documented subject, although literature focuses on the application toward the DRM reaction. However, the literature concentrates largely on the use of Ni and SiO_2 with minimal variation from this combination. Several works concerning the use of Ni@SiO_2 yolk shell materials for the DRM reaction have been reported in recent years [7, 30, 31, 56 57], demonstrating a high level of effectiveness for this material combination. Notably, Lu et al. [57] detail a simple synthesis for a multicore yolk–shell material that displayed no sign of deactivation over 50 hours of continuous exposure to the DRM reaction, although their technique produced a wide distribution of particle sizes. In Li et al.'s 2014 work [31], they produced a Ni@Ni@SiO_2 material that involved not only a central Ni core but other Ni cores dispersed throughout the SiO_2 sphere. This material performed extremely effectively within the DRM, displaying high conversion with negligible deactivation and carbon deposition, after over 80 hours of continuous reaction. This serves to demonstrate the longevity provided by the porous shell and its associated confinement effects.

Although both Ni and SiO_2 provide numerous reasons for their success in this field, for example, the thermal and chemical resistivity of SiO_2 and high activity of Ni toward CO_2 reduction, this shows clear room for diversification within the catalytic composition.

Careful selection of promoters or support media greatly improves the life span of a catalyst. For example, the utilization of materials such as CeO_2 provide redox properties [58, 59] to increase the resistance to carbon formation, while materials such as SiO_2 or Al_2O_3 provide thermal resistivity and high dispersion to prevent sintering. Although these inclusions are relatively simple for SMP catalysts, their inclusion in the YS morphology can be more complex, as the main active sites are contained within the shell. Core size is very much a constraining factor. There are many reports on YS materials ranging in size from 200 to 800 nm [7, 33, 60, 61], while SMP materials often feature sizes in the range of microns. Therefore, the incorporation of synergistic materials can be achieved through the adaptation of the shell material. There are examples of metal@CeO_2 YS particles [62] that are particularly resistant to coking within the DRM, while maintaining high surface area. It can be inferred, therefore, that the support design considerations that SMP materials enjoy are analogous toward the shell material of encapsulated catalysts.

15.3.2 Photochemical Reactions with CO_2

Hollow shell and multi-shell materials represent a significant advancement in photocatalytic applications owing to increased light absorption afforded to the morphology through massively increased surface area. This section will briefly cover the basic principles of photocatalysis before expanding on the uses, variations, and how the physical characteristics of YS materials are applied in this field.

15.3.2.1 Principles of Photocatalysis

Following the discovery of the photoelectric effect, some of the first work to be undertaken was toward "artificial photosynthesis," building upon the principles of naturally occurring photosynthesis to produce more complex chemicals from CO_2 and light. Some of the earliest reports of these phenomena utilize semiconductor materials to achieve their goals, utilizing light to supplement or replace electrical input [63–65]. The basic mechanism of photocatalysis can be applied to most photocatalytic reactions and can be described in three steps [66–68]: (i) light adsorption, (ii) charge separation and transfer, and (iii) surface reaction. Semiconductor materials are used for photocatalysis as, according to band theory, they possess a quantum mechanically forbidden zone between the low energy orbital (valence band) and the high energy orbital (conduction band), known as the bandgap (E_g). Metals, however, do not possess this gap and instead have an overlap between these two levels, allowing for electrical conductance with minimal loss or activation energy [69, 70]. For electronic movement between the two bands in a semiconductor material, E_g must be overcome through the excitation of an electron in the valence band by a photon of energy $h\nu \geq E_g$. When an electron is excited between the bands in this manner, it leaves behind a positively charged "hole" and both the hole and the excited electron are key to subsequent reactions. Following photoexcitation, the produced charge carriers diffuse toward the surface of the particles; during this process, the charge carriers tend to recombine either in the bulk or on the surface. It is therefore of great importance to the materials effectiveness that the electron and hole are prevented from recombination for as long as possible so that they can be used in reaction.

Therefore, to be an effective photocatalyst, a material must possess the ability to undergo five key physiochemical processes: (i) light absorption, (ii) charge separation, (iii) charge migration, (iv) charge recombination, and (v) surface redox reactions [71–74].

To improve the performance of intrinsic (charge carriers are equal) semiconductor materials, dopants are introduced within their crystal lattice to produce more holes (positive charges, or p-type materials) or electrons (negative charges, or n-type materials)

While both types of semiconductors rely on charge carriers within the crystal lattice, a p-type semiconductor has been doped with electron acceptor atoms to produce more "free" positive charges within the lattice. N-type semiconductors have been doped with electron-donating atoms to increase the number of "free" electrons and these negative charges within the lattice. Both types prolong the separation of the electron–hole pairing and can alter numerous other properties of the semiconductor. These include extending the light absorption range without changing the bandgap [75, 76], reducing or enlarging the bandgap energy, or reducing the bandgap while not introducing intermediate states [77].

15.3.2.2 Prominent Materials

The most prominent material in this field is TiO_2, owing to its high stability under photocatalytic conditions, abundancy, and reaction versatility [78]. Further to this, the bandgaps displayed by the two crystal structures of TiO_2 (anatase and rutile) also overlap with that of the visible spectrum (c. 1.75–3 eV). Anatase is selected preferentially to rutile, despite a slightly increased bandgap energy of 3.2 eV compared to 3.0 eV found in rutile. The justification for this preference is the more desirable ability to prevent electron–hole recombination [79, 80]. However, it has been shown that a 70 : 30 mix of anatase:rutile, respectively, has superior photocatalytic properties to pure structures composed of either component [81–83].

Further to TiO_2, there are a number of other effective heterogeneous materials for the photochemical reduction of CO_2 (Figure 15.5). These include Cu_2O [84–86], WO_3 [87, 88], and even Cu-doped graphene [89]. However, TiO_2 retains preference in the field, albeit with promotion/doping by other elements, as it is highly tunable toward a variety of photocatalytic reactions with high conversions and yields.

15.3.2.3 Benefits of YS in Photocatalysis

As with all catalysts, the structure has a profound effect on the performance of the photocatalyst. For example, one-dimensional (1D) nanowire structures have been found to contribute to electron transport in the axial direction, transporting holes in the radial direction and substantially improving the charge carrier separation [90]. 2D structures offer increased surface area, accommodating more active sites and loading positions than traditional structures. Three-dimensional (3D) structures, such as YS or hollow shell (HS) materials, offer even more benefits for photocatalysis. As discussed earlier, this morphology offers significant variability with the associated advantages of large surface areas, tunable shell thickness, and variable void space.

The efficiency of a semiconductor photocatalyst is determined by its ability to harvest light. When light hits a surface, it is absorbed according to Beer's law (Eq. (15.1)). A flat or solid material will strongly reflect light. However, when light hits a thin hollow structure, it will penetrate and refract according to the optical

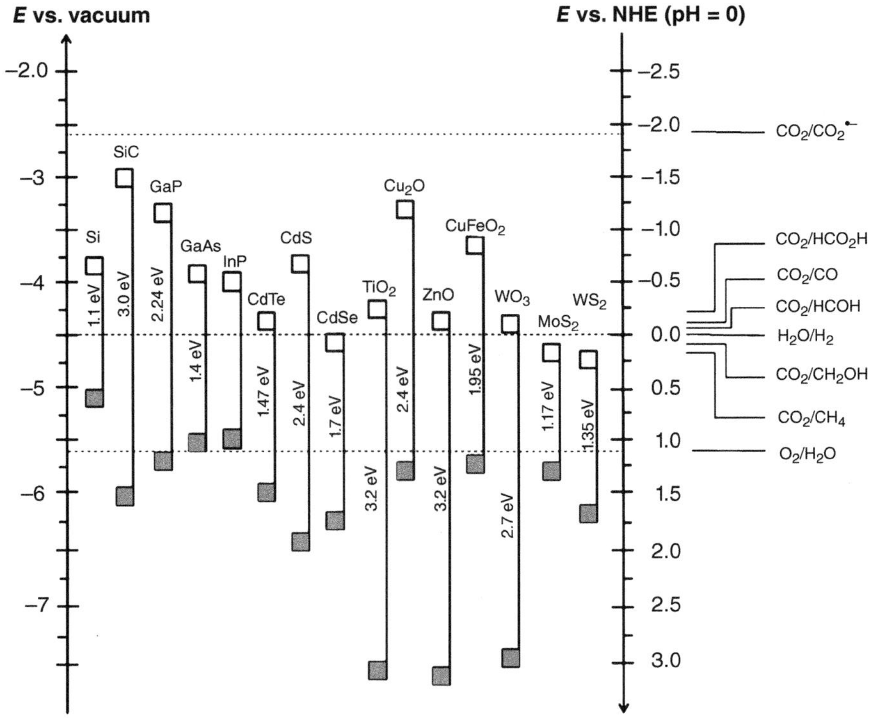

Figure 15.5 Diagram of the conduction band (white squares) and valence band (gray squares) potentials of several popular semiconductor materials, with the associated potentials for CO_2 and water redox couples at pH 0. Left axis versus vacuum, while the right axis is against a normal hydrogen electrode (NHE). Source: From White et al. [78]. Copyright (2015) American Chemical Society.

properties of the material. This latter example is demonstrated in Figure 15.6, which also demonstrates the importance of the void space. In a YS or hollow sphere, the light is refracted multiple times within the shell, once past the surface layer. These subsequent refractions of the original incident photons increase the amount of energy transferred to the material, thus increasing the photocatalytic output of the catalyst [91].

$$A = -\log \frac{I}{I_0} = \varepsilon LC \tag{15.1}$$

where A is adsorption, I and I_0 are the intensity of penetrated and source light, ε is molar absorptivity, L is the sample thickness, and C is the concentration, which for a solid material is a constant.

Although traditional structured photocatalysts can improve on their absorption with plasmonic dopants or through careful combination of materials, the characteristic void space enhances this property through the light-scattering effect within HS and YS materials. Combining the HS/YS morphology with plasmonic materials

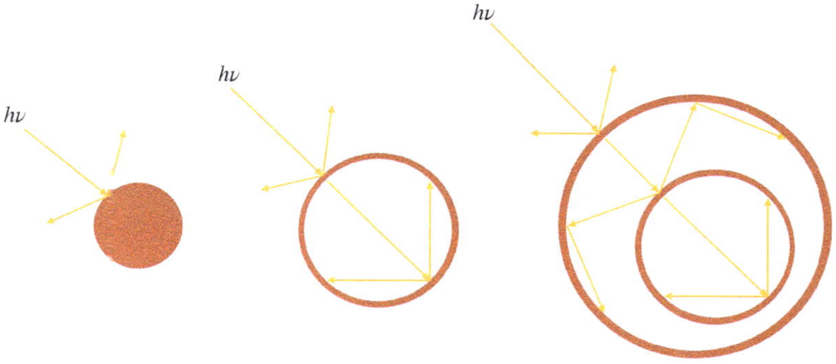

Figure 15.6 Diagram of how light interacts and refracts with solid, hollow, and layered hollow media.

not only increases visible light absorption but also strengthens the light-scattering effect [91].

A further benefit of employing a HS is that this structure can be applied to the production of photonic crystals [92]. These crystals, when combined with the scattering effect, produce a further phenomenon known as the slow-photon effect. This radically improves the light absorption of the material when the bandgap of the semiconductor overlaps with its absorption range. The effect has only been found in HS structures of TiO_2 [93], ZnO [94], WO_3 [95], and TaON [96], where it is notable that both the size of the void space and the angle of incident light are crucial to the successful generation of the effect [94].

When attempting to circumvent the problem of rapid charge recombination within a material, the most effective methodologies rely on reducing the migration distance, extending the carrier diffusion length or by establishing an electric field. The hierarchal morphologies, especially HS, solve this issue in two ways. Firstly, the thin shells offer a reduced migration distance, improving the chances of the charge carriers being used within the reaction rather than recombine. Secondly, directed charge separation occurs within the hollow structure heterojunctions, on both the inner and outer surfaces [91], as a result of the quantum confinement effect.

There have been several reports of both TiO_2 YS [97, 98] and CS [99, 100] materials used for CO_2 reduction with a variety of core materials. Commonly, both the CS and YS materials used for photocatalytic reactions involve the use of noble metals in the core or shell, owing to the favorable reaction performance offered by these metals and their advantageous local surface plasmonic resonance (LSPR). This enhances the photocatalytic properties of the catalyst through the generation of a strong electromagnetic field that serves to prolong electron–hole pair recombination times, extending the absorbable wavelengths of light into the visible spectrum [97], and further developing the light-scattering effect within the particle. There are few current examples of photocatalytic materials that use the hierarchal morphology which do not include a noble metal (e.g. Au and Pt) [98] to enhance the material with LSPR.

A very recent work details the use of Al@MIL-53 yolk–shell particles within the RWGS reaction, exploiting the LSPR properties of Al nanocrystals to enhance the photocatalytic efficiencies of the YS material within both the RWGS and hydrogen-deuterium exchange reactions. The YS catalyst detailed significantly increased CO_2 uptake with an associated increase in the amount of CO produced, when compared to non-encapsulated Al nanocrystals [101].

Much current literature on the topic is centered on pure or modified TiO_2 as both the shell and core, or even mono layer/multilayer hollow spheres [102]. These iterations benefit from increased surface area and thus increased internal refraction and adsorption of light.

The most recent work in this field was published in 2019 and concerns the use of a TiO_{2-x} YS material modified with a CoAl-layered double hydroxide (LDH) for the conversion of CO_2 [103]. This material was further designed to include Ti^{3+} species that further enhance the visible light activity of the YS material, while simultaneously increasing the n-type nature of the semiconductor. LDH materials have been shown to possess increased activity within the visible spectrum and increased quantum yield over TiO_2 alone, although they suffer from limited photoconversion efficiency and charge separation performance [104, 105].

This work detailed extremely high production rates of methanol (251 μmol $g_{cat}{}^{-1}$ h) from the reduction of CO_2 over the TiO_{2-x}/LDH YS, comparable to an H-P25-supported CoAl LDH material that was also tested, substantially surpassing the production rates reported for this latter material. In fact, the results for the titular YS material far exceeded that of any of the other materials tested, clearly demonstrating the beneficial synergy between the LDH and TiO_2 in addition to the Ti^{3+} defects toward the photocatalytic reduction of CO_2, further demonstrating the efficacy of a hierarchal morphology.

15.4 Synthesis Techniques for CS/YS: A Brief Overview

The routes available to produce CS/YS nanoparticles are numerous and range in complexity, particularly for the YS variant. Some recent reviews [1, 33] cover a significant number of synthetic routes for YS materials in great detail. These range from developing methods (e.g. selective etching) to more popular methods such as soft and hard templating. These latter techniques offer high degrees of control of end product characteristics (from porosity to particle size) [7, 30, 106, 107]. This section will introduce and discuss a small selection of the most common techniques in the field to provide background into method selection and form a comparative basis to other methods.

The single core@shell structure of the YS variant of the encapsulated material is the simplest in this hierarchal morphology. Several more advanced variations exist in literature, such as multiple core contained within a single shell, multiple shells, cores attached to the shell, and a single core within (Figure 15.7) [1, 23, 31]. These advanced variations can be more time consuming to produce owing to an extended synthetic procedure, although they continue to offer increasing benefits over the simplest YS morphology and allow for even more diverse potential variation with little added complexity.

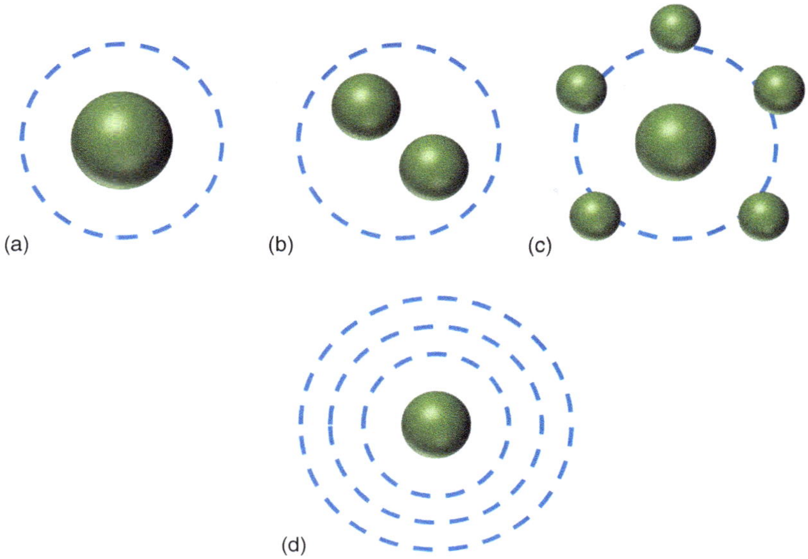

Figure 15.7 Schematic illustration of several YS variations: (a) standard single core–yolk shell, (b) multi-core–yolk–shell, (c) single core–yolk–shell with cores embedded in the shell, and (d) single core–yolk–shell with multiple shells. Source: Based on Purbia and Paria [1].

15.4.1 Soft Templating Techniques

One method used to produce mesoporosity and high surface area is the use of a soft template, consisting of surfactants or polymers, i.e. cetyl-trimethyl-ammonium bromide (CTAB), to produce mesoporous shells (IUPAC defines mesoporosity as being between 2 and 50 nm). This method has been demonstrated to produce extremely high surface areas of 700 $m^2\,g^{-1}$ [7] and even as high as 1309 $m^2\,g^{-1}$ [108]. Figure 15.8 illustrates the general procedure of this method. This technique is commonly used to produce silica shells, using surfactants, cosurfactants, or polymers. The hydrophobic chains and a hydrophilic head act as the soft template, spontaneously creating micelles either around the core material (Figure 15.8a) or droplets of solvent (Figure 15.8b). This forms either yolk–shell capsules or hollow spheres, respectively. When the silica precursor is added, hydrolysis induces polymerization at the interface between the micelle and water because of the Coulombic interactions of the anionic shell precursor and the cationic N ions (hydrophilic heads) on the micelle surface [108]; the surfactant forcibly creates both a void space and mesoporous channels within the shell media during its formation. Subsequent hydrolysis and condensation of the precursor mixture produce the shell, with the final thermal treatment or washing [109] removing the soft template material from the channels of the shell to produce mesopores.

This modified Stöber technique offers a simple and effective method of encapsulation [110, 111] that provides the further benefit of producing mesoporous shell media. The procedure is sensitive to pH, hydrothermal conditions, solvent ratio, and

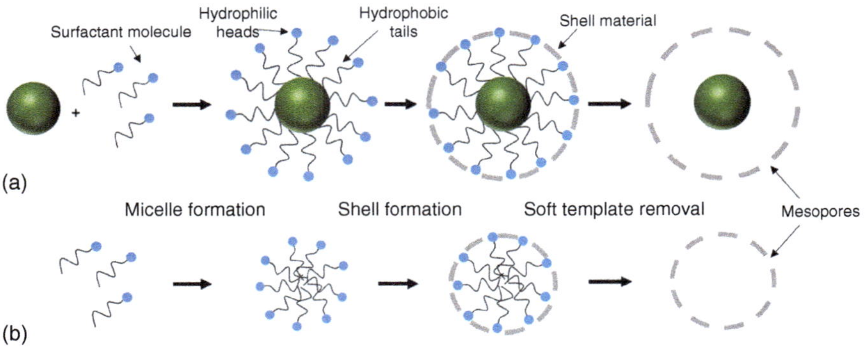

Figure 15.8 Schematic illustration of how soft templating can achieve either mesoporous (a) yolk–shell materials or (b) hollow–shell materials.

core material particle size. Commonly, cationic surfactants are used for this synthetic method, as they also act as a structural directing agent within the system. A diverse study into the effect of varying surfactant combinations and usage to produce silica spheres was undertaken in 2010, largely concerning the use of CTAB with other surfactants and polymers, i.e. polyvinyl alcohol (PVA) and polyvinylpyrrolidone (PVP) [108]. This work found that there is significant variation of the final product depending on preparation materials and conditions used. For example, while the addition of CTAB produced mesoporous silica spheres of double the size compared to the control experiment (no surfactant or polymer), using both CTAB and 5 wt% PVA produced silica spheres with the highest surface area (1309 m^2 g^{-1}), although this incurred the largest produced average particle size. The study also detailed the effect of other conditions, i.e. increasing the alcohol/silica precursor ratio reduces the size of the spheres (1.5 to 0.01 μm), while varying the silica precursor/surfactant ratio controls can increase the size of the spheres (65–740 nm). A later 2012 study made use of a water-in-oil emulsion and a nonionic surfactant as the soft template. The authors detailed how various different shapes of YS can be obtained through the control of the aging process, temperature, and surfactant chain length, providing further evidence of morphological control unique to the soft templating method [112].

15.4.2 Hard Templating Techniques

Hard templating is usually analogous to soft templating, although the polymers/surfactants are replaced with more rigid materials that serve as a sacrificial layer between the core material and the shell. This layer is removed chemically (e.g. etching) or thermally (e.g. calcination). This technique is effective for producing metal oxide shells [111] and is used commonly to produce TiO_2 HS or YS materials. Materials like SiO_2 make highly effective hard templates as SiO_2 can easily be removed with solvents. One such work details the synthesis of Au@TiO_2 YS particles through the use of a carbon hard template upon which the TiO_2 was deposited, prior to its removal [97]. Another work details the synthesis of several variations of *m*-ZrO_2-TiO_2 HS and YS particles through the use of a SiO_2 hard template, which was removed solvothermally with 0.5 M ammonia [113]. These works serve to support the flexibility that a hard template offers when considering the synthesis of certain YS materials.

Another sacrificial layer-based methodology known as galvanic replacement also allows for the efficient production toward highly tunable morphology, especially in terms of material porosity. The technique involves the exchange of base metals from a metal salt with more noble metal ions from a solution; the more basic metal acts as both the reducing agent and the sacrificial layer. This exchange of base metal and noble metal ions then achieves electrochemical equilibrium, forming small void spaces that get larger toward the center of the particle. Although a viable method, because of the reliance on noble metals, the technique is not widely used.

These examples of synthetic techniques serve to demonstrate the numerous avenues that produce this morphology, while allowing for complete control of the shell thickness, core size and shape, void space, and porosity [1].

15.4.2.1 Metal Oxide/Carbide Shells

Hard templating methodologies are being used to produce novel shell media for this morphology toward more advanced applications. As discussed previously, a popular example of an active metal oxide shell is TiO_2 or Titania, which features heavily in photocatalytic applications owing to its low bandgap energy and ability to absorb in the visible spectrum [10].

Other novel shells have also begun to appear in the literature. With the advancement of the Stöber process to produce carbon spheres [114], work is beginning to appear with further advancements that can produce hollow metallic carbide [52, 115, 116] or hetero-atom-doped carbon spheres [117–119], all detailing promising performance in CO_2 recycling. These demonstrate catalytic activity in other applications, ranging from photocatalytic applications [52] to oxygen reduction [120, 121].

As the shell material can be thought of as a re-engineering of the support phase of a catalyst, its importance is made even greater as the shell can now be designed to participate in the reaction, rather than be merely supportive of the active phase. However, the requirements expected of support phases remain and the choice of material, with respect to deactivation resistances, is vital to the success of the material.

15.5 Future Advancement

This chapter has introduced the concept of a highly effective hierarchal morphology, along with several routes for CO_2 recycling, to which they have been applied. Although this structure has clear advantages over their predecessors, their development and utilization remain very much open avenues for research.

The key to the prolonged success and implementation of this morphology relies on the continued development of the materials and their synthetic pathways to optimize the cost/benefit trade-off that commercially viable catalysts must satisfy.

When considering demanding reactions, such as those concerned with chemical CO_2 reduction, and the application of hierarchal morphologies toward them, it is crucial to consider the benefit afforded by a synergistic shell over an inert one, or whether perhaps the development of the core media is where the true value lies.

The development of other shell media is another wide field in itself, as etching methodologies become increasingly popular and effective, as they would allow for new shell media that is otherwise impossible to achieve. Only through continued research and understanding can we hope to define the limits of this impressive morphology. As it stands, the future is bright with potential for this family of materials, although their perfection remains to the future.

References

1 Purbia, R. and Paria, S. (2015). Yolk/shell nanoparticles: classifications, synthesis, properties, and applications. *Nanoscale* 7: 19789–19873.

2 Stöber, W. and Fink, A. (1968). Controlled growth of monodisperse silica spheres in the micron size range. *J. Colloid Interface Sci.* 26: 62–69.

3 Liu, J., Qiao, S.Z., Hartono, S.B., and Lu, G.Q. (2010). Monodisperse yolk-shell nanoparticles with a hierarchical porous structure for delivery vehicles and nanoreactors. *Angew. Chem. Int. Ed.* 49: 4981–4985.

4 Chen, C., Hsieh, M.-F., Fang, X. et al. (2012). Hollow mesoporous aluminosilica spheres with perpendicular pore channels as catalytic nanoreactors. *ACS Nano* 6: 4434–4444.

5 Wang, S., Zhang, M., and Zhang, W. (2011). Yolk-shell catalyst of single Au nanoparticle encapsulated within hollow mesoporous silica microspheres. *ACS Catal.* 1: 207–211.

6 Wu, S.-H., Tseng, C.-T., Lin, Y.-S. et al. (2011). Catalytic nano-rattle of Au@hollow silica: towards a poison-resistant nanocatalyst. *J. Mater. Chem.* 21: 789–794.

7 Price, C.-A.H., Pastor-Pérez, L., Ramirez Reina, T., and Liu, J. (2018). Robust mesoporous bimetallic yolk–shell catalysts for chemical CO_2 upgrading via dry reforming of methane. *React. Chem. Eng.* 3: 433–436.

8 Lin, C.H., Liu, X., Wu, S.H. et al. (2011). Corking and uncorking a catalytic yolk-shell nanoreactor: stable gold catalyst in hollow silica nanosphere. *J. Phys. Chem. Lett.* 2: 2984–2988.

9 Wang, L. and Yamauchi, Y. (2013). Metallic nanocages: synthesis of bimetallic Pt–Pd hollow nanoparticles with dendritic shells by selective chemical etching. *J. Am. Chem. Soc.* 135: 16762–16765.

10 Li, G., Liao, C., and Jiang, G. (2018). Hollow TiO_2 spheres with improved visible light photocatalytic activity synergistically enhanced by multi-stimulative: morphology advantage, carbonate-doping and the induced Ti_{3+}. *J. Environ. Sci.* 72: 153–165.

11 Yin, G., Yuan, X., Du, X. et al. (2018). Efficient reduction of CO_2 to CO using cobalt–cobalt oxide core-shell catalysts. *Chem. A Eur. J.* 24: 2157–2163.

12 Gu, W. and Yushin, G. (2013). Review of nanostructured carbon materials for electrochemical capacitor applications: advantages and limitations of activated carbon, carbide-derived carbon, zeolite-templated carbon, carbon aerogels, carbon nanotubes, onion-like carbon, and graphene. *WIREs Energy Envrion.* 3: 424–473.

13 Pakhare, D. and Spivey, J. (2014). A review of dry (CO_2) reforming of methane over noble metal catalysts. *Chem. Soc. Rev.* 43: 7813–7837.

14 Olah, G.A., Goeppert, A., Czaun, M., and Prakash, G.K.S. (2013). Bi-reforming of methane from any source with steam and carbon dioxide exclusively to metgas ($CO–2H_2$) for methanol and hydrocarbon synthesis. *J. Am. Chem. Soc.* 135: 648–650.

15 Du, X., Zhang, D., Gao, R. et al. (2013). Design of modular catalysts derived from NiMgAl-LDH@m-SiO_2 with dual confinement effects for dry reforming of methane. *Chem. Commun.* 49: 6770.

16 Kim, J.H., Suh, D.J., Park, T.J., and Kim, K.L. (2000). Effect of metal particle size on coking during CO_2 reforming of CH_4 over Ni-alumina aerogel catalysts. *Appl. Catal., A* 197: 191–200.

17 Bartholomew, C.H. (2001). Mechanisms of catalyst deactivation. *Appl. Catal., A* 212: 17–60.

18 Zhang, M., Wang, C., Liu, C. et al. (2018). Metal–organic framework derived Co_3O_4/C@SiO_2 yolk–shell nanoreactors with enhanced catalytic performance. *J. Mater. Chem. A* 6: 11226–11235.

19 Fu, Q. Li, W.X., Yao, Y. et al. (2010). Interface-confined ferrous centers for catalytic oxidation. *Science* 328: 1141–1144.

20 Yang, F., Deng, D., Pan, X. et al. (2015). Understanding nano effects in catalysis. *Natl. Sci. Rev.* 2: 183–201.

21 Zicovich-Wilson, C.M., Corma, A., and Viruela, P. (1994). Electronic confinement of molecules in microscopic pores. A new concept which contributes to the explanation of the catalytic activity of zeolites. *J. Phys. Chem.* 98: 10863–10870.

22 Sastre, G. and Corma, A. (2009). The confinement effect in zeolites. *J. Mol. Catal.* 305: 3–7.

23 Pan, X. and Bao, X. (2011). The effects of confinement inside carbon nanotubes on catalysis. *Acc. Chem. Res.* 44: 553–562.

24 Choi, K.M., Na, K., Somorjai, G.A., and Yaghi, O.M. (2015). Chemical environment control and enhanced catalytic performance of platinum nanoparticles embedded in nanocrystalline metal-organic frameworks. *J. Am. Chem. Soc.* 137: 7810–7816.

25 Li, H., Xiao, J., Fu, Q., and Bao, X. (2017). Confined catalysis under two-dimensional materials. *Proc. Natl. Acad. Sci. U.S.A* 114: 5930–5934.

26 Derouane, E.G. (1986). Shape selectivity in catalysis by zeolites: the nest effect. *J. Catal.* 100: 541–544.

27 Algara-Siller, G., Lehtinen, O., Wang, F.C. et al. (2015). Square ice in graphene nanocapillaries. *Nature* 519: 443–445.

28 Vasu, K.S., Prestat, E., Abraham, J. et al. (2016). Van der Waals pressure and its effect on trapped interlayer molecules. *Nat. Commun.* 7: 12168.

29 Zhao, X., Li, H., Zhang, J. et al. (2016). Design and synthesis of NiCe@m-SiO_2 yolk-shell framework catalysts with improved coke- and sintering-resistance in dry reforming of methane. *Int. J. Hydrogen Energy* 41: 2447–2456.

30 Price, C., Earles, E., Pastor-Pérez, L. et al. (2018). Advantages of yolk shell catalysts for the DRM: a comparison of Ni/ZnO@SiO_2 vs. Ni/CeO_2 and Ni/Al_2O_3. *Chemistry (Easton)* 1: 3–16.

31 Li, Z., Mo, L., Kathiraser, Y., and Kawi, S. (2014). Yolk-satellite-shell structured Ni-Yolk@Ni@SiO_2 nanocomposite: superb catalyst toward methane CO_2 reforming reaction. *ACS Catal.* 4: 1526–1536.

32 Li, Z., Li, M., Bian, Z. et al. (2016). Design of highly stable and selective core/yolk-shell nanocatalysts-review. *Appl. Catal., B* 188: 324–341.

33 Wang, M., Boyjoo, Y., Pan, J. et al. (2017). Advanced yolk-shell nanoparticles as nanoreactors for energy conversion. *Chin. J. Catal.* 38: 970–990.

34 Saha, A., Mondal, A., Maiti, S. et al. (2017). A facile method for the synthesis of a C@MoO_2 hollow yolk–shell structure and its electrochemical properties as a faradaic electrode. *Mater. Chem. Front.* 1: 1585–1593.

35 Tanksale, A., Beltramini, J.N., and Lu, G.M. (2010). A review of catalytic hydrogen production processes from biomass. *Renewable Sustainable Energy Rev.* 14: 166–182.

36 Li, X. (2005). Diversification and localization of energy systems for sustainable development and energy security. *Energy Policy* 33: 2237–2243.

37 Fayaz, F., Danh, H.T., Nguyen-Huy, C. et al. (2016). Promotional effect of Ce-dopant on Al_2O_3-supported Co catalysts for syngas production via CO_2 reforming of ethanol. *Proc. Procedia Eng.* 148: 646–653.

38 United Nations Adoption of the Paris Agreement (2015). In *Proceedings of the Framework Convention on Climate Change, Paris*, 30 November 2015.

39 Abdullah, B., Abd Ghani, N.A., and Vo, D.V.N. (2017). Recent advances in dry reforming of methane over Ni-based catalysts. *J. Cleaner Prod.* 162: 170–185.

40 de Falco, M., Iaquaniello, G., and Centi, G. (2013). CO_2: a valuable source of carbon. *Green Energy Technol.* 137: 194.

41 Ganesh, I. (2014). Conversion of carbon dioxide into methanol – a potential liquid fuel: Fundamental challenges and opportunities (a review). *Renewable Sustainable Energy Rev.* 31: 221–257.

42 Bui, M., Adjiman, C.S., Bardow, A. et al. (2018). Carbon capture and storage (CCS): the way forward. *Energy Environ. Sci.* 11: 1062–1176.

43 Billig, E., Decker, M., Benzinger, W. et al. (2019). Non-fossil CO_2 recycling – the technical potential for the present and future utilization for fuels in Germany. *J. CO2 Util.* 30: 130–141.

44 Bian, Z., Das, S., Wai, M.H. et al. (2017). A review on bimetallic nickel-based catalysts for CO_2 reforming of methane. *ChemPhysChem* 18: 3117–3134.

45 Liu, C., Ye, J., Jiang, J., and Pan, Y. (2011). Progresses in the preparation of coke resistant Ni-based catalyst for steam and CO_2 reforming of methane. *ChemCatChem* 3: 529–541.

46 Kaw , S., Kathiraser, Y., Ni, J. et al. (2015). Progress in synthesis of highly active and stable nickel-based catalysts for carbon dioxide reforming of methane. *ChemSusChem* 8: 3556–3575.

47 Rostrup-Nielsen, J.R. and Bak Hansen, J.H. (1993). CO_2-reforming of methane over transition metals. *J. Catal.* 144: 38–49.

48 Seo, H. (2018). Recent scientific progress on developing supported Ni catalysts for dry (CO_2) reforming of methane. *Catalysts* 8: 110.

49 Li, W., Wang, H., Jiang, X. et al. (2018). A short review of recent advances in CO_2 hydrogenation to hydrocarbons over heterogeneous catalysts. *RSC Adv.* 8: 7651–7669.

50 Zhang, S., Muratsugu, S., Ishiguro, N., and Tada, M. (2013). Ceria-doped Ni/SBA-16 catalysts for dry reforming of methane. *ACS Catal.* 3: 1855–1864.

51 Pereñiguez, R., Gonzalez-delaCruz, V.M., Caballero, A., and Holgado, J.P. (2012). $LaNiO_3$ as a precursor of Ni/La_2O_3 for CO_2 reforming of CH_4: Effect of the presence of an amorphous NiO phase. *Appl. Catal., B* 123–124: 324–332.

52 Wang, Y., Zhang, L., Zhang, X. et al. (2017). Openmouthed β-SiC hollow-sphere with highly photocatalytic activity for reduction of CO_2 with H_2O. *Appl. Catal., B* 206: 158–167.

53 Alper, E. and Yuksel Orhan, O. (2017). CO_2 utilization: developments in conversion processes. *Petroleum* 3: 109–126.

54 Boyjoo, Y., Merigot, K., Lamonier, J.F. et al. (2015). Synthesis of $CaCO_3$@C yolk-shell particles for CO_2 adsorption. *RSC Adv.* 5: 24872–24876.

55 Boyjoo, Y., Pareek, V.K., and Liu, J. (2014). Synthesis of micro and nano-sized calcium carbonate particles and their applications. *J. Mater. Chem. A* 2: 14270–14288.

56 Yang, W., Liu, H., Li, Y. et al. (2016). Properties of yolk–shell structured Ni@SiO_2 nanocatalyst and its catalytic performance in carbon dioxide reforming of methane to syngas. *Catal. Today* 259: 438–445.

57 Lu, Y., Guo, D., Ruan, Y. et al. (2018). Facile one-pot synthesis of Ni@HSS as a novel yolk-shell structure catalyst for dry reforming of methane. *J. CO2 Util.* 24: 190–199.

58 Price, C., Pastor-Pérez, L., le Saché, E. et al. (2017). Highly active Cu-ZnO catalysts for the WGS reaction at medium–high space velocities: effect of the support composition. *Int. J. Hydrogen Energy* 42: 10747–10751.

59 Tamura, M. and Tomishige, K. (2015). Redox properties of CeO_2 at low temperature: the direct synthesis of imines from alcohol and amine. *Angew. Chem. Int. Ed.* 54: 864–867.

60 Park J.C. and Song, H. (2011). Metal@Silica yolk-shell nanostructures as versatile bifunctional nanocatalysts. *Nano Res.* 4: 33–49.

61 He, L., Li, L., Zhang, L. et al. (2014). ZIF-8 templated fabrication of rhombic dodecahedron-shaped ZnO@SiO_2, ZIF-8@SiO_2 yolk–shell and SiO_2 hollow nanoparticles. *CrystEngComm*: 6534–6537.

62 Das, S., Ashok, J., Bian, Z. et al. (2018). Silica–Ceria sandwiched Ni core–shell catalyst for low temperature dry reforming of biogas: coke resistance and mechanistic insights. *Appl. Catal., B* 230: 220–236.

63 Russell, P.G., Kovac, N., Srinivasan, S., and Steinberg, M. (1977). The electrochemical reduction of carbon dioxide, formic acid, and formaldehyde. *J. Electrochem. Soc.* 124: 1329.

64 Halmann, M. (1978). Photoelectrochemical reduction of aqueous carbon dioxide on p-type gallium phosphide in liquid junction solar cells. *Nature* 275: 115–116.

65 Hori, Y. (2008). Electrochemical CO_2 reduction on metal electrodes. In: *Modern Aspects of Electrochemistry* (eds. C.G. Vayenas, R.E. White and M.E. Gamboa-Aldeco), 89–189. New York, NY: Springer New York.

66 Luo, B., Liu, G., and Wang, L. (2016). Recent advances in 2D materials for photocatalysis. *Nanoscale* 8: 6904–6920.

67 Kudo, A. and Miseki, Y. (2009). Heterogeneous photocatalyst materials for water splitting. *Chem. Soc. Rev.* 38: 253–278.

68 Ong, W.J., Tan, L.L., Ng, Y.H. et al. (2016). Graphitic carbon nitride (g-C_3N_4)-based photocatalysts for artificial photosynthesis and environmental remediation: are we a step closer to achieving sustainability? *Chem. Rev.* 116: 7159–7329.

69 Bocarsly, A.B., Tachikawa, H., and Faulkner, L.R. (1996). *Laboratory techniques in electroanalytical chemistry, Revised and Expanded*, 2e (eds. P. Kissinger and W.R. Heineman). CRC Press. ISBN: 9780824794453.

70 Bard, A.J. and Faulkner, L.R. (2001). *Electrochemical Methods, Fundamentals and Applications*, 2e. Wiley. ISBN: 9780471043720.

71 Linsebigler, A.L., Lu, G., and Yates, J.T. (1995). Photocatalysis on TiO_2 surfaces: principles, mechanisms, and selected results. *Chem. Rev.* 95: 735–758.

72 Fujishima, A., Zhang, X., and Tryk, D.A. (2008). TiO_2 photocatalysis and related surface phenomena. *Surf. Sci. Rep.* 63: 515–582.

73 Mills, A. and Le Hunte, S. (1997). An overview of semiconductor photocatalysis. *J. Photochem. Photobiol., A* 108: 1–35.

74 Hoffmann, M.R., Martin, S.T., Choi, W., and Bahnemann, D.W. (1995). Environmental applications of semiconductor photocatalysis. *Chem. Rev.* 95: 69–96.

75 Li, J. and Wu, N. (2015). Semiconductor-based photocatalysts and photoelectrochemical cells for solar fuel generation: a review. *Catal. Sci. Technol.* 5: 1360–1384.

76 Wang, J., Tafen, D.N., Lewis, J.P. et al. (2009). Origin of photocatalytic activity of nitrogen-doped TiO_2 nanobelts. *J. Am. Chem. Soc.* 131: 12290–12297.

77 Meng, F., Hong, Z., Arndt, J. et al. (2012). Visible light photocatalytic activity of nitrogen-doped $La_2Ti_2O_7$ nanosheets originating from band gap narrowing. *Nano Res.* 5: 213–221.

78 White, J.L., Baruch, M.F., Pander, J.E. et al. (2015). Light-driven heterogeneous reduction of carbon dioxide: photocatalysts and photoelectrodes. *Chem. Rev.* 115: 12888–12935.

79 Liu, L., Zhao, H., Andino, J.M., and Li, Y. (2012). Photocatalytic CO_2 reduction with H_2O on TiO_2 nanocrystals: comparison of anatase, rutile, and brookite polymorphs and exploration of surface chemistry. *ACS Catal.* 2: 1817–1828.

80 Schulte, K.L., DeSario, P.A., and Gray, K.A. (2010). Effect of crystal phase composition on the reductive and oxidative abilities of TiO_2 nanotubes under UV and visible light. *Appl. Catal., B* 97: 354–360.

81 Wang, W.-N., An, W.-J., Ramalingam, B. et al. (2012). Size and structure matter: enhanced CO_2 photoreduction efficiency by size-resolved ultrafine Pt nanoparticles on TiO_2 single crystals. *J. Am. Chem. Soc.* 134: 11276–11281.

82 Wang, P.-Q., Bai, Y., Liu, J.-Y. et al. (2012). One-pot synthesis of rutile TiO_2 nanoparticle modified anatase TiO_2 nanorods toward enhanced photocatalytic reduction of CO_2 into hydrocarbon fuels. *Catal. Commun.* 29: 185–188.

83 Hurum, D.C., Agrios, A.G., Gray, K.A. et al. (2003). Explaining the enhanced photocatalytic activity of degussa P25 mixed-phase TiO_2 using EPR. *J. Phys. Chem. B.* 107: 4545–4549.

84 Ba, X., Yan, L.L., Huang, S. et al. (2014). New way for CO_2 reduction under visible light by a combination of a Cu electrode and semiconductor thin film: Cu_2O conduction type and morphology effect. *J. Phys. Chem. C* 118: 24467–24478.

85 Tennakone, K., Jayatissa, A.H., and Punchihewa, S. (1989). Selective photoreduction of carbon dioxide to methanol with hydrous cuprous oxide. *J. Photochem. Photobiol., A* 49: 369–375.

86 Handoko, A.D. and Tang, J. (2013). Controllable proton and CO_2 photoreduction over Cu_2O with various morphologies. *Int. J. Hydrogen Energy* 38: 13017–13022.

87 Aurian-Blajeni, B., Halmann, M., and Manassen, J. (1980). Photoreduction of carbon dioxide and water into formaldehyde and methanol on semiconductor materials. *Sol. Energy* 25: 165–170.

88 Chen, X., Zhou, Y., Liu, Q. et al. (2012). Ultrathin, single-crystal WO_3 nanosheets by two-dimensional oriented attachment toward enhanced photocatalystic reduction of CO_2 into hydrocarbon fuels under visible light. *ACS Appl. Mater. Interfaces* 4: 3372–3377.

89 Shown, I., Hsu, H.C., Chang, Y.C. et al. (2014). Highly efficient visible light photocatalytic reduction of CO_2 to hydrocarbon fuels by cu-nanoparticle decorated graphene oxide. *Nano Lett.* 14: 6097–6103.

90 Dong, F., Zhao, W., and Wu, Z. (2008). Characterization and photocatalytic activities of C, N and S co-doped TiO_2 with 1D nanostructure prepared by the nano-confinement effect. *Nanotechnology* 19.

91 Xiao, M., Wang, Z., Lyu, M. et al. (2018). Hollow nanostructures for photocatalysis: advantages and challenges. *Adv. Mater.* 31.

92 Wu, M., Jin, J., Liu, J. et al. (2013). High photocatalytic activity enhancement of titania inverse opal films by slow photon effect induced strong light absorption. *J. Mater. Chem. A* 1: 15491–15500.

93 Liao, G., Chen, S., Quan, X. et al. (2010). Photonic crystal coupled TiO_2/polymer hybrid for efficient photocatalysis under visible light irradiation. *Environ. Sci. Technol.* 44: 3481–3485.

94 Liu, J., Jin, J., Li, Y. et al. (2014). Tracing the slow photon effect in a ZnO inverse opal film for photocatalytic activity enhancement. *J. Mater. Chem. A* 2: 5051–5059.

95 Chen, X., Ye, J., Ouyang, S. et al. (2011). Enhanced incident photon-to-electron conversion efficiency of tungsten trioxide photoanodes based on 3d-photonic crystal design. *ACS Nano* 5: 4310–4318.

96 Rugge, A., Park, J.S., Gordon, R.G., and Tolbert, S.H. (2005). Tantalum(V) nitride inverse opals as photonic structures for visible wavelengths. *J. Phys. Chem. B* 109: 3764–3771.

97 Tu, W., Zhou, Y., Li, H. et al. (2015). Au@TiO_2 yolk-shell hollow spheres for plasmon-induced photocatalytic reduction of CO_2 to solar fuel via a local electromagnetic field. *Nanoscale* 7: 14232–14236.

98 Wang, W., Xu, D., Cheng, B. et al. (2017). Hybrid carbon@TiO_2 hollow spheres with enhanced photocatalytic CO_2 reduction activity. *J. Mater. Chem. A* 5: 5020–5029.

99 Tsai, C.-W., Chen, H.M., Liu, R.-S. et al. (2011). Ni@NiO core–shell structure-modified nitrogen-doped $InTaO_4$ for solar-driven highly efficient CO_2 reduction to methanol. *J. Phys. Chem. C* 115: 10180–10186.

100 Zhai, Q., Xie, S., Fan, W. et al. (2013). Photocatalytic conversion of carbon dioxide with water into methane: platinum and Copper(I) oxide co-catalysts with a core-shell structure. *Angew. Chem. Int. Ed.* 52: 5776–5779.

101 Robatjazi, H., Weinberg, D., Swearer, D.F. et al. (2019). *Metal-organic frameworks tailor the properties of aluminum nanocrystals*, vol. 5.

102 Feng, S., Wang, M., Zhou, Y. et al. (2015). Double-shelled plasmonic Ag-TiO_2 hollow spheres toward visible light-active photocatalytic conversion of CO_2 into solar fuel. *APL Mater.* 3: 104416.

103 Ziarati, A., Badiei, A., Grillo, R., and Burgi, T. (2019). 3D Yolk@Shell TiO_2-x/LDH architecture: tailored structure for visible light CO_2 conversion. *ACS Appl. Mater. Interfaces* 11: 5903–5910.

104 Zhao, Y., Jia, X., Waterhouse, G.I.N. et al. (2016). Layered double hydroxide nanostructured photocatalysts for renewable energy production. *Adv. Energy Mater.* 6: 1501974.

105 Mohapatra, L. and Parida, K. (2016). A review on the recent progress, challenges and perspective of layered double hydroxides as promising photocatalysts. *J. Mater. Chem. A* 4: 10744–10766.

106 Yu, M., Liang, C., Lui, M. et al. (2014). Yolk–shell Fe_3O_4 @ZrO_2 prepared by a tunable polymer surfactant assisted sol–gel method for high temperature stable microwave absorption. *J. Mater. Chem. C* 2: 7275–7283.

107 Wu, X.-J. and Xu, D. (2010). Soft template synthesis of yolk/silica shell particles. *Adv. Mater.* 22: 1516–1520.

108 Ahmed, A., Clowes, R., Willneff, E. et al. (2010). Synthesis of uniform porous silica microspheres with hydrophilic polymer as stabilizing agent. *Ind. Eng. Chem. Res.* 49: 602–608.

109 Qiu, P., Kang, K., Kim, K. et al. (2015). Facile synthesis of uniform yolk-shell structured magnetic mesoporous silica as an advanced photo-Fenton-like catalyst for degrading rhodamine B. *RSC Adv.*

110 Liu, L., Yang, H.Q., Kleitz, F. et al. (2012). Yolk-shell hybrid materials with a periodic mesoporous organosilica shell: ideal nanoreactors for selective alcohol oxidation. *Adv. Funct. Mater.* 22: 591–599.

111 Li, G.D. and Tang, Z.Y. (2014). Noble metal nanoparticle@metal oxide core/yolk-shell nanostructures as catalysts: recent progress and perspective. *Nanoscale* 6: 3995–4011.

112 Dahlberg, K.A. and Schwank, J.W. (2012). Synthesis of Ni@SiO_2 nanotube particles in a water-in-oil microemulsion template. *Chem. Mater.* 24: 2635–2644.

113 Guan, B., Wang, T., Zeng, S. et al. (2014). A versatile cooperative template-directed coating method to synthesize hollow and yolk-shell mesoporous zirconium titanium oxide nanospheres as catalytic reactors. *Nano Res.* 7: 246–262.

114 Liu, J., Qiao, S.Z., Liu, H. et al. (2011). Extension of the stöber method to the preparation of monodisperse resorcinol-formaldehyde resin polymer and carbon spheres. *Angew. Chem. Int. Ed.* 50: 5947–5951.

115 Pan, Y.-X., Zhuang, H.-Q., Ma, H. et al. (2019). Tungsten carbide hollow spheres flexible for charge separation and transfer for enhanced visible-light-driven photocatalysis. *Chem. Eng. Sci.* 194: 71–77.

116 Lin, L., Chen, M., and Wu, L. (2018). Synthesis of molybdenum–tungsten bimetallic carbide hollow spheres as pH-Universal electrocatalysts for efficient hydrogen evolution reaction. *Adv. Mater. Interfaces* 5: 1801302.

117 Wenelska, K., Ottmann, A., Moszyński, D. et al. (2018). Facile synthesis N-doped hollow carbon spheres from spherical solid silica. *J. Colloid Interface Sci.* 511: 203–208.

118 Wang, D.-W., Zhou, R., Liu, J. et al. (2014). Hierarchical mesoporous yolk–shell structured carbonaceous nanospheres for high performance electrochemical capacitive energy storage. *Chem. Commun.* 51: 2518–2521.

119 Tang, J., Liu, J., Salunkhe, R.R. et al. (2015). Nitrogen-doped hollow carbon spheres with large mesoporous shells engineered from diblock copolymer micelles. *Chem. Commun.* 52: 505–508.

120 Yang, T., Liu, J., Zhou, R. et al. (2014). N-doped mesoporous carbon spheres as the oxygen reduction reaction catalysts. *J. Mater. Chem. A* 2: 18139–18146.

121 Hu, Y., Jensen, J.O., Zhang, W. et al. (2014). Hollow spheres of iron carbide nanoparticles encased in graphitic layers as oxygen reduction catalysts. *Angew. Chem. Int. Ed.* 53: 3675–3679.

16

Aliphatic Polycarbonates Derived from Epoxides and CO_2

Sebastian Kernbichl and Bernhard Rieger

Technical University Munich, WACKER-Chair of Macromolecular Chemistry, Department of Chemistry, Lichtenbergstr. 4, 85748 Garching, Germany

Abbreviations

BDI	β-diiminate
BPA-PC	bisphenol-A polycarbonate
CCU	carbon capture and utilization
CHO	cyclohexene oxide
Đ	polydispersity index
DFT	density functional theory
DMAP	4-(dimethylamino)pyridine
HPPO	hydrogen peroxide to propylene oxide
LO	limonene oxide
MW	molecular weight
NBS	*N*-bromosuccinimide
PCHC	poly(cyclohexene carbonate)
PLC	poly(limonene carbonate)
PO	propylene oxide
PPC	poly(propylene carbonate)
PPNCl	bis(triphenylphosphine)iminium chloride
TMS	trimethylsilyl
TOF	turnover frequency
TPP	tetraphenylporphyrine
ZnGA	zinc glutarate
ZnSA	zinc succinate

16.1 Introduction

In light of its increasing concentration in the atmosphere and its industrial emergence, carbon dioxide is clearly a topic of intense research. Carbon dioxide capture

Engineering Solutions for CO_2 Conversion, First Edition.
Edited by Tomas R. Reina, José A. Odriozola, and Harvey Arellano-Garcia.

and utilization (CCU) deals with the aim of using industrially produced CO_2 as a C1-feedstock for high-quality chemical products. Because carbon dioxide is chemically rather inert, high energy barriers must be overcome [1]. The coupling of CO_2 with an epoxide offers the possibility of using carbon dioxide for the preparation of cyclic carbonates or polycarbonates. Cyclic carbonates are applied as electrolytes in lithium ion batteries or as high boiling aprotic polar solvents [2, 3]. Depending on their molecular weight (MW) and their kind of end groups, polycarbonates are used in the polyurethane production (low MW, hydroxyl terminated) or in the fabrication of rigid plastics (high MW) as a possible alternative to bisphenol-A polycarbonate (BPA-PC) [4]. Polycarbonates derived from epoxides and CO_2 offer promising properties such as high transparency, UV stability, and high Young's moduli [5]. The coupling of propylene oxide (PO) and CO_2 offers the most promising polymer poly(propylene carbonate) (PPC) because PO is industrially available in large quantities and a higher CO_2 content is generated compared to other aliphatic polycarbonates (Figure 16.1a). The widespread commercialization of PPC although was hindered because of the low glass transition temperature of 30–40 °C. Polyurethane foams from Bayer based on poly(ether carbonates) made of PO and CO_2 evidence the potential use of carbon dioxide in large-scale polymerizations [7]. The industrial feasibility of such processes always depends on the activity of the employed catalyst. Because the initial discovery in 1969 of diethylzinc and water serving as a heterogeneous catalyst in the synthesis of PPC, much research has been made on the question of how the activity of the catalysts can be improved [8]. Efforts mainly concentrate on homogeneous systems and often initially use cyclohexene oxide (CHO) as it can be copolymerized more easily because of its ring strain to poly(cyclohexene carbonate) (PCHC) (Figure 16.1b). Poly(limonene carbonate) (PLC) was reported for the first time in 2004 but gained attention only very recently (Figure 16.1c) [6]. The bio-based origin of limonene oxide (LO) and the polymers' very promising properties expand the class of the well-known aliphatic polycarbonates PPC and PCHC. This review gives an overview of the catalytic copolymerization of various epoxides and CO_2 and focuses on the latest findings especially in the field of PLC.

16.2 Aliphatic Polycarbonates

16.2.1 Synthesis of the Monomers

During recent decades, several processes for the production of PO have been established, starting with the hydrochlorination route in 1930 [9]. Because of the high costs of the starting materials and arising by-products, the preparation has been improved several times. The hydrogen peroxide to propylene oxide (HPPO) process works with the aid of a titanium-doped silicate catalyst and produces water as the only by-product [10]. 7.5 mio. $t\,a^{-1}$ PO were produced this way in 2015 [11].

Despite the industrial abundance of PO, cyclohexene oxide is often used in an initial attempt at testing a catalyst. Because of the ring strain of the cyclohexane ring, this epoxide can be ring-opened more easily after coordination to a suitable catalyst.

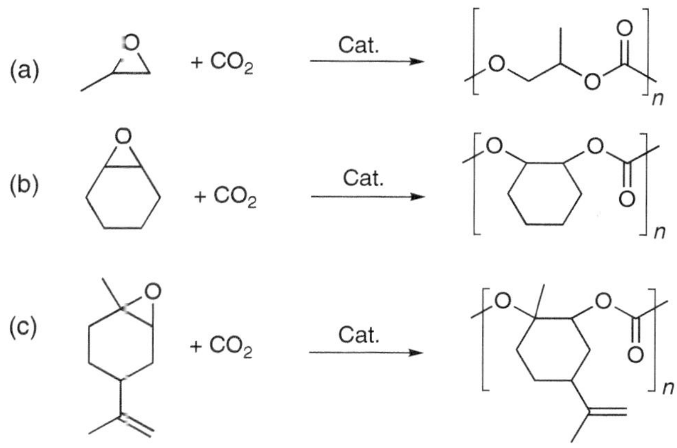

Figure 16.1 Alternating copolymerization of CO_2 with (a) propylene oxide, (b) cyclohexene oxide, and (c) limonene oxide. Source: (c) Adapted from Byrne et al. [6].

Cyclohexene oxide itself is usually prepared through epoxidation with peracids or even catalytically with manganese porphyrins and oxygen [12, 13].

Carbon dioxide used in the copolymerization process in academic research is usually taken from gas bottles in high purity. Williams and coworkers reported in 2015 whether CO_2 from a nearby power station can be copolymerized with epoxides using established catalysts [14]. Three homogeneous dinuclear Zn/Mg systems showed a high tolerance toward gaseous impurities. Polyols from CHO and CO_2 were obtained in a controlled manner, demonstrating the possibility to apply industrially emitted carbon dioxide without further purification.

The third type of epoxide discussed in this chapter is limonene oxide. It is obtained via the epoxidation of limonene. One enantiomer, (*R*)-limonene, can be found in orange peels and gets extracted at a volume of 70 kt a^{-1} [15]. Categorized as a cyclic monoterpene, limonene bears one chiral C-atom and two unsaturated carbon bonds prone to chemical modification. This renewable non-food resource represents a promising class of feedstocks because the 1,2 epoxidation products display an auspicious candidate for the copolymerization with CO_2 [16]. In addition, it shows strong resemblance to the well-studied epoxide cyclohexene oxide. The topic of sustainable epoxide/CO_2 copolymerization was summarized by Darensbourg and coworker in 2017 [17]. Limonene oxide is commercially available in a 46 : 54 cis : trans mixture usually obtained via acid-catalyzed oxidation of (*R*)-limonene (Figure 16.2a) [18]. Because of the fact that the most active catalysts almost exclusively polymerize the trans-diastereomer, a route toward a higher trans-content was established in 2016 by first treating limonene with *N*-bromosuccinimide (NBS) to enable the more selective ring closure in the second step (Figure 16.2b) [16].

16.2.2 Mechanistic Aspects of the Copolymerization of Epoxides and CO_2

The process of the CO_2/epoxide copolymerization can be divided into three catalytic steps, namely, initiation, propagation, and termination (Scheme 16.1). During

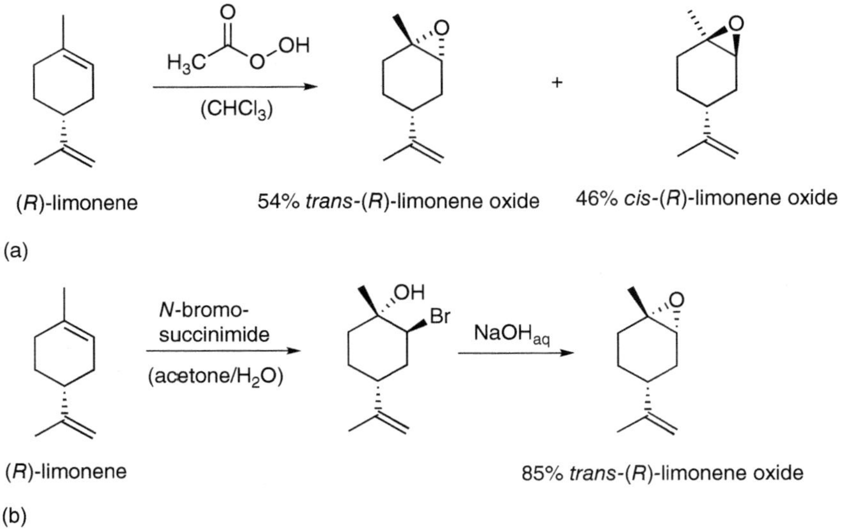

Figure 16.2 Oxidation of (*R*)-limonene via two different routes to different *trans*-(*R*)-limonene oxide contents. Source: (a) Adapted from Greenspan and Linder [18], (b) Adapted from Hauenstein et al. [16].

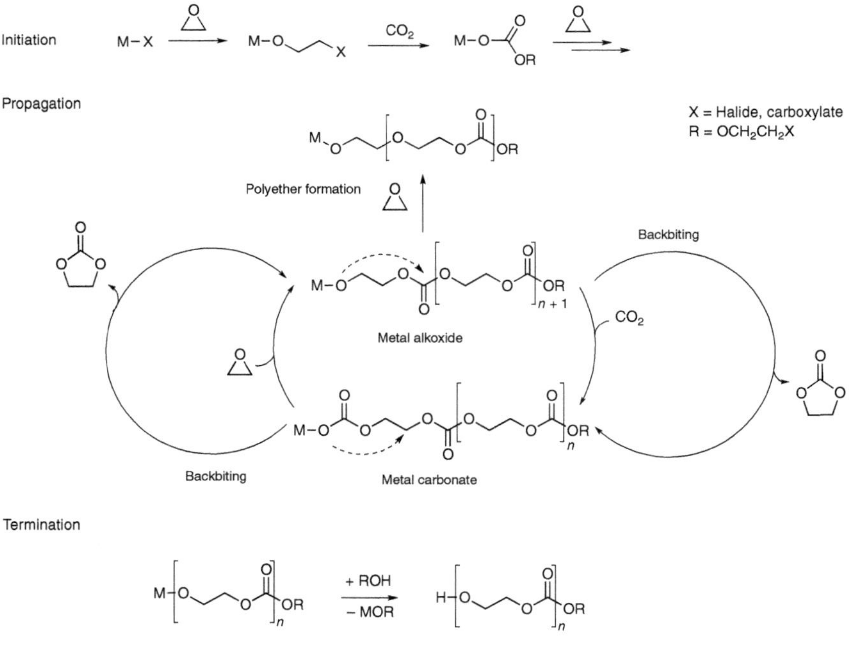

Scheme 16.1 Copolymerization of CO_2 and epoxides in three elementary steps: initiation, propagation, and termination.

initiation, either carbon dioxide or the epoxide inserts into the M—X bond, creating a metal carboxylate or a metal alkoxide species. The nature of the catalyst decides which of the two monomers is more likely to be inserted in the first step. The alternating incorporation of CO_2 and epoxide marks the period of propagation, in which the metal alkoxide and the metal carbonate are interconverted. There are two important side reactions during propagation that occur to a varying extent, which depend on the type of epoxide and the catalyst, as well as the general polymerization conditions such as reaction temperature or CO_2 pressure. The consecutive insertion of two (or more) epoxide molecules results in the formation of polyether units. In contrast to that, the so-called backbiting reaction to a five-membered carbonate ring is the thermodynamically favored product. It is the predominant side reaction for the copolymerization of PO and CO_2.

For most catalysts, the copolymerization reaction can be described as an immortal reaction. On the one hand, this allows the synthesis of polymers in well-defined molecular weights because an inverse relationship exists between the catalyst concentration and the resulting molecular weight. On the other hand, this living character also enables the formation of block copolymers by the addition of a second type of epoxide. The polymerizations are usually terminated by the addition of a protic compound such as methanol. The underlying mechanism is a chain transfer reaction between the polymeric chain end and the proton. This chain transfer can also occur during the polymerization if any protic impurities such as water are present in the reaction mixture. This automatically leads to polymers of lower molecular weight. In the case of the cyclohexene oxide chain transfer, it also explains the often observed bimodal molecular weight distribution because of the in situ formed cyclohexanediol that again acts as a chain transfer agent.

Applying the same reaction conditions for the two epoxide types, CHO and PO, in the copolymerization with CO_2 results in a major difference regarding reactivity and selectivity. The reason was investigated more closely by the group of Darensbourg in 2003 using a chromium salen catalyst [19]. Activation energies were determined by in situ infrared (IR) kinetic measurements showing that, for both monomers, the respective cyclic product requires a higher activation energy than the polycarbonate (Figure 16.3). However, the decisive difference is the smaller energy gap between the two products in the case of PO (33 $kJ\,mol^{-1}$), explaining the lower selectivity for polycarbonate formation at elevated temperatures.

16.2.3 Thermal Stability and Possible Degradation Pathways

Processability of polymers is always strongly linked to the question of thermal stability. A multitude of works is dealing with the thermal degradation of aliphatic polycarbonates, especially PPC [20–25]. Two main decomposition pathways have been observed thereby. On the one hand, the intramolecular backbiting reaction to cyclic carbonates is facilitated by OH-end groups (Scheme 16.2a). These hydroxy end groups are usually obtained if the catalytic copolymerization is terminated by the addition of acidified methanol. On the other hand, chain unzipping occurs if a nucleophile is located at a polymer chain end (Scheme 16.2b). Catalyst residues accelerate

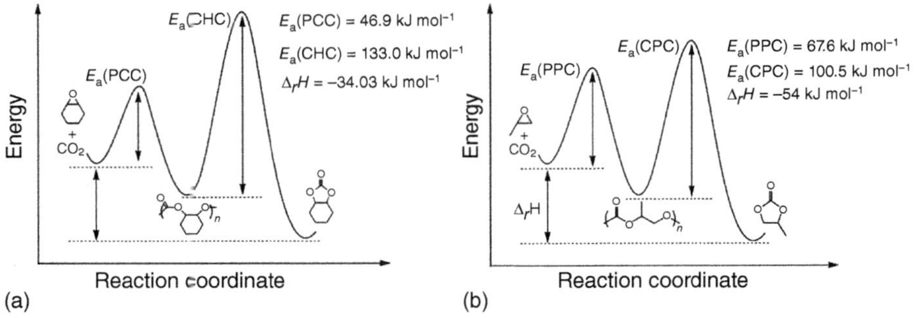

Figure 16.3 Reaction coordinate diagram for the coupling of (a) CHO and (b) PO with CO_2 using a chromium salen catalyst. Source: Darensbourg et al. [19].

the backbiting reaction severely in the temperature range of 150–180 °C [26]. At higher temperatures (~200 °C), random chain scission has been mostly observed, leading to the splitting of a polymer chain with the release of CO_2.

Decomposition temperatures reported in the literature are difficult to compare because of varying conditions, the type of analysis, and the kind of polymer. Most publications indicate 180 °C as the upper processability limit for PPC. Different attempts have been made to further increase the thermal stability of aliphatic polycarbonates. The backbiting reaction can be prevented to some extent via end-capping of the polymer chain ends [20]. Anhydrides seem to be the most promising candidates along with acetates and chlorophosphates as they are less reactive than the hydroxy functionality.

Thermogravimetric analysis of PLC revealed a $T_{5\%}$ of 225 °C, indicating that, at 225 °C, 5% of the sample was converted to volatile compounds [16]. Endcapping with maleic anhydride further increased the decomposition temperature to a $T_{5\%}$ of 240 °C. Gas chromatography mass spectrometry (GC-MS) measurement of a hydroxy-terminated PLC sample heated to 230 °C for 20 minutes did not show any cyclic limonene carbonate, exhibiting instead a variety of different decomposition products including LO [16]. This circumstance was transferred to the base-initiated depolymerization of PLC by Koning and coworkers [27] A similar base-assisted depolymerization was previously observed for poly(cyclopentene carbonate), yielding both the epoxide and the cyclic carbonate [28]. In the case of PLC, the strong organic base triazabicyclodec-5-ene leads to the deprotonation of the hydroxyl polymer chain end, enabling fast backbiting to limonene oxide and CO_2 at 110 °C. The controlled depolymerization of aliphatic polycarbonates represents an interesting approach in terms of recyclability and sustainability.

16.2.4 Mechanical Properties

Figure 16.4 illustrates the three copolymers derived from CO_2 and the respective epoxide. The main advantages and disadvantages are clearly visible: Although PCHC struggles with brittleness and PPC exhibits a soft, deformable property, PLC is best suited for solvent casting along with a high transmission of the polymer film.

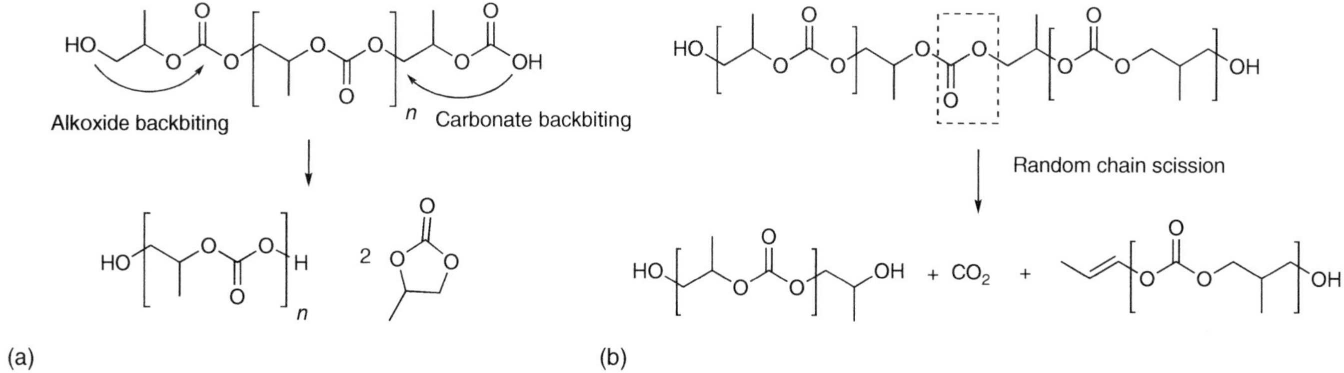

Scheme 16.2 Degradation mechanism for PPC: chain scission (a) and backbiting (b). Source: Phillips et al. [20].

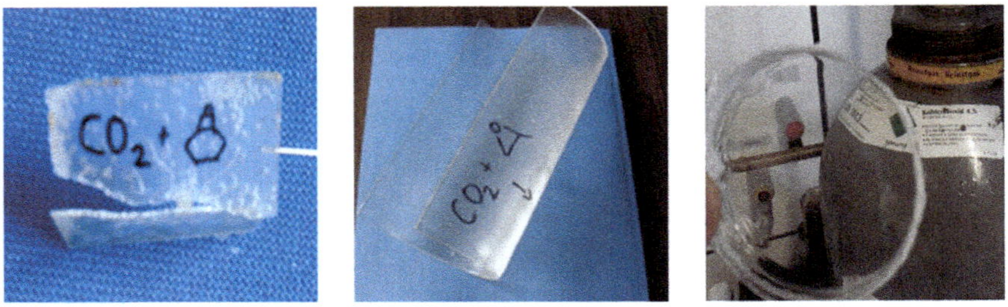

(a) (b) (c)

Figure 16.4 Poly(cyclohexene carbonate) (a), poly(propylene carbonate) (b), and poly(limonene carbonate) (c).

Similar to the determination of the thermal stability of aliphatic polycarbonates, mechanical properties cannot be easily compared because they are strongly dependent on the respective microstructure, the molecular weight, and cyclic carbonate impurities. In the case of PPC, the glass transition temperature is highly dependent on the molecular weight of the polymer and ranges between 30 and 40 °C. PPC with a T_g of 30 °C exhibits Young's moduli of 200–1000 MPa, whereas the polymer with a higher T_g shows an increased stiffness of 700–1400 MPa [5, 26, 29]. Aliphatic polycarbonates are often compared with the aromatic bisphenol-A-based polycarbonate. Apart from a Young's modulus of 2400 ± 400 MPa, the aromatic polymer shows an elongation at a break of $40 \pm 35\%$, opening up a multitude of different areas of application. PCHC is a much stiffer material than PPC and resembles BPA-PC in some aspects. The main disadvantage of PCHC thus far arises from its brittleness reflected in a Young's modulus of 3600 ± 100 MPa and an elongation at break of $1.7 \pm 0.6\%$ [30].

Efforts to overcome the main problems in terms of material properties for both PPC and PCHC were made by using a second type of epoxide in the polymerization. The so-called terpolymerization enables the synthesis of polymers in an adjustable composition and hence in a tunable glass transition temperature [31]. Terpolymerization of CHO, PO, and CO_2 was thought to improve the brittleness of PCHC, but it only resulted in a reduced E-modulus (2100 MPa) with an otherwise unchanged elongation at break [32]. Other epoxides such as hexene oxide or styrene oxide were also successfully polymerized with the established monomers PO and CHO [33, 34].

The mechanical behavior of PLC was tested for the first time in 2016 [16]. Hot-pressed polymer specimen could be produced at a thickness of 300 μm and tested using a stress–strain machine. A Young's modulus of 950 MPa, a tensile strength of 55 MPa, and an elongation at break of 15% were obtained. These promising properties indicate that PLC is a possible alternative to BPA-PC.

16.3 Catalyst Systems for the CO_2/Epoxide Copolymerization

For other classes of polymerization, early catalytic attempts started with heterogeneous systems. Although the number and the kind of active sites is difficult to determine for such systems, they have already been industrially applied for centuries in PPC production [26, 35, 36]. In contrast, homogeneous catalysts are the subject of intense research because of their solubility, tunability, and the possibility of determining the underlying mechanism in situ. The viscosity of the polymerization solution plays an important role regarding polymer processability and activity. A low amount of solvent enables high activity but reduces the diffusion of the monomer, whereas an increased dilution lowers the activity, to some extent, because of spatially separated metal centers, but allows higher yields and improved processability [37].

16.3.1 Heterogeneous Catalysts

The development of catalysts that are active in the copolymerization of epoxides and CO_2 has been excellently reviewed by a number of groups [37–47]. Following a brief historical overview of the most important milestones in the catalysis of copolymerization, the focus of this chapter will be on the possibility for the terpolymerization of epoxides with other types of monomers. In addition, the coupling of limonene oxide and CO_2 giving the fully bio-based polymer PLC will be discussed in detail. The copolymerization of carbon dioxide and epoxides was observed for the very first time when Inoue et al. mixed diethylzinc and water with PO and CO_2 in 1969 [8, 48]. Despite the low activity of $0.12\,h^{-1}$ as well as low molecular weights, this finding marks the establishment of an intensively studied research field. Soga et al. changed to a system of $Zn(OH)_2$ and various dicarboxylic acids and in this way created the first well-defined heterogeneous catalyst [49]. Glutaric acid turned out to be the most active partner to $Zn(OH)_2$, enabling an activity of $1.1\,h^{-1}$. The easy preparation along with the nontoxicity of zinc made this system industrially relevant for the preparation of PPC. Nevertheless, the reaction mechanism remained unknown for a long time. Single-crystal X-ray spectroscopy revealed a tetrahedral coordination of four carboxylic groups around the zinc center, which prevents monomer diffusion, thus limiting the catalyst's activity [50]. Attempts to increase the ZnGA (zinc glutarate) surface area by ball milling, modification of the stirring procedure, or the use of additives such as SiO_2 did not lead to significantly higher activity [51–53].

In 2011, Rieger and coworkers demonstrated the importance of the Zn–Zn distance by varying the chain length of the carboxylic acid ranging from succinic acid (C_4) to pimelic acid (C_7) [51]. Interestingly, zinc succinate (ZnSA) indeed showed nearly no activity in the copolymerization of PO and CO_2, whereas ZnGA, zinc adipate, and zinc pimelate exhibited similar activity. The reason could be addressed by measuring the solid-state structures, indicating a Zn–Zn distance only for the latter three systems of 4.6–4.8 Å, whereas ZnSA did not show such a defined spatial proximity (Figure 16.5). This principle of two interacting zinc centers was later transferred to the development of homogeneous catalysts.

16.3.2 Overview of the Homogeneous Catalytic Systems

Although heterogeneous catalysts display numerous advantages, including sufficient activities in the copolymerization, defining the active sites in these systems is challenging. As a result, improvements to the catalyst's structure as well as the

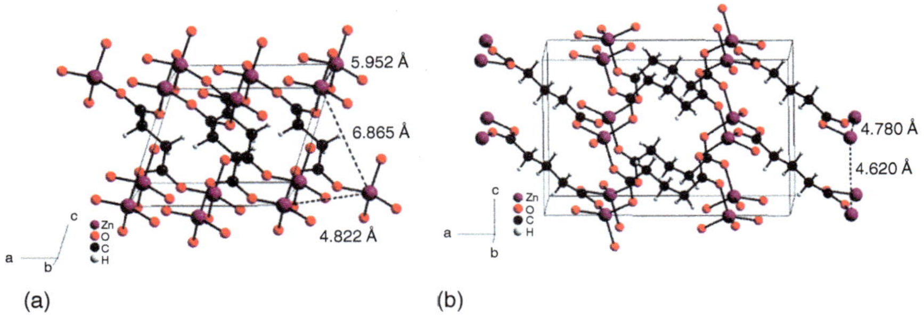

Figure 16.5 Representation of part of the repeating unit in two different zinc carboxylates: (a) zinc succinate and (b) zinc glutarate. Source: Klaus et al. [51]. © 2011 American Chemical Society.

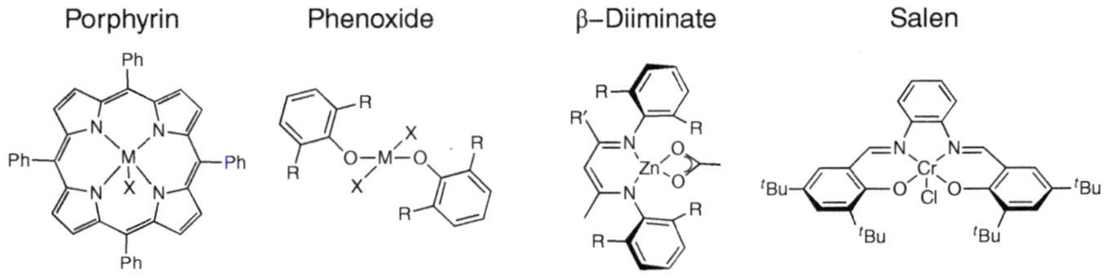

Figure 16.6 Types of homogeneous catalysts active in the copolymerization of CO_2 and epoxides. Source: Adapted from Lu [4].

study of reaction kinetics are difficult to address. Regarding polymer chemistry, the variety of active sites also causes broad polydispersity indices (*Đ*) with severe effects on thermal and mechanical properties. In 1978, Inoue and coworkers was again first in developing an aluminum tetraphenylporphyrin (TPP) complex (**1a**, M = Al, X = OMe) [54–56]. By using $EtPh_3PBr$ as a cocatalyst, the complex successfully catalyzed the CHO/CO_2 and PO/CO_2 polymerization. Figure 16.6 illustrates the four main complex structures used in the copolymerization of CO_2 and epoxides.

When Kruper and Dellar used chromium as the central metal for porphyrin based complexes in combination with the cocatalysts 4-(dimethylamino)pyridine (DMAP) and bis(triphenylphosphine)iminium chloride (PPNCl), they observed mostly cyclic carbonate formation both for cyclohexene oxide and PO [57]. In contrast, switching to cobalt (**1a**, M = Co, X = halide) led to a decrease in activity because of the reduced Lewis acidity of the metal, but it yielded high polycarbonate selectivities, especially for PO/CO_2 (99% polycarbonate, turnover frequency (TOF) = 188 h^{-1}) [58]. Various porphyrin-based catalysts were synthesized and tested in the copolymerization of PO and CO_2 [59–66].

The first aryl–oxide zinc complex paved the way for a novel class of catalysts, namely, the phenoxide system (**1b**) bearing the advantage of being soluble and as a result more easily characterized and improved [67]. Their activity in the copolymerization reaction is mainly dependent on the aryl ligand framework and the base

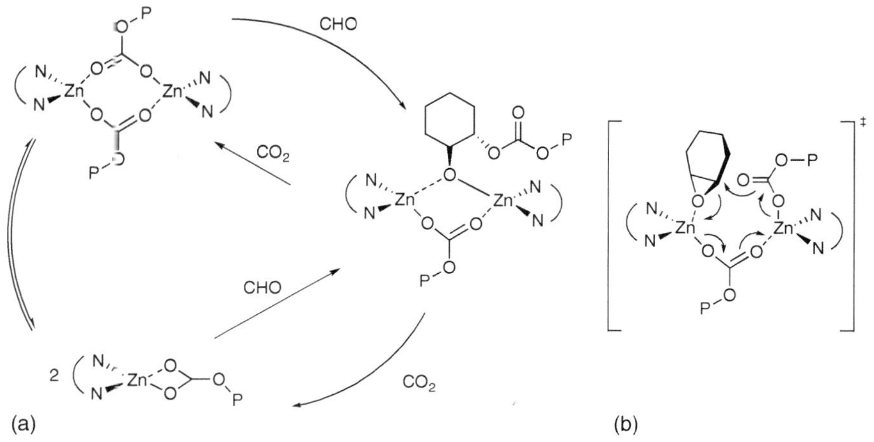

Scheme 16.3 Proposed copolymerization mechanism for BDI–Zn catalysts (a) and transition state of the epoxide ring-opening step (b). Source: Based on Coates et al. [39], Moore et al. [76].

ligand at the metal center [68]. Substituents in ortho position were screened systematically, which show the trend of increased activity along with decreasing substituent size and increasing electronegativity (F > Cl > Br, X = thf) [69]. Apart from one of the first successful terpolymerizations from CHO, PO, and CO_2, this class of catalysts also indicated that epoxide activation is more decisive for high activities than the carbon dioxide insertion step [30, 34, 70].

A breakthrough in the copolymerization of epoxides and CO_2 was achieved by Coates in 1998 with the complexation of a β-diiminate (BDI) ligand to zinc (**1c**) [71]. A variety of modifications were performed at the pentane backbone as well as at the aniline moieties and the nucleophilic initiating group [39, 72–75]. Having a closer look at the reaction kinetics, again a bimetallic mechanism was observed with a reaction order of $[Zn]_{tot} = 1.73$ [76]. The complexes are usually in a monomer–dimer equilibrium and form these dimeric species via a bridging of the initiating group. Scheme 16.3 shows the ring-opening step of CHO in a bimetallic transition state and the consecutive insertion of CO_2 into the metal alkoxide bond.

The synthetic realization of a flexibly tethered bimetallic catalyst was done by Rieger and coworkers (Figure 16.7). In the copolymerization of CHO with CO_2, the bimetallic catalyst showed a TOF of 23 000 h^{-1} at 100 °C and 30 bar [77]. Further increasing the Lewis acidity at the zinc center by introducing CF_3-groups at the pentane backbone managed an improvement in the activity to the highest reported TOF for CHO with 155 000 h^{-1} [77]. The two zinc centers seem to have the perfect distance for enabling a rapid CHO coordination and ring opening. The reaction orders have been determined for catalyst **2a** and a switch in the rate-determining step observed [78]. At CO_2 pressures below 25 bar, the insertion of CO_2 becomes rate limiting, whereas the insertion of CHO is no longer bearing the reaction order of 1, but shows zero-order dependency.

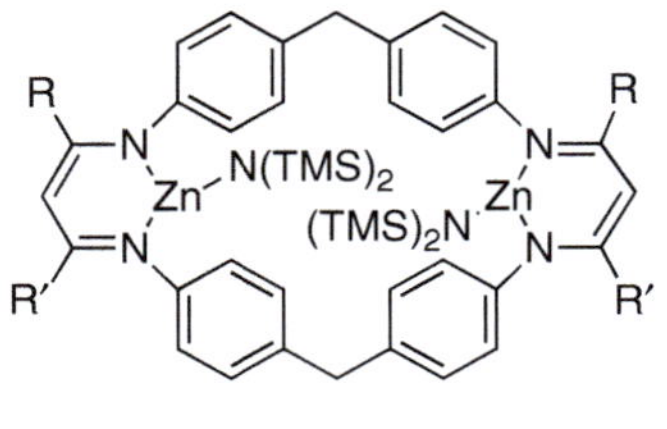

R = Me; R′ = Me **2a**
R = CF_3; R′ = Ph **2b**

2a TOF = 23 000h^{-1}
2b TOF = 155 000 h^{-1}

Figure 16.7 Flexibly tethered bimetallic zinc complex. Source: Kissling et al. [77].

Applying catalyst **2a** in the copolymerization of PO and CO_2 resulted in very low activities and selectivities toward polycarbonate formation [79, 80]. Density functional theory (DFT) calculations indicate a thermodynamically stable six-membered transition state consisting of an alkoxide-terminated chain end, the catalyst, and a carbonate moiety with 104.8 kJ mol^{-1} to be overcome in the case of PO as an epoxide. The incorporation of a carbon dioxide molecule in the polymerization of CHO as an epoxide is less hindered, explaining this discrepancy in the activity for both epoxides. A great deal of effort was put into the modification of dimeric complexes, but no major improvements were achieved for PO. Returning to the mononuclear structure and introducing two electron-withdrawing CF_3-groups at the pentane backbone revealed an "all-rounder" for the copolymerization of various epoxides and CO_2 (Figure 16.8) [81]. In the case of CHO, high turnover frequencies (TOF = 5520 h^{-1}) were again observed. In addition, however, challenging epoxides such as PO (TOF = 53 h^{-1}), styrene oxide (TOF = 26 h^{-1}), or the less explored

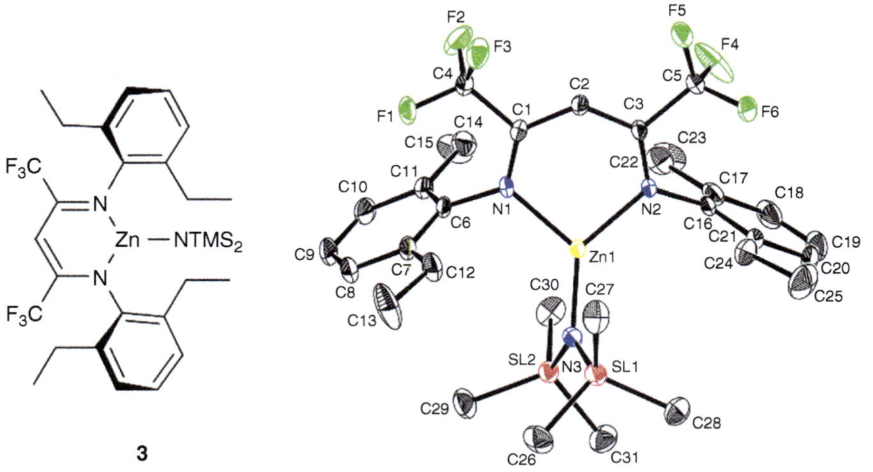

Figure 16.8 Structure of the β-diiminato-zinc-$NTMS_2$ complex **3** and its ORTEP style representation with ellipsoids drawn at the 50% probability level. Source: Reiter et al. [81]. © The Royal Society of Chemistry.

limonene oxide (TOF = 310 h^{-1}) were copolymerized to high molecular weights (M_n = 21.0–134 kg mol^{-1}) in a controlled manner ($Đ$ = 1.2–1.8). A more detailed overview of the copolymerization of limonene oxide with CO_2 is presented in Section 16.3.4.

The most intensively studied system for the copolymerization of epoxides and CO_2 is the metal salen system (**1d**). It has been excellently reviewed by a number of groups and will only be discussed briefly within this chapter [41, 42, 82]. Although suffering from overall lower activities compared to BDI complexes, salen ligands coordinated to a suitable transition metal (Cr, Co, etc.) allow high polymer selectivity under mild conditions and the introduction of regioselectivity. The first cobalt salen complex was published by Coates and coworkers in 2003 [83]. For the coupling of PO and CO_2 high polymer selectivities (>99%) for polymerization temperatures between 15 and 40 °C and molecular weights of up to 21.7 kg mol^{-1} were achieved. The *N*,*N*,*O*,*O*-tetradentate ligand structure has been optimized, but the cocatalyst plays an important role as well. The axial X group at the metal center should be an electrophile of poor leaving ability (e.g. 2,4-dinitrophenoxy or halide), whereas bulky nucleophiles with low coordination ability are good candidates for a cocatalyst (e.g. Bu_4NCl and PPNCl). Head-to-tail connectivities of >95% were able to be realized [84]. Based on these results, the salen systems have been improved in terms of activity, stereoselectivity, and regioselectivity [85–88].

A series of bimetallic complexes supported by a macrocyclic ligand system based on diphenolates were prepared by Williams and coworkers (Figure 16.9) [90]. The complexes were active in the copolymerization of CHO and CO_2, even at very low pressures (p_{CO_2} = 1 bar), with the polycarbonate selectivity remaining high. The introduction of defined equivalents of H_2O acting as chain transfer agents allowed the synthesis of polyols in controlled molecular weights. These polyols represent an interesting group of candidates for the production of polyurethanes [89]. Synthetic modifications both at the metal center and at the nucleophilic leaving group led to improvements in terms of activity and selectivity [91–94].

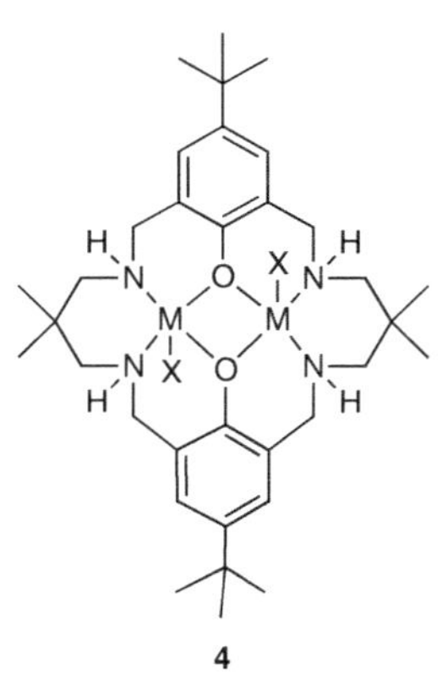

4a M = Zn, X = OAc
4b M = Mg, X = OAc
4c M = Mg, X = Br
4d M = Zn, X = CF_3CO_2
4e M = Zn, X = Ph

4

Figure 16.9 Macrocyclic diphenolate complexes with Zn or Mg as a central metal. Source: Kember and Williams [89].

16.3.3 Terpolymerization Pathways

The class of macrocyclic diphenolate complexes was also able to catalyze the ring-opening polymerization of other types of monomers such as lactones and anhydrides in combination with epoxides. In a first example, terpolymers from CHO, CO_2, and ε-caprolactone were synthesized to ABA block copolymers consisting of polyester and polycarbonate units. The catalyst exclusively initiated the copolymerization of CHO and CO_2, followed by the ring-opening polymerization of ε-CL only after releasing CO_2 completely from the reaction atmosphere [95]. In this case, CHO served as an situ formed alkoxide initiator enabling the ring-opening step of the lactone. The monomer scope was even extended to anhydrides, allowing the chemoselective synthesis of copolyesters and copolyester carbonates [96]. Using mixtures of three different monomers limit the synthesis of polymers to ABA block copolymers [97]. Pentablocks were realized by activating a similar dizinc catalyst with cyclohexanediol [98]. The diol enables the in situ formation of two alkoxide initiators, allowing the combination of three different catalytic cycles in a one-pot polymerization of four different types of monomers, namely, phthalic anhydride, ε-decalactone, CHO, and CO_2. Another approach for the combination of different catalytic cycles was presented by Wang et al. in 2018 [99]. The terpolymerization of CO_2, epoxides, and vinyl monomers by a porphyrin aluminum complex is presented. The advantage of this catalyst is its compatibility with the reversible addition fragmentation chain transfer (RAFT) polymerization, which enables the synthesis of PPC-*b*-PMMA copolymers to $M_n = 11.9–21.6\,kg\,mol^{-1}$ in narrow dispersities (1.09–1.14). The switch in the catalytic cycle was realized by a bifunctional chain transfer agent. A different approach was done by our group in order to realize both consecutively and simultaneously proceeding catalytic cycles [100]. BDI–Zn–$NTMS_2$ complex **3** shows high activity for both the ring-opening polymerization of β-butyrolactone and the copolymerization of CO_2 and epoxides [81, 101]. CO_2-controlled AB and BA block copolymers were prepared in one-pot polymerizations. Specifically, the polymerization procedure starting with the copolymerization of CO_2 and CHO ran in a highly selective manner because no incorporation of β-butyrolactone (BBL) was observed at that stage, and the ROP of BBL did not start until the excess of the carbon dioxide pressure was released. By lowering the CO_2 pressure to 3 bar, similar reaction rates for the two types of polymerization led to a statistical composition of the terpolymers. Molecular weights between 34 and $166\,kg\,mol^{-1}$ were obtained. Interestingly, the block copolymers exhibit two glass transition temperatures ($T_{g,\,PHB} \sim 2\,°C$, $T_{g,\,PCHC} \sim 115\,°C$) because of phase separation, whereas the mix-T_g of the statistical terpolymer ranges between 36 and 91 °C. Mechanistic insights could be gained using in situ IR spectroscopy, 2D NMR spectroscopy, and polarimetry, thus suggesting two zinc–alkoxide intermediates that can coordinate either BBL or CO_2 (Figure 16.10). Following carbon dioxide insertion, the respective zinc carbonate facilitates the ring opening of CHO.

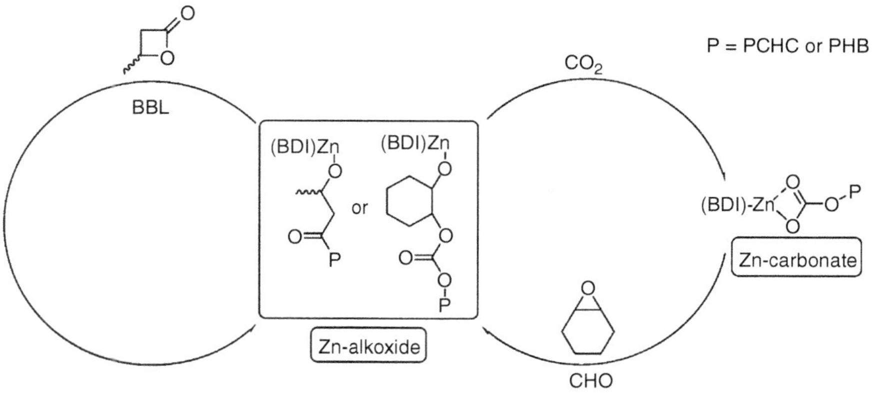

Figure 16.10 Postulated mechanism for the terpolymerization reaction of BBL, CHO, and CO_2 under 3 bar CO_2 pressure with catalyst **3**. Source: Kernbichl et al. [100]. © 2017 American Chemical Society.

16.3.4 Limonene Oxide: Recent Advances in Catalysis and Mechanism Elucidation

The bio-based limonene oxide was successfully copolymerized with CO_2 by Coates et al. in 2004 using the Lewis acidic zinc catalyst **5** (Figure 16.11) [6]. This catalyst bears one electron-withdrawing CF_3-group in the pentane backbone and an acetate moiety serving as a nucleophilic initiating group. A maximum TOF of $37\,h^{-1}$ was achieved at a reaction temperature of 25 °C. Surprisingly, the activity decreased drastically for temperatures above 35 °C. The reason for this was later explained by a relatively low ceiling temperature of PLC in the presence of a catalyst [81]. The second interesting observation was the exclusive incorporation of the *trans*-limonene oxide into the polymer even though the initial monomer consisted of a 46 : 54 cis : trans mixture. This fact renders the synthesis of PLC with BDI–Zn catalysts economically unfriendly. One approach was achieved by the group of Kleij with the introduction of an amino-triphenolate aluminum complex **6** with PPNCl as a cocatalyst [102]. Based on DFT calculations, it was shown that a precoordinated LO is prone to a nucleophilic attack by the chloride of PPNCl. This ring-opening step of the epoxide can occur on both positions of the oxirane ring. Surprisingly, the attack on the α-C-atom (position of methyl-group) requires the lowest energy, but overall, both diastereomers are incorporated into the polymer. In the same year, Coates changed the nucleophilic initiating group of system **5** from acetate to a trimethylsilylamide group [103]. The crystal structure of a 1 : 1 mixture of amorphous poly(*R*-limonene carbonate) and poly(*S*-limonene carbonate) indicates a semicrystalline nature of the packed polymer chains [104].

A second approach was demonstrated by Greiner and Rieger [16] The stereoselective epoxidation route from (+)-limonene with NBS enables the synthesis of 90%

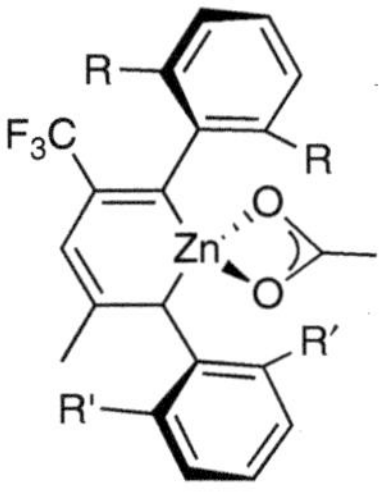

R = Et, R′ = Et **5b**
R = Et, R′ = ^{i}Pr **5b**

5

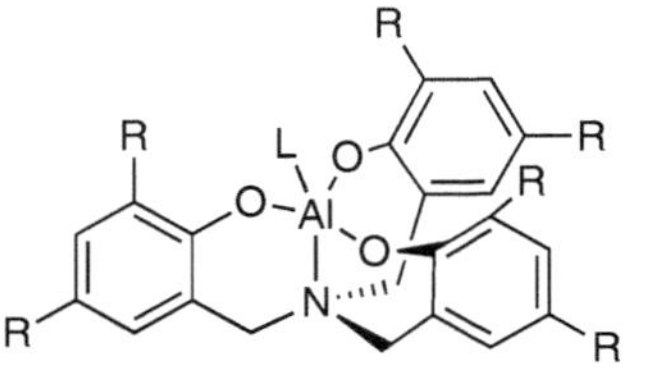

R = Me
L = THF

6

Figure 16.11 Active initiators for the copolymerization of CO_2 and limonene oxide: BDI–Zn–OAc **5** and the amino-triphenolate aluminum complex **6**. Source: (a) Adapted from Byrne et al. [6], (b) Peña Carrodeguas et al. [102].

trans-limonene oxide, allowing the use of BDI–Zn catalysts in a more cost-effective way. In addition, higher molecular weights (>100 kDa) were achieved, and the most important thermal, optical, and mechanical properties were determined. A glass transition temperature of 130 °C along with a transmission of a polymer film of 97% and a Young's modulus of 0.95 GPa with an elongation at break of 15% demonstrate the high potential of PLC for use as a rigid plastic alternative. Within this work, the reaction orders were determined for the polymerization with a BDI–Zn–OAc catalyst, and a possible mechanism could be postulated. In the case of CHO as an epoxide, the overall rate equation is $-d[CHO]/dt = k{\cdot}[CO_2]^0{\cdot}[CHO]^1{\cdot}[Zn]^1$ for the same catalyst [76]. This demonstrates that the coordination of CHO is the rate-determining step, whereas the incorporation of CO_2 is very fast. The decisive difference in the rate law for limonene oxide was observed for the order with respect to limonene oxide: $-d[LO]/dt = k{\cdot}[CO_2]^0{\cdot}[LO]^2{\cdot}[Zn]^1$. Because of the higher steric demand of the limonene oxide molecule, it is assumed that a nucleophilic attack of the acetate group is not possible within the first step (Figure 16.12). It rather requires the coordination of a second LO molecule to spatially separate the two zinc centers. After a rearrangement, the initiating group can finally successfully ring open the epoxide, and the alternating addition of carbon dioxide and limonene oxide can take place.

Modifying the BDI structure by the introduction two electron-withdrawing groups and an $NTMS_2$-initiating group led to the highest activity in the copolymerization of LO and CO_2 at a TOF of 310 h^{-1} [81]. In this work, the reason for low conversions at elevated temperatures was addressed in more detail. As previously observed by Coates, the polymerization to PLC hardly reaches high conversions and actually results in lower yields when polymerized at 60 °C instead of at 40 °C. It was assumed that either *cis*-LO that stays in the reaction mixture and is not polymerized or the double bond of LO might act as a catalyst poison, preventing the polymerization from proceeding to high conversion. Both using 99% *trans*-LO and fully hydrated LO, also called menthene oxide, indeed showed the same polymerization behavior and no increase in yield at 60 °C. Using in situ IR spectroscopy, a deeper insight into polymerization activity at certain temperatures could be gained, indicating a relatively low ceiling temperature at 60 °C for LO concentrations of 3.5 mol l^{-1}. [81] Until

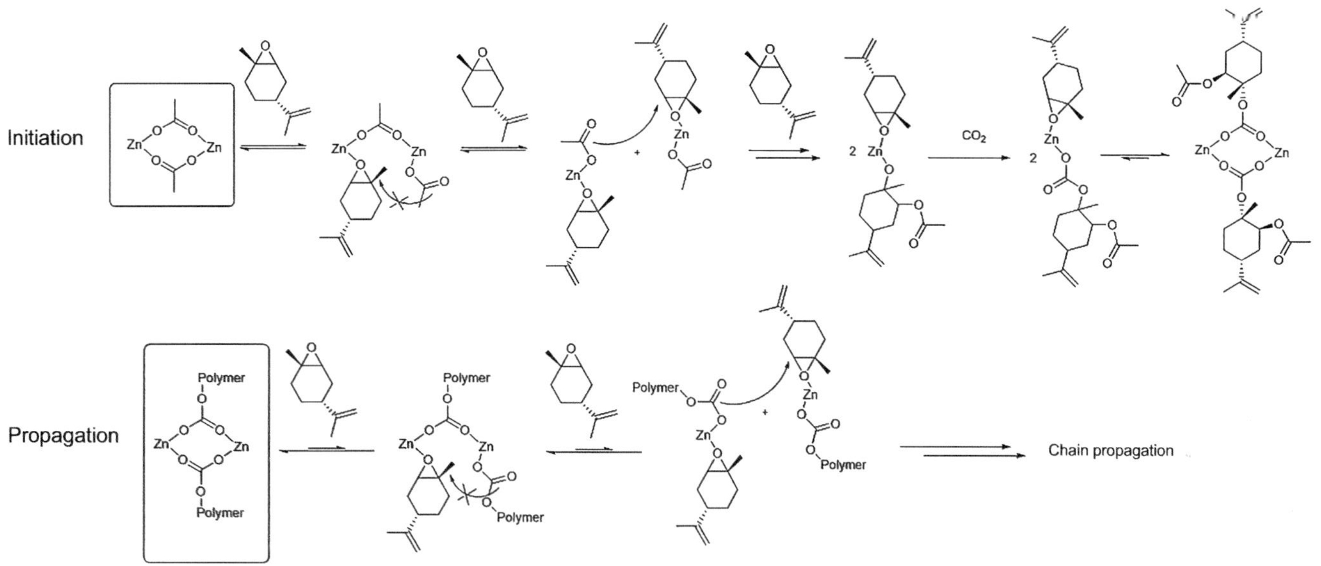

Figure 16.12 Postulated mechanism for the copolymerization of limonene oxide and CO_2 with BDI−Zn−OAc. Source: Hauenstein et al. [16], © Royal Society of Chemistry.

now, a multitude of different synthetic modifications were performed and possible applications were investigated [27, 105–110]. Kleij and coworkers developed a stepwise approach to poly(limonene dicarbonate) via PLC, followed by the epoxidation of the isopropenyl group that again can be transformed into the respective carbonate with the help of CO_2 and PPNCl [111]. Glass transition temperatures of up to 180 °C were obtained for molar masses of 15 kg mol^{-1} ($Đ = 1.31$). The group of Koning also investigated the synthesis and use of poly(limonene-8,9-oxide carbonate) [112]. They obtained this aliphatic epoxy polycarbonate through the alternating copolymerization of limonene dioxide and CO_2 with a BDI–Zn–$NTMS_2$ catalyst. The epoxy group opens a magnitude of possibilities for post-modification: Thiols, carboxylic acids, or CO_2 again afford functional polymers for different areas of application such as epoxy resins or non-isocyanate polyurethanes.

16.4 Conclusion

Aliphatic polycarbonates have been known for more than 50 years but remain highly interesting in a variety of research areas. Although PPC, derived from PO and CO_2, has already found its industrial application in small-scale processes, PCHC continues to serve as an academic benchmark for catalysis research. Although heterogeneous systems dominate the industrial PO/CO_2 copolymerization, dramatic improvements become possible with homogeneous catalysts in the past decade. BDI zinc systems seem to be the most active catalysts for the copolymerization of cyclohexene oxide and CO_2. Besides BDI complexes, macrocyclic diphenolate systems allow the coupling of CO_2 and epoxides as well as the incorporation of another type of monomer such as cyclic esters or anhydrides. These one-pot terpolymerizations can be nicely controlled via the carbon dioxide pressure. The combination of various polymerization cycles by a single catalyst represents an interesting tool for chemoselectively building up block copolymers with different features. BDI complexes also enable the coupling of the bio-based epoxide limonene oxide and CO_2 to the very promising material PLC. Intense research in recent years has revealed possible areas of application for PLC: In addition to the use as a non-isocyanate building block in polyurethane production or in coating applications, the detailed determination of its thermal, optical, and mechanical properties has made it a potential rigid plastic alternative to BPA-PC. However, turnover numbers still need to be increased for limonene oxide, and it will be very interesting to see whether this limitation can be solved in the near future in order to make an industrial process possible.

References

1 Bhanage, B.M. and Arai, M. (2014). *Transformation and Utilization of Carbon Dioxide*. Berlin/Heidelberg, Germany: Springer.

2 Sakakura, T. and Kohno, K. (2009). *Chem. Comm.*: 1312–1330.

3 Büttner, H., Longwitz, L., Steinbauer, J. et al. (2017). *Top. Curr. Chem.* 375: 50.
4 Lu, X.B. (2015). *Carbon Dioxide and Organometallics*. Springer International Publishing.
5 Luinstra, G.A. (2008). *Polym. Rev.* 48: 192–219.
6 Byrne, C.M., Allen, S.D., Lobkovsky, E.B., and Coates, G.W. (2004). *J. Am. Chem. Soc.* 126: 11404–11405.
7 Langanke, J., Wolf, A., Hofmann, J. et al. (2014). *Green Chem.* 16: 1865–1870.
8 Inoue, S., Koinuma, H., and Tsuruta, T. (1969). *Die Makromol. Chem.* 130: 210–220.
9 H. Baer, M. Bergamo, A. Forlin, L. H. Pottenger, J. Lindner, in *Ullmann's Encyclopedia of Industrial Chemistry*, 2012.
10 Yilmaz, B. and Müller, U. (2009). *Top. Catal.* 52: 888–895.
11 Kreimeyer, A., Eckes, P., Fischer, C. et al. (2015). *Angew. Chem. Int. Ed.* 54: 3178–3195.
12 Zhou X.-T., Ji, H.-B., Xu, J.-C. et al. (2007). *Tetrahedron Lett.* 48: 2691–2695.
13 R. J. Gall, F. P. Greenspan (1956). Preparation of cyclohexene oxide, Google Patents US2745848 A.
14 Chapman, A.M., Keyworth, C., Kember, M.R. et al. (2015). *ACS Catal.* 5: 1581–1588.
15 Ciriminna, R., Lomeli-Rodriguez, M., Demma Carà, P. et al. (2014). *Chem. Commun.* 50: 15288–15296.
16 Hauenstein, O., Reiter, M., Agarwal, S. et al. (2016). *Green Chem.* 18: 760–770.
17 Poland, S.J. and Darensbourg, D.J. (2017). *Green Chem.* 19: 4990–5011.
18 F. P. Greenspan, S. M. Linder, (1961). Method of making carvomenthene oxide, Google Patents US3014929 A.
19 Darensbourg, D.J., Yarbrough, J.C., Ortiz, C., and Fang, C.C. (2003). *J. Am. Chem. Soc.* 125: 7586–7591.
20 Phillips, O., Schwartz, J.M., and Kohl, P.A. (2016). *Polym. Degrad. Stab.* 125: 129–139.
21 Chisholm, M.H., Navarro-Llobet, D., and Zhou, Z. (2002). *Macromolecules* 35: 6494–6504.
22 Liu, B., Zhao, X., Wang, X., and Wang, F. (2003). *J. Appl. Polym. Sci.* 90: 947–953.
23 Varghese, J.K., Na, S.J., Park, J.H. et al. (2010). *Polym. Degrad. Stab.* 95: 1039–1044.
24 Lin, S., Yu, W., Wang, X., and Zhou, C. (2014). *Ind. Eng. Chem. Res.* 53: 18411–18419.
25 Darensbourg, D.J. (2018). *Polym. Degrad. Stab.* 149: 45–51.
26 Rieger, B., Künkel, A., and Coates, G.W. (2012). *Synthetic Biodegradable Polymers*. Berlin/Heidelberg, Germany: Springer.
27 Li, C. Sablong, R.J., van Benthem, R.A.T.M., and Koning, C.E. (2017). *ACS Macro Lett.* 6: 684–688.
28 Darensbourg, D.J., Wei, S.-H., Yeung, A.D., and Ellis, W.C. (2013). *Macromolecules* 46: 5850–5855.
29 Qin, Y. and Wang, X. (2010). *Biotechnol. J.* 5: 1164–1180.

30 Koning, C., Wildeson, J., Parton, R. et al. (2001). *Polymer* 42: 3995–4004.
31 Seong, J.E., Na, S.J., Cyriac, A. et al. (2010). *Macromolecules* 43: 903–908.
32 Thorat, S.D., Phillips, P.J., Semenov, V., and Gakh, A. (2003). *J. Appl. Polym. Sci.* 89: 1163–1176.
33 Shi, L., Lu, X.-B., Zhang, R. et al. (2006). *Macromolecules* 39: 5679–5685.
34 Darensbourg, D.J. and Holtcamp, M.W. (1995). *Macromolecules* 28: 7577–7579.
35 Demirel, Y. (2015). *J. Chem. Eng. Process Technol.* 6: 236.
36 Wang, X.L., Du, F.G., Jiao, J. et al. (2007). *J. Biomed. Mater. Res. Part B* 83B: 373–379.
37 Klaus, S., Lehenmeier, M.W., Anderson, C.E., and Rieger, B. (2011). *Coord. Chem. Rev.* 255: 1460–1479.
38 Darensbourg, D.J. and Holtcamp, M.W. (1996). *Coord. Chem. Rev.* 153: 155–174.
39 Coates, G.W. and Moore, D.R. (2004). *Angew. Chem. Int. Ed.* 43: 6618–6639.
40 Sugimoto, H. and Inoue, S. (2004). *J. Polym. Sci. Pol. Chem.* 42: 5561–5573.
41 Darensbourg, D.J., Mackiewicz, R.M., Phelps, A.L., and Billodeaux, D.R. (2004). *Acc. Chem. Res.* 37: 836–844.
42 Darensbourg, D.J. (2007). *Chem. Rev.* 107: 2388–2410.
43 Kember, M.R., Buchard, A., and Williams, C.K. (2011). *Chem. Commun.* 47: 141–163.
44 Darensbourg, D.J. and Wilson, S.J. (2012). *Green Chem.* 14: 2665–2671.
45 Childers, M.I., Longo, J.M., Van Zee, N.J. et al. (2014). *Chem. Rev.* 114: 8129–8152.
46 Wang, Y. and Darensbourg, D.J. (2018). *Coord. Chem. Rev.* 372: 85–100.
47 Kozak, C.M., Ambrose, K., and Anderson, T.S. (2018). *Coord. Chem. Rev.* 376: 565–587.
48 Inoue, S., Koinuma, H., and Tsuruta, T. (1969). *J. Polym. Sci.* 7: 287–292.
49 Soga, K., Imai, E., and Hattori, I. (1981). *Polym. J.* 13: 407–410.
50 Ree, M., Hwang, Y., Kim, J.-S. et al. (2006). *Catal. Today* 115: 134–145.
51 Klaus, S., Lehenmeier, M.W., Herdtweck, E. et al. (2011). *J. Am. Chem. Soc.* 133: 13151–13161.
52 Meng, Y.Z., Du, L.C., Tiong, S.C. et al. (2002). *J. Polym. Sci. Pol. Chem.* 40: 3579–3591.
53 Gao, L.J., Xiao, M., Wang, S.J. et al. (2007). *J. Appl. Polym. Sci.* 104: 15–20.
54 Takeda, N. and Inoue, S. (1978). *Die Makromol. Chem.* 179: 1377–1381.
55 Aida, T. and Inoue, S. (1985). *J. Am. Chem. Soc.* 107: 1358–1364.
56 Kojima, F., Aida, T., and Inoue, S. (1986). *J. Am. Chem. Soc.* 108: 391–395.
57 Kruper, W.J. and Dellar, D.D. (1995). *J. Org. Chem.* 60: 725–727.
58 Qin, Y., Wang, X., Zhang, S. et al. (2008). *J. Polym. Sci. Pol. Chem.* 46: 5959–5967.
59 Chisholm, M.H. and Zhou, Z. (2004). *J. Am. Chem. Soc.* 126: 11030–11039.
60 Chen, P., Chisholm, M.H., Gallucci, J.C. et al. (2005). *Inorg. Chem.* 44: 2588–2595.
61 Anderson, C.E., Vagin, S.I., Xia, W. et al. (2012). *Macromolecules* 45: 6840–6849.
62 Chatterjee, C. and Chisholm, M.H. (2012). *Inorg. Chem.* 51: 12041–12052.

63 Anderson, C.E., Vagin, S.I., Hammann, M. et al. (2013). *ChemCatChem* 5: 3269–3280.

64 Xia, W., Vagin, S.I., and Rieger, B. (2014). *Chem. Eur. J.* 20: 15499–15504.

65 Sheng X., Wang, Y., Qin, Y. et al. (2014). *RSC Adv.* 4: 54043–54050.

66 Wu, W., Sheng, X., Qin, Y. et al. (2014). *J. Polym. Sci. Pol. Chem.* 52: 2346–2355.

67 Geerts, R.L., Huffman, J.C., and Caulton, K.G. (1986). *Inorg. Chem.* 25: 1803–1805.

68 Darensbourg, D.J., Niezgoda, S.A., Draper, J.D., and Reibenspies, J.H. (1998). *J. Am. Chem. Soc.* 120: 4690–4698.

69 Darensbourg, D.J., Wildeson, J.R., Yarbrough, J.C., and Reibenspies, J.H. (2000). *J. Am. Chem. Soc.* 122: 12487–12496.

70 Darensbourg, D.J., Holtcamp, M.W., Struck, G.E. et al. (1999). *J. Am. Chem. Soc.* 121: 107–116.

71 Cheng, M., Lobkovsky, E.B., and Coates, G.W. (1998). *J. Am. Chem. Soc.* 120: 11018–11019.

72 Cheng, M., Moore, D.R., Reczek, J.J. et al. (2001). *J. Am. Chem. Soc.* 123: 8738–8749.

73 Moore, D.R., Cheng, M., Lobkovsky, E.B., and Coates, G.W. (2002). *Angew. Chem. Int. Ed.* 41: 2599–2602.

74 Bourget-Merle, L., Lappert, M.F., and Severn, J.R. (2002). *Chem. Rev.* 102: 3031–3066.

75 Cheng, M., Darling, N.A., Lobkovsky, E.B., and Coates, G.W. (2000). *Chem. Commun.*: 2007–2008.

76 Moore, D.R., Cheng, M., Lobkovsky, E.B., and Coates, G.W. (2003). *J. Am. Chem. Soc.* 125: 11911–11924.

77 Kissling, S., Lehenmeier, M.W., Altenbuchner, P.T. et al. (2015). *Chem. Commun.* 51: 4579–4582.

78 Lehenmeier, M.W., Kissling, S., Altenbuchner, P.T. et al. (2013). *Angew. Chem. Int. Ed.* 52: 9821–9826.

79 Kissling, S., Altenbuchner, P.T., Lehenmeier, M.W. et al. (2015). *Chem. Eur. J.* 21: 8148–8157.

80 Allen, S.D., Moore, D.R., Lobkovsky, E.B., and Coates, G.W. (2002). *J. Am. Chem Soc.* 124: 14284–14285.

81 Reiter, M., Vagin, S., Kronast, A. et al. (2017). *Chem. Sci.* 8: 1876–1882.

82 Lu, X.-B. and Darensbourg, D.J. (2012). *Chem. Soc. Rev.* 41: 1462–1484.

83 Qin, Z., Thomas, C.M., Lee, S., and Coates, G.W. (2003). *Angew. Chem. Int. Ed.* 42: 5484–5487.

84 Lu, X.-B., Shi, L., Wang, Y.-M. et al. (2006). *J. Am. Chem. Soc.* 128: 1664–1674.

85 Darensbourg, D.J. and Yarbrough, J.C. (2002). *J. Am. Chem. Soc.* 124: 6335–6342.

86 Cohen, C.T., Chu, T., and Coates, G.W. (2005). *J. Am. Chem. Soc.* 127: 10869–10878.

87 Lu, X.-B. and Wang, Y. (2004). *Angew. Chem. Int. Ed.* 43: 3574–3577.

88 Paddock, R.L. and Nguyen, S.T. (2005). *Macromolecules* 38: 6251–6253.

89 Kember, M.R. and Williams, C.K. (2012). *J. Am. Chem. Soc.* 134: 15676–15679.

90 Buchard, A., Kember, M.R., Sandeman, K.G., and Williams, C.K. (2011). *Chem. Commun.* 47: 212–214.

91 Kember, M.R., Knight, P.D., Reung, P.T.R., and Williams, C.K. (2009). *Angew. Chem. Int. Ed.* 48: 931–933.

92 Kember, M.R., White, A.J.P., and Williams, C.K. (2010). *Macromolecules* 43: 2291–2298.

93 Buchard, A., Jutz, F., Kember, M.R. et al. (2012). *Macromolecules* 45: 6781–6795.

94 Jutz, F., Buchard, A., Kember, M.R. et al. (2011). *J. Am. Chem. Soc.* 133: 17395–17405.

95 Paul, S., Romain, C., Shaw, J., and Williams, C.K. (2015). *Macromolecules* 48: 6047–6056.

96 Romain, C., Zhu, Y., Dingwall, P. et al. (2016). *J. Am. Chem. Soc.* 138: 4120–4131.

97 Stößer, T. and Williams, C.K. (2018). *Angew. Chem. Int. Ed.* 57: 6337–6341.

98 Chen, T.T.D., Zhu, Y., and Williams, C.K. (2018). *Macromolecules* 51: 5346–5351.

99 Wang, Y., Zhao, Y., Ye, Y. et al. (2018). *Angew. Chem. Int. Ed.* 57: 3593–3597.

100 Kernbichl, S., Reiter, M., Adams, F. et al. (2017). *J. Am. Chem. Soc.* 139: 6787–6790.

101 Kronast, A., Reiter, M., Altenbuchner, P.T. et al. (2016). *Organometallics* 35: 681–685.

102 Peña Carrodeguas, L., González-Fabra, J., Castro-Gómez, F. et al. (2015). *Chem. Eur. J.* 21: 6115–6122.

103 Auriemma, F., De Rosa, C., Di Caprio, M.R. et al. (2015). *Angew. Chem. Int. Ed.* 54: 1215–1218.

104 Auriemma, F., De Rosa, C., Di Caprio, M.R. et al. (2015). *Macromolecules* 48: 2534–2550.

105 Hauenstein, O., Agarwal, S., and Greiner, A. (2016). *Nat. Commun.* 7: 11862.

106 Li, C., Sablong, R.J., and Koning, C.E. (2015). *Eur. Polym. J.* 67: 449–458.

107 Li, C., Johansson, M., Sablong, R.J., and Koning, C.E. (2017). *Eur. Polym. J.* 96: 337–349.

108 Stößer, T., Li, C., Unruangsri, J. et al. (2017). *Polym. Chem.* 8: 6099–6105.

109 Hauenstein, O., Rahman, M.M., Elsayed, M. et al. (2017). *Adv. Mater. Technol.* 2: 1700026.

110 Parrino, F., Fidalgo, A., Palmisano, L. et al. (2018). *ACS Omega* 3: 4884–4890.

111 Kindermann, N., Cristòfol, À., and Kleij, A.W. (2017). *ACS Catal.* 7: 3860–3863.

112 Li, C., Sablong, R.J., and Koning, C.E. (2016). *Angew. Chem. Int. Ed.* 128: 11744–11748.

17

Metal–Organic Frameworks (MOFs) for CO_2 Cycloaddition Reactions

Ignacio Campello, Antonio Sepúlveda-Escribano, and Enrique V. Ramos-Fernández

Universidad de Alicante, Instituto Universitario de Materiales, Departamento de Química Inorgánica, Laboratorio de Materiales Avanzados, Ctra. San Vicente, Alicante s/n, E-03690 San Vicente del Raspeig, Spain

17.1 Introduction to MOF

The so-called metal–organic frameworks (MOFs) are relatively new materials, which now represent a very wealthy and productive research field [1]. They are hybrid materials that lay at the interface of organic and inorganic compounds. However, because of their interesting properties, they overpass the frontier of a simply uncommon compounds to become materials with a myriad of potential applications [2–4]. The potential of these materials lays in the almost infinite combination of different compounds, which can be synthetized. That is the reason why the amount of publications about MOFs has increased exponentially in the past decade.

MOFs are hybrid systems with one or more metallic atoms as inorganic part (metallic centers), bonded to an organic molecule via coordination bond (linker). This later plays the role of the ligand in coordination chemistry. The periodic table limits the metallic centers able to participate in the compounds, but there are MOFs made of almost all of them, going from alkaline to rare earth metals. However, the number of organic ligands is much larger because a small change in the ligand turns the MOF toward one or another final structure. The vast amount of possible materials results in a huge amount of opportunities for research. Figure 17.1 schematizes an MOF structure constructed by an iron-based center and two different types of linkers. More than 5000 different structures of MOFs [1, 6–8] have been discovered, pointing the structural diversity that can be achieved with this type of materials (Figure 17.2).

17.2 MOFs as Catalysts

The use of MOFs as catalysts has been deeply revised in many interesting reviews [10, 11]. In this chapter, we will focus on particular catalytic active sites in MOFs

Engineering Solutions for CO_2 Conversion, First Edition.
Edited by Tomas R. Reina, José A. Odriozola, and Harvey Arellano-Garcia.

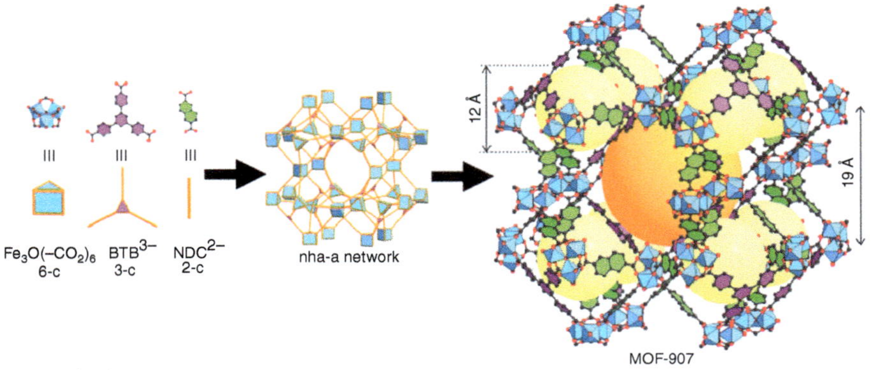

Figure 17.1 Schematic representation of the MOF structural levels and components. Source: Nguyen et al. [5].

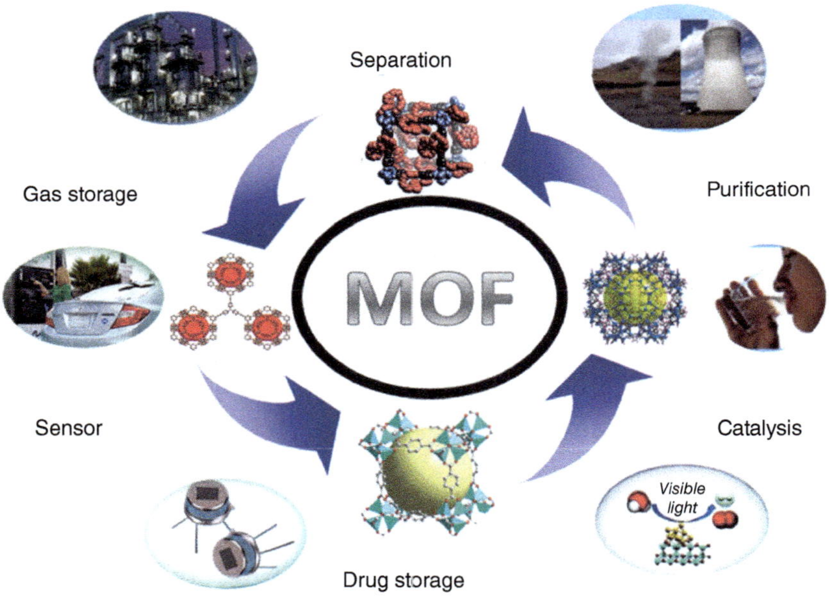

Figure 17.2 Examples of different applications involving MOF materials. Source: Chaemchuen et al. [9].

that can be tuned to enhance their reactivity toward CO_2 conversion reactions. The utilization of MOFs as catalysts for CO_2 transformations is plenty of possible applications. However, they drastically decrease when the thermal stability of MOFs is considered because, for example, CO_2 hydrogenations require high temperatures that would make MOF unsuitable. Thus, the applications of MOFs as catalysts for CO_2 transformations are restricted to three main groups of reactions: photocatalysis, electroreduction of CO_2, and CO_2 cycloadditions. The two first applications, which are out the scope of this chapter and will not further be considered, have been recently the object of excellent reviews [12–14]. Regarding CO_2 additions, it is worth to mention that it is by far the most studied reaction using MOFs as catalysts because MOFs have several unique properties that make them important for this type of reactions, as it would be further explained in this chapter.

The use of MOFs as catalysts support is also a very developed field in MOF research. Actually, many important achievements have been reached. Thus, the Pardo's group has developed the synthesis of sub-nanometric metal particles with very unique properties that can only be achieved in MOFs structures that are able to catalyze the water gas shift (WGS) reaction at room temperature [15]. Saying so, the discussion of metal nanoparticles supported on MOFs will be omitted in this chapter because often the catalytic performance of metal nanoparticles supported on classical supports overpasses the performance of the ones supported on MOFs. Furthermore, the focus on this chapter will be the intrinsic catalytic behavior of MOFs for CO_2 activation.

17.2.1 Active Sites in MOFs: Lewis Acid Sites

17.2.1.1 Historical Overview

When some MOFs are synthetized, the coordination sites of the metal centers may not be fully filled out with the functional groups of the linkers. These unsaturated sites can be occupied by some coordinating guest species such as water, hydroxyl groups, or solvent molecules used in the synthesis. If the MOF is treated at moderate or high temperature and/or under vacuum, the coordinating guest molecules can be removed. This generates the so-called open metal sites, OMSs (also referred to as unsaturated metal sites, coordinatively unsaturated metal sites, or exposed metal centers), that are electron deficient; in other words, they are Lewis acid centers. Figure 17.3 shows how OMSs in the inorganic cluster of MIL-101(Cr) are generated. In the past few years, OMSs have been used as a catalyst in Lewis acid catalyzed

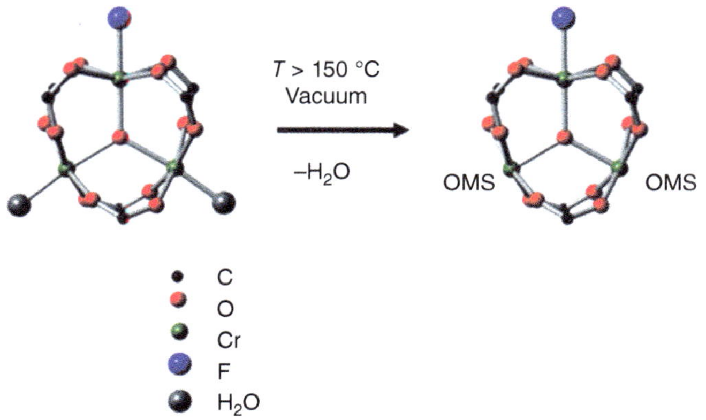

Figure 17.3 Generation of open metal site (OMS) illustrative example. Source: Rogge et al.[16]. Published by The Royal Society of Chemistry.

reactions such as acetalization of aldehydes [17], cyanosilylation of aldehydes [18], and in oxidation reactions [19–21].

The OMSs are extremely important for CO_2 conversion, as it would be explained below. Therefore, it is worth detailing how OMSs in MOFs were discovered, the most important MOFs having OMSs, and how OMSs can be intentionally created, incorporated, and modified in MOFs. The first MOF discovered having OMSs was the so-called HKUST-1 [22] (HKUST = Hong Kong University of Science and Technology). After this discovery, the catalysis community started to use this material as a catalyst in simple reactions [17]. The applicability of the HKUST-1 in catalysis is very limited because it is not stable in the presence of water, and water is involved in many reactions. Thus, HKUST-1 is a nice lab case but is far from being applicable. Saying so, it is important to remark that HKUST-1 was an important MOF for the catalysis community because it was the first MOF discovered with intrinsic catalytic properties.

Later in 2005, the group of Gerard Ferey discovered two new MOFs called MIL-100(Cr) and MIL-101(Cr) (MIL = Matériaux de l'Institut Lavoisier) [23, 24]. These materials contain one OMS per metal atom in the structure and have a great thermal and chemical stability. Furthermore, both materials have a very developed mesoporosity that help to avoid internal mass transfer limitations, making the OMSs more accessible to the reactant. These materials were first applied in selective sulfoxidation of aryl sulfides using hydrogen peroxide as the oxidant, showing the ability of the OMS to activate peroxide [25]. Later on, Kholdeeva and coworkers [20, 26] studied this material as a catalyst in several oxidation reactions of organic compounds, also using peroxides as the oxidant. It was later when MIL-100,1 material was prepared with iron. This metal not only provides acid centers but also redox centers. In this sense, this material can catalyze oxidation reactions using molecular oxygen as an oxidant, as Llabrés i Xamena and coworkers has demonstrated [21]. Continuing with the MIL-100,1 family, Mitchell et al. demonstrated that the Sc-containing version of these MOFs outperformed the catalytic behavior of their counterparts with Fe, Cr in several reactions catalyzed by Lewis acid [27].

Almost at the same time, in Norway, two new families of materials having strong Lewis acidity were developed: UiO-66 [2, 28–31] (Univeresitetet i Oslo) and CPO-27 [32–39] (Coordination Polymer of Oslo), the latter one being also known as MOF-77. It has been deeply studied by the Long's group [40–44] at the University of Berkeley. These materials have been a breakthrough in the applications of MOF in catalysis because their Lewis acidity is especially strong and tunable and the density of Lewis acid sites is very high. The case of UiO-66(Zr) is the most relevant one, and since its discovery, more than 3000 publications have been reported studying its potential applications as an adsorbent [2], catalyst [28–31, 45–47], etc. Some good examples of the application of UiO-66 as a catalyst are the work carried out by de Vos and coworkers [30, 48–50]. They demonstrated that the Lewis acidity of UiO-66 could be modified as desired not only by choosing the proper linker but also by modifying the metal cluster with another metal, changing the electronic configuration to a large extent [50, 51].

17.2.1.2 Tunability of the Lewis Acid Sites

17.2.1.2.1 Nature of the Metal Cluster

The presence of the Lewis acid sites may be finely tuned for catalytic applications. The strength of the Lewis acid centers in an MOF depends on the electronic properties of the metal cluster. Hence, the most important factor controlling the strength of acidity is the nature of the metal cluster and the coordinating topology. Thus, MOFs have been created having open metal sites with metals such as Mn, Ni [27], Cu [36], Zr [31], Fe [52], Zn, Sc [27], etc. [39, 53]. In this way, it is possible to prepare MOFs having the same topology and types of OMSs but constructed with different metals. Thus, the strength of the OMSs can be compared in the same conditions, as it has been done by Wright and coworkers [27], showing that in many cases, the scandium-based MOF outperformed the catalytic behavior of their counterpart.

Another approach to modify the Lewis acidity in MOFs is the preparation of materials having two different metals in the same cluster. Especially relevant are the works published by Stock and coworkers [54]; they demonstrated that the catalytic behavior of the MOF was strongly modified by substituting some zirconium atoms in the metallic centers by cerium atoms, enhancing the catalytic performance. These authors developed an one-pot hydrothermal synthesis method to control the degree of substitution. The metal centers in the MOF can also be substituted by other metal cations by a post-synthetic methodology. In this sense, metal centers without OMSs in an MOF can be transformed in an MOF with OMSs. It can be done by submitting the parent MOF to a post-synthetic treatment where the second metal can exchange the metal present in the MOF. The work by Dinca [53] fully demonstrated that the Zn^{2+} cation in MOF-5 can be substituted by several M^{3+} cations.

Another way to modify the acidity of the MOF is the utilization of a proper linker. MOFs having the same topology and metal centers can be prepared with linkers having different functional groups that might vary from electron-withdrawing to electron-donating groups. This can alter the electronic properties of the metal centers in such way that their acidity would be modified as desired. This was demonstrated by the de Vos's group in several works [49, 50]. They found that there was a strong correlation between the chemical nature of the functional groups in the linker and the Lewis acidity, and in consequence, the catalytic performance can be tailored.

17.2.1.2.2 Defect Generation

In some MOFs, the OMSs are inherent to the structure (see Figure 17.3); in other MOFs, however, it must be induced. One way to generate OMS sites is by inducing defects in the MOF structure. As it was explained above, the Lewis acid sites in MOFs are due to the presence of free coordination sites. MOFs are crystals that can be defined as an infinite periodic repetition of fragments in space. However, actual crystals are very far for being perfect and many defects can appear. If the number of defects is not very large, they would not affect the overall properties of the crystal. In some cases, the number of defects may be so large that the overall properties are defined by the defects.

If during the synthesis of an MOF some of the linkers are missed, this will generate extra OMSs because the metal would not be fully coordinated. These types of defects appear in all the crystallization processes in general, but they can also be intentionally created.

Surface Defects The first type of defect one can find in a crystal is the crystal boundary. MOF crystals can terminate exposing metal centers, the linkers, or even a combination of both. This would depend on the MOF topology, structure, chemistry, and crystal orientation. Several works have already been reported in which the surface imperfections of the MOFs have been used for catalytic applications. In 2010, Chizallet et al. [55] published a work demonstrating that surface defects in ZIF-8 (ZIF means zeolitic imidazolate framework) could catalyze transesterification reactions in a very efficient way. This remarkable result was explained on the basis of the amount of surface defects. Thus, the authors showed that the number of imidazole moieties as well as Lewis acid sites is crucial for the catalytic performance. In this sense, it is reasonable to think that the crystal size is very important for the catalytic performance because the number of exposed active sites would depend on the surface-to-volume ratio. In fact, this has been demonstrated in several studies [53, 56]. However, the surface-to-volume ratio is not the only factor affecting the catalytic activity for this type of materials. Thus, it has been demonstrated that the exposed crystal face is also a key point. Hence, Chen et al. [57] prepared ZIF-8 with different morphologies, from rhombic dodecahedral to cubes. They found that ZIF-8 cubes were one order of magnitude more active than their counterparts, indicating that the surface defects can be tailored not only in number (surface–volume ratio) but also in strength.

Bulk Defects The bulk defects are those inside the MOF crystal. For using these defects as active sites, the reactants must diffuse in through the crystal, and the produced product diffuses out. Thus, MOFs should have the proper porosity to allow reactants to diffuse faster than reaction takes place in the active sites. Bulk defects are also present in many MOFs after the synthesis; however, after a typical synthesis, the number of bulk defects in the MOF is not very high. The number of defects must be increased by using synthetic and post-synthetic approaches.

Bulk defects can be created during the synthesis. A very simple way is to speed up the crystallization rate. A fast crystallization process produces crystals with very poor quality in terms of crystallinity. This provokes a large number of defects in the lattice that might be used in catalysis. This was first reported by Ravon et al. [58] and further developed by Ramos-Fernandez et al. [59, 60] and Llabrés i Xamena and coworkers [57]. A second approach is the use of *dummy linkers* or modulators during the synthesis. In a typical synthesis of an MOF, a bidentate (terephthalic acid) linker is used to build up the frameworks. If a portion of monodentate linker (such as benzoic acid) is added to the synthesis mixture, both linkers will compete for the coordination with the metal. When the MOF crystallizes, the presence of monodentate linkers in the final structure produces defects that are normally called "*truncated missing linkers*." This approach is very promising not only to create defects, OMS, but also to control the strength of the Lewis acid sites.

Post-synthetic approaches to generate OMSs are also very important. Thus, a prepared MOF can be submitted to different treatments to extract or leach out linkers from the framework, generating in that way the defects. This was firstly demonstrated by Rosseinsky and coworkers [61]; they prepared an amino acid-based MOF that was later treated with diluted HCl solutions. They found that the treated samples were far more active in the methanolysis of *cis*-2,3-epoxybutane than the untreated one. Later on, Vermoortele et al. [48] followed the same approach, but in this case, with MIL-100(Fe), finding out that the generation of Lewis acid sites was very remarkable.

17.2.1.3 Active Sites in MOFs: Lewis Basic Sites

Lewis basic sites correspond to species with pairs of free electrons in their configuration. They are defined as the electron donor specie that bonds to a Lewis acid in order to form a so-called adduct or complex [62]. MOFs can have intrinsic Lewis basicity because oxygen atoms forming part of likers (i.e. terephthalate) may have a basic character [63–65]. However, this Lewis basicity is partially neutralized by the bonding with the metal center. The strength and types of Lewis base sites can also be tuned and modified by choosing the proper synthetic approaches.

The most used way to induce Lewis basicity in an MOF is the functionalization of the linker with basic sites. If linker containing nitrogen functional groups are used (for example, $-NH_2$), an MOF having basicity will be obtained. Nitrogen atom in the amine group has five electrons in the valance band, three of them being used for bonding and the other two being unpaired. This causes heavy repulsions with the other bonds of the molecule and, in order to find a more favorable energy, these groups try to share these electrons with Lewis acids. Because nitrogen functionalities are very commonly found in the organic linkers of MOFs, nitrogen groups are the most common atoms referred when it turns to Lewis basicity. The basicity of amines and amides is a largely studied topic in organic chemistry [63], and, similarly with Lewis basic groups, they are in a close relationship with substituents and the surrounding environment of the nitrogen groups. Because MOFs are formed with a large variety of linkers, the possibility of functionalization of the basic groups opens vast possibilities. This trend has been largely studied [64, 65] because of the interesting applications that functionalized nitrogen groups present.

The interaction of basic sites with acid sites in the same MOF was demonstrated by Serra-Crespo et al. [66]. In this work, the authors demonstrated that amine groups and OMS of the lattice interact in such a way that the flexibility of MIL-53 is strongly modified because of these interactions.

Another way to introduce Lewis basicity is by creating defects in the frameworks, as it was explained in Section 17.2.1.2.2. In many cases, upon defect generation, not only Lewis acid sites are produced but also Lewis conjugated bases are also created. An illustrative example of this is the surface defects that can contain uncoordinated metals and undercoordinated linkers with different acid–base properties.

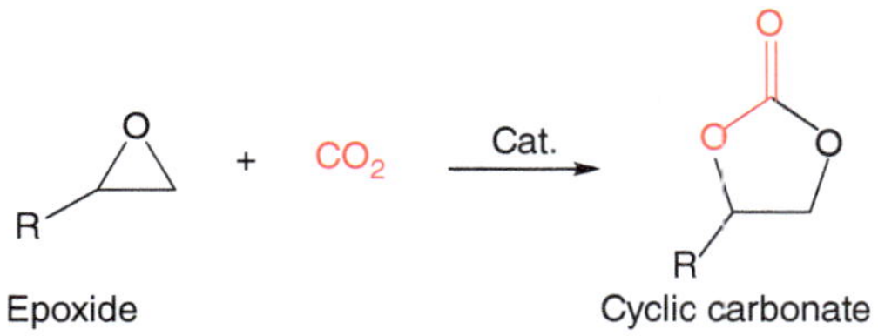

Scheme 17.1 Synthesis of cyclic carbonates from an epoxide.

17.3 CO_2 Cycloadditions

The cycloaddition reaction is one of the most efficient ways for CO_2 chemical fixation. The reaction consists basically in the opening of an epoxide ring forming cyclic carbonates upon reaction with CO_2. This reaction does not involve the reduction of the carbon atom in CO_2, which makes it less energetically demanding that other CO_2 conversion reactions. Furthermore, this reaction is also more environmentally friendly and less toxic than the traditional methods to produce cyclic carbonates that involve the use of phosgene. Cyclic carbonates are valuable products that can be used as electrolytes [61] or as a monomer in the production of polycarbonate that are produced at massive scale (Scheme 17.1) [61].

17.3.1 Reaction Mechanism

There are many possible mechanisms involved in this reaction, and it remains very difficult to clarify the exact mechanism occurring because even using powerful synchrotron techniques, only few clues can be extracted. This is due to the high number of intermediate species and interactions that take place, which complicates to totally discard a given reaction pathway. Saying so, it is generally accepted that Lewis acid sites are the main active sites in this reaction. Thus, the epoxide first binds to the Lewis acid site through the basic oxygen; then, a nucleophilic attack by a cocatalyst is needed (typically an ammonium halide) to activate the epoxide. The nucleophilic attack is typically carried out by the halide, forming in that way a halo-alkoxide. This intermediate can further react with the CO_2 through a cycloaddition reaction forming the cyclic carbonate and regenerating the halide [61]. In most cases, the nucleophilic attack is very fast, and the rate-limiting step is the coordination of the epoxide to the Lewis acid site. Most of scientific efforts have been devoted to produce solid Lewis acid materials with the proper strength to activate the epoxide. Figure 17.4 shows the reaction mechanism explained above, with a cobalt-based MOF being used as a catalyst [61].

Cycloaddition reactions proceed in the absence of cocatalysts when some catalysts are used. This has been ascribed to the presence of both acid and basic sites in the catalysts. The epoxide is coordinated to the Lewis acid site and the created reaction intermediate can be stabilized by the presence of the basic centers that might act as nucleophiles. The presence of basic sites may make unnecessary the use of a cocatalyst because the nucleophilic specie would be the Lewis basic site. In Section 17.3.2,

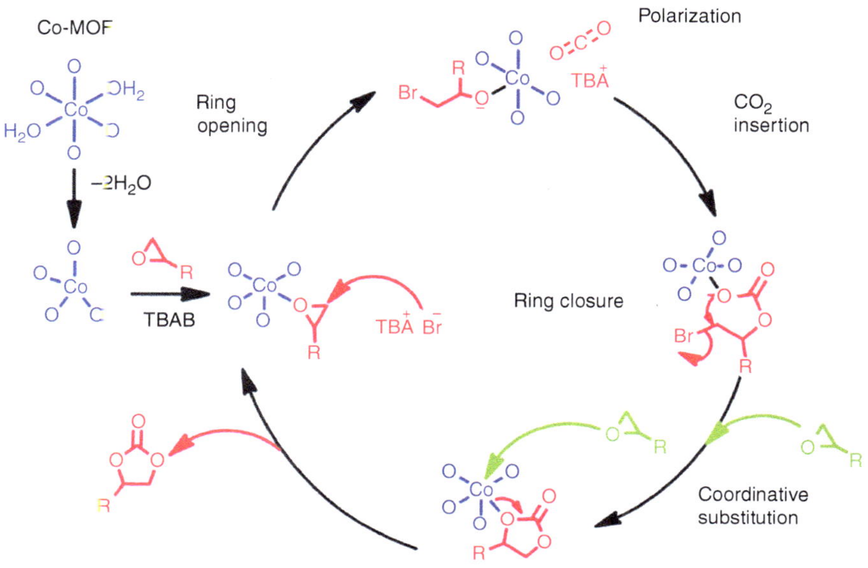

Figure 17.4 Proposed mechanism of CO_2 cycloaddition. Source: Dhankhar and Nagaraja [67]. © Royal Society of Chemistry, 2019.

details will be given on how MOFs can catalyze the cycloaddition reactions in the presence and absence of cocatalysts.

Another mechanism proposed by Kim et al. [68] describes the activation of the CO_2 molecule on the basic sites of the linkers, and the activation of the epoxide due to the acidic sites corresponding to the metallic centers, which are then more suitable for producing the reaction.

17.3.2 CO_2 Cycloadditions Reactions Catalyzed by Lewis Acid MOFs

MOFs having only Lewis acid sites cannot catalyze the reaction if a cocatalyst is not present in the reaction media, as it has been explained above. There are many examples reporting the catalytic activity of MOFs having Lewis acid sites. Table 17.1 summarizes the works published in CO_2 cycloadditions catalyzed by MOFs with and without cocatalysts. It is very difficult to establish relationships between Lewis acidity and conversion. It is also difficult to compare different MOFs because in many cases, the reaction conditions are different. Regarding the Lewis acidity, MOFs constructed from early transition metals (Zr and Hf) should be more active than mid (Cr and Fe) and late (Cu and Zn) transition metals because of the strength of their acidic character [41–43, 74]. Two good examples of MOFs based on early transition metal used as catalysts for CO_2 cycloaddition is the one reported by Guillerm and coworkers [61] (yttrium-based MOF) and the one published by Farha and coworkers [61] (hafnium-based MOF). The work by Guillerm shows that the catalytic activity of the yttrium-based MOF cannot reach their homogeneous counterpart catalyst, indicating that mass transport problems are also a key factor when MOFs are used for catalysis. The work by Farha shows that the Hf-based MOF is exceptionally good for this reaction. They ascribed this good behavior to the presence of a small amount of Brönsted acid sites that work hand in hand with the Lewis acid sites. One example of CO_2 cycloaddtions catalyzed by MOF constructed with mid transition metals is the MIL-101(Cr), firstly reported by Ferey et al. [24] and later used in catalysis by Zalomaeva et al. [77]. Furthermore, they compared the catalytic behavior of MIL-101(Cr) against the same framework constructed with iron (MIL-101(Fe)) [20] (Table 17.1, entry 11). They found that the Fe-based MOF was more effective than the Cr-based one

MOFs with late transition metals would be in principle less active because their Lewis acidity should be lower. As it has been explained above, this is not always straightforward. The advantage of using weak Lewis acid sites is that the epoxide will not be strongly adsorbed, decreasing the possibility of deactivation. Thus, it has been reported that HKUST-1 can be used for CO_2 cycloadditions. The main problem of this MOF is the lack of thermal and chemical stability [78].

17.3.3 CO_2 Cycloaddition Reactions Catalyzed by Lewis Acid and Basic MOFs

As it has been mentioned above, CO_2 cycloaddition reactions can be catalyzed in the absence of cocatalysts. To do it, the catalyst must be bifunctional: it must contain Lewis acid sites together with Lewis base sites that work as nucleophiles. These MOFs can be intentionally designed for that purpose. There are two main families of MOFs having Lewis acidity–basicity: those having OMSs and nitrogen-containing linkers and defective MOFs having both functionalities.

17.3.4 Defective MOFs for CO_2 Cycloaddition Reactions

It has been demonstrated that MOFs having coordinatively saturated metal centers (not having OMSs) and inert linkers (no basic function) can show a remarkable catalytic activity in CO_2 cycloadditions. Thus, Miralda et al. [69] (Table 17.1, entry 1[a]) showed that ZIF-8 was active in the absence of a cocatalyst. They showed that the presence of defects in the ZIF-8, imidazole moieties which are strong Lewis bases, as well as undercoordinated Zn atoms, were the responsible of the activity. Other types of ZIF materials have also shown catalytic activity in CO_2 cycloadditions [72]. The works about ZIF-8 and this reaction are of especial interest because ZIF-8 is very easily prepared and tuned (Table 17.1, entry 4). ZIF-8 has very narrow pores, and its catalytic activity is maintained even when sterically hindered reactants are used. Therefore, this shows that the catalytic activity mainly takes place at the external crystal surface, and this is a catalysis process assisted by surface defects. Miralda et al. [69] decided to functionalize the surface of the ZIF-8 crystals with amine groups in order to improve the interaction between CO_2 and the catalyst (Table 17.1, entry 1[b]). The functionalized materials catalyzed the reaction more efficiently than the unfunctionalized one. This example points out the importance of defects in MOFs for this type of transformations.

Table 17.1 List of different MOFs and their performance in cycloaddition reactions.

Entry	MOF	Substrate	Temp (°C)	P (bar)	Conversion (%)	Selectivity (%)	Time (h)	Cocatalyst	Ref
1[a]	ZIF-8	EP	80	7	84	52	4	Yes	Miralda et al. [69]
2[a]	ZIF-8	EP	100	7	75	72	5	No	Zhu et al. [70]
2[b]	ZIF-8	SO	100	7	54	100	5	No	Zhu et al. [70]
3	ZIF-8	SO	100	20	53	100	4	No	Kim et al. [71]
1[b]	NR2-ZIF-8	EP	80	7	100	73	4	Yes	Miralda et al. [69]
4	ZIF-68	SO	120	10	93	100	4	No	Yang et al. [72]
5	gea-MOF-1	SO,EP	120	20	85.8	—	6	Yes	Guillerm et al. [73]
6	Hf-NU-1000	SO	25	1	100	100	48	Yes	Beyzavi et al. [74]
7	Co-MOF-74	SO	100	20	96	100	4	No	Cho et al. [75]
8	Mg-MOF-74	SO	100	20	94	100	4	No	Yang et al. [76]
9[a]	Mg-MOF-74	SO	100	20	96	100	4	No	Kim et al. [68]
10[a]	Cr-MIL-101	SO	70	8	95	60	7	No	Zalomaeva et al. [77]
10[b]	Cr-MIL-101	SO	25	8	49	33	24	Yes	Zalomaeva et al. [77]
9[b]	Cr-MIL-101	SO	100	20	63	100	4	Yes	Kim et al. [68]
11	Fe-MIL-101	SO	25	8	93	100	48	Yes	Maksimchuk et al. [20]
12	HKUST-1	EP	100	7	64	52	4	No	Macias et al. [78]
13	HKUST-1	SO	100	20	63	100	4	Yes	Kim et al. [68]
14[a]	Ni(salphen)-MOF	SO,EP	80	20	84	—	4	Yes	Ren et al. [79]
14[b]	Ni(salphen)-MOF	SO,EP	80	20	81	—	4	Yes	Ren et al. [79]
15[a]	MIL-68(In)	SO	150	8	39	—	8	Yes	Lescouet [80]
15[b]	MIL-68(In)-NH_2	SO	150	8	71	—	8	Yes	Lescouet [80]

Table 17.1 (Continued)

Entry	MOF	Substrate	Temp (°C)	P (bar)	Conversion (%)	Selectivity (%)	Time (h)	Cocatalyst	Ref
9^c	UIO-66	SO	100	20	48	100	1	Yes	Kim et al. [68]
9^d	UIO-66-NH_2	SO	100	20	70	100	1	Yes	Kim et al. [68]
16	F-IRMOF-3	SO,EP	140	20	8480	100	5, 1.5	Yes	Zhou et al. [81]
9^e	IR-MOF-3	SO	100	20	33	100	4	Yes	Kim et al. [68]
16	int-MOF-5/*n*-Bu4-NBr	SO	50	4	85	100	24	Yes	H. Kim et al. [82]
17	Zn-DABCO	EP	100	8	100	100	12	No	Mousavi et al. [83]
18	Mn-MOF-1	EP	25	1	98	100	48	Yes	Sharma et al. [84]
19	Cu-MOF-1	EP	80	25	96	100	5	Yes	Guo et al. [85]
20	Im-MnF	SO,EP	140 100	15	58/96	97/99	6	Yes	Wu et al. [86]
21	{[Ni3HL(μ3-OH)(H_2O)$_2$]·3(H_2O)·DMA}n	SO,EP	100	10	90 100	100	12, 6	Yes	Yang et al. [87]
22	Ni-TCPE-1	SO	100	10	100	100	12	Yes	Zhou et al. [88]
23	UiO-66/Cu-BTC	SO,EP	60	12	2691	100 100	8	Yes	Park et al. [89]
24	Zn-BTC-2MeIm	EP	100	30	100	99	6	No	Xiao et al. [90]
25	Li-UTSA-16	SO,EP	120	12	26 100	100 100	6	No	Zhang et al. [91]
26	Ni-MOF-1	SO,EP	60	1	9497	100 100	6	Yes	Li et al. [92]
27	Mn–BTC	EP	105	30	98	96	9	No	Xiao et al. [93]
28	Co(tp)(bpy)	EP	100	10	96	98	8	No	Xiao et al. [94]
29	UIO-66-OH	SO,EP	140	10	9690	100 100	2	Yes	Noh et al. [95]
30	UMCM-15	SO,EP	25	12	4580	100 100	18	Yes	Park et al. [96]
31	MIL-101-IMBr-6	SO,EP	80	8	5592	8796	4	No	Li et al. [97]
32	ZIF-8-90	EP	100	20	96	99	4	No	Lui et al. [98]
33	ZIF-22	SO,EP	120	12	7599	9498	2	No	Park et al. [99]

Yang et al. [72] utilized ZIF-68 as a catalyst (Table 17.1, entry 4). The main differences between ZIF-8 and ZIF-68 is that ZIF-68 has a different topology with larger pores, and it is functionalized with a nitro group where ZIF-8 has a methyl group. This study demonstrated that ZIF-68 behaves better, and it was ascribed to the presence of the nitro group and to the best accessibility of reactants to the bulk defects.

17.3.5 MOFs Having Functional Linkers for CO_2 Cycloaddition Reactions

One of the most reported ways to introduce Lewis base functionalities in MOFs is the utilization of amino functionalized linkers as building blocks. In this way, the amine pendant groups will provide weak Lewis basicity [80]. The combination of this Lewis basicity with the OMSs will produce bifunctional MOFs able to catalyze CO_2 cycloadditions. Thus, the Lewis acid sites must activate the epoxide while the basic sites may have two roles as nucleophiles or as active centers for CO_2 molecules. The first example of bifunctional acid–base MOF catalyst for CO_2 additions was reported by Lescoutet et at [80]. (Table 17.1, entry 15[a,b]). In this work, the authors used MIL-68(In) and the amine-functionalized version, the MIL-68(In)-NH_2. They found that both materials were catalytically active in CO_2 cycloadditions. The catalytic activity was ascribed to the presence of OMSs and a small number of defects in the unfunctionalized MOF. The amino version was far more active because of the presence of Lewis basic sites in the form of amine groups. Another example of MOFs having bifunctional acid–base character is the UiO-66(Zr) and its amino functionalized version UiO-66(Zr) [68] (Table 17.1, entry 9[c,d]). It is known that strong Lewis basic sites [100] do not improve the catalytic activity, actually they can even be detrimental.

Besides the amino base Lewis acid sites, it is also possible to immobilize or to use other functional groups to avoid the use of cocatalysts. In this sense, the works proposing the grafting of the cocatalyst molecule to the MOF are very interesting. The cocatalysts are normally quaternary ammonium salts with ionic–liquid character. Hence, it is important to remark that the MOF should have a mesoporous structure because ionic–liquid moieties are very bulky. Dee Won, in a pioneering work [101], reported the immobilization of an ammonium salt in IRMOF-3 and the remarkable behavior of this material in CO_2 cycloadditions. The main problem was its lack of stability; however, this study can be considered as an elegant proof of concept. Ma et al. [102] reported the immobilization of a quaternary ammonium salt in a amino-functionalized mesoporous MIL-101(Cr) structure. They found that this composite material could be used and recycled and that the amine groups in the MIL-101(Cr) served as anchoring sites for the ammonium salts. If an unfunctionalized material is used, the ammonium salts are not efficiently anchored. Later on, Sun et al. [103] followed a similar approach also using Mil-101(Cr) structure but, in this case, it was functionalized with sulfonic groups that were used as anchoring sites. We would like to point out that these types of approaches are very promising because the degree of freedom to develop new catalytic materials is very high. Actually, since 2017, many new research studies have been carried out in this topic, and many new achievements are expected.

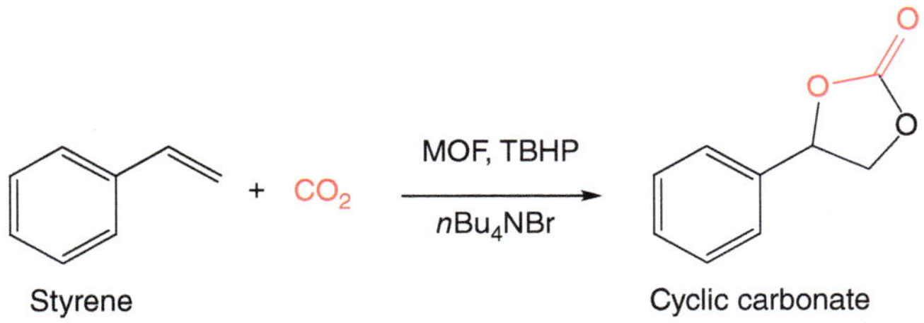

Scheme 17.2 Oxidative carboxylation reaction of styrene.

17.4 Oxidative Carboxylation

As it has been mentioned in Section 17.2, Lewis acid sites present in MOFs are not only good catalysts for Lewis acid catalyzed reactions but also for oxidation reactions when a peroxide is used as an oxidant. Actually, it has been demonstrated that Lewis acid sites of ML-101(Cr or Fe) may catalyze the olefin oxidation to produce their corresponding aldehydes, alcohols, etc. [59, 104]. Moreover, it has been shown that finely tuning the reaction conditions, this type of materials can also be used for olefin epoxidation. The epoxidation of styrene using MOFs as a catalyst has been recently studied by Sun et al. [105]. They discovered that selectivity toward the styrene epoxide was around 95% at half conversion and using *tert*-butyl hydroperoxide as an oxidant. The work by Dinca and coworkers [104] is also remarkable. The authors used a bimetallic version of MOF-5 a catalyst for epoxidation and explained the catalytic behavior in terms of the presence of redox pairs.

In this context, one-pot oxidative carboxylation of styrene with CO_2 would be an interesting concept to go from olefins to cyclic carbonates. The idea consists of using styrene instead of styrene epoxide as the reactant. This change is interesting because epoxides are usually expensive and explosive. The study of the performance of some MOFs for this reaction has been recently reported [106]. By applying mild conditions (1 bar and 80 °C), the authors were able to obtain values of 87% of conversion and selectivity of 99% using the so-called MOF-590 as a catalyst. However, it was necessary to add an oxidant cocatalyst (TBHP) *tert*-butyl hydroxide. Because this new approach has not been widely studied, it remains difficult to propose a mechanism (Scheme 17.2).

The authors also used a large library of different MOFs as a catalyst for the one-pot synthesis of cyclic carbonates, finding out that promising MOFs should contain Lewis acid sites.

References

1 Férey, G. (2008). Hybrid porous solids: past, present, future. *Chem. Soc. Rev.* 37: 191–214.

2 Cunha, D., Ben Yahia, M., Hall, S. et al. (2013). Rationale of drug encapsulation and release from biocompatible porous metal-organic frameworks. *Chem. Mater.* 25: 2767–2776.

3 Homayoonnia, S. and Zeinali, S. (2016). Design and fabrication of capacitive nanosensor based on MOF nanoparticles as sensing layer for VOCs detection. *Sensors Actuators, B Chem.* 237: 776–786.

4 Bagheri, A., Taghizadeh M., Behbahani M., Asgharinezhad A.A. et al. (2012). Synthesis and characterization of magnetic metal-organic framework (MOF) as a novel sorbent, and its optimization by experimental design methodology for determination of palladium in environmental samples. *Talanta* 99: 132–139.

5 Nguyen, H.L. et al. (2018). A complex metal-organic framework catalyst for microwave-assisted radical polymerization. *Commun. Chem.* 1: 1–7.

6 Eddaoudi, M. and Eubank, J.F. (2010). Insight into the development of metal-organic materials (MOMs): at zeolite-like metal-organic frameworks (ZMOFs). In: *Metal-Organic Frameworks: Design and Application* (ed. L.R. MacGillivray), 37–89. Wiley https://doi.org/10.1002/9780470606858.ch2.

7 Stassen, I. et al. (2017). An updated roadmap for the integration of metal-organic frameworks with electronic devices and chemical sensors. *Chem. Soc. Rev.* 46: 3185–3241.

8 Czaja, A., Leung, E., Trukhan, N., and Müller, U. (2011). Industrial MOF synthesis. In: *Metal-Organic Frameworks: Applications from Catalysis to Gas Storage*, 337–352. Wiley-VCH https://doi.org/10.1002/9783527635856.ch14.

9 Chaemchuen, S., Kabir, N.A., Zhou, K., and Verpoort, F. (2013). Metal-organic frameworks for upgrading biogas via CO_2 adsorption to biogas green energy. *Chem. Soc. Rev.* 42: 9304–9332.

10 Wang, C., Liu, D., and Lin, W. (2013). Metal-organic frameworks as a tunable platform for designing functional molecular materials. *J. Am. Chem. Soc.* 135: 13222–13234.

11 Gascon, J., Corma, A., Kapteijn, F. et al. (2014). Metal organic framework catalysis: quo vadis? *ACS Catalysis* 4: 361–378.

12 Dhakshinamoorthy, A., Li, Z., and Garcia, H. (2018). Catalysis and photocatalysis by metal organic frameworks. *Chem. Soc. Rev.* 47: 8134–8172.

13 Li, Y., Xu, H., Ouyang, S., and Ye, J. (2016). Metal–organic frameworks for photocatalysis. *Phys. Chem. Chem. Phys.* 18: 7563–7572.

14 Al-Rowaili, F.N., Jamal, A., Ba Shammakh, M.S., and Rana, A. (2018). A review on recent advances for electrochemical reduction of carbon dioxide to methanol using metal–organic framework (MOF) and non-MOF catalysts: challenges and future prospects. *ACS Sustain. Chem. Eng.* 6: 15895–15914.

15 Rivero-Crespo, M.A. et al. (2018). Confined Pt1 1+ water clusters in a MOF catalyze the low-temperature water–gas shift reaction with both CO_2 oxygen atoms coming from water. *Angewandte Chemie - International Edition* (59, 52): 17094–17099. https://doi.org/10.1002/anie.201810251.

16 Rogge, S.M.J. et al. (2017). Metal-organic and covalent organic frameworks as single-site catalysts. *Chem. Soc. Rev.* 46: 3134–3184.

17 Schlichte, K., Kratzke, T., and Kaskel, S. (2004). Improved synthesis, thermal stability and catalytic properties of the metal-organic framework compound $Cu_3(BTC)_2$. *Microporous Mesoporous Mater.* 73: 81–88.

18 Gómez-Lor, B. et al. (2005). Novel 2D and 3D indium metal-organic frameworks: topology and catalytic properties. *Chem. Mater.* 17: 2568–2573.

19 Skobelev, I.Y., Sorokin, A.B., Kovalenko, K.A. et al. (2013). Solvent-free allylic oxidation of alkenes with O_2 mediated by Fe- and Cr-MIL-101. *J. Catal.* 298: 61–69.

20 Maksimchuk, N.V., Zalomaeva, O.V., Skobelev, I.Y. et al. (2012). Metal–organic frameworks of the MIL-101 family as heterogeneous single-site catalysts. *Proc. R. Soc. A Math. Phys. Eng. Sci.* 468: 2017–2034.

21 Santiago-Portillo, A. et al. (2015). MIL-101 as reusable solid catalyst for autoxidation of benzylic hydrocarbons in the absence of additional oxidizing reagents. *ACS Catal.* 5: 3216–3224.

22 Chui, S.S.-Y., Lo, S.M.-F., Charmant, J.P.H. et al. (1999). A chemically functionalizable nanoporous material $[Cu_3(TMA)_2\ (H_2O)_3](n)$. *Science* 283 (5405): 1148–1150.

23 Vimont, A. et al. (2006). Investigation of acid sites in a zeotypic giant pores chromium(III) carboxylate. *J. Am. Chem. Soc.* 128: 3218–3227.

24 Férey, G. et al. (2005). A chromium terephthalate-based solid with unusually large pore volumes and surface area. *Science* 309: 2040–2042.

25 Hwang, Y.K. et al. (2009). Selective sulfoxidation of aryl sulfides by coordinatively unsaturated metal centers in chromium carboxylate MIL-101. *Appl. Catal. A Gen.* 358: 249–253.

26 Hwang, Y.K., Ferey, G., Lee, U.-H., and Chang, J.-S. (2013). Liquid phase oxidation of organic compounds by metal-organic frameworks. In: *Liquid Phase Oxidation via Heterogeneous Catalysis* (eds. M.G. Clerici and O.A. Kholdeeva), 371–409. Wiley https://doi.org/10.1002/9781118356760.ch8.

27 Mitchell, L. et al. (2013). Remarkable Lewis acid catalytic performance of the scandium trimesate metal organic framework MIL-100(Sc) for C—C and C—N bond-forming reactions. *Catal. Sci. Technol.* 3: 606–617.

28 DeStefano, M.R., Islamoglu, T., Garibay, S.J. et al. (2017). Room-temperature synthesis of UiO-66 and thermal modulation of densities of defect sites. *Chem. Mater.* 29: 1357–1361.

29 Valenzano, L. et al. (2011). Disclosing the complex structure of UiO-66 metal organic framework: a synergic combination of experiment and theory. *Chem. Mater.* 23: 1700–1718.

30 Smolders, S. et al. (2017). Unravelling the redox-catalytic behavior of Ce_{4+} metal-organic frameworks by X-ray absorption spectroscopy. *ChemPhysChem* 19: 373–378.

31 Van de Voorde, B. et al. (2015). Improving the mechanical stability of zirconium-based metal–organic frameworks by incorporation of acidic modulators. *J. Mater. Chem. A* 3: 1737–1742.

32 Bonino, F. et al. (2008). Local structure of CPO-27-Ni metallorganic framework upon dehydration and coordination of NO. *Chem. Mater.* 20: 4957–4968.

33 Chavan, S. et al. (2009). Response of CPO-27-Ni towards CO, N_2 and C_2H_4. *Phys. Chem. Chem. Phys.* 11: 9811–9822.

34 Tagliabue, M. et al. (2011). Methane storage on CPO-27-Ni pellets. *J. Porous Mater.* 18: 289–296.

35 Chavan, S. et al. (2012). Functionalization of CPO-27-Ni through metal hexacarbonyls: the role of open Ni_{2+} sites. *Microporous Mesoporous Mater.* 157: 56–61.

36 Märcz, M., Johnsen, R.E., Dietzel, P.D.C., and Fjellvåg, H. (2012). The iron member of the CPO-27 coordination polymer series: synthesis, characterization, and intriguing redox properties. *Microporous Mesoporous Mater.* 157: 62–74.

37 Guasch, J., Dietzel, P.D.C., Collier, P., and Acerbi, N. (2015). The effect of solvent and temperature in the synthesis of CPO-27-Ni by reflux. *Microporous Mesoporous Mater.* 203: 238–244.

38 Drenchev, N. et al. (2018). Open metal sites in the metal-organic framework CPO-27-Cu: detection of regular and defect copper species by CO and NO probe molecules. *J. Phys. Chem. C* 122: 17238–17249.

39 Rosnes, M.H., Nesse, F.S., Opitz, M., and Dietzel, P.D.C. (2019). Morphology control in modulated synthesis of metal-organic framework CPO-27. *Microporous Mesoporous Mater.* 275: 207–213.

40 Bachman, J.E., Smith, Z.P., Li, T. et al. (2016). Enhanced ethylene separation and plasticization resistance in polymer membranes incorporating metal-organic framework nanocrystals. *Nat. Mater.* 15: 845–849.

41 Xiao, D.J., Oktawiec, J., Milner, P.J., and Long, J.R. (2016). Pore environment effects on catalytic cyclohexane oxidation in expanded Fe_2 (dobdc) analogues. *J. Am. Chem. Soc.* 138: 14371–14379.

42 Borycz, J. et al. (2016). Structural and electronic effects on the properties of Fe_2 (dobdc) upon oxidation with N_2O. *Inorg. Chem.* 55: 4924–4934.

43 Carraro, F. et al. (2017). Catalysis in MOFs: general discussion. *Faraday Discuss.* 201: 369–394.

44 Bachman, J.E. and Long, J.R. (2016). Plasticization-resistant Ni_2(dobdc)/polyimide composite membranes for the removal of CO_2 from natural gas. *Energy Environ. Sci.* 9: 2031–2036.

45 Nouar, F. et al. (2015). Tuning the properties of the UiO-66 metal organic framework by Ce substitution. *Chem. Commun.* 51: 14458–14461.

46 Wang, A. et al. (2016). Titanium incorporated with UiO-66(Zr)-type metal–organic framework (MOF) for photocatalytic application. *RSC Adv.* 6: 3671–3679.

47 Deleu, W.P.R., Stassen, I., Jonckheere, D. et al. (2016). Waste PET (bottles) as a resource or substrate for MOF synthesis. *J. Mater. Chem. A* 4: 9519–9525.

48 Vermoortele, F. et al. (2012). Tuning the catalytic performance of metal-organic frameworks in fine chemistry by active site engineering. *J. Mater. Chem.* 22: 10313–10321.

49 Vermoortele, F. et al. (2013). Synthesis modulation as a tool to increase the catalytic activity of metal-organic frameworks: the unique case of UiO-66(Zr). *J. Am. Chem. Soc.* 135: 11465–11468.

50 Vermoortele, F. et al. (2012). Electronic effects of linker substitution on Lewis acid catalysis with metal-organic frameworks. *Angew. Chemie - Int. Ed.* 51: 4887–4890.

51 Lammert, M. et al. (2015). Cerium-based metal organic frameworks with UiO-66 architecture: synthesis, properties and redox catalytic activity. *Chem. Commun.* 51: 12578–12581.

52 Casco, M.E. et al. (2016). Paving the way for methane hydrate formation on metal-organic frameworks (MOFs). *Chem. Sci.* 7: 3658–3666.

53 Hu, Z. and Zhao, D. (2017). Metal–organic frameworks with Lewis acidity: synthesis, characterization, and catalytic applications. *CrystEngComm* 19: 4066–4081.

54 Lammert, M., Glißmann, C., and Stock, N. (2017). Tuning the stability of bimetallic Ce(iv)/Zr(iv)-based MOFs with UiO-66 and MOF-808 structures. *Dalt. Trans.* 46: 2425–2429.

55 Chizallet, C., Lazare, S., Bazer-Bachi, D. et al. (2010). Catalysis of transesterification by a nonfunctionalized metal–organic framework: acido-basicity at the external surface of ZIF-8 probed by FTIR and *ab initio* calculations. *J. Am. Chem. Soc.* 132 (35): 12365–12377. https://doi.org/10.1021/ja103365s.

56 Yang, Q., Xu, Q., and Jiang, H.-L. (2017). Metal–organic frameworks meet metal nanoparticles: synergistic effect for enhanced catalysis. *Chem. Soc. Rev.* 46: 4774–4808.

57 Chen, L. et al. (2016). Facet-dependent catalytic activity of ZIF-8 nanocubes and rhombic dodecahedra based on tracing substrate diffusion in pores by SERS: a case study for surface catalysis of MOFs. *Catal. Sci. Technol.* 6: 1616–1620.

58 Ravon, U. et al. (2010). Engineering of coordination polymers for shape selective alkylation of large aromatics and the role of defects. *Microporous Mesoporous Mater.* 129: 319–329.

59 Ramos-Fernandez, E.V., Garcia-Domingos, M., Juan-Alcañiz, J. et al. (2011). MOFs meet monoliths: hierarchical structuring metal organic framework catalysts. *Appl. Catal. A Gen.* 391: 261–267.

60 Juan-Alcañiz, J., Goesten, M.G., Ramos-Fernandez, E.V. et al. (2012). Towards efficient polyoxometalate encapsulation in MIL-100(Cr): Influence of synthesis conditions. *New J. Chem.* 36: 977–987.

61 Ingleson, M.J. et al. (2008). Generation of a solid Brønsted acid site in a chiral framework. *Chem. Commun.* 11: 1287–1289. https://doi.org/10.1039/B718443C.

62 Pearson, R.G. (1938). Acids and bases. *Science* 151: 172.

63 North, M. (1994). Amines and amides. *Contemp. Org. Synth.* 1: 475–494.

64 Verma, A., De, D., Tomar, K., and Bharadwaj, P.K. (2017). An amine functionalized metal-organic framework as an effective catalyst for conversion of CO_2 and Biginelli reactions. *Inorg. Chem.* 56: 9765–9771.

65 Yang, Y. et al. (2015). Synthesis and characterization of three amino-functionalized metal-organic frameworks based on the 2-aminoterephthalic ligand. *Dalt. Trans.* 44: 8190–8197.

66 Serra-Crespo, P. et al. (2012). Interplay of metal node and amine functionality in NH_2 -MIL-53: modulating breathing behavior through intra-framework interactions. *Langmuir* 28: 12916–12922.

67 Dhankhar, S.S. and Nagaraja, C.M. (2019). Construction of a 3D porous Co(ii) metal-organic framework (MOF) with Lewis acidic metal sites exhibiting selective CO_2 capture and conversion under mild conditions. *New J. Chem.* 43: 2163–2170.

68 Kim, J., Kim, S.N., Jang, H.G. et al. (2013). CO_2 cycloaddition of styrene oxide over MOF catalysts. *Appl. Catal. A Gen.* 453: 175–180.

69 Miralda, C.M., Macias, E.E., Zhu, M. et al. (2012). Zeolitic imidazole framework-8 catalysts in the conversion of CO_2 to chloropropene carbonate. *ACS Catal.*: 8–11.

70 Zhu, M. (2013). ZIF-8: novel catalytic material for the conversion of CO_2 to cyclic carbonates. ThinkIR: The University of Louisville's Institutional Repository.

71 Zhu, M., Srinivas, D., Bhogeswararao, S. et al. (2013). Catalytic activity of ZIF-8 in the synthesis of styrene carbonate from CO_2 and styrene oxide. *Catal. Commun.* 32: 36–40.

72 Yang, L. et al. (2014). Zeolitic imidazolate framework-68 as an efficient heterogeneous catalyst for chemical fixation of carbon dioxide. *J. Mol. Catal. A Chem.* 392: 278–283.

73 Guillerm, V. et al. (2014). Discovery and introduction of a (3,18)-connected net as an ideal blueprint for the design of metal–organic frameworks. *Nat. Chem.* 6: 673.

74 Beyzavi, M. H. *et al.* (2014). A Hafnium-based metal – organic framework as an efficient and multifunctional catalyst for facile CO_2 fixation and regioselective and enantioretentive epoxide activation.

75 Cho, H.Y., Yang, D.A., Kim, J. et al. (2012). CO_2 adsorption and catalytic application of Co-MOF-74 synthesized by microwave heating. *Catal. Today* 185: 35–40.

76 Yang, D.A., Cho, H.Y., Kim, J. et al. (2012). CO_2 capture and conversion using Mg-MOF-74 prepared by a sonochemical method. *Energy Environ. Sci.* 5: 6465–6473.

77 Zalomaeva, O.V. et al. (2013). Cyclic carbonates synthesis from epoxides and CO_2 over metal-organic framework Cr-MIL-101. *J. Catal.* 298: 179–185.

78 MacIas, E.E., Ratnasamy, P., and Carreon, M.A. (2012). Catalytic activity of metal organic framework $Cu_3(BTC)_2$ in the cycloaddition of CO_2 to epichlorohydrin reaction. *Catal. Today* 198: 215–218.

79 Ren, Y. et al. (2013). Ni(salphen)-based metal-organic framework for the synthesis of cyclic carbonates by cycloaddition of CO_2 to epoxides. *RSC Adv.* 3: 2167–2170.

80 Lescouet, T., Chizallet, C., and Farrusseng, D. (2012). The origin of the activity of amine-functionalized metal–organic frameworks in the catalytic synthesis of cyclic carbonates from epoxide and CO_2. *ChemCatChem* 4: 1725–1728.

81 Zhou, X., Zhang, Y., Yang, X. et al. (2012). Functionalized IRMOF-3 as efficient heterogeneous catalyst for the synthesis of cyclic carbonates. *J. Mol. Catal. A Chem.* 361–362: 12–16.

82 Kim, H. et al. (2019). Synthesis of cyclic carbonate by CO_2 fixation to epoxides using interpenetrated MOF-5/n-Bu_4 NBr. *J. Mater. Sci.* 54 https://doi.org/10.1007/s10853-019-03702-6.

83 Mousavi, B., Chaemchuen, S., Moosavi, B., and Zhou, K. (2017). CO_2 cycloaddition to epoxides by using M-DABCO metal–organic frameworks and the influence of the synthetic method on catalytic reactivity. *ChemistryOpen* 6: 674–680. https://doi.org/10.1002/open.201700060.

84 Sharma, N., Dhankhar, S.S., and Nagaraja, C.M. (2019). Microporous and mesoporous materials a Mn(II) -porphyrin based metal-organic framework (MOF) for visible-light- assisted cycloaddition of carbon dioxide with epoxides. *Microporous Mesoporous Mater.* 280: 372–378.

85 Sun, X. et al. (2017). Metal–organic framework mediated cobalt/nitrogen-doped carbon hybrids as efficient and chemoselective catalysts for the hydrogenation of nitroarenes. *ChemCatChem* 9: 1854–1862.

86 Wu, Y. et al. (2019). Mn-based MOFs as efficient catalysts for catalytic conversion of carbon dioxide into cyclic carbonates and DFT studies. *Chem. Eng. Sci.* 201: 288–297.

87 Yang, Y., Guo, Z., Chen, X., and Liu, J. (2019). A Ni_3O-cluster based porous MOF for catalytic conversion of CO_2 to cyclic carbonates. *J. Solid State Chem.* 276: 190–193.

88 Zhou, Z., He, C., Xiu, J. et al. (2015). Metal–organic polymers containing discrete single-walled nanotube as a heterogeneous catalyst for the cycloaddition of carbon dioxide to epoxides. *J. Am. Chem. Soc.* 137: 3–6. https://doi.org/10.1021/jacs.5b07925.

89 Kurisingal, J.F., Rachuri, Y., Gu, Y. et al. (2019). General binary metal-organic frameworks: catalysts for the efficient solvent-free CO_2 fixation reaction via cyclic carbonates synthesis. *Appl. Catal. A, Gen.* 571: 1–11.

90 Wu, Y. et al. (2018). Zn_2 $(C_9H_3O_6)(C_4H_5N_2)(C_4H_6N_2)_3$ MOF as a highly efficient catalyst for chemical fixation of CO_2 into cyclic carbonates and kinetic studies. *Chem. Eng. Res. Des.* 140: 273–282.

91 Zhang, X. et al. (2018). Microporous and mesoporous materials the fixation of carbon dioxide with epoxides catalyzed by cation-exchanged metal-organic framework. *Microporous Mesoporous Mater.* 258: 55–61.

92 Li, J. et al. (2019). Porous metal-organic framework with Lewis acid – base bifunctional sites for high efficient CO_2 adsorption and catalytic conversion to cyclic carbonates. 106: 70–75.

93 Wu, Y. et al. (2018). 3D-monoclinic M–BTC MOF (M = Mn, Co, Ni) as highly efficient catalysts for chemical fixation of CO_2 into cyclic carbonates. *Journal of Industrial and Engineering Chemistry* 58: 296–303.

94 Song, X. et al. (2018). Dual-linker metal-organic frameworks as efficient carbon dioxide conversion catalysts. *Appl. Catal. A, Gen.* 566: 44–51.

95 Noh, J. et al. (2018). Functional group effects on a metal-organic framework catalyst for CO_2 cycloaddition. *J. Ind. Eng. Chem.* 64: 478–483.

96 Babu, R. et al. (2017). Aqueous microwave-assisted synthesis of non-interpenetrated metal-organic framework for room temperature cycloaddition of CO_2 and epoxides. *Appl. Catal. A, Gen.* 544: 126–136.

97 Liu, D., Li, G., and Liu, H. (2018). Applied surface science functionalized MIL-101 with imidazolium-based ionic liquids for the cycloaddition of CO_2 and epoxides under mild condition. *Appl. Surf. Sci.* 428: 218–225.

98 Xiang, W., Sun, Z., Wu, Y. et al. (2019). Enhanced cycloaddition of CO_2 to epichlorohydrin over zeolitic imidazolate frameworks with mixed linkers under solventless and co-catalyst-free condition. *Catal. Today* 339 https://doi.org/10.1016/j.cattod.2019.01.050.

99 Hwang, G. et al. (2016). A highly efficient zeolitic imidazolate framework catalyst for the co-catalyst and solvent free synthesis of cyclic carbonates from CO_2. *Biochem. Pharmacol.* 15: 123–130.

100 Tu, M. and Davis, R.J. (2001). Cycloaddition of CO_2 to epoxides over solid base catalysts. *J. Catal.* 199: 85–91.

101 Kim, Y.-J. and Park, D.-W. (2013). Functionalized IRMOF-3: an efficient heterogeneous catalyst for the cycloaddition of allyl glycidyl ether and CO_2. *J. Nanosci. Nanotechnol.* 13: 2307–2312.

102 Ma, D. et al. (2015). Bifunctional MOF heterogeneous catalysts based on the synergy of dual functional sites for efficient conversion of CO_2 under mild and co-catalyst free conditions. *J. Mater. Chem. A* 3: 23136–23142.

103 Sun, Y. et al. (2018). Facile approach to graft ionic liquid into MOF for improving the efficiency of CO_2 chemical fixation. *ACS Appl. Mater. Interfaces* 10: 27124–27130.

104 Kholdeeva, O.A. et al. (2014). Hydrocarbon oxidation over Fe- and Cr-containing metal-organic frameworks MIL-100 and MIL-101 – a comparative study. *Catal. Today* 238: 54–61.

105 Sun, J., Yu, G., Huo, Q. et al. (2014). Epoxidation of styrene over Fe(Cr)-MIL-101 metal–organic frameworks. *RSC Adv.* 4: 38048–38054.

106 Nguyen, H.T.D. et al. (2018). A series of metal-organic frameworks for selective CO_2 capture and catalytic oxidative carboxylation of olefins. *Inorg. Chem.* 57: 13772–13782.

18

Plasma-Assisted Conversion of CO_2

Kevin H. R. Rouwenhorst[1], Gerard J. van Rooij[2], and Leon Lefferts[1]

[1] *University of Twente, MESA+ Institute for Nanotechnology, Faculty of Science and Technology, Catalytic Processes and Materials, Drienerlolaan 5, 7500 AE Enschede, Netherlands*
[2] *Dutch Institute for Fundamental Energy Research (DIFFER), Nonequilibrium Fuel Conversion, De Zaale 20, 5612 AJ Eindhoven, Netherlands*

18.1 Introduction

The nonequilibrium or nonthermal chemistry in the plasma phase forms a promising approach toward direct conversion of CO_2 [1]. Experimental results obtained in the 1970s in the former USSR indicate that the net reaction $CO_2 \rightarrow CO + \frac{1}{2}O_2$ can be achieved with over 80% energy efficiency [2, 3]. It may be of no surprise that such unprecedented efficiency has been food for inspiration and is being considered interesting for the field of CO_2 reuse. As this research was carried out during the cold war for a two-step hydrogen production process and perhaps also for CO_2 emission mitigation in a military context, it remained relatively unknown to the rest of the world until it was summarized in the book on *Plasma Chemistry* by Fridman in 2008 [4]. The past decade, the subject has been well recognized by the international plasma chemistry community and a number of groups started investigating the maximally achievable energy efficiency for the reduction of CO_2 in plasma, both numerically and experimentally, pure and mixed with other molecules, in homogeneous plasma configurations as well as coupled to catalysis [1, 5, 6].

The desire to couple plasma and catalysis resonates in both communities. Although the plasma phase is extremely promising for homogeneous conversion of CO_2, its high reactivity causes loss of selectivity when applied to molecules or mixtures that can form multiple products, e.g. methane, hydrogen, and water. The product slate is determined by quenching reactions and radical chain reactions with poor control. Catalysis following plasma activation is pursued to gain selectivity to targeted compounds by enhancing specific reaction pathways [7]. The field of catalysis also recognizes the potential of enhancing reactivity by plasma for reducing demands on operation temperature and/or pressure or breaking scaling relations based on the Evans–Polanyi relationships [8].

In this chapter, the main concepts and mechanisms underlying the plasma approach will be introduced, as well as a review on the recent advancements of

Engineering Solutions for CO_2 Conversion, First Edition.
Edited by Tomas R. Reina, José A. Odriozola, and Harvey Arellano-Garcia.

the field. Furthermore, the coupling of a plasma and a catalyst is discussed with reported experimental performance, followed by the perspective of plasma-assisted CO_2 conversion.

18.1.1 What Is a Plasma?

The word plasma is generally used to denote gas-phase systems in which an appreciable amount of free charge is present in the form of free electrons and ionized atoms and/or molecules but as a whole is electrically neutral. In 1879, Sir William Crookes termed this ionized gas as *the fourth state of matter* [9]. The term *plasma* was first introduced by Langmuir and Tonks in the 1920s to describe the glowing region of a discharge [4, 9]. Characteristic features of plasma are its collective behavior because of the long-range action of the Coulomb force between the charged particles and its ability to conduct current.

Plasma is often considered as the fourth state of matter. Indeed, heating up a gas to high temperatures will first make collisions between the particles sufficiently energetic to dissociate any molecule into its constituent atoms. Heating even further will reach a point at which one of the outer valence electrons may be knocked out to create free charge and to form a plasma. In this picture, all species in the system are in thermal equilibrium and only (ionized) atoms and electrons are present.

Alternatively, plasma can form because of nonthermal ionization. As an example, astrophysical plasma often forms because of photoionization under extremely low-temperature conditions [10]. Laboratory discharges, subject of the present, are most commonly electric discharges, in which ionization is sustained by application of electric fields. These fields energize in first instance the free electrons as these are most mobile in view of their small mass. Collisions of the electrons with heavy atoms and molecules are not efficient in transferring momentum because of the large mass difference. Consequently, the temperature of the electron population will remain much higher (typically >10.000 K) than the heavy particles (i.e. molecules) that might even stay at room temperature. Under such conditions, one speaks of nonthermal plasma (NTP) [11]. The high electron energy opens energy transfer to internal excitation and ionization of the heavy particles.

18.1.2 History

The field of plasma chemistry emerged from an application for which the capabilities provided by the plasma phase are also nowadays unequivocal: ozone production. The formation ozone was discovered in electric sparks in 1785 by Martinus van Marum and identified as a new chemical compound in 1839 by Schonberg [4, 12]. In 1857, Werner Siemens achieved a breakthrough toward reliable ozone formation with a dielectric barrier discharge (DBD), which is still the NTP discharge of choice for industrial ozone production [13].

The first large-scale commercial application of plasma concerned the Birkeland–Eyde process for nitrogen fixation via nitric oxide production [14, 15]. A truly large factory was built in Notodden around 1906: 32 thermal arc furnaces were installed, each of 1000 HP (750 kW) [16]. The electricity was supplied from a waterfall power station that was built at the same time as the factory. Evidently, cheap electricity was essential for the economic viability of the process. The alternative technology for ammonia production developed by Haber and Bosch turned out to be less energy intensive and forced the Notodden factory to shut down [17].

Current industrial applications in which plasma technology has become established for its unique capabilities include material processing such as deposition of coating and high aspect ratio etching, cleaning of exhaust gasses and water, and light sources [18].

18.1.3 Electrification

Comparison of those cases in which industrial application of plasma has demonstrated its unique capabilities with the ones it has lost the competition with alternative approaches underlines its traditional weakness as an expensive technology. Plasma requires often power supplies that are expensive and difficult to scale up to take advantage of economies of scale. Moreover, input energy is in the form of electricity, which is traditionally expensive compared to chemical (i.e. fossil) energy input [19]. Recent developments in technology as well as society may provide a second chance to plasma chemistry. Power supply technology has tremendously advanced and is becoming cheaper and cheaper [20]. For example, where MW (microwave) power generators used to be exclusively magnetron based, developments in telecom are driving cheaper and robust semiconductor amplifiers. The societal awareness of the risks imposed by climate change is driving an energy transition and, intimately related, an interest for CO_2 chemistry. The continuously increasing deployment of sustainable energy sources is shifting the paradigm of electricity as most expensive form of energy and already examples are seen where solar energy is cheaper than fossil fuel-based energy.

At the same time, the introduction of sustainable energy sources has made clear that new technologies are required to address intermittency and to ensure that demand and supply are matched. Energy storage and conversion will thus become increasingly important, which essentially will have to follow the intermittency of the sustainable energy production. Particularly in this respect, plasma may well turn out to be advantageous as it is easily switchable, and with cheaper power supply equipment, it may be economically feasible not to run the process continuously [21].

18.1.4 Thermodynamics

The production of ozone and nitric oxide has the common feature of being thermodynamically uphill reactions with a high reaction barrier. Consequently, all chemistry will be frozen under room temperature conditions, and heating is the obvious and traditional way forward to drive the chemistry via the $T\Delta S$ term. However, as the backward reaction will inherently involve a lower barrier for uphill reactions, it is generally difficult to realize appreciable conversion yields. Thus, in situ removal of the product from the plasma volume is then the solution.

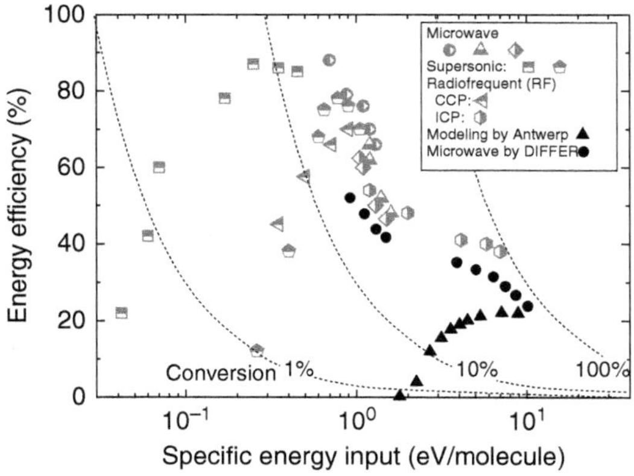

Figure 18.1 Energy efficiency of plasma-assisted CO_2 reduction as a function of the specific energy input. The gray-and-white markers are reported by the former Soviet Union work [2, 3]. The solid markers show initial modeling results of vibrational ladder climbing and early experimental microwave results since the revival of the field. Contours of conversion degree are indicated by the dashed lines. Source: Adapted from Refs. [2, 3, 22, 23].

The NTP phase offers an alternative via internal excitation of the reagents, effectively acting on the ΔH term. At the one hand, it means that the reactions will run at lower gas kinetic temperatures. Moreover, it may provide opportunities to break micro-reversibility if it were possible to control the excitation of reactants versus products or choose a catalyst that is only acting on excited reagents.

18.1.5 Homogeneous Plasma Activation of CO_2

An overview of the aforementioned high energy efficiencies is reported from the experiments in the 80s [4]. An overview of these early results on CO_2 dissociation in plasma is shown in Figure 18.1 together with some first results since the revival of the field [22, 23]. The plot summarizes the following key aspects. Firstly, as said, the energy efficiency reached values close to 90%. Energy efficiency in this respect refers to the fraction of the input plasma discharge power that is converted into enthalpy of formation of CO in the effluent. It thus refers to the net reaction $CO_2 \rightarrow CO + ½O_2$. It neglects power losses in power supplies and other equipment (e.g. 20–30% in the case of microwave generation).

Secondly, the plot reflects that microwave discharges have demonstrated to be superior for the purpose of CO_2 dissociation. The prevalent explanation has been the NTP nature of microwave plasma because of which vibrational excitation is the dominant energy transfer mechanism between electrons and CO_2 molecules [4]. We will come back to the mechanism that takes advantage of vibrational excitation below. For now, we would like to emphasize the keyword "energy transfer": the plasma is used here as an energy transfer medium, transferring the input microwave power into (internal) heating of the CO_2 molecules.

The third aspect hinted at is that recent experiments such as from the work within DIFFER have not been capable of reproducing the earlier record energy efficiencies. Indeed, present-day experiments have not exceeded values of ~50%. Such a value is nevertheless impressive as it equals the limit of thermodynamic equilibrium conversion. Here, the (temperature-dependent) thermodynamic conversion efficiency is calculated as the fraction of CO molecules in the equilibrium composition multiplied with the formation enthalpy per molecule and divided by the total enthalpy change of the mixture. It yields for a pressure of 0.1 bar (a "typical" pressure in the experiments reported in the graph) a maximum of just over 50% at a temperature of ~3200 K [24]. In practice, it would require ideal quenching of the mixture preventing any back reactions.

The fourth and final feature is the clear discrepancy of the modeling results with the experiments. The modeling accounted solely for the NTP effects (i.e. dissociation via vibrational excitation) [22] and hence indicate that also other pathways must have been contributing in the experiments.

18.1.6 Mechanisms

Let us for a moment return to the concept of energy transfer in plasma. Again, the free plasma electrons are accelerated because of the externally applied electric fields and their collisions with the atoms and molecules brings about an electric resistivity that constitutes the power input to the system. However, the nature of the electron-induced dissociation reactions is inherently different from dissociation resulting from energetic collisions between CO_2 molecules. In the latter case, momentum can be directly transferred to one of the O atoms to give it sufficient energy to overcome the binding potential. In the case of electron collisions, these cannot transfer momentum to the atoms but will exchange energy with the bound electrons and induce transition as sketched in Figure 18.2. Key here is that these transitions follow the Franck Condon principle, i.e. the atoms have not the time to move at the timescale of the electron impact energy transfer and correspond to vertical transitions in the potential energy diagram. The consequence is that reaction barriers are involved, where the excess energy is converted into heat or internal excitation. In thermal plasma, this simply contributes to the overall heating of the system and thus to driving the thermal chemistry in the discharge. However, in NTP conditions, it may be that the overall temperature remains close to room temperature. Under such conditions, all heat generation is non-recoverable and must be considered as heat loss.

On the basis of the transitions sketched in between the potential energy curves, we summarize the following main dissociation mechanisms because of electron impact reactions. The occurrence of such mechanisms for various reduced electric fields is shown in Figure 18.3. These involve different energy thresholds E_{th} that are to be compared with the 2.9 eV CO formation enthalpy [4]:

i. Electron impact ionization followed by dissociative recombination (not indicated in Figure 18.2),

$$e + CO_2 \rightarrow CO_2{}^+ + 2e \rightarrow CO + O + e \qquad E_{th} > 13.8\,\text{eV} \tag{18.1}$$

ii. Electron impact dissociative excitation (red arrow in Figure 18.2),

$$e + CO_2 \rightarrow CO + O(^1S) + e \qquad E_{th} \sim 11.5\,\text{eV} \tag{18.2}$$

iii. Vibrationally enhanced electron impact dissociative excitation (blue arrow, a single vibrational excitation reduces the barrier significantly),

$$e + CO_2(v \geq 1) \rightarrow CO + O(^3P) + e \qquad E_{th} \sim 5.5\,\text{eV} \tag{18.3}$$

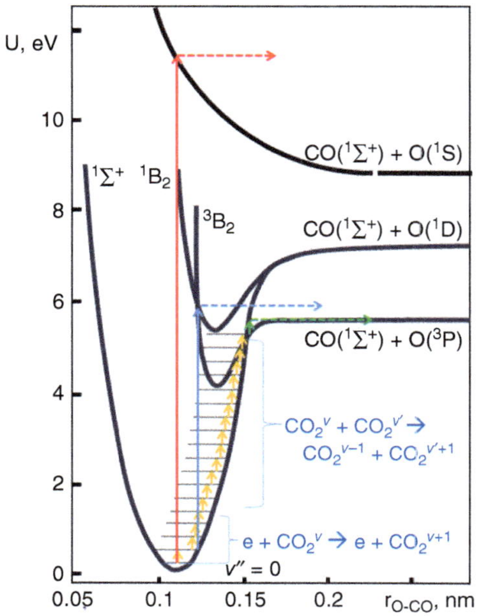

Figure 18.2 Illustration of different electron impact driven dissociation processes on the basis of the ground-state potential energy diagram with vibrational levels indicated and some electronic terms of CO_2 [25]. Indicated is the so-called ladder-climbing process (yellow arrows). It is initiated by electron impact excitation of the lower vibrational levels, which is carried by vibrational exchange up to the point of dissociation. Opposed to barrier-less ladder climbing is dissociative excitation, which involves a large activation barrier (red arrow). Also indicated is how vibrational excitation may lower such an activation barrier (blue arrow). Source: Adapted from van Rooij et al. [25].

iv. Vibrational ladder climbing (yellow/green arrows), where initial vibrational excitation by electron impact causes overpopulation that is transferred up the ladder up to the point of dissociation by excitation exchange in CO_2–CO_2 collisions,

$$e + CO_2(v \geq 1) \rightarrow CO + O(^3P) + e \qquad E_{th} \sim 0.3\ \text{eV} \tag{18.4}$$

Here, the threshold energy of process 4 is assumed to be the asymmetric stretch vibrational quantum. Furthermore, it is important to mention that process 3 is only considered as a concept and has not yet been included in any modeling as an overlap between states is unknown. Incorporation of the other pathways in numerical modeling is pursued by a number of groups [22, 26, 27] and results in significant dissociation under truly NTP conditions in which gas temperatures remain significantly lower than the vibrational temperatures. However, as said, performance is predicted to be lower than what has recently been achieved experimentally.

Evidently, the pathways benefiting from vibrational excitation have the lowest energy threshold. This feature of essentially channeling the input power via vibrational excitation toward the desired reaction, in this case, CO_2 dissociation, is at the heart of the concept of nonequilibrium plasma chemistry. In effect, it offers (nearly) a barrier-free reaction channel via electron impact reactions. Under ideal NTP conditions, these are at play while the gas kinetic temperature remains as low as at room temperature! This is in contrast with thermally driving CO_2 dissociation, which produces maximum amounts of CO at temperatures of 3000–3500 K. It also means that the NTP process does not require quenching in view of its low gas kinetic temperature.

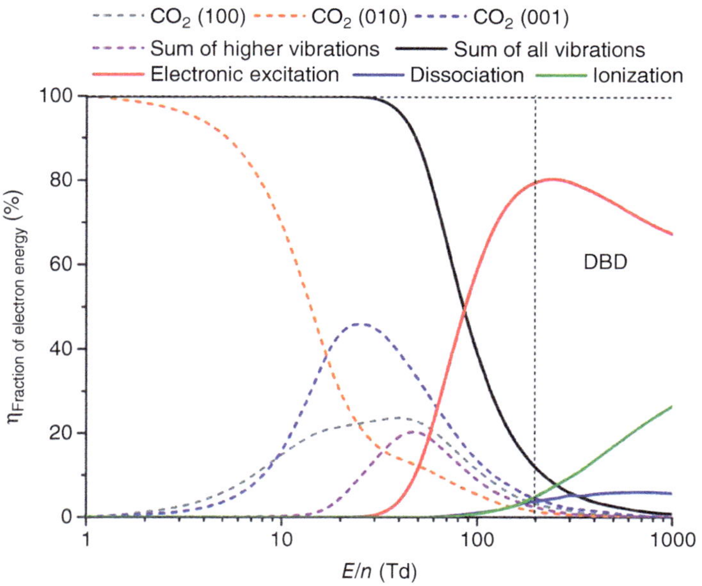

Figure 18.3 Channels for activation of CO_2 as a function of the reduced electric field (E/n). Source: Snoeckx and Bogaerts [5].

One may have noticed that oxygen radicals are produced. These must be involved in a follow-up reaction with CO_2 to realize the maximum energy efficiency and prevent the (three-body) back reaction. The follow-up reaction is also endothermic ($\Delta H = 0.3$ eV) and may again be stimulated with vibrational excitation:

$$O + CO_2(v \geq 1) \rightarrow CO + O_2 \tag{18.5}$$

As the power density in the plasma discharge is increased, the gas kinetic temperature will increase. Thermal dissociation will set in at temperatures exceeding 1700 K and will reach the aforementioned optimum of just over 50% at 3200 K [24]. In fact, also here vibrational nonequilibrium may be beneficial by quenching the thermally produced atomic oxygen via production of additional CO in (18.5). This is referred to as super-ideal quenching and would bring the efficiency limit of thermal conversion in plasma beyond 60% [4].

18.1.7 Plasma Reactors

Various plasma reactors (Figure 18.4) have been used for plasma-assisted CO_2 conversions, varying in temperature, pressure, plasma intensity, and frequency of discharges [1, 28]. The most used reactor is the DBD reactor, which can be operated at atmospheric pressures and at temperatures in the range 300–700 K, typically with a frequency of tens of kHz. A DBD reactor is often used because of its simple configuration (i.e. an AC potential applied in between an inner electrode and an outer

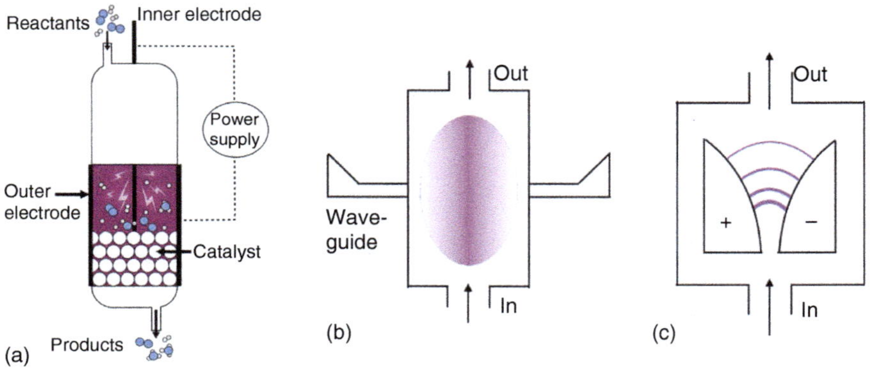

Figure 18.4 Various plasma reactors used for chemical conversions: dielectric barrier discharge (a), microwave reactor (b), and cylindrical gliding arc (c). Source: Mehta et al. [6]; Bogaerts and Neyts [1].

electrode with a dielectric barrier) and its proven viability for industrial ozone formation. For the context of plasma catalysis, it is especially attractive as it provides the possibility of implementing a packed bed or a catalytic wall reactor configuration.

Microwave plasmas are generated by electromagnetic radiation in a waveguide, without using electrodes. Microwave (MW) reactors and radiofrequency (RF) reactors both require reduced pressures (typically 10^{-5}–10^{-2} bar). The temperature within the plasma increases with increasing pressure, thereby transitioning from a NTP to a thermal plasma. A main difference between a microwave plasma and a radiofrequency plasma is the frequency. It ranges from 0.3 to 10 GHz for microwave reactors, while radiofrequency reactors operate at frequencies varying from Hz of MHz.

Gliding arcs are generated between two flat diverging electrodes, at the shortest distance between two electrodes. Afterward, the arc "glides" toward the larger inter-electrode distance until the plasma extinguishes and a new gliding arc is generated at the shortest distance between the two electrodes. A gliding arc reactor operates at atmospheric pressures, but temperatures within the plasmas are typically high (1000–3000 K), making this reactor unsuitable in combination with a catalyst.

18.1.8 Performance in Various Plasma Reactors

Recently, Snoeckx et al. [5] reviewed the state of the art of plasma conversion of CO_2. Many studies involved DBDs in view of their inherent NTP character and ease of operation at atmospheric pressure. However, their performance in terms of energy efficiency appears typically to be limited to below 20%. Their conversion can reach up to 40%, which is controlled via the specific energy input (or, equivalently, the residence time of the gas in the reactor) and goes on the expense of energy efficiency. The limited efficiency is interpreted to be due to a relatively high average electron energy that drives (18.1, 18.2) as the main dissociation pathways.

Microwave discharges cover a large parameter range in terms of conversion and efficiency and reach the highest energy efficiencies. The discharge type is identified to be best suitable to channel discharge power to vibrational modes and is thus best equipped to benefit from vibrational excitation. As said before, the best results (efficiency >60%) date from the early work [2, 3, 29–32] and have not been reproduced in recent years. All recent work is within the range of thermal equilibrium conversion [33–37], and gas temperature measurements have demonstrated gas temperatures exceeding 3000 K and thus thermal conversion to be significant [24, 25]. As such high temperature regions have also been observed in the early work, it seems that a combination of thermal conversion with nonequilibrium chemistry in the periphery is an interesting route for further optimization.

The other type of plasma discharge that is promising with respect to providing high levels of vibrational excitation is the so-called gliding arc discharge. Indeed, energy efficiencies up to 50% are common [38] and also record values of 65% have been reported [39]. Model calculations revealed that GA discharges also induce elevated temperatures, and thus, thermal conversion is likely to play a significant role [40]. The key to success may be in combining thermal conversion in combination with vibrationally induced super-ideal quenching, as was recognized for microwave discharges

18.2 Plasma-catalytic CO_2 Conversion

The control of the selectivity toward the desired products is limited for the conversion of CO_2 (in the presence of, e.g., H_2, CH_4, or H_2O) in a plasma environment [6, 18, 41, 42]. A catalyst may be used to overcome this selectivity issue. The introduction of a catalyst may change the product distribution [6, 18, 41] even though the CO_2 conversion is not necessarily increased [5, 43–47]. In the Section 18.2.1, a general introduction is given regarding the complexity of coupling a plasma and a catalyst, after which catalyst development and the experimental performance of various CO_2 conversion reactions are discussed. Lastly, a perspective is given on plasma-assisted CO_2 conversions.

18.2.1 Introduction

The energy efficiency of CO_2 conversion may be enhanced by using a packing material in the reactor [48, 49]. Furthermore, some catalysts allow for the formation of products, which cannot be produced in the plasma environment, such as formaldehyde [43] and C4 hydrocarbons [50]. Another benefit of plasma catalysis as compared to thermal catalysis is the reduced coke formation because of the reverse-Boudouard reaction ($CO_2 + C \rightleftharpoons 2CO$) upon CO_2 activation [6, 51].

Plasma catalysis is generally defined as the synergistic effect of combining a plasma and a catalyst, in which the result of combining the plasma and the catalyst is larger than the sum of the individual contributions of the plasma and the catalyst [52]. Often, synergy is claimed upon combining plasma with a catalyst.

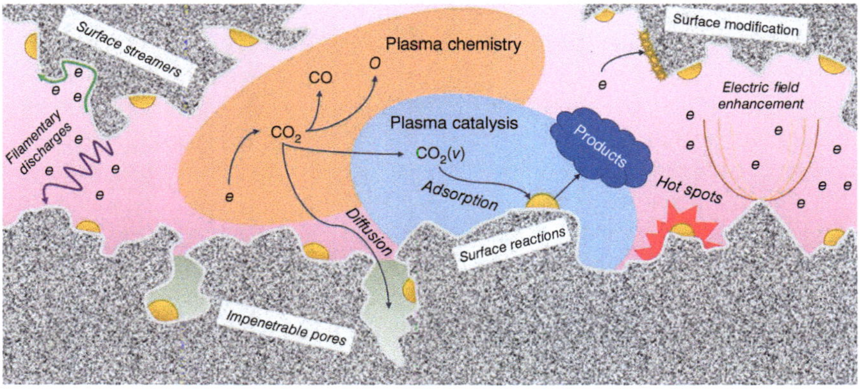

Figure 18.5 Schematic representation of various plasma–catalyst interactions as recognized by the field (see e.g. [51]).

However, plasma catalyst synergies are not trivial to uncover, as the underlying principles of plasma catalysis are not fully understood [7, 53, 54]. Various possible plasma–catalyst interactions that have been proposed are shown in Figure 18.5.

The various species present in the plasma environment include electrons, positive and negative ion photons, and neutral atoms and molecules in ground state or internally excited [9]. These may interact with the catalyst in various manners. A catalyst may influence the conversion of CO_2 in the presence of a plasma in three manners, namely, (i) by quenching of radicals from the plasma environment, (ii) by surface reactions to catalytically produce complex products, and (iii) by modifying the plasma characteristics. The first function (i.e. the quenching of radicals from the plasma environment) comes from the ionized nature of the plasma. Dissociated atoms in the plasma environment are highly reactive. Therefore, these radicals are easily adsorbed at catalyst surfaces [9]. Surface reactions of plasma-activated molecules (the second function) are complex and not well understood, especially when the surface catalyzes the dissociation (of activated molecules) or surface reactions. Quenching and surface reactions contribute in the case of plasma activation near the surface as will be discussed later, while the plasma chemistry far from the surface is also altered in the case of modification of the plasma environment by the catalyst acting as a dielectric packing material (third function).

Plasma catalytic reactors can be classified as in-plasma catalysis reactors or post-plasma catalysis reactors [9, 55]. In the case of in-plasma catalytic reactors, also termed as plasma-driven catalytic reactors, the catalyst and plasma are located in the same position in the reactor [55]. In the case of post-plasma catalytic reactors, also termed as plasma-assisted catalytic reactors, the plasma generation and the catalyst bed are spatially separated [55]. Although in-plasma catalysis can be used to activate short-lived species over a catalyst, post-plasma catalysis is only relevant for long-lived species [55]. A complication of in-plasma catalysis is the mutual influence of the plasma and the catalyst on one another, as discussed in Section 18.2.2.

18.2.2 Mutual Influence of Plasma and Catalyst

A multidisciplinary approach is required to understand the mutual influence of the plasma and the catalyst [55], covering various length scales and timescales [9, 54, 56, 57]. For now, most information is obtained by macroscopic performance testing of plasma reactor. It is difficult to draw conclusions on reaction mechanisms of the processes proceeding within the plasma reactor. Techniques to probe events inside the plasma reactor and especially at the reaction site would help plasma catalysis to develop further as a field of science. Fundamental understanding of surface processes in plasma catalysis therefore requires the development of in situ and in operando characterization techniques [54].

The mutual influence of a plasma and a catalyst was first systematically categorized by Van Durme et al. [58] and heavily discussed thereafter. Possible mutual influences of a plasma and a catalyst have been reviewed by Neyts and coworkers [18, 28, 51] and Whitehead [7, 53, 54]. Mutual influences of a plasma and a catalyst relevant for plasma-driven CO_2 conversions are discussed hereafter, with experimental examples. The complexity is even more severe because any change in, e.g., the catalyst will induce changes in the plasma, which again influences the catalyst. In short, effects can generally not be isolated [7, 59].

For discussing the synergy between plasma and catalysis, it is useful to distinguish between physical (or plasma) versus chemical (or catalysis) effects regarding the mutual influence. The physical effects are generally better characterized and understood compared to the chemical effects [18, 53]. A commonly encountered effect is electric field enhancement and consequent decreased plasma onset voltage, as was substantiated by simulations [60–63]. At some point, the field enhancement may be so large that microdischarges will form on the surface or even the discharge may entirely move from the gas phase to the catalyst surface [53], as imaged by Kim and coworkers [64]. Another example of how the discharge nature can be changed is the transition from a filamentary discharge to a glow discharge [41, 51, 59]. All these cases are essentially catalyst-enhanced plasma generation that alters the distribution of active species in the plasma environment [59].

The chemical effects are firstly introduced by the activation of the reagents. For example, a higher adsorption probability of species on the catalyst may be achieved because of plasma excitation (yielding a lower apparent activation barrier). Such enhanced adsorption may be rooted in the observed orders of magnitude increase upon vibrational excitation of a molecule [65]. Furthermore, the apparent activation barrier for the reaction may decrease upon plasma activation of molecules when the activation of reactants is rate limiting in the overall reaction [66, 67]. The production of excited states or radicals may also open alternative surface reaction pathways [41, 51].

On the other hand, as opposed to effects from the presence of non-ground-state reagent molecules, the surface properties of the catalyst might be altered because of the presence of plasma. This may occur via physical modifications of the catalyst surface because of bombardment with energetic particles, via reduction of the catalyst, coke formation, changes in the physicochemical properties of the catalyst (such as the catalyst work function), and the formation of hot spots. It was postulated that the physicochemical properties of the catalyst may be altered by surface charging, thereby altering the work function and consequently the activity of the surface [68, 69]. Furthermore, it was reported that surface cracking and peeling of metal nanoparticles can occur in a DBD reactor in the presence of CH_4, while this is not significant for pure CO_2 feeds [70]. Similarly, changes in the surface morphology of a metal oxide catalyst after plasma treatment were reported after treatment in a DBD reactor [71].

A prerequisite for an effective interaction between the plasma and the catalyst is that the plasma-activated species can reach the catalytic surface before decaying to the ground-state molecules [53]. Plasma-activated species have a limited lifetime, giving a limited traveling distance by diffusion [55]. A direct interaction between the plasma and the catalyst surface is obtained when the distance between the plasma and the catalyst surface is smaller than the maximum traveling distance of plasma-activated species [55]. This is especially relevant when the catalyst is porous (see Figure 18.5). Large pores allow generation of plasma, while inside pores in the sub-micron range plasma cannot be generated [63, 72, 73]. Thus, only plasma-activated species with a long lifetime can penetrate into the pores, while most other species decay to the ground state before reaching the active catalyst surface within the pores. It should be noted that the majority of the pores in a typical support material are smaller than 50 nm or so. In many cases, the contribution of the external surface area is likely to dominate.

18.2.3 Catalyst Development

The choice for the catalyst depends on the type of plasma activation of molecules and which activated species are utilized for the chemical reaction [74]. From an energy efficiency point of view, it may be more efficient to use vibrationally excited molecules as (i) these make up a major portion of the species formed in a CO_2 plasma [75] and (ii) this allows for a mild activation of CO_2 in the plasma phase. Typical activation barriers for CO_2 dissociation over transition metals are in the range 100–300 kJ mol^{-1} [76], while the dissociation energy for CO_2 to O and CO 531 kJ mol^{-1} in the plasma environment. Furthermore, using vibrationally excited CO_2 molecules with subsequent dissociative adsorption on the surface allows for better control over surface reactions. Snoeckx et al. [5] put the efficiency target for commercial applications at 60% (about 887 kJ mol^{-1}). Plasma-generated radicals are very reactive, and chemical reactions may already occur before the species have diffused to the catalyst surface, causing nonselective recombinations, thereby lowering the overall efficiency. Therefore, mildly activated CO_2 molecules with subsequent reaction over a catalytic surface are the most energy-efficient pathway for the selective formation of products (see Figure 18.6).

The operating conditions of the plasma reactor (temperature, pressure, frequency, etc.) have a profound influence on the species formed [5]. Therefore, the choice for the optimum catalyst depends on the type of the plasma reactor and the conditions used. When looking at the processing conditions and heat integration required,

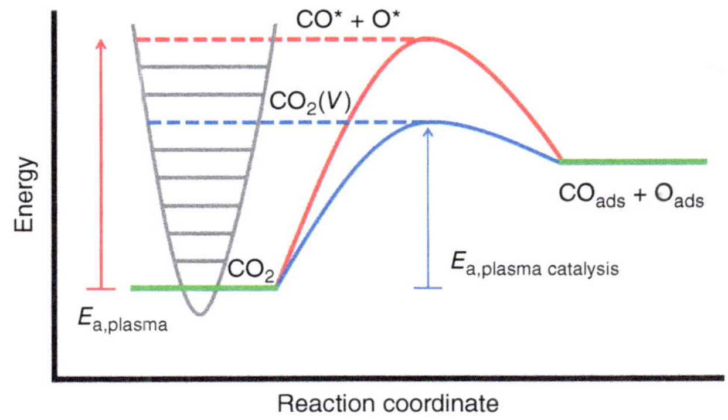

Figure 18.6 Schematic of the dissociative surface adsorption of various activated CO_2 species. Blue: vibrationally excited CO_2 molecules. Red: CO and O radical. X_{ads} represents an adsorbed species X. X^* represents a plasma-excited radical X.

NTPs are preferred for plasma-driven CO_2 conversion processes. Thermal plasmas operate at high temperatures, where there are essentially no molecules adsorbed on a solid surface. Furthermore, catalysts deactivate at high temperatures because of sintering effects. Thermal plasmas are fully composed of radicals, implying that the reaction pathway is via radicals. On the other hand, excited molecules are present in NTPs. Atmospheric pressure plasma reactors are beneficial over low-pressure plasma reactors for practical applications.

In case vibrationally excited molecules are aimed for, catalysts similar to thermal catalysts can be used. Vibrationally excited or electronically excited molecules are often the dominant plasma-activated species in NTPs (see Figure 18.3) [75]. Plasma activation is used to bring down the operating temperature for dissociative adsorption-limited catalytic processes. Typically, the optimum active metal for plasma catalysis is a late transition metal, which desorb reaction products easily and thus the operating temperature can be low. For noble metals, the adsorption of reactants is usually limiting the reaction rate. Through vibrational excitation of molecules, part of the activation barrier can be overcome, thereby allowing for plasma-enhanced catalytic activity over a noble metal. The active metal still has an activating role in this case, as the CO_2 is not dissociated in the plasma environment. Apart from the choice for the active metal, the choice for the support and promoter composition has an influence on the activity in some cases [67]. The choice of the support depends on the pore size distribution, as well as potential electronic active metal–support interactions. Metal–support interactions can influence the binding strength of surface-adsorbed species on the metal and thereby the reactivity of surface-adsorbed species. As plasmas can only be generated in pores of typically >1 μm [77] as discussed above, oxides with highly porous structures with pore sizes below 1 μm and large internal surface areas (such as γ-Al_2O_3 with ~100 $m^2\ g^{-1}$) are not beneficial for the maximum interaction between the active metal catalyst and the plasma (see Figure 18.5) [63].

In case radicals are aimed for, the ideal catalysts will likely be very different from those used in thermal catalysis. Radicals do not need to be dissociated over a surface as dissociation already occurred in the plasma. Thus, the surface has a recombination function, which can be a simple quenching reaction or, e.g., a more elaborate hydrogenation reaction. The quenching function can be performed by various types of materials. On the other hand, transition metals may be used to hydrogenate the adsorbed radicals to more complex H-, C-, and O-containing molecules (see Section 18.2.4). For these catalysts, the efficient plasma activation of reactants toward radicals is important [78]. Therefore, the choice of the catalyst (support) is mostly determined by properties influencing the discharge characteristics (e.g. dielectric properties and morphology), thereby influencing the plasma-phase dissociation of activated molecules. Recently, Akay and coworkers [79–81] proposed to mix a dielectric material with a supported transition metal catalyst in the plasma reactor. Herein, the dielectric material acts as a plasma catalyst promoter (PCP), which modifies the plasma characteristics, whereas the transition metal catalyst performs the catalytic function.

In case localized hot spots are formed near the catalyst surface, radicals may be formed as well and the local thermodynamic equilibrium conditions are altered significantly. Heating effects were reported to be a function of the specific energy input [55]. Therefore, catalysts active for the formation of CO_2 via oxidation or reforming of product species should not be used in this case, as heating would obviously enhance these reactions.

18.2.4 Experimental Performance

Various authors reviewed CO_2 conversion in the presence of a plasma and a catalyst recently [1, 5, 6, 18, 41, 78, 82–85]. Most authors used DBD reactors, as these operate at atmospheric pressure, have a low cost, and a simple operation [41]. By now, there are hundreds of articles describing the activation of CO_2 by the combination of a plasma and a catalyst [82], and in recent years, there is an increased attention for plasma catalytic CO_2 conversions [53, 82]. Examples of CO_2 conversion can be categorized into CO_2 splitting to CO and O_2 (CO_2 feed), dry reforming of methane (CO_2 and CH_4 feed), hydrogenation of CO_2 (CO_2 and H_2 feed), and artificial photosynthesis (CO_2 and H_2O feed). Additional feeds (H_2, CH_4, and H_2O) are added to change the product distribution and to overcome the issue of forming an explosive mixture composed of CO and O_2 [86]. For example, the selectivity toward O_2 can be completely suppressed upon addition of CH_4, as the oxygen is converted to oxygenates or H_2O [87]. However, the addition of such feeds may also cause an explosive mixture containing CO, O_2, and H_2. Depending on the feed composition, typical products include syngas, formaldehyde, hydrocarbons, and oxygenates [18]. The aim here is not to give an extensive review of the research performed, but to give an overview of plasma–catalyst synergies in various CO_2 conversion processes with examples from experimental research.

Three types of catalysts are typically used for the plasma–catalytic conversion of CO_2, namely, (i) oxide-supported metal catalysts, (ii) zeolites, and (iii) dielectric materials. The latter class of materials mostly modifies the plasma characteristics, rather than fulfilling a surface catalytic function (although surface processes were claimed to happen for these materials as well [44]). Another manner to modify the plasma characteristics is by changing the surface morphology [18]. Sharp edges cause an enhancement in the electric field, thereby altering the activation of species in the vicinity of the surface [18, 28, 51].

18.2.4.1 CO_2 Dissociation

Various plasma reactor types have been used for plasma catalytic CO_2 splitting, with most research focusing on DBD reactors, MW reactors, and gliding arcs. CO_2 splitting in various plasma reactors is extensively discussed in Refs. [5, 88]. Empty DBD reactors typically have conversions up to 40% with energy efficiencies of only 5–10% (see Figure 18.7) [89–92], as also discussed in the first part of this chapter. By tweaking the plasma in a burst mode, energy efficiencies of up to 20% may be attained [92, 93]. The presence of atmospheric N_2 in the gas mixture does not negatively affect the CO_2 conversion and energy efficiency in a DBD reactor [94]. Recent results with MW reactors indicate energy efficiencies of 10–50% at conversions of 10–80% at reduced pressures (up to few 100 mbar) [23, 25, 95]. At atmospheric pressures, energy efficiencies of 5–20% are attained at conversions of 10–50% for MW reactors [35]. Energy efficiencies of 20–40% (and occasionally 60%) at 0–20% conversion are reported for gliding arcs [38, 96–100].

The introduction of a packing material (Al_2O_3, $BaTiO_3$, TiO_2, and ZrO_2) can increase the CO_2 conversion and energy efficiency by a factor two for DBD reactors [1, 44, 45, 91, 101]. However, not all packed DBD reactors enhance the yield as compared to empty DBD reactors [5, 44–47, 102]. The introduction of a catalyst (such as

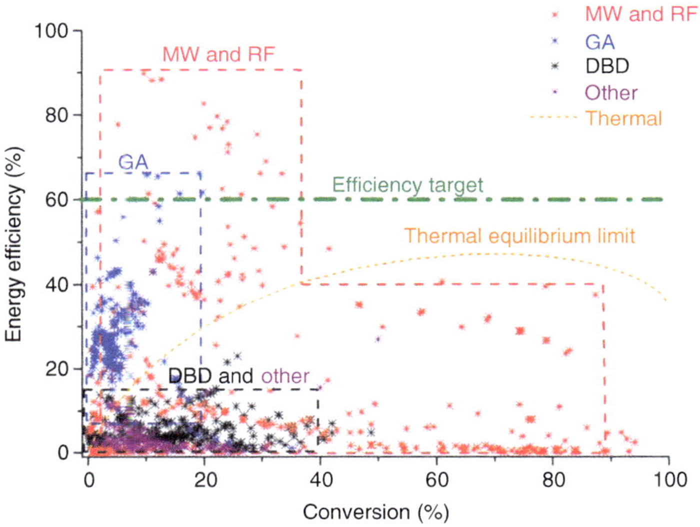

Figure 18.7 Comparison of literature data for CO_2 splitting in various plasma reactors. Source: Snoeckx and Bogaerts [5].

Rh) may facilitate the reverse reaction [35]. A photocatalytic effect by the formation of electron–hole pairs was proposed for $BaTiO_3$ and TiO_2 packings, causing oxygen vacancies as an active site for adsorption and activation of CO_2 [44]. Furthermore, the support particle size has an effect on the CO_2 splitting reactions [101], which may be attributed to plasma modification, because of, e.g., the number of contact points between particles and sharp-edged features. The dielectric constant of the packing material determines the plasma characteristics, such as the ignition voltage and the streamer onset voltage [103]. Decreasing the discharge barrier gap increases the CO_2 conversion at the cost of a lower energy efficiency [102]. Concluding, the efficiency of plasma catalytic CO_2 splitting depends on the reactor geometry, packing material, and particle size [47]. To the best of our knowledge, there are no examples of catalyst that are able to enhance reaction of O radicals with CO_2, without enhancing recombination reactions forming CO_2.

18.2.4.2 Dry Reforming of Methane

Plasma-based dry reforming of methane has attracted interest in recent years and a wide range of plasma reactors have been investigated [5, 82, 88, 104–106]. The benchmark for commercializing plasma-based dry reforming is about 4.27 eV per molecule CO_2 and CH_4 for the formation of syngas (see Figure 18.8), with a potential lower benchmark on energy efficiency in case of formation of higher value products (C2–C4 hydrocarbons) [5]. The main products are CO and H_2. The formation of oxygenates such as formic acid, acetic acid, propionic acid, butyric acid, methanol, ethanol, acetone, and formaldehyde and C2–C4 hydrocarbons have also been reported in low concentrations (<1%) [43, 107–109]. At high $CH_4 : CO_2$ ratios, the formation of C2–C4 hydrocarbons such as C_2H_2 becomes dominant [87]. The $CH_4 : CO_2$ ratio determines the degree of carbon deposition (dominant at high

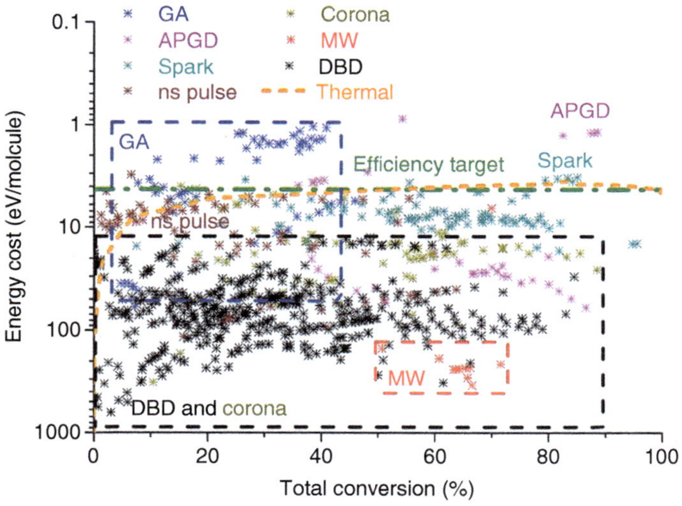

Figure 18.8 Comparison of literature data for the dry reforming of methane in various plasma reactors. Source: Snoeckx and Bogaerts [5].

ratios) and the formation of H_2O from H (dominant at low ratios) [85]. Empty DBD reactors typically have conversions of up to 60% at an energy cost of 20–100 eV per molecule [5]. Even upon optimization, the lowest predicted energy cost is about 16.9 eV per molecule (at low conversions of 0.01%) [110]. Thus, a catalyst is required to increase the energy efficiency in a DBD reactor. High conversions of 50–70% can also be obtained for MW reactors, albeit with energy costs of over 100 eV per molecule [5]. Energy costs of down to 0.75–2.6 eV per molecule with conversions up to 40–80% have been reported for gliding arcs and gliding arc plasmatrons [5, 87, 111–115]. The CO_2 conversion increases with increasing CH_4 fraction (up to 25%) [87], which may be attributed to the formation of CH_2 and CH_3 radicals, which are formed upon the dissociation of CO_2 [116]. Furthermore, the addition of O_2 to the feed mixture (oxidative dry reforming of methane) can be beneficial for the conversion of methane [112]. Promising results have also been reported for other plasma reactors (ns-pulsed discharges, spark discharges, and atmospheric pressure glow discharges), with conversions of up to 40–90% at energy costs of about 1.2–10 eV per molecule [107, 108, 117, 118].

The addition of a dielectric material as a catalyst [119] to a DBD reactor can increase the conversion to up to 80% [5] and the product distribution can be tuned [43]. A dielectric material ($BaTiO_3$, SiO_2, and TiO_2) can be used to shift the discharge mode and thereby the activation of CO_2 [120]. Dielectric materials as catalysts (barium zinc titanate [BZT] and barium iron niobate [BFN]) in a DBD reactor have yielded CO_2 conversions of 45–60% at an energy cost of about 13–16 eV per molecule [121]. It should be noted that these materials do not truly act as catalysts as discussed above.

Various supported metal catalysts have been introduced into DBD reactors [85]. Even though the catalysts can tune the selectivity of products formed, metal catalysts may also facilitate the conversion of products back to CO_2 [43]. For most catalysts used so far, syngas is the major product and the fraction of oxygenates it typically only up to 1% [43]. Thus, a more rational catalyst design is required to form higher value products, which have a high barrier to be formed in the gas phase. However, the selective production of oxygenates is also reported [107]. The choice of the active metal (Au, Cu, and Pt) determines the distribution of oxygenates formed [43]. For example, formaldehyde is only formed over Au and Pt, and not over Cu-based catalysts [43]. Furthermore, the conversion of CO_2 over Ni-based catalysts is influenced by the choice of support supports (γ-Al_2O_3, MgO, SiO_2, and TiO_2), which may be attributed to the number of basic sites for CO_2 adsorption [49]. CO_2 is a Lewis acid and a high density of basic sites can increase the residence time on the catalyst surface [49]. The increased residence time of CO_2 on the surface may lead to a lower carbon deposition [49], which is primarily caused by methane on the surface.

Zeolites have also been used by various authors. CO_2 conversions of up to 37% at an energy cost of about 24 eV per molecule have been reported for zeolites [122]. As compared to a blank reactor, the introduction of zeolites can tune the product selectivity toward C2–C4 hydrocarbons [123, 124] and specifically unsaturated C2 hydrocarbons [109]. The formation of higher hydrocarbons (C5+) and oxygenates can be suppressed upon introduction of a zeolite [123, 124]. [123, 124]. A possible

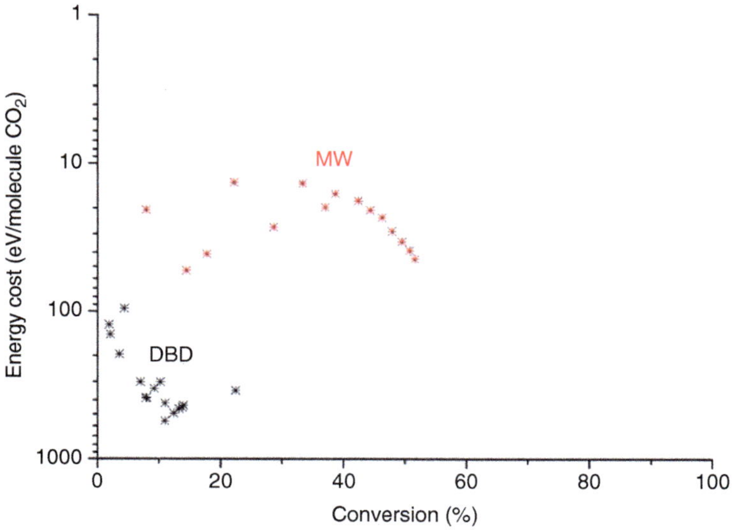

Figure 18.9 Comparison of literature data for CO_2 hydrogenation in various plasma reactors. Source: Bogaerts and Snoeckx [85].

explanation for the hydrocarbon selectivities is the shape selectivity provided by the zeolites [109, 123, 124].

18.2.4.3 CO_2 Hydrogenation

The hydrogenation of CO_2 has not been studied extensively (see Figure 18.9). Note that this reaction is exothermic and that a plasma can only help to master activation barriers, rather than storing energy as explained in Section 18.2.3. The type of plasma reactors used include DBD reactors, MW reactors, RF plasmas, corona discharge, and surface discharge. The primary products are CO and H_2O [85]. Other reported products are CH_4, CH_3OH, CH_5OH, CH_2O, CH_2O_2, C_2H_6O, C_2H_2, C_2H_4, and C_2H_6 [85]. Low energy efficiencies (caused by high energy consumption of 100 eV per molecule CO_2) and low conversions (0–20%) are typically achieved in a DBD reactor. The reactor wall (Al_2O_3, SiO_2) can affect the conversion of CO_2 [125]. Higher conversions of up to over 50% at down to 10 eV per molecule CO_2 are attainable in MW reactors at reduced pressures. As compared to the dry reforming of methane, low concentrations of CH_2 and CH_3 are present in the case of CO_2 and H_2 feeds. CH_2 and CH_3 radicals are beneficial for the conversion of CO_2 [116]. Furthermore, the formation of CHO radicals in the case of CO_2 and H_2 feeds promotes the reaction back to CO_2 [116]. Thus, the hydrogenation of CO_2 without a catalyst does not appear attractive.

Selective formation of CO from CO_2 and H_2 via the reversed water gas sift (R-WGS) reaction (80%) feeds is achieved by introducing various metal catalysts (Cu/Al_2O_3, $Cu\text{-}Mn/Al_2O_3$, and Mn/Al_2O_3) in a DBD reactor [48]. Furthermore, the introduction of a catalyst (Cu/Al_2O_3, $Cu/ZnO/Al_2O_3$, and Pt/Al_2O_3) can enhance the CO_2 conversion to methanol by up to an order of magnitude in DBD reactors [126, 127]. CO_2 conversions of up to 20% and selectivities toward CH_3OH of 55% are attainable in a DBD reactor [127], while the operating temperature can also be decreased as compared to thermal catalysis (although it is difficult to disentangle whether heating on the surface effects play a crucial role) [126]. Furthermore, the methanation and CO formation over Ni-based catalysts can be enhanced by plasma activation of the C=O bond in CO_2 [128–131], which is the rate-determining step in thermal catalysis. Furthermore, the addition of argon can decrease the breakdown voltage, thereby enhancing the charge transfer through the reactor and the methanation rate by CO_2 activation in the plasma phase [130]. CO_2 conversions of up to 80–85% and CH_4 selectivities of >95% are reported in a DBD reactor [129, 131]. Upon introduction of a catalyst (metal-loaded La-ZrO_2) in a corona discharge, CO_2 conversions of 40% and CO selectivities of >95% are attainable [132].

18.2.4.4 Artificial Photosynthesis

Plasma-driven artificial photosynthesis (with CO_2 and H_2O reactants) has not been studied extensively as any conversion is challenging, considering thermodynamics. The main products are H_2, CO, and O_2, while H_2O_2, $C_2H_2O_4$, CH_2O_2, CH_4, C_2H_6O, CH_3OH, C_2H_5OH, C_2H_2, C_3H_4, and carbon nanofibers have also been reported [5]. Hydrogen-rich syngas ratios (H_2 : CO) of up to 8.6 can be achieved [133]. Most studies have been performed with a DBD reactor, MW reactor, gliding arc, and corona discharge [85]. For both DBD reactors and gliding arcs, water molecules quench activated CO_2 molecules, thereby limiting the energy efficiency of converting CO_2 [38, 85, 98]. In the plasma phase, H radicals easily recombine with OH to water, making the formation of complex oxygenates difficult [133]. Furthermore, OH radicals may recombine with CO to CO_2, lowering the observed CO_2 conversion upon water addition [133]. As shown in Figure 18.10, conversions of up to 50% at an energy cost of 20 eV per CO_2 molecule are attainable for MW plasma reactors operating

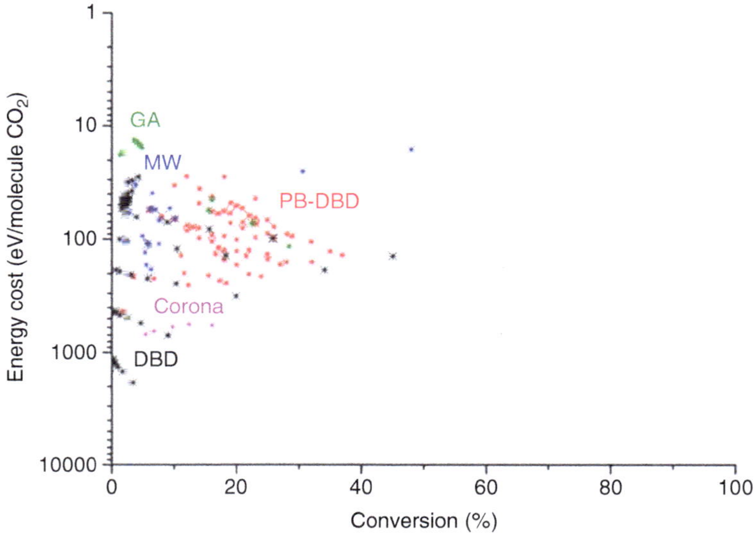

Figure 18.10 Comparison of literature data for artificial photosynthesis in various plasma reactors. Source: Bogaerts and Snoeckx [85].

at reduced pressures (40–80 mbar) [134], which may be due to reduced quenching possibilities at low pressures [85].

The direct formation of oxygenates at significant yields can only be achieved in the presence of a catalyst, as the formation of these products involves many multiple steps. The CO_2 conversion can be increased from about 5% for a blank reactor to up to 20% upon introducing a catalyst (Al_2O_3, Ni/Al_2O_3, NiO/Al_2O_3, NiO/TiO_2, and TiO_2) in a DBD reactor [134–136]. However, the conversion of water is not necessarily increased upon the introduction of a catalyst [136]. The adsorption of plasma-activated species can limit the recombination to CO_2 and H_2O [133], which are the thermodynamically most stable molecules. Possibly, oxygen vacancies in the catalyst lower the activation barrier for dissociative CO_2 adsorption [134]. Upon introduction of a catalyst ($Cu/ZnO/Al_2O_3$) in a negative corona discharge, the selectivity toward ethanol is enhanced and the CO_2 conversion is increased by a factor three [137].

18.3 Perspective

As follows from the experimental observations, the combination of a plasma with the right catalyst appears to be the way forward for plasma-driven CO_2 conversion. The key challenge is to increase the energy efficiency of plasma catalysis for CO_2 conversions. So far, most research followed a trial-and-error approach, rather than a more rigorous science-based approach, which is currently being developed [6, 69]. However, even in such cases, the type of molecule or radical that is involved in plasma catalytic conversions, e.g. vibrational excitation, is generally postulated and experimental evidence is lacking.

18.3.1 Models Describing Plasma Catalysis

Plasma–catalyst interactions occur at various length scales with different time constants [53, 54, 56, 57]. Kim et al. [9] described typical reactions occurring in the plasma and on the catalyst surface, as shown in Figure 18.11. Thus, various types of models describing plasma–catalyst interactions are required for describing the microscopic and macroscopic behavior of plasma catalysis. Models concerning plasma-phase activation and plasma–catalyst interactions are discussed in detail elsewhere [51, 57, 138, 139].

Even though models exist for plasma-phase CO_2 conversions [75, 116], models describing the interactions of plasma-activated species with a catalyst surface have not been fully developed for CO_2 conversions [140]. Recently, density functional theory (DFT) calculations were used for plasma-enhanced catalytic ammonia synthesis, based on vibrational excitations in N_2 [8]. Similar studies for CO_2 activation can provide insights into catalyst design for plasma-enhanced CO_2 conversions, as there is still a lack of insight in which catalyst should perform best in a plasma environment [18]. Inspiration can be obtained from the extensive DFT calculations performed for methane, methanol, and ethanol synthesis from CO_2 and H_2 with

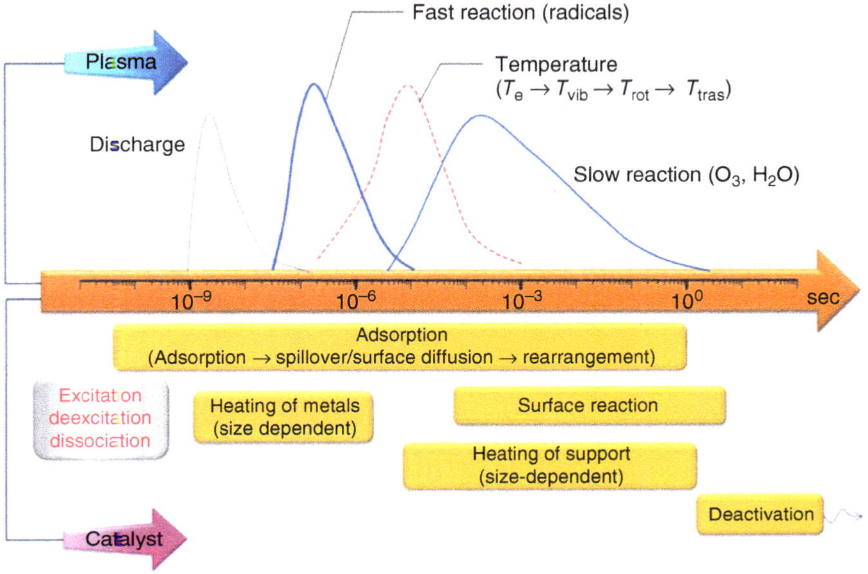

Figure 18.11 Typical timescales of plasma-phase reactions and catalytic surface reactions. Source: Kim et al. [9].

thermal catalysis [141, 142]. Furthermore, kinetic analyses for plasma catalysis may aid further understanding of mechanisms involved [66, 67, 143, 144]. All in all, insights from heterogeneous catalysis are required to develop a fundamental description of plasma catalytic CO_2 conversions [41].

18.3.2 Scale-Up and Process Considerations

Although energy efficiencies have remained too low for practical applications, process development is important for determining the requirements for scale up (such as energy efficiency and conversion). Although not extensively studied, some process concepts have been developed based on plasma-driven CO_2 conversions. An example of such a process concept is the combination of water electrolysis with oxidative dry reforming of methane in a gliding arc plasma, operating at an overall energy efficiency of about 80% [115]. This can serve as a novel process for syngas production for methanol synthesis. Integrated pilots including a plasma reactor and separation steps are required to assess the technological viability of plasma-driven CO_2 conversions. Product separation is often crucial because of thermodynamic limitations, efficiency losses, and safety issues (such as formation of an explosive mixture in case of the CO_2 splitting reaction). Plasma membrane reactors are currently investigated as an alternative to provide enhanced mass transfer and separation [145, 146]. Another option for product removal is the use of oxygen-scavenging materials, such as base metals. Inert gases such as argon and helium may also be added to reduce the explosion risks, albeit at the cost of more expensive separation and recuperation steps, as well as a higher energy input [85].

Figure 18.12 Industrial plasma-driven ozone generators (DBD reactors) for 60 kg h^{-1} ozone production. Source: Kogelschatz [147].

In some cases, plasma-driven conversion processes have developed into mature industrial processes. Such an example is ozone formation [147]. The earliest investigation by Martinus van Marum and Werner von Siemens into ozone formation in electrical sparks and DBD reactors dates from 1785 to 1857, respectively [12, 13]. Nowadays, production rates of hundreds of kg h^{-1} are attainable in a single installation, with single-pass conversions of 5–15% [147]. By using a filtration as a separation step, the overall oxygen conversion can be close to 100% [148]. In an industrial ozone generator, typical reactor lengths are 1–3 m with reactor diameters ranging from 20 to 50 mm (see Figure 18.12) [147]. The inner electrode is typically 0.5–1 mm in width [147]. Thus, industrially applied DBD reactors are similar to those used in a laboratory setup and scale-up is mainly provided by parallelization. It should be noted that ozone generation is operated without a catalyst [148], while plasma-driven CO_2 conversions may require a catalyst in the discharge gap.

18.4 Conclusion

This chapter has illustrated that it is possible to dissociate CO_2 with reasonable efficiency at low conversion without the application of catalysts. In particular, when the reagent mixture becomes more complex, for example, when methane or hydrogen is admixed to target higher value-added products, it becomes useful to introduce catalysts to improve especially the process selectivity. Although synthesis gas remains the main product during dry reforming (CH_4 and CO_2), catalysts enhance the formation of oxygenates and even hydrocarbons. Catalysts can also selectively enhance the formation of CO, methanol, or methane during CO_2 hydrogenation in plasma reactors.

The energy efficiency of the homogeneous CO_2 plasma conversion process can become as high as, or may even exceed, the efficiency of the thermal conversion process. However, large-scale thermal-catalytic processes are generally highly heat-integrated and operate at still higher overall efficiency. The concept of heat integration is not trivial in the plasma process and is to be taken into account for an objective assessment of the economic viability.

There are several reasons why plasma chemistry might become important for future technologies, even in cases that the energy efficiency remains lower than for the traditional processes. Firstly, sustainable resources (wind, solar, and biomass) are available in dispersed mode, requiring scale-down of processes. The consequence is that heat integration and compression become more expensive because of the lacking economy of scale. Furthermore, thermal processes are mostly not capable of following rapid variation in loads from intermittent sources, such as solar power or wind power. Therefore, lower energy efficiency may be acceptable when cheap distributed electrical power is available, such that plasma-driven CO_2 conversions may compete with conventional thermal approaches. Transportation of chemicals can be minimized by the distributed production with plasma-driven processes. All in all, plasma-driven CO_2 conversion processes may provide a solution for intermittent, small-scale applications where mild process operations are important, such as small-scale energy storage from surplus renewable energy. Plasma-driven CO_2 conversion is a potential on-demand technology [149]. The current focus for plasma-driven CO_2 conversion is on better understanding of plasma chemistry as well as plasma catalysis, which may lead to an enhanced energy efficiency through catalyst optimization and plasma optimization.

References

1 Bogaerts, A. and Neyts, E.C. (2018). Plasma technology: an emerging technology for energy storage. *ACS Energy Lett.* 3 (4): 1013–1027. https://doi.org/10.1021/acsenergylett.8b00184.

2 Rusanov, V.D., Fridman, A.A., and Sholin, G.V. (1981). The physics of a chemically active plasma with nonequilibrium vibrational excitation of molecules. *Phys. Usp.* 24 (6): 447–474.

3 Azizov, R.I., Vakar, A.K., Zhivotov, V.K. et al. (1983). Non-equilibrium plasma-chemical process of CO_2 decomposition in a supersonic microwave discharge. *Proc. USSR Acad. Sci.* 271: 94–97.

4 Fridman, A. (2008). *Plasma Chemistry*. Cambridge University Press.

5 Snoeckx, R. and Bogaerts, A. (2017). Plasma technology – a novel solution for CO_2 conversion? *Chem. Soc. Rev.* 46 (19): 5805–5863. https://doi.org/10.1039/c6cs00066e.

6 Mehta, P., Barboun, P., Go, D.B. et al. (2019). Catalysis enabled by plasma activation of strong chemical bonds: a review. *ACS Energy Lett.* 4 (5): 1115–1133. https://doi.org/10.1021/acsenergylett.9b00263.

7 Whitehead, J.C. (2016). Plasma-catalysis: the known knowns, the known unknowns and the unknown unknowns. *J. Phys. D: Appl. Phys.* 49 (24): 243001. https://doi.org/10.1088/0022-3727/49/24/243001.

8 Mehta, P., Barboun, P., Herrera, F.A. et al. (2018). Overcoming ammonia synthesis scaling relations with plasma-enabled catalysis. *Nat. Catal.* 1 (4): 269–275. https://doi.org/10.1038/s41929-018-0045-1.

9 Kim, H.-H., Teramoto, Y., Ogata, A. et al. (2016). Plasma catalysis for environmental treatment and energy applications. *Plasma Chem. Plasma Process.* 36 (1): 45–72. https://doi.org/10.1007/s11090-015-9652-7.

10 Eddington, A.S. (1928). Monthly notice. *Roy. Astron. Soc.* 88: 352.

11 Von Engel, A. and Cozens, J.R. (1965). Flame plasmas. *Adv. Electron. Electron Phys.* 20 (C): 99–146. https://doi.org/10.1016/S0065-2539(08)60495-3.

12 Hackmann, W.D. (1972). The researches of Dr. Martinus Van Marum (1750–1837) on the influence of electricity on animals and plants. *Med.History* 16 (1): 11–26. https://doi.org/10.1017/S0025727300017221.

13 von Siemens, W. (1857). Ueber die elektrostatische Induction und die Verzögerung des Stroms in Flaschendrähten. *Ann. Phys.* 178 (9): 66–122. https://doi.org/10.1002/andp.18571780905.

14 Patil, B.S., Hessel, V., Seefeldt, L.C. et al. (2017). Nitrogen fixation. In: *Ullmann's Encyclopedia of Industrial Chemistry*, 1–21. Wiley-VCH Verlag GmbH & Co. KGaA https://doi.org/10.1002/14356007.a17_471.pub2.

15 Patil, B.S., Wang, Q., Hessel, V., and Lang, J. (2015). Plasma N2-fixation: 1900–2014. *Catal. Today* 256: 49–66. https://doi.org/10.1016/j.cattod.2015.05.005.

16 Ernst, F.A. (1928). *Industrial Chemical Monographs: Fixation of Atmospheric Nitrogen*. London: Chapman & Hall.

17 Smil, V. (2004). *Enriching the Earth: Fritz Haber, Carl Bosch, and the Transformation of World Food Production*. Cambridge, MA: MIT Press.

18 Neyts, E.C., Ostrikov, K., Sunkara, M.K., and Bogaerts, A. (2015). Plasma catalysis: synergistic effects at the nanoscale. *Chem. Rev.* 115 (24): 13408–13446. https://doi.org/10.1021/acs.chemrev.5b00362.

19 Van Geem, K.M., Galvita, V.V., and Marin, G.B. (2019). Making chemicals with electricity. *Science* 364: 734–735.

20 van Rooij, G.J., Akse, H.N., Bongers, W.A., and van de Sanden M.C.M. (2018). Plasma for electrification of chemicalindustry: a case study on CO_2 reduction. *Plasma Phys. Control. Fusion* 60: 014019.

21 Bogaerts, A. and Neyts, E.C. (2018). Plasma technology: an emerging technology for energy storage. *ACS Energy Lett.* 3: 1013–1027.

22 Kozàk, T. and Bogaerts, A. (2014). Splitting of CO_2 by vibrational excitation in non-equilibrium plasmas: a reaction kinetics model. *Plasma Sources Sci. Technol.* 23 (4) https://doi.org/10.1088/0963-0252/23/4/045004.

23 Bongers, W., Bouwmeester, H., Wolf, B. et al. (2017). Plasma-driven dissociation of CO_2 for fuel synthesis. *Plasma Processes Polym.* 14 (6): 1–8. https://doi.org/10.1002/ppap.201600126.

24 den Harder, N., van den Bekerom, D.C.M., Al, R.S. et al. (2017). Homogeneous CO_2 conversion by microwave plasma: wave propagation and diagnostics. *Plasma Processes Polym.* 14 (6): 1–24. https://doi.org/10.1002/ppap.201600120.

25 van Rooij, G.J., van den Bekerom, D.C.M., den Harder, N. et al. (2015). Taming microwave plasma to beat thermodynamics in CO_2 dissociation. *Faraday Discuss.* 183: 233–248. https://doi.org/10.1039/c5fd00045a.

26 Diomede, P., Van De Sanden, M.C.M., and Longo, S. (2017). Insight into CO_2 dissociation in plasma from numerical solution of a vibrational diffusion equation. *J. Phys. Chem. C* 121 (36): 19568–19576. https://doi.org/10.1021/acs.jpcc.7b04896.

27 Silva, T., Grofulović, M., Klarenaar, B.L.M. et al. (2018). Kinetic study of low-temperature CO_2 plasmas under non-equilibrium conditions. I. Relaxation of vibrational energy. *Plasma Sources Sci. Technol.* 27 (1) https://doi.org/10.1088/1361-6595/aaa56a.

28 Neyts, E.C. (2016). Plasma-surface interactions in plasma catalysis. *Plasma Chem. Plasma Process.* 36 (1): 185–212. https://doi.org/10.1007/s11090-015-9662-5.

29 Semiokhin, I., Andreev, Y.P., and Panchenkov, G. (1964). Dissociation of carbon dioxide circulating in a silent electric Discharge. *Russ. J. Phys. Chem.* 38 (8): 1126.

30 Maltsev, A.N., Eremin, E.N., and Ivanter, L.V. (1967). Dissociation of carbon dioxide in glow discharge. *Russ. J. Phys. Chem.* 41: 633.

31 Andreev, Y.P., Semiokhin, I.A., Voronkov, Y. et al. (1971). Dissociation of carbon dioxide in pulse discharge. *Russ. J. Phys. Chem.* 45 (11): 1587.

32 Asisov, R. F., Givotov, V. K., Krasheninnikov, E. G., Petrushev, B. I., Potapkin, B. V., Rusanov, V. D., … Kurchatov, I. V. (1981). Carbon dioxide dissociation in non-equilibrium plasma. In *5th International Symposium on Plasma Chemistry* (p. 774). Edinburgh, 10–14 August 1981.

33 Tsuji, M., Tanoue, T., Nakano, K., and Nishimura, Y. (2001). Decomposition of CO_2 into CO and O in a microwave-excited discharge flow of CO_2/He or CO_2/Ar mixtures. *Chem. Lett.* 1 (1): 22–23. https://doi.org/10.1246/cl.2001.22.

34 Vesel, A., Mozetic, M., Drenik, A., and Balat-Pichelin, M. (2011). Dissociation of CO_2 molecules in microwave plasma. *Chem. Phys.* 382 (1–3): 127–131. https://doi.org/10.1016/j.chemphys.2011.03.015.

35 Spencer, L.F. and Gallimore, A.D. (2013). CO_2 dissociation in an atmospheric pressure plasma/catalyst system: A study of efficiency. *Plasma Sources Sci. Technol.* 22 (1) https://doi.org/10.1088/0963-0252/22/1/015019.

36 Chen, G., Georgieva, V., Godfroid, T. et al. (2016). Plasma assisted catalytic decomposition of CO_2. *Appl. Catal., B* 190: 115–124. https://doi.org/10.1016/j.apcatb.2016.03.009.

37 Goede, A.P.H., Bongers, W.A., Graswinckel, M.F. et al. (2014). Production of solar fuels by CO_2 plasmolysis. *EPJ Web Conf.* 79: 1–5. https://doi.org/10.1051/epjconf/20137901005.

38 Nunnally, T., Gutsol, K., Rabinovich, A. et al. (2011). Dissociation of CO_2 in a low current gliding arc plasmatron. *J. Phys. D: Appl. Phys.* 44 (27) https://doi.org/10.1088/0022-3727/44/27/274009.

39 Kim, S.C., Lim, M.S., and Chun, Y.N. (2014). Reduction characteristics of carbon dioxide using a plasmatron. *Plasma Chem. Plasma Process.* 34 (1): 125–143. https://doi.org/10.1007/s11090-013-9499-8.

40 Wang, W., Berthelot, A., Kolev, S. et al. (2016). CO_2 conversion in a gliding arc plasma: 1D cylindrical discharge model. *Plasma Sources Sci. Technol.* 25 (6) https://doi.org/10.1088/0963-0252/25/6/065012.

41 Liu, M., Yi, Y., Wang, L. et al. (2019). Hydrogenation of carbon dioxide to value-added chemicals by heterogeneous catalysis and plasma catalysis. *Catalysts* 9: 275. https://doi.org/10.3390/catal9030275.

42 Liu, C.-J., Xu, G.-H., and Wang, T. (1999). Non-thermal plasma approaches in CO_2 utilization. *Fuel Process. Technol.* 58 (2): 119–134. https://doi.org/10.1016/S0378-3820(98)00091-5.

43 Wang, L., Yi, Y., Wu, C. et al. (2017). One-step reforming of CO_2 and CH_4 into high-value liquid chemicals and fuels at room temperature by plasma-driven catalysis. *Angew. Chem. Int. Ed.* 56 (44): 13679–13683. https://doi.org/10.1002/anie.201707131.

44 Mei, D., Zhu, X., Wu, C. et al. (2016). Plasma-photocatalytic conversion of CO_2 at low temperatures: understanding the synergistic effect of plasma-catalysis. *Appl. Catal., B* 182: 525–532. https://doi.org/10.1016/j.apcatb.2015.09.052.

45 Van Laer, K. and Bogaerts, A. (2015). Improving the conversion and energy efficiency of carbon dioxide splitting in a Zirconia-packed dielectric barrier discharge reactor. *Energy Technology* 3 (10): 1038–1044. https://doi.org/10.1002/ente.201500127.

46 Duan, X., Hu, Z., Li, Y., and Wang, B. (2015). Effect of dielectric packing materials on the decomposition of carbon dioxide using DBD microplasma reactor. *AlChE J.* 61 (3): 898–903. https://doi.org/10.1002/aic.14682.

47 Michielsen, I., Uytdenhouwen, Y., Pype, J. et al. (2017). CO_2 dissociation in a packed bed DBD reactor: first steps towards a better understanding of plasma catalysis. *Chem. Eng. J.* 326: 477–488. https://doi.org/10.1016/j.cej.2017.05.177.

48 Zeng, Y. and Tu, X. (2016). Plasma-catalytic CO_2 hydrogenation at low temperatures. *IEEE Trans. Plasma Sci.* 44 (4): 1–7. https://doi.org/10.1109/TPS.2015.2504549.

49 Mei, D., Ashford, B., He, Y., and Tu, X. (2017). Plasma-catalytic reforming of biogas over supported Ni catalysts in a dielectric barrier discharge reactor: effect of catalyst supports. *Plasma Processes Polym.* 14 (6) https://doi.org/10.1002/ppap.201600076.

50 Zeng, Y.X., Wang, L., Wu, C.F. et al. (2018). Low temperature reforming of biogas over K-, Mg- and Ce-promoted Ni/Al_2O_3 catalysts for the production of hydrogen rich syngas: understanding the plasma-catalytic synergy. *Appl. Catal., B* 224: 469–478. https://doi.org/10.1016/j.apcatb.2017.10.017.

51 Neyts, E.C. and Bogaerts, A. (2014). Understanding plasma catalysis through modelling and simulation – a review. *J. Phys. D: Appl. Phys.* 47 (22): 1–18. https://doi.org/10.1088/0022-3727/47/22/224010.

52 Whitehead, J.C. (2010). Plasma catalysis: a solution for environmental problems *Pure Appl. Chem.* 82 (6): 1329–1336. https://doi.org/10.1351/PAC-CON-10-02-39.

53 Whitehead, J.C. (2018). Plasma-catalysis: introduction and history. In: *Plasma Catalysis: Fundamentals and Applications* (eds. X. Tu, T. Nozaki and J.C. Whitehead), 1–19. Springer International Publishing AG.

54 Whitehead, J.C. (2019). Plasma-catalysis: is it just a question of scale? *Front. Chem. Sci. Eng.* 13: 264–273. https://doi.org/10.1007/s11705-019-1794-3.

55 Kim, H.-H., Teramoto, Y., Negishi, N., and Ogata, A. (2015). A multidisciplinary approach to understand the interactions of nonthermal plasma and catalyst: a review. *Catal. Today* 256 (Part 1): 13–22. https://doi.org/10.1016/j.cattod.2015.04.009.

56 Ostrikov, K. (2019). Plasma-nano-interface in perspective: from plasma-for-nano to nano-plasmas. *Plasma Phys. Controlled Fusion* 61 (1): 014028. https://doi.org/10.1088/1361-6587/aad770.

57 Neyts, E.C. and Brault, P. (2017). Molecular dynamics simulations for plasma-surface interactions. *Plasma Processes Polym.* 14 (1–2): 1–18. https://doi.org/10.1002/ppap.201600145.

58 Van Durme, J., Dewulf, J., Leys, C., and Van Langenhove, H. (2008). Combining non-thermal plasma with heterogeneous catalysis in waste gas treatment: a review. *Appl. Catal., B* 78 (3–4): 324–333. https://doi.org/10.1016/j.apcatb.2007.09.035.

59 Wang, B., Xu, X., Xu, W. et al. (2018). The mechanism of non-thermal plasma catalysis on volatile organic compounds removal. *Catal. Surv. Asia* 22 (2): 73–94. https://doi.org/10.1007/s10563-018-9241-x.

60 Van Laer, K. and Bogaerts, A. (2015). Fluid modelling of a packed bed dielectric barrier discharge plasma reactor. *Plasma Sources Sci. Technol.* 25 (1): 015002. https://doi.org/10.1088/0963-0252/25/1/015002.

61 Van Laer, K. and Bogaerts, A. (2017). Influence of gap size and dielectric constant of the packing material on the plasma behaviour in a packed bed DBD reactor: a fluid modelling study. *Plasma Processes Polym.* 14 (4–5) https://doi.org/10.1002/ppap.201600129.

62 Van Laer, K. and Bogaerts, A. (2017). How bead size and dielectric constant affect the plasma behaviour in a packed bed plasma reactor: a modelling study. *Plasma Sources Sci. Technol.* 26 (8) https://doi.org/10.1088/1361-6595/aa7c59.

63 Bogaerts, A., Zhang, Q., Zhang, Y. et al. (2019). Burning questions of plasma catalysis: answers by modeling. *Catal. Today* 337: 3–14. https://doi.org/10.1016/j.cattod.2019.04.077.

64 Wang, W., Kim, H., Van Laer, K., and Bogaerts, A. (2018). Streamer propagation in a packed bed plasma reactor for plasma catalysis applications G RA P H I C A L AB S T R A C T. *Chem. Eng. J.* 334: 2467–2479. https://doi.org/10.1016/j.cej.2017.11.139.

65 Rettner, C.T. and Stein, H. (1987). Effect of vibrational energy on the dissociative chemisorption of N2 on Fe(111). *J. Chem. Phys.* 87 (1): 770–771. https://doi.org/10.1063/1.453575.

66 Kim, J., Go, D.B., and Hicks, J.C. (2017). Synergistic effects of plasma-catalyst interactions for CH_4 activation. *Phys. Chem. Chem. Phys.* 19 (20): 13010–13021. https://doi.org/10.1039/c7cp01322a.

67 Rouwenhorst, K.H.R., Kim, H.-H., and Lefferts, L. (2019). Vibrationally excited activation of N2 in plasma-enhanced catalytic ammonia synthesis: a kinetic analysis. *ACS Sustainable Chem. Eng.* 7 (20): 17515–17522. https://doi.org/10.1021/acssuschemeng.9b04997.

68 Bal, K.M. and Neyts, E.C. (2019). Overcoming old scaling relations and establishing new correlations in catalytic surface chemistry: combined effect of charging and doping. *J. Phys. Chem. C* 123: 6141–6147. https://doi.org/10.1021/acs.jpcc.9b01216.

69 Bal, K.M., Huygh, S., Bogaerts, A., and Neyts, E.C. (2018). Effect of plasma-induced surface charging on catalytic processes: application to CO_2 activation. *Plasma Sources Sci. Technol.* 27 (2): 024001. https://doi.org/10.1088/1361-6595/aaa868.

70 Lee, C.J., Lee, D.H., and Kim, T. (2017). Modification of catalyst surface from interaction between catalysts and dielectric barrier discharge plasma. *J. Nanosci. Nanotechnol.* 17 (4): 2707–2710. https://doi.org/10.1166/jnn.2017.13352.

71 Guo, Y., Ye, D., Chen, K. et al. (2006). Toluene decomposition using a wire-plate dielectric barrier discharge reactor with manganese oxide catalyst in situ. *J. Mol. Catal. A: Chem.* 245 (1–2): 93–100. https://doi.org/10.1016/j.molcata.2005.09.013.

72 Roland, U., Holzer, F., and Kopinke, F.D. (2002). Improved oxidation of air pollutants in a non-thermal plasma. *Catal. Today* 73 (3–4): 315–323. https://doi.org/10.1016/S0920-5861(02)00015-9.

73 Zhang, Y.R., Van Laer, K., Neyts, E.C., and Bogaerts, A. (2016). Can plasma be formed in catalyst pores? A modeling investigation. *Appl. Catal., B* 185: 56–67. https://doi.org/10.1016/j.apcatb.2015.12.009.

74 Rouwenhorst, K.H.R., Engelmann, Y., van ‘t Veer, K., Postma, R.S., Bogaerts, A., and Lefferts, L. (2020). Plasma-driven catalysis: green ammonia synthesis with intermittent electricity. *Green Chem.* 22: 6258–6287.

75 Alliati, M., Mei, D., and Tu, X. (2018). Plasma activation of CO_2 in a dielectric barrier discharge: a chemical kinetic model from the microdischarge to the reactor scales. *J. CO2 Util.* 27: 308–319. https://doi.org/10.1016/j.jcou.2018.07.018.

76 Hummelshøj, J.S., Abild-pedersen, F., Studt, F. et al. (2012). CatApp: a web application for surface chemistry and heterogeneous catalysis. *Angew. Chem. Int. Ed.* 51 (1): 272–274. https://doi.org/10.1002/anie.201107947.

77 Whitehead, J.C. (2020). Advances in plasmas for a sustainable future: roadmap paper on plasma catalysis. *J. Phys. D: Appl. Phys.*

78 Ashford, B. and Tu, X. (2017). Non-thermal plasma technology for the conversion of CO_2. *Curr. Opin. Green Sustainable Chem.* 3: 45–49. https://doi.org/10.1016/j.cogsc.2016.12.001.

79 Akay, G. and Zhang, K. (2017). Process intensification in ammonia synthesis using novel coassembled supported microporous catalysts promoted by nonthermal plasma. *Ind. Eng. Chem. Res.* 56 (2): 457–468. https://doi.org/10.1021/acs.iecr.6b02053.

80 Akay, G. (2017). Sustainable ammonia and advanced symbiotic fertilizer production using catalytic multi-reaction-zone reactors with nonthermal plasma and simultaneous reactive separation. *ACS Sustainable Chem. Eng.* 5 (12): 11588–11606. https://doi.org/10.1021/acssuschemeng.7b02962.

81 Chiremba, E., Zhang, K., Kazak, C., and Akay, G. (2017). Direct nonoxidative conversion of methane to hydrogen and higher hydrocarbons by dielectric barrier discharge plasma with plasma catalysis promoters. *AIChE J.* 63 (10): 4418–4429. https://doi.org/10.1002/aic.15769.

82 Puliyalil, H., Lašič Jurković, D., Dasireddy, V.D.B.C., and Likozar, B. (2018). A review of plasma-assisted catalytic conversion of gaseous carbon dioxide and methane into value-added platform chemicals and fuels. *RSC Adv.* 8 (48): 27481–27508. https://doi.org/10.1039/c8ra03146k.

83 Bogaerts, A., Kozák, T., Van Laer, K., and Snoeckx, R. (2015). Plasma-based conversion of CO_2: current status and future challenges. *Faraday Discuss.* 183: 217–232. https://doi.org/10.1039/c5fd00053j.

84 Hussain, A., Tahir, M., Aishah, N., and Amin, S. (2019). Recent developments in non-thermal catalytic DBD plasma reactor for dry reforming of methane. *Energy Convers. Manage.* 183: 529–560. https://doi.org/10.1016/j.enconman.2018.12.112.

85 Bogaerts, A. and Snoeckx, R. (2019). Plasma-based CO_2 conversion. In: *An Economy Based on Carbon Dioxide and Water* (eds. M. Aresta, I. Karimi and S. Kawi), 287–325. Springer, Cham https://doi.org/10.1007/978-3-030-15868-2_8.

86 Aerts, R., Snoeckx, R., and Bogaerts, A. (2014). In-situ chemical trapping of oxygen in the splitting of carbon dioxide by plasma. *Plasma Processes Polym.* 11 (10): 985–992. https://doi.org/10.1002/ppap.201400091.

87 Cleiren, E., Heijkers, S., Ramakers, M., and Bogaerts, A. (2017). Dry reforming of methane in a gliding arc plasmatron: towards a better understanding of the plasma chemistry. *ChemSusChem* 10 (20): 4025–4036. https://doi.org/10.1002/cssc.201701274.

88 Lebouvier, A., Iwarere, S.A., Argenlieu, P. et al. (2013). Assessment of carbon dioxide dissociation as a new route for syngas production: a comparative review and potential of plasma-based technologies. *Energy Fuels* 27 (5): 2712–2722. https://doi.org/10.1021/ef301991d.

89 Aerts, R., Somers, W., and Bogaerts, A. (2015). Carbon dioxide splitting in a dielectric barrier discharge plasma: a combined experimental and computational study. *ChemSusChem* 8 (4): 702–716. https://doi.org/10.1002/cssc.201402818.

90 Paulussen, S., Verheyde, B., Tu, X. et al. (2010). Conversion of carbon dioxide to value-added chemicals in atmospheric pressure dielectric barrier discharges. *Plasma Sources Sci. Technol.* 19 (3): 034015. https://doi.org/10.1088/0963-0252/19/3/034015.

91 Yu, Q., Kong, M., Liu, T. et al. (2012). Characteristics of the decomposition of CO_2 in a dielectric packed-bed plasma reactor. *Plasma Chem. Plasma Process.* 32 (1): 153–163. https://doi.org/10.1007/s11090-011-9335-y.

92 Ozkan, A., Bogaerts, A., and Reniers, F. (2017). Routes to increase the conversion and the energy efficiency in the splitting of CO_2 by a dielectric barrier discharge. *J. Phys. D: Appl. Phys.* 50 (8): 084004. https://doi.org/10.1088/1361-6463/aa562c.

93 Ozkan, A., Dufour, T., Silva, T. et al. (2016). DBD in burst mode: solution for more efficient CO_2 conversion? *Plasma Sources Sci. Technol.* 25 (5): 055005. https://doi.org/10.1088/0963-0252/25/5/055005.

94 Snoeckx, R., Heijkers, S., Van Wesenbeeck, K. et al. (2016). CO_2 conversion in a dielectric barrier discharge plasma: N2 in the mix as a helping hand or problematic impurity? *Energy Environ. Sci.* 9 (3): 999–1011. https://doi.org/10.1039/c5ee03304g.

95 Silva, T., Britun, N., Godfroid, T., and Snyders, R. (2014). Optical characterization of a microwave pulsed discharge used for dissociation of CO_2. *Plasma Sources Sci. Technol.* 23 (2): 025009. https://doi.org/10.1088/0963-0252/23/2/025009.

96 Ramakers, M., Trenchev, G., Heijkers, S. et al. (2017). Gliding arc plasmatron: providing an alternative method for carbon dioxide conversion. *ChemSusChem* 10 (12): 2642–2652. https://doi.org/10.1002/cssc.201700589.

97 Sun, S.R., Wang, H.X., Mei, D.H. et al. (2017). CO_2 conversion in a gliding arc plasma: performance improvement based on chemical reaction modeling. *J. CO2 Util.* 17: 220–234. https://doi.org/10.1016/j.jcou.2016.12.009.

98 Indarto, A., Yang, D.R., Choi, J.-W. et al. (2007). Gliding arc plasma processing of CO_2 conversion. *J. Hazard. Mater.* 146 (1–2): 309–315. https://doi.org/10.1016/j.jhazmat.2006.12.023.

99 Liu, J.L., Park, H.W., Chung, W.J., and Park, D.W. (2016). High-efficient conversion of CO_2 in AC-pulsed Tornado gliding arc plasma. *Plasma Chem. Plasma Process.* 36 (2): 437–449. https://doi.org/10.1007/s11090-015-9649-2.

100 Wang, W., Mei, D., Tu, X., and Bogaerts, A. (2017). Gliding arc plasma for CO_2 conversion: better insights by a combined experimental and modelling approach. *Chem. Eng. J.* 330: 11–25. https://doi.org/10.1016/j.cej.2017.07.133.

101 Butterworth, T., Elder, R., and Allen, R. (2016). Effects of particle size on CO_2 reduction and discharge characteristics in a packed bed plasma reactor. *Chem. Eng. J.* 293 (February): 55–67. https://doi.org/10.1016/j.cej.2016.02.047.

102 Uytdenhouwen, Y., Van Alphen, S., Michielsen, I. et al. (2018). A packed-bed DBD micro plasma reactor for CO_2 dissociation: does size matter? *Chem. Eng. J.* 348 (May): 557–568. https://doi.org/10.1016/j.cej.2018.04.210.

103 Butterworth, T. and Allen, R.W.K. (2017). Plasma-catalyst interaction studied in a single pellet DBD reactor: dielectric constant effect on plasma dynamics.

Plasma Sources Sci. Technol. 26 (6): 065008. https://doi.org/10.1088/1361-6595/aa6c35.

104 Lavoie, J.-M. (2014). Review on dry reforming of methane, a potentially more environmentally-friendly approach to the increasing natural gas exploitation. *Front. Chem.* 2: 024001. https://doi.org/10.3389/fchem.2014.00081.

105 Tao, X., Bai, M., Li, X. et al. (2011). CH_4-CO_2 reforming by plasma – challenges and opportunities. *Prog. Energy Combust. Sci.* 37 (2): 113–124. https://doi.org/10.1016/j.pecs.2010.05.001.

106 Istadi and Amin, N.A.S. (2006). Co-generation of synthesis gas and C_{2+} hydrocarbons from methane and carbon dioxide in a hybrid catalytic-plasma reactor: a review. *Fuel* 85 (5–6): 577–592. https://doi.org/10.1016/j.fuel.2005.09.002.

107 Scapinello, M., Martini, L.M., and Tosi, P. (2014). CO_2 hydrogenation by CH_4 in a dielectric barrier discharge: catalytic effects of nickel and copper. *Plasma Processes Polym.* 11 (7): 624–628. https://doi.org/10.1002/ppap.201400023.

108 Ghorbanzadeh, A.M., Norouzi, S., and Mohammadi, T. (2005). High energy efficiency in syngas and hydrocarbon production from dissociation of CH_4-CO_2 mixture in a non-equilibrium pulsed plasma. *J. Phys. D: Appl. Phys.* 38 (20): 3804–3811. https://doi.org/10.1088/0022-3727/38/20/006.

109 Gallon, H.J., Tu, X., and Whitehead, J.C. (2012). Effects of reactor packing materials on H2 production by CO_2 reforming of CH_4 in a dielectric barrier discharge. *Plasma Processes Polym.* 9 (1): 90–97. https://doi.org/10.1002/ppap.201100130.

110 Snoeckx, R., Zeng, Y.X., Tu, X., and Bogaerts, A. (2015). Plasma-based dry reforming: improving the conversion and energy efficiency in a dielectric barrier discharge. *RSC Adv.* 5 (38): 29799–29808. https://doi.org/10.1039/c5ra01100k.

111 Wu, A., Yan, J., Zhang, H. et al. (2014). Study of the dry methane reforming process using a rotating gliding arc reactor. *Int. J. Hydrogen Energy* 39 (31): 17656–17670. https://doi.org/10.1016/j.ijhydene.2014.08.036.

112 Li, K., Liu, J.L., Li, X.S. et al. (2016). Warm plasma catalytic reforming of biogas in a heat-insulated reactor: dramatic energy efficiency and catalyst auto-reduction. *Chem. Eng. J.* 288: 671–679. https://doi.org/10.1016/j.cej.2015.12.036.

113 Tu, X. and Whitehead, J.C. (2014). Plasma dry reforming of methane in an atmospheric pressure AC gliding arc discharge: co-generation of syngas and carbon nanomaterials. *Int. J. Hydrogen Energy* 39 (18): 9658–9669. https://doi.org/10.1016/j.ijhydene.2014.04.073.

114 Allah, Z.A. and Whitehead, J.C. (2015). Plasma-catalytic dry reforming of methane in an atmospheric pressure AC gliding arc discharge. *Catal. Today* 256 (P1): 76–79. https://doi.org/10.1016/j.cattod.2015.03.040.

115 Li, K., Liu, J.L., Li, X.S. et al. (2018). Novel power-to-syngas concept for plasma catalytic reforming coupled with water electrolysis. *Chem. Eng. J.* 353: 297–304. https://doi.org/10.1016/j.cej.2018.07.111.

116 De Bie, C., Van Dijk, J., and Bogaerts, A. (2016). CO_2 hydrogenation in a dielectric barrier discharge plasma revealed. *J. Phys. Chem. C* 120 (44): 25210–25224. https://doi.org/10.1021/acs.jpcc.6b07639.

117 Chung, W.C. and Chang, M.B. (2016). Review of catalysis and plasma performance on dry reforming of CH_4 and possible synergistic effects. *Renewable Sustainable Energy Rev.* 62: 13–31. https://doi.org/10.1016/j.rser.2016.04.007.

118 Li, D., Li, X., Bai, M. et al. (2009). CO_2 reforming of CH_4 by atmospheric pressure glow discharge plasma: a high conversion ability. *Int. J. Hydrogen Energy* 34 (1): 308–313. https://doi.org/10.1016/j.ijhydene.2008.10.053.

119 Michielsen, I., Uytdenhouwen, Y., Bogaerts, A., and Meynen, V. (2019). Altering conversion and product selectivity of dry reforming of methane in a dielectric barrier discharge by changing the dielectric packing material. *Catalysts* 9 (1) https://doi.org/10.3390/catal9010051.

120 Mei, D., Zhu, X., He, Y. et al. (2015). Plasma-assisted conversion of CO_2 in a dielectric barrier discharge reactor: understanding the effect of packing materials. *Plasma Sources Sci. Technol.* 24 (1): 015011. https://doi.org/10.1088/0963-0252/24/1/015011.

121 Chung, W.-C., Pan, K.-L., Lee, H.-M., and Chang, M.-B. (2014). Dry reforming of methane with dielectric barrier discharge and ferroelectric packed-bed reactors. *Energy Fuels* 28 (12): 7621–7631. https://doi.org/10.1021/ef5020555.

122 Krawczyk, K., Młotek, M., and Ulejczyk, B. (2014). Methane conversion with carbon dioxide in plasma-catalytic system. *Fuel* 117: 608–617. https://doi.org/10.1016/j.fuel.2013.08.068.

123 Eliasson, B., Liu, C.J., and Kogelschatz, U. (2000). Direct conversion of methane and carbon dioxide to higher hydrocarbons using catalytic dielectric-barrier discharges with zeolites. *Ind. Eng. Chem. Res.* 39 (5): 1221–1227. https://doi.org/10.1021/ie990804r.

124 Zhang, K., Eliasson, B., and Kogelschatz, U. (2002). Direct conversion of greenhouse gases to synthesis gas and C4 hydrocarbons over Zeolite HY promoted by a dielectric-barrier discharge. *Ind. Eng. Chem. Res.* 41 (6): 1462–1468. https://doi.org/10.1021/ie0105021.

125 Mora, E.Y., Sarmiento, A., and Vera, E. (2016). Alumina and quartz as dielectrics in a dielectric barrier discharges DBD system for CO_2 hydrogenation. *J. Phys. Conf. Ser.* 687 (1): 012020. https://doi.org/10.1088/1742-6596/687/1/012020.

126 Eliasson, B., Kogelschatz, U., Xue, B., and Zhou, L.M. (1998). Hydrogenation of carbon dioxide to methanol with a discharge-activated catalyst. *Ind. Eng. Chem. Res.* 37 (8): 3350–3357. https://doi.org/10.1021/ie9709401.

127 Wang, L., Yi, Y., Guo, H., and Tu, X. (2018). Atmospheric pressure and room temperature synthesis of methanol through plasma-catalytic hydrogenation of CO_2. *ACS Catal.* 8 (1): 90–100. https://doi.org/10.1021/acscatal.7b02733.

128 Nizio, M., Albarazi, A., Cavadias, S. et al. (2016). Hybrid plasma-catalytic methanation of CO_2 at low temperature over ceria zirconia supported Ni catalysts. *Int. J. Hydrogen Energy* 41 (27): 11584–11592. https://doi.org/10.1016/j.ijhydene.2016.02.020.

129 Jwa, E., Lee, S.B., Lee, H.W., and Mok, Y.S. (2013). Plasma-assisted catalytic methanation of CO and CO_2 over Ni – zeolite catalysts. *Fuel Process. Technol.* 108: 89–93. https://doi.org/10.1016/j.fuproc.2012.03.008.

130 Zeng, Y. and Tu, X. (2017). Plasma-catalytic hydrogenation of CO_2 for the cogeneration of CO and CH_4 in a dielectric barrier discharge reactor: effect of argon addition. *J. Phys. D: Appl. Phys.* 50 (18): 184004. https://doi.org/10.1088/1361-6463/aa64bb.

131 Mikhail, M., Wang, B., Jalain, R. et al. (2019). Plasma-catalytic hybrid process for CO_2 methanation: optimization of operation parameters. *React. Kinet. Mech. Catal.* 126: 629–643. https://doi.org/10.1007/s11144-018-1508-8.

132 Oshima, K., Shinagawa, T., Nogami, Y. et al. (2014). Low temperature catalytic reverse water gas shift reaction assisted by an electric field. *Catal. Today* 232: 27–32. https://doi.org/10.1016/j.cattod.2013.11.035.

133 Snoeckx, R., Ozkan, A., Reniers, F., and Bogaerts, A. (2017). The quest for value-added products from carbon dioxide and water in a dielectric barrier discharge: a chemical kinetics study. *ChemSusChem* 10 (2): 409–424. https://doi.org/10.1002/cssc.201601234.

134 Chen, G., Britun, N., Godfroid, T. et al. (2017). An overview of CO_2 conversion in a microwave discharge: the role of plasma-catalysis. *J. Phys. D: Appl. Phys.* 50 (8) https://doi.org/10.1088/1361-6463/aa5616.

135 Mahammadunnisa, S., Reddy, E.L., Ray, D. et al. (2013). CO_2 reduction to syngas and carbon nanofibres by plasma-assisted in situ decomposition of water. *Int. J. Greenhouse Gas Control* 16: 361–363. https://doi.org/10.1016/j.ijggc.2013.04.008.

136 Ma, X., Li, S., Ronda-Lloret, M. et al. (2018). Plasma assisted catalytic conversion of CO_2 and H_2O over Ni/Al_2O_3 in a DBD reactor. *Plasma Chem. Plasma Process.* 39 (1): 109–124. https://doi.org/10.1007/s11090-018-9931-1.

137 Zhao, B., Liu, Y., Zhu, Z. et al. (2018). Highly selective conversion of CO_2 into ethanol on $Cu/ZnO/Al_2O_3$ catalyst with the assistance of plasma. *J. CO2 Util.* 24 (October 2017): 34–39. https://doi.org/10.1016/j.jcou.2017.10.013.

138 Neyts, E.C. (2018). Atomistic simulations of plasma catalytic processes. *Front. Energy Res.* 12 (1): 145–154. https://doi.org/10.1007/s11705-017-1674-7.

139 Bogaerts, A., De Bie, C., Snoeckx, R., and Kozák, T. (2017). Plasma based CO_2 and CH_4 conversion: a modeling perspective. *Plasma Processes Polym.* 14 (6): 1–21. https://doi.org/10.1002/ppap.201600070.

140 Tan, X. and Go, D.B. (2019). Rational design of plasma-enhanced catalysis at microscale dimensions for the dissociation of CO_2. *J. Electrostat.* 97 (January): 71–74. https://doi.org/10.1016/j.elstat.2018.12.003.

141 Studt, F., Sharafutdinov, I., Abild-Pedersen, F. et al. (2014). Discovery of a Ni-Ga catalyst for carbon dioxide reduction to methanol. *Nat. Chem.* 6 (4): 320–324. https://doi.org/10.1038/nchem.1873.

142 Medford, A.J., Lausche, A.C., Abild-Pedersen, F. et al. (2014). Activity and selectivity trends in synthesis gas conversion to higher alcohols. *Top. Catal.* 57 (1–4): 135–142. https://doi.org/10.1007/s11244-013-0169-0.

143 Giammaria, G., van Rooij, G., and Lefferts, L. (2019). Plasma catalysis: distinguishing between thermal and chemical effects. *Catalysts* 9 (2): 185. https://doi.org/10.3390/catal9020185.

144 Giammaria, G. and Lefferts, L. (2020). Synergy between dielectric barrier discharge plasma and calcium oxide for reverse water gas shift. *Int. J. Chem. Eng.* https://doi.org/10.1016/j.cej.2019.123806.

145 Hessel, V., Anastasopoulou, A., Wang, Q. et al. (2013). Energy, catalyst and reactor considerations for (near)-industrial plasma processing and learning for nitrogen-fixation reactions. *Catal. Today* 211: 9–28. https://doi.org/10.1016/j.cattod.2013.04.005.

146 Mori, S., Matsuura, N., Tun, L.L., and Suzuki, M. (2016). Direct synthesis of carbon nanotubes from only CO_2 by a hybrid reactor of dielectric barrier discharge and solid oxide electrolyser cell. *Plasma Chem. Plasma Process.* 36 (1): 231–239. https://doi.org/10.1007/s11090-015-9681-2.

147 Kogelschatz, U. (2003). Dielectric barrier discharge: their history, discharge physic and Industrial Applications. *Plasma Chem. Plasma Process.* 23 (1): 1–46. https://doi.org/10.1023/A:1022470901385.

148 Hong, J., Prawer, S., and Murphy, A.B. (2018). Plasma catalysis as an alternative route for ammonia production: status, mechanisms, and prospects for progress. *ACS Sustainable Chem. Eng.* 6 (1): 15–31. https://doi.org/10.1021/acssuschemeng.7b02381.

149 Brandenburg, R., Bogaerts, A., Bongers, W. et al. (2018). White paper on the future of plasma science in environment, for gas conversion and agriculture. *Plasma Processes Polym.* 16 (1): 1–18. https://doi.org/10.1002/ppap.201700238.

Index

Engineering Solutions for CO_2 Conversion, First Edition.
Edited by Tomas R. Reina, José A. Odriozola, and Harvey Arellano-Garcia.

p

q

r